Poly(Ethylene Glycol) Chemistry

Biotechnical and Biomedical Applications

TOPICS IN APPLIED CHEMISTRY

Series Editors: Alan R. Katritzky, FRS
Kenan Professor of Chemistry
University of Florida, Gainesville, Florida

Gebran J. Sabongi
Laboratory Manager, Encapsulation Technology Center
3M, St. Paul, Minnesota

BIOCATALYSTS FOR INDUSTRY
Edited by Jonathan S. Dordick

CHEMICAL TRIGGERING
Reactions of Potential Utility in Industrial Processes
Gebran J. Sabongi

THE CHEMISTRY AND APPLICATION OF DYES
Edited by David R. Waring and Geoffrey Hallas

HIGH-TECHNOLOGY APPLICATIONS OF ORGANIC COLORANTS
Peter Gregory

INFRARED ABSORBING DYES
Edited by Masaru Matsuoka

POLY(ETHYLENE GLYCOL) CHEMISTRY
Biotechnical and Biomedical Applications
Edited by J. Milton Harris

RADIATION CURING
Science and Technology
Edited by S. Peter Pappas

STRUCTURAL ADHESIVES
Edited by S. R. Hartshorn

TARGET SITES FOR HERBICIDE ACTION
Edited by Ralph C. Kirkwood

A Continuation Order Plan is available for this series. A continuation order will bring delivery of each new volume immediately upon publication. Volumes are billed only upon actual shipment. For further information please contact the publisher.

Poly(Ethylene Glycol) Chemistry

Biotechnical and Biomedical Applications

Edited by

J. Milton Harris

Department of Chemistry
University of Alabama in Huntsville
Huntsville, Alabama

Plenum Press • New York and London

1992

Library of Congress Cataloging-in-Publication Data

Poly(ethylene glycol) chemistry : biotechnical and biomedical applications / edited by J. Milton Harris.
p. cm. -- (Topics in applied chemistry)
Includes bibliographical references and index.
ISBN 0-306-44078-4
1. Polyethylene glycol--Biotechnology. I. Harris, J. Milton. II. Title: Polyethylene glycol chemistry. III. Series.
TP248.65.P58P65 1992
610'.28--dc20 92-16319
CIP

ISBN 0-306-44078-4

A Division of Plenum Publishing Corporation
233 Spring Street, New York, N.Y. 10013

Printed in the United States of America

Contributors

Abraham Abuchowski, Enzon, Inc., South Plainfield, New Jersey 07080

J. D. Andrade, Center for Biopolymers at Interfaces, Department of Bioengineering, University of Utah, Salt Lake City, Utah 84112

Kris P. Antonsen, Center for Bioengineering, University of Washington, Seattle, Washington 98195; *present address*: Miles, Inc., Berkeley, California 94701

Ernst Bayer, Institute for Organic Chemistry, University of Tübingen, D-7400 Tübingen, Germany

Karin Bergström, Berol Nobel, S-44485 Stenungsund, Sweden

Donald E. Brooks, Departments of Pathology and Chemistry, University of British Columbia, Vancouver, British Columbia V6T 2B5, Canada

P. Caliceti, Department of Pharmaceutical Sciences (Centro di Studio di Chimica del Farmaco e dei Prodotti Biologicamente Attivi del CNR), University of Padua, 35100 Padua, Italy

J. Carlino, Celtrix Laboratories, Palo Alto, California 94303

S. Chu, Celtrix Laboratories, Palo Alto, California 94303

P. Claesson, The Surface Force Group, The Royal Institute of Technology, S-10044 Stockholm; and Institute for Surface Chemistry, S-11486 Stockholm, Sweden

Frank F. Davis, Enzon, Inc., South Plainfield, New Jersey 07080

T. M. Fyles, Department of Chemistry, University of Victoria, Victoria, British Columbia V8W 2A2, Canada

C.-G. Gölander, Institute for Surface Chemistry, S-11486 Stockholm, Sweden

Wayne R. Gombotz, Bristol-Myers Squibb, Pharmaceutical Research Institute, Seattle, Washington 98121

N. B. Graham, Department of Pure and Applied Chemistry, University of Strathclyde, Glasgow G1 1XL, Scotland, U.K.

Wang Guanghui, Center for Bioengineering, University of Washington, Seattle, Washington 98195

J. Milton Harris, Department of Chemistry, University of Alabama in Huntsville, Huntsville, Alabama 35899

James N. Herron, Center for Biopolymers at Interfaces, and Departments of Bioengineering and Pharmaceutics, University of Utah, Salt Lake City, Utah 84112

Howard Higley, Celtrix Laboratories, Palo Alto, California 94303

Yoshihiko Hirose, Amano Pharmaceutical Company, Ltd., Nagoya 481, Japan

Allan S. Hoffman, Center for Bioengineering, University of Washington, Seattle, Washington 98195

Krister Holmberg, Berol Nobel, S-44485 Stenungsund, Sweden; *present address*: Institute for Surface Chemistry, S-11486 Stockholm, Sweden

Thomas A. Horbett, Center for Bioengineering, University of Washington, Seattle, Washington 98195

Hitoshi Ishida, Faculty of Engineering, Kumamoto University, Kurokami, Kumamoto 860, Japan

Göte Johansson, Department of Biochemistry, Chemical Center, University of Lund, S-22100 Lund, Sweden

Sung Wan Kim, Department of Pharmaceutics, Center for Controlled Chemical Delivery, University of Utah, Salt Lake City, Utah 84108

Masami Kimura, Kumamoto University Medical School, Kumamoto 860, Japan

Toshimitsu Konno, Kumamoto University Medical School, Kumamoto 860, Japan

Chyi Lee, Enzon, Inc., South Plainfield, New Jersey 07080

Kap Lim, Center for Biopolymers at Interfaces, and Departments of Bioengineering and Pharmaceutics, University of Utah, Salt Lake City, Utah 84112

Hiroshi Maeda, Kumamoto University Medical School, Kumamoto 860, Japan

Edward W. Merrill, Department of Chemical Engineering, Massachusetts Institute of Technology, Cambridge, Massachusetts 02139

Kwang Nho, Enzon, Inc., South Plainfield, New Jersey 07080

Katsutoshi Ohkubo, Faculty of Engineering, Kumamoto University, Kurokami, Kumamoto 860, Japan

Hirosuke Okada, Department of Fermentation Technology, Faculty of Engineering, Osaka University, 2-1 Yamada-oka, Suita-shi, Osaka 565, Japan

Ki Dong Park, Department of Pharmaceutics, Center for Controlled Chemical Delivery, University of Utah, Salt Lake City, Utah 84108

Wolfgang Rapp, Institute for Organic Chemistry, University of Tübingen, D-7400 Tübingen, Germany

W. Rhee, Research and Development, Collagen Corporation, Palo Alto, California 94303

Takashi Sagawa, Faculty of Engineering, Kumamoto University, Kurokami, Kumamoto 860, Japan

L. Sartore, Department of Pharmaceutical Sciences (Centro di Studio di Chimica del Farmaco e dei Prodotti Biologicamente Attivi del CNR), University of Padua, 35100 Padua, Italy

Ikuharu Sasaki, Amano Pharmaceutical Company, Ltd., Nagoya 481, Japan

P. J. Sather, Departments of Chemistry and Pathology, University of British Columbia, Vancouver, British Columbia V6T 2B5, Canada

O. Schiavon, Department of Pharmaceutical Sciences (Centro di Studio di Chimica del Farmaco e dei Prodotti Biologicamente Attivi del CNR), University of Padua, 35100 Padua, Italy

M. R. Sedaghat-Herati, Department of Chemistry, Southwest Missouri State University, Springfield, Missouri 65804-0089

Alec Sehon, MRC Group for Allergy Research, Department of Immunology, Faculty of Medicine, The University of Manitoba, Winnipeg, Manitoba R3E 0W3, Canada

Kim A. Sharp, Departments of Pathology and Chemistry, University of British Columbia, Vancouver, British Columbia V6T 2B5, Canada; *present address*: Department of Biochemistry and Molecular Biophysics, College of Physicians and Surgeons of Columbia University, New York, New York 10032

Maj-Britt Stark, Department of Histology, University of Göteborg, S-40033 Göteborg, Sweden; *present address*: Astra Hässle AB, Kärragatan 5, S-43183 Mölndal, Sweden

P. Stenius, Institute for Surface Chemistry, S-11486 Stockholm, Sweden

S. Jill Stocks, Departments of Pathology and Chemistry, University of British Columbia, Vancouver, British Columbia V6T 2B5, Canada; *present address*: Biotechnology Processes Services Ltd., Durham DL1 6YL, England

Folke Tjerneld, Deparment of Biochemistry, Chemical Center, University of Lund, S-22100 Lund, Sweden

Itaru Urabe, Department of Fermentation Technology, Faculty of Engineering, Osaka University, 2-1 Yamada-oka, Suita-shi, Osaka 565, Japan

James M. Van Alstine, Departments of Pathology and Chemistry, University of British Columbia, Vancouver, British Columbia V6T 2B5, Canada; *present address*: Department of Biological Sciences, The University of Alabama in Huntsville, Huntsville, Alabama 35899

F. M. Veronese, Department of Pharmaceutical Sciences (Centro di Studio di Chimica del Farmaco e dei Prodotti Biologicamente Attivi del CNR), University of Padua, 35100 Padua, Italy

Tetsuya Yomo, Department of Fermentation Technology, Faculty of Engineering, Osaka University, 2-1 Yamada-oka, Suita-shi, Osaka 565, Japan

Kohji Yoshinaga, Faculty of Engineering, Kyushu Institute of Technology, Sensui, Tobata-ku, Kitakyushu 804, Japan

Samuel Zalipsky, Enzon, Inc., South Plainfield, New Jersey 07080; *present address*: Department of Chemistry, Rutgers–The State University of New Jersey, Piscataway, New Jersey 08855

Preface

The idea for this book came from discussions among participants in a symposium on biotechnical applications at the "Pacifichem 89" meeting in Honolulu. It was the majority opinion of this group that a volume dedicated to biotechnical and biomedical applications of PEG chemistry would enhance research and development in this area. Though the book was conceived at the Honolulu meeting, it is not a proceedings of this symposium. Several groups who did not participate in this meeting are represented in the book, and the book incorporates much work done after the meeting. The book does not include contributions in all related areas to which PEG chemistry has been applied. Several invited researchers declined to participate, and there is not enough space in this single volume to properly cover all submissions. Chapter 1—an overview of the topic—discusses in brief applications not given detailed coverage in specifically devoted chapters.

The following topics are covered: introduction to and fundamental properties of PEG and derivatives in Chapters 1–3; separations using aqueous polymer two-phase partitioning in Chapters 4–6; PEG-proteins as catalysts in biotechnical applications in Chapters 7 and 8; biomedical applications of PEG-proteins in Chapters 9–13; PEG-modified surfaces for a variety of biomedical and biotechnical applications in Chapters 14–20; and synthesis of new PEG derivatives in Chapters 21 and 22.

I thank the authors who have tolerated the glacial process that leads to a bound book. I thank Alice, Adele, and Bert for taking care of the frozen pipes back home while I was otherwise engaged in Honolulu.

J. Milton Harris

Huntsville, Alabama

Contents

3. Molecular Simulation of Protein–PEG Interaction

Kap Lim and James N. Herron

4. PEG-Derivatized Ligands with Hydrophobic and Immunological Specificity: Applications in Cell Separation

Donald E. Brooks, James M. Van Alstine, Kim A. Sharp, and S. Jill Stocks

5. Affinity Partitioning in PEG-Containing Two-Phase Systems

Göte Johansson

6. Aqueous Two-Phase Partitioning on an Industrial Scale

Folke Tjerneld

11. Toxicity of Bilirubin and Detoxification by PEG–Bilirubin Oxidase Conjugate: A New Tactic for Treatment of Jaundice

Hiroshi Maeda, Masami Kimura, Ikuharu Sasaki, Yoshihiko Hirose, and Toshimitsu Konno

12. PEG-Modified Hemoglobin as an Oxygen Carrier

Kwang Nho, Samuel Zalipsky, Abraham Abuchowski, and Frank F. Davis

17. Poly(Ethylene Glycol) Gels and Drug Delivery

N. B. Graham

18. PEO-Modified Surfaces—*In Vitro*, *Ex Vivo*, and *In Vivo* Blood Compatibility

Ki Dong Park and Sung Wan Kim

19. Immobilization of Proteins via PEG Chains

Krister Holmberg, Karin Bergström, and Maj-Britt Stark

20. Polystyrene-Immobilized PEG Chains: Dynamics and Application in Peptide Synthesis, Immunology, and Chromatography

Ernst Bayer and Wolfgang Rapp

21. Use of Functionalized Poly(Ethylene Glycol)s for Modification of Polypeptides

Samuel Zalipsky and Chyi Lee

22. Synthesis of New Poly(Ethylene Glycol) Derivatives

J. Milton Harris, M. R. Sedaghat-Herati, P. J. Sather, Donald E. Brooks, and T. M. Fyles

1

Introduction to Biotechnical and Biomedical Applications of Poly(Ethylene Glycol)

J. MILTON HARRIS

1.1. INTRODUCTION

At first glance, the polymer known as poly(ethylene glycol) or PEG appears to be a simple molecule. It is a linear or branched, neutral polyether, available in a variety of MWs, and soluble in water and most organic solvents. Despite its apparent simplicity,

$$HO\text{—}(CH_2CH_2O)_n\text{—}CH_2CH_2OH$$

poly(ethylene glycol)

this molecule is the focus of much interest in the biotechnical and biomedical communities. Primarily this is because PEG is unusually effective at excluding other polymers from its presence when in an aqueous environment. This property translates into protein rejection, formation of two-phase systems with other polymers, nonimmunogenicity, and nonantigenicity. In addition, the polymer is nontoxic and does not harm active proteins or cells although it interacts with cell membranes. It is subject to ready chemical modification and attachment to other molecules and surfaces, and when attached to other molecules it has little effect on their chemistry but controls their solubility and increases their size. These properties, which are described in more detail below, have led to a variety of important biotechnical and biomedical applications, a summary of which is also presented below.

Five key early works set the stage for the applications described in this volume; without regard to any order these are: (1) the observation that PEG can be used to

---

J. MILTON HARRIS • Department of Chemistry, University of Alabama in Huntsville, Huntsville, Alabama 35899.

Poly(Ethylene Glycol) Chemistry: Biotechnical and Biomedical Applications, edited by J. Milton Harris. Plenum Press, New York, 1992.

drive proteins and nucleic acids from solution for purification and crystal growth[1–3]; (2) Albertsson's discovery that PEG and dextran, when mixed with buffer, form aqueous polymer two-phase systems, which are hospitable to biological materials and are extremely useful for purification of these biological materials[4]; (3) the finding that PEG interacts with cell membranes to give cell fusion, a key process in biotechnology[5–7]; (4) Davis and Abuchowski's observation that covalent attachment of PEG to proteins gives active conjugates that are nonimmunogenic and nonantigenic and have greatly increased serum lifetimes[8]; and (5) Nagaoka's finding that covalent attachment of PEG to surfaces greatly retards protein adsorption to these surfaces.[9]

These early works have led to several active areas of investigation, which have produced hundreds of research publications. The present volume is the first to bring these diverse applications together in one source. A major goal of this book is to show the interrelationships among these different areas and thus stimulate new areas of investigation.

1.2. PROPERTIES OF PEG

At molecular weights less than 1000, PEGs are viscous, colorless liquids; higher molecular weight PEGs are waxy, white solids.[10,11] The melting point of the solid is proportional to molecular weight, approaching a plateau at about 67 °C. The molecular weights commonly used in biomedical and biotechnical applications range from a few hundred to approximately 20,000. Since PEG is usually prepared by an anionic initiation process with few chain-transfer and termination steps, molecular weight distributions are generally observed to be narrow.[9] It should be noted, however, that the commonly used monomethyl ethers of PEG exhibit a rather broad molecular weight distribution because of the presence of high molecular weight PEG (i.e., the polymer with two hydroxyl terminal groups).[12] This PEG is produced from trace hydroxide acting as an initiator, and since this difunctional polymer grows at both ends, it has a higher molecular weight than the monomethyl ether, which grows at only one end.

PEGs are also sometimes referred to as poly(ethylene oxide) (PEO), poly(oxyethylene) (POE), and polyoxirane. In general usage, poly(ethylene glycol) refers to polyols of molecular weights below about 20,000, poly(ethylene oxide) refers to higher molecular weight polymers, and poly(oxyethylene) and poly(oxirane) are not specific in this regard. The *Chemical Abstracts* registry number for PEG is 25322-68-3.

PEGs possess a variety of properties pertinent to biomedical and biotechnical applications. A summary of these properties is given here, with applications described in Section 1.3. The following is a brief listing of some properties of interest.

1. Soluble in water, toluene, methylene chloride, many organic solvents.
2. Insoluble in ethyl ether, hexane, ethylene glycol.

3. Insoluble in water at elevated temperature.
4. Solubility and partitioning controlled by making derivatives.
5. Forms complexes with metal cations.
6. Highly mobile; large exclusion volume in water.
7. Can be used to precipitate proteins and nucleic acids.
8. Forms two-phase systems with aqueous solutions of other polymers.
9. Nontoxic; FDA approved for internal consumption.
10. Hospitable to biological materials.
11. Causes cell fusion (in high concentration).
12. Weakly immunogenic.

If covalently linked PEG will:

13. Solubilize other molecules.
14. Render proteins nonimmunogenic and toleragenic.
15. Reduce rate of clearance through kidney.
16. Render surfaces protein-rejecting.
17. Alter electroosmotic flow.
18. Move molecules across cell membranes.
19. Alter pharmacokinetics.

1.2.1. Solubility and Partitioning

PEG exhibits a bizarre solubility pattern, as it is soluble both in water and (to varying degrees) in many organic solvents including toluene, methylene chloride, ethanol, and acetone (hence PEG is frequently described as amphiphilic).[10] Interestingly, the closely related poly(methylene oxide), poly(propylene oxide), and isomeric polyacetaldehyde are not soluble in water. PEG is insoluble in hexane and similar aliphatic hydrocarbons and in ethyl ether and ethylene glycol, molecules which closely resemble PEG. This solubility pattern is of much use in synthesis of PEG derivatives since reactions can be conducted in an organic solvent, such as toluene, and the product isolated by addition to a nonsolvent such as hexane or ethyl ether. The Hildebrand solubility parameter is 10.3 $(\mathrm{cal\,cm^{-3}})^{0.5}$ for PEG.[11]

Since PEG is soluble in both organic and aqueous media, it is apparent that the polymer will be present to some extent in both phases of an organic-water, two-phase system. Although this partitioning has not been studied in great detail, it is known that PEG will partition in favor of water in a water–benzene system and in favor of methylene chloride in a water–methylene chloride system. A common isolation procedure in synthesis of PEG derivatives is to extract the derivative into water and then into methylene chloride.

In biological systems, it appears that PEG partitions between aqueous medium and cell membranes. Evidence for this lies in the observation that PEG induces cell fusion, a property used to great benefit in production of hybridomas and monoclonal antibodies.[5–7] This topic is discussed in Section 1.3.5.

PEG solubility and partitioning patterns can be altered by attachment of hydrophobic tails, as in the common PEG-based surfactants, and by including hydrophobic comonomers in the polymer backbone, as in the common ethylene oxide–propylene oxide copolymers (inclusion of greater than 50% propylene oxide will make the polymer water-insoluble).[10] This alteration in solubility can be used to control partitioning of PEG derivatives in two-phase systems, such as benzene versus water, where attachment of hydrocarbon tails shifts partitioning in favor of the organic layer.[13]

PEGs also have the unusual property of possessing a lower consolute temperature (LCT), or cloud point, of approximately 100 °C in water; that is, raising the temperature above 100 °C will result in insolubility and formation of two phases.[10,11] The LCT for PEG varies somewhat depending on molecular weight, concentration, and pH. Increasing salt concentration can greatly lower the LCT. Also, inclusion of propylene oxide comonomer lowers the LCT proportionately, until the polymer with 60% ethylene oxide and 40% propylene oxide becomes insoluble at 37 °C. Attachment of hydrophobic end groups, as in the large number of PEG-based surfactants, has a similar effect. This inverse solubility–temperature relationship in water has several practical applications.

Finally, PEG solubility properties are revealed in the many uses of PEG as a copolymer in a variety of block copolymers (see Chapter 14 by Merrill on PEG-polyurethanes) and in polymer blends.[11] In the block copolymers the PEG can form an immiscible "soft segment" phase which becomes the phase in contact with an aqueous medium. Similarly, the blends can be miscible or immiscible blends. Polymer miscibility is a complex phenomenon in which changes in thermodynamic parameters are important. A discussion of this subject is given by Bailey and Koleske.[11]

1.2.2. Metal Binding

PEGs form complexes with metal cations.[10,11] This property is demonstrated by application of PEGs as "phase transfer agents," a process of much importance in organic chemistry.[13–15] In this application the PEG transfers a salt from solid phase or aqueous phase to organic phase by complexing or coordinating with the metal cation and assisting its partition into the organic phase. The corresponding anion has to follow along to maintain charge neutrality, and in the process the anion becomes much more reactive because it is poorly solvated or "naked"; hence, this process is frequently referred to as phase transfer catalysis. Examples include movement of $KMnO_4$ from solid phase into benzene (using an alkyl-PEG) and transfer of metal picrates from water into methylene chloride (using PEG itself).[13–15]

It is noteworthy that phase transfer catalysis was first described for "crown ethers," which are cyclic ethers resembling rings made of PEG of approximately six ethylene oxide units.[16] The molecular structures of these ethers resemble crowns in which oxygens lie at the points of the crown. Complexation occurs when an electron-deficient metal cation coordinates with the electron-rich oxygens. After complexation, the metal is rendered hydrophobic and organic soluble, because it is hidden inside the hydrocarbon portion of the crown ether. Crown ethers show selectivity in metal binding related to the size of the cavity in the center of the crown. Interestingly, PEGs also show selectivity in metal binding, apparently because the PEGs adopt helical conformations with cavities of preferred sizes.[11,14]

1.2.3. Aqueous Solutions

The behavior of PEG in aqueous environments is the key to its importance in biomedical and biotechnical applications. In simple terms, PEG in aqueous solution acts as a highly mobile molecule with a large exclusion volume. Relaxation-time studies show rapid motion of the polymer chain[17] and gel chromatography shows that PEGs are much larger in solution than many other molecules (e.g., proteins) of comparable molecular weight.[18,19] Interesting consequences of this property are that PEG excludes other polymers and, if the concentration of PEG is high enough, will form two-phase systems with other polymers. Applications that result include protein and nucleic acid precipitation and two-phase partitioning and, when the PEG is covalently bound, protein-rejecting molecules and surfaces. Also, as noted in the discussion of solubility (Section 1.2.1), incorporation of hydrophobic end groups, inclusion of hydrophobic propylene oxide as comonomer, and addition of salts lead to lowering of the lower consolute temperature; this phenomenon has been utilized to produce two-phase systems for partitioning purifications.

Molecular modeling and theoretical investigations of PEG solution properties and interactions with other polymers are discussed in Chapters 2, 3, and 15. Theoretical investigation of formation of two-phase systems has also been the subject of several recent studies.[20–23]

1.2.4. Covalently Bound PEGs

The terminal hydroxyl groups of the PEG molecule provide a ready site for covalent attachment to other molecules and surfaces.[24] Molecules to which PEG is attached usually remain active, demonstrating that bound PEG does not denature proteins or hinder the approach of other small molecules. Bound PEG does, however, retain its ability to repel other large molecules, and thus PEG-modified surfaces and PEG-modified proteins are protein-rejecting. Consequently, there are many examples of PEG-enzymes with small substrates that remain active.[24,25] Although PEG-protein interactions with large molecules have not yet been investigated in great detail, it is known that some interactions can occur. For example, we have shown that the PEG-modified antibody against alkaline phosphatase will continue to bind with alkaline phosphatase[26] and that the PEG-modified antibody against red blood cells will continue to bind with red blood cells.[27] Similarly, a variety of PEG-bound affinity ligands (both protein and nonprotein) used in two-phase partitioning are known to continue binding with their substrates (see Chapters 4, 5, and 6). On the other hand, the well-known reduction of immunogenicity and antigenicity for PEG-modified proteins and the nonfouling nature of PEG-modified surfaces must derive in large part from the inability of nonbound proteins to approach these materials. Obviously more work is needed to give a better understanding of interaction of PEG-modified moieties with other molecules.

Workers investigating protein-rejecting surfaces have conducted experiments to elucidate better the nonfouling nature of PEG-surfaces (see Chapters 14–16, and 18). We give here a qualitative view drawn from this volume and from our recent work.[28] As noted above, PEG in aqueous solution is a highly mobile molecule with a large exclusion volume and, of course, the molecule is neutral and possesses no acidic sites (excluding the hydroxyl group which acts as a weak hydrogen-bond acid) and only weakly basic ether linkages. The molecule is also heavily hydrated. One immediate conclusion is that there are few sites to which proteins can bind. Moreover, the rapid motion of the molecule gives an approaching protein little time in which to form a positive interaction.[17]

Additionally, consider what happens if a protein senses and binds with the surface beneath the PEG. There can be two consequences. First, if the PEG molecules are compressed against the surface, without losing water molecules, there will be an unfavorable negative entropy change resulting from the reduced motion of the highly mobile, hydrated chain. This entropic disadvantage can be overcome if water molecules are stripped from the PEG chains, but this in turn is enthalpically unfavorable. Since protein adsorption to PEG-modified surfaces is minimal, it is apparent that neither of these consequences obtains.

Thus PEG molecules bound to surfaces can be viewed as being mobile, heavily hydrated, elastic spheres. A protein interacting with these spheres can temporarily deform the sphere but cannot attach to or significantly compress the sphere. A dependence of this phenomenon on molecular weight has been observed and is discussed in Chapter 2.

Molecules bound to PEG have altered solubility properties. Thus, PEG attachment may be used to improve water solubility. Similarly, PEG binding can improve solubility of molecules in organic solvents, and this property has been utilized to make PEG-enzymes that are active in dry organic solvents (Section 1.3.3). In view of these applications, and the solubility of PEGs and PEG conjugates in water and organic solvents, it would be expected that PEG could be used to move molecules across cell membranes. Preliminary evidence that this can in fact be accomplished is presented in Section 1.3.5.

Covalent linkage of PEGs also increases the size of the molecule to which the PEG is bound, and this property has been utilized to reduce the rate of clearance of molecules through the kidney (Section 1.3.2). Finally, covalent linkage of PEGs alters the electrical nature of surfaces since charges on the surface become buried beneath a viscous, hydrated, neutral layer. This property has been utilized in capillary electrophoresis to control electroosmotic flow (Section 1.3.3).

1.2.5. Toxicity

Of much interest in the biomedical area is the fact that PEG is nontoxic and has been approved by the FDA for internal consumption.[29–31] PEG is used in large quantities for drug compounding and for a wide variety of cosmetic and personal care products, and PEG-proteins have been cleared for clinical trials in humans (Section 1.3.2). It is of interest that small PEGs of molecular weight less than 400 may exhibit some toxicity.[29] Free PEG administered intravenously to humans is readily excreted through the kidney.[32]

PEG is poorly immunogenic, which, of course, is crucial to the development of PEG-proteins as drugs (Section 1.3.2).[33] Richter and Akerblom have studied antibody formation in humans exposed to PEG and PEG-proteins, and find that PEG is poorly immunogenic, while PEG-proteins can elicit a mild anti-PEG response.[34,35] Interestingly, this response dwindles with increased exposure time, and is sufficiently weak to be of no clinical significance.

Finally, it is noteworthy that aqueous solutions of PEG are hospitable to living cells. This fact has long been appreciated by experimentalists dealing with aqueous polymer two-phase systems,[4] and PEG is utilized in tissue culture media and for organ preservation.

1.3. BIOTECHNICAL AND BIOMEDICAL APPLICATIONS OF PEG

The purpose of this section is to provide a brief review of applications and especially to show how the different chapters of this book are related. Obviously this short review cannot be exhaustive, but an attempt has been made to provide leading references to primary sources, and of course many references can be found in the individual chapters. A recent review by Topchieva contains many additional references.[36]

1.3.1. Biological Separations

This volume includes three chapters on the "phase partitioning" technique first developed by Albertsson.[4] The formation of aqueous polymer two-phase systems has been known since the 19th century, but it remained for Albertsson to utilize the phase systems for biological purifications. The procedure involves purification of biological materials (cells, cell fragments, viruses, proteins, nucleic acids) by partitioning between the two phases formed by solution of a pair of polymers (typically PEG and dextran) in aqueous buffer. It is indeed interesting that immiscible layers can be formed, although each layer is composed of more than 90% water. As noted earlier, this incompatibility of PEG with other polymers is one of the key properties of PEG.

The physical mechanism for this incompatibility is not clear (it has been examined in terms of Flory–Huggins theory),[20–23] but the incompatibility of PEG with neutral polymers, such as dextran, is clearly related to its incompatibility with proteins and nucleic acids that makes possible nonimmunogenic enzymes and nonfouling surfaces. It should be noted that PEG is not unusual in formation of two-phase systems; many pairs of polymers exhibit this phenomenon, and it is even possible to form multilayered systems by simultaneous solution of several polymers. Salt solutions also form two-phase systems with PEG and water (the salt essentially lowers the lower consolute temperature), and PEG will form a two-phase system with pure water above its lower consolute temperature.

A detailed discussion of many aspects of phase partitioning follows in Chapters 3–6, but it is pertinent to summarize a few key points regarding the technique. Partitioning of materials between the two phases, or between the phases and the phase interface in the case of particles, can be controlled by varying phase components including salt concentration, salt identity, pH, polymer identity, polymer MW, and phase-volume ratio. It is especially noteworthy that increasing PEG MW acts to drive proteins to the dextran-rich phase, while increasing dextran MW acts to drive proteins to the PEG-rich phase. Also, certain salts give "neutral" phase systems in which purifications are not sensitive to the charge of the desired protein or cell but depend rather on subtle factors such as variation in membrane lipids. Other salts, however, are not partitioned equally between the phases and thus give "charged" phase systems in which purifications are dependent on charge of the target moiety.

Affinity partitioning can be performed by covalent attachment of affinity ligands to PEG. Ligands that have been utilized include dyes, antibodies, enzyme inhibitors, metals, and hydrophobic alkyl chains. The target substance can be separated from the affinity ligand by the usual pH and salt variations or, if an ethylene oxide–propylene oxide copolymer is used, by heating above the lower consolute point of the polymer.[37] Affinity partitioning separations offer powerful competition for other affinity techniques such as cell sorting and chromatography. Frequently, the technique is applied by use of multiple separations in the form of countercurrent distribution (repeated single-tube separations) or countercurrent chromatography (in which the two immiscible liquids flow past each other).

Finally, it is important to note that phase partitioning offers many advantages for large-scale, industrial purifications. The technique can be scaled up directly, and many techniques are available for handling liquid–liquid, two-phase systems. The inexpensive polymers can be recovered and recycled after use. An important advantage for work with cell homogenates (e.g., in recombinant DNA technology) is that cell debris collects at the phase interface while nucleic acids can be collected in the bottom phase. Thus partitioning is especially attractive as a first step in large-scale recovery of proteins.

There are several variations on this two-phase partitioning theme. For example, PEG bis-copper-chelates can be used for metal-affinity precipitation of proteins.[38] The PEG-derived detergent Triton X-114 can be dissolved in buffer, mixed with a crude protein mixture, and heated above the cloud point to form a two-phase system and partition the proteins.[39] Also, a chromatographic form of partitioning can be derived by immobilizing the dextran–water phase on a chromatographic support and eluting with the PEG-water phase.[40]

1.3.2. PEG-Proteins and PEG-Peptides for Medical Applications

In 1977, Davis, Abuchowski, and co-workers demonstrated that covalent attachment of PEG to a protein gives minimal loss of activity and renders the protein nonimmunogenic and nonantigenic, thus imparting greatly increased serum lifetime.[8,33] This observation is highly significant because it opens the way to intravenous administration of proteins for treatment of inborn errors of metabolism and other disorders. Additional benefits of protein "pegylation" are reduced rate of kidney clearance and enhanced association with cells (see Section 1.3.5).[41] Examples of medically useful PEG-modified proteins include PEG-asparaginase for treatment of acute leukemias,[42] PEG-adenosine deaminase for treatment of severe combined immunodeficiency disease,[43] and PEG-superoxide dismutase and PEG-catalase for limiting tissue injury resulting from reactive oxygen species associated with ischemia and related pathological events.[41] In the present volume, Chapter 21 describes synthetic methods for attaching PEG to proteins, Chapter 9 describes several useful applications of PEG-proteins including PEG-superoxide dismutase, Chapter 11 describes use of PEG-bilirubin oxidase for treatment of jaundice, Chapter 12 describes PEG-hemoglobin as a whole blood substitute, and Chapter 13 describes use of PEG-collagen for soft tissue replacement.

Lightly pegylated protein allergens can be used in hyposensitization of allergic patients.[44] Work in this area is described in Chapter 10.

There is also much pharmaceutical interest in PEG modification of small drug molecules (penicillin, procaine, aspirin, etc.) because of enhanced water solubility, altered pharmacokinetics, and greatly increased size and reduced rate of renal clearance.[45] For example, there is interest in using PEG-peptides, such as PEG-interleukin 2, for stimulation of the immune system.[46]

1.3.3. PEG-Proteins for Chemical/Biotechnical Applications

In addition to the medical and separations applications of PEG-proteins described above, these conjugates have other applications in biotechnology. In particular it is noteworthy that PEG attachment induces solubility of PEG-enzymes in dry organic solvents, such as chloroform, where they exhibit interesting properties.[47]

Another application of PEGs is in synthesizing artificial enzymes by connecting binding sites and catalytic sites via PEG linkers. The PEG linker is a key aspect of this application because of its length, hydrophilicity, and flexibility. In Chapter 8, Yomo, Urabe, and Okada describe application of this concept in preparing semisynthetic oxidases. In a related work Yoshinaga, Ishida, Sagawa, and Ohkubo (Chapter 7) have coupled catalytically active metal–porphyrin complexes to PEG-modified BSA, which acts as a binding site, to obtain hybrid catalysts that are active in organic media.

1.3.4. PEG on Surfaces

1.3.4.1. Nonfouling Surfaces

There is a great deal of current interest in using surface-bound PEG for preventing protein adsorption. This topic is discussed in Chapters 14–16 and a qualitative discussion of the mechanism for protein rejection was given above in Section 1.2.4. Attaching a monolayer of PEG to a surface can reduce the likelihood of many medically undesirable processes. Drug delivery from PEG-based gels, which also exhibit excellent protein-rejecting properties, shows much promise as well; this topic is discussed in Chapter 17.

In a related application, recent work has shown that incorporation of PEG-lipids into the surface of liposomes gives increased serum lifetime and altered pharmacokinetics.[48,49] Thus it appears that PEG on the surface of the liposome acts, as in the case of PEG-proteins, to shield the small particles from the immune system.

1.3.4.2. As a Tether between Molecules and Surfaces

PEG exhibits interesting properties when used as a tether or linker to couple an active molecule to a surface. In this application the PEG acts to inhibit nonspecific protein adsorption on the surface. Additionally, the tethered molecules have been shown to be highly active, behaving essentially as free molecules in solution. This observation can be rationalized by assuming that the unusually long, well-hydrated PEG linker moves the active molecule well out into solution, some distance from the surface.

Applications of this concept to extracorporeal devices, peptide synthesis (see Bonora *et al.* for related oligonucleotide synthesis),[50] diagnostics, and blood contacting materials are described in Chapters 18–20.

1.3.4.3. For Control of Electroosmosis

Electrosmosis is the fluid flow adjacent to a charged surface that results (for example, in capillary electrophoresis) when an electrical potential is applied, causing soluble counterions to migrate toward the oppositely charged electrode. Workers in capillary electrophoresis have long been interested in eliminating or controlling electroosmosis. One approach to solving this problem has been to adsorb polysaccharides onto the charged surface to produce a viscous layer which impedes counterion movement. A problem with this approach is that the polysaccharides tend to desorb, especially in the presence of proteins. We showed in 1986 that covalently bound PEG-5000 essentially eliminates electroosmotic flow, while lower molecular weight PEGs reduce such flow.[51,52] Since our early work there has been a resurgence of interest in capillary electrophoresis as an analytically technique and other workers have examined PEG coatings.[53] An additional benefit of the PEG coatings is that protein adsorption to the capillary surface is also greatly reduced.

1.3.5. Cell Membrane Interactions

The amphiphilic nature of PEG indicates that PEG might interact with cell membranes but the details of these interactions remain relatively poorly understood. Evidence that this interaction occurs is the well-known PEG-induced fusion of cells and liposomes.[5–7,54] Although the mechanism of this process remains debatable, it is clear that membranes absorb large quantities of PEG and that PEG–membrane interaction is probably involved in membrane fusion.[54,55]

This membrane-associative behavior of PEG encouraged Beckman *et al.* to examine endothelial cell uptake of PEG-superoxide dismutase and catalase.[41] These workers found that the PEG-enzymes were in fact taken up by the cells and remained active within the cells. Three mechanisms were proposed for this uptake: (1) direct penetration of membranes, (2) binding of PEG-enzymes to membrane surfaces, and (3) uptake by endocytosis. The first mechanism was considered to be unlikely because PEG is insoluble in alkanes, and thus the PEG-enzyme would not be expected to be soluble in the aliphatic core of phospholipid membranes. However, PEG is known to associate with the phospholipid head-group region of membranes.[54] This fact plus the observed kinetics of uptake and the enhancement of uptake following mechanical injury of the endothelial cell membrane are consistent with membrane binding followed by endocytosis.

In a related work, Veronese and co-workers noted that PEG-superoxide dismutase, unlike free enzyme, could not be recovered completely from plasma after addition to whole blood.[56] These workers suggest that the PEG-enzyme is bound to the blood cells.

This cellular association of PEG-proteins is significant in predicting effects of PEG-protein therapy and in interpreting observed results. Thus it is probable that PEG-proteins associate with cells and do not simply act in plasma.[41,56]

An especially intriguing example of PEG interaction with membranes is the observation by Bittner and colleagues that PEG can be used to fuse the severed halves of invertebrate nerve cells.[57,58] Morphological continuity is demonstrated by transfer of dyes between the fused segments. This work raises the exciting possibility that PEG might be used to fuse severed nerve cells in humans. In a related work, Geron and Meiri have shown that PEG induces fusion of synaptic vesicles and surface membranes of nerve cells.[59]

Clifton and co-workers made the intriguing observation that PEG assists penetration of the blood–brain barrier by vitamin E succinate.[60] This property was used effectively to reduce the damage resulting from mechanical trauma to the brains of rats. If this observation may be generalized to include other animal systems, the use of PEGs could be highly significant for assisting transport of a variety of materials across the blood–brain barrier. Two studies, however, indicate a lack of generality. In the first, Bundgaard and Cserr found that hagfish cerebral capillaries were impermeable to PEG,[61] and in the second, Spigelman *et al.* found that a complex solvent mixture containing PEG 300 had no effect on permeability of the blood–brain barrier.[62] It is important to note, however, that studies of the effects of PEG variables (molecular weight, end groups, copolymers, covalent linkage, etc.) on this intriguing process have not been conducted.

Finally, we note that the interpretation of many two-phase partitioning results requires the concept that PEG interacts with cell membranes (see Chapters 4–6).

1.3.6. Summary

It is apparent that PEG is of immense utility in addressing a variety of needs in biotechnology and medicine. In addition to the "mainline" applications discussed here there are several others that probably are related to the current undertaking but which, as yet, have not received extensive consideration in the literature. These include use of PEGs in cryoprotection, pharmaceutical preparations, intestinal lavage, tissue culture, and organ protection. In any event, the pertinence of PEG and PEG chemistry to medicine and biology is clear, and it is this author's opinion that the next few years will see a rapid growth in related science and applications. It is hoped that the following chapters will assist in this growth.

ACKNOWLEDGMENTS

The author gratefully acknowledges the many helpful discussions with Allan Hoffman and the financial support of this work by the National Aeronautics and Space Administration and the National Institutes of Health.

REFERENCES

1. A. Polson, G. M. Potgieter, J. F. Largier, G. E. Mears, and F. J. Joubert, *Biochim. Biophys. Acta 82*, 463 (1964).
2. M. Zeppezauer and S. Brishammar, *Biochim. Biophys. Acta 94*, 581 (1965).
3. P. W. Chun, M. Fried, and E. F. Ellis, *Anal. Biochem. 19*, 481 (1967).
4. P.-Å. Albertsson, *Partition of Cell Particles and Macromolecules*, 3rd ed., Wiley, New York (1986).
5. K. N. Kao, F. Constabel, M. R. Michayluck, and O. L. Gamborg, *Planta 120*, 215 (1974).
6. Q. F. Ahkong, D. Fisher, W. Tampion, and J. A. Lucy, *Nature 253*, 194 (1975).
7. G. Pontecorvo, *Somat. Cell Genet. 1*, 397 (1975).
8. A. Abuchowski, T. van Es, N. C. Palczuk, and F. F. Davis, *J. Biol. Chem. 252*, 3578 (1977).
9. Y. Mori, S. Nagaoka, H. Takiuchi, T. Kikuchi, N. Noguchi, H. Tanzawa, and Y. Noishiki, *Trans. Am. Soc. Artif. Internal Organs 28*, 459 (1982).
10. F. E. Bailey, Jr. and J. V. Koleske, *Poly(Ethylene Oxide)*, Academic Press, New York (1976).
11. F. E. Bailey, Jr. and J. V. Koleske, *Alkylene Oxides and Their Polymers*, Marcel Dekker, New York (1991).
12. J. M. Dust, Z.-H. Fang, and J. M. Harris, *Macromolecules 23*, 3742 (1990).
13. J. M. Harris and M. G. Case, *J. Org. Chem. 48*, 5390 (1983).
14. J. M. Harris, N. H. Hundley, T. G. Shannon, and E. C. Struck, *J. Org. Chem. 47*, 4789 (1982).
15. J. M. Harris, N. H. Hundley, T. G. Shannon, and E. C. Struck, in: *Crown Ethers and Phase Transfer Catalysis in Polymer Science* (L. Mathias and C. E. Carreher, eds.), p. 371, Plenum Press, New York (1984).
16. C. Starks and C. Liotta, *Phase Transfer Catalysis*, Academic Press, New York (1978).

17. S. Nagaoka, Y. Mori, H. Takiuchi, K. Yokota, H. Tanzawa, and S. Nishiumi, in: *Polymers as Biomaterials* (S. W. Shalaby, A. S. Hoffman, B. D. Ratner, and T. A. Horbett, eds.), p. 361, Plenum Press, New York (1985).
18. K. Hellsing, *J. Chromatogr. 36*, 270 (1968).
19. A. P. Ryle, *Nature 206*, 1256 (1965).
20. D. E. Brooks, K. A. Sharp, and D. Fisher, in: *Partitioning in Aqueous Two-Phase Systems* (H. Walter, D. E. Brooks, and D. Fisher, eds.), Chap. 2, Academic Press, New York (1985).
21. A. Gustafsson, H. Wennerstrom, and F. Tjerneld, *Polymer 27*, 1768 (1986).
22. J. N. Baskir, T. A. Hatton, and U. W. Suter, *J. Phys. Chem. 93*, 2111 (1989).
23. C. A. Haynes, R. A. Beynon, R. S. King, H. W. Blanch, and J. M. Prausnitz, *J. Phys. Chem. 93*, 5612 (1989).
24. J. M. Harris, *J. Macromol. Sci. Rev. Macromol. Chem. Phys. C25*, 325 (1985).
25. K. Yoshinaga, S. G. Shafer, and J. M. Harris, *J. Bioact. Compatible Polym. 2*, 49 (1987).
26. L. J. Karr, P. A. Harris, D. M. Donnely, and J. M. Harris, unpublished results.
27. L. J. Karr, J. M. Van Alstine, R. S. Snyder, S. G. Shafer, and J. M. Harris, *J. Chromatogr. 442*, 219 (1988).
28. K. Bergström, K. Holmberg, A. Safranj, A. S. Hoffman, M. J. Edgell, B. A. Hovanes, and J. M. Harris, *J. Biomed. Mater. Res.* (in press).
29. D. A. Herold, K. Keil, and D. E. Bruns, *Biochem. Pharmacol. 38*, 73 (1989).
30. H. F. Smyth, Jr., C. P. Carpenter, and C. S. Weil, *J. Am. Pharm. Assoc. 39*, 349 (1950).
31. A. J. Johnson, M. H. Darpatkin, and J. Newman, *Br. J. Hematol. 21*, 21 (1971).
32. C. B. Shaffer and F. H. Critchfield, *J. Am. Pharm. Assoc. 36*, 152 (1947).
33. A. Abuchowski and F. F. Davis, in: *Enzymes as Drugs* (J. Holsenberg and J. Roberts, eds.), p. 367, Wiley, New York (1981).
34. A. W. Richter and E. Åkerblom, *Int. Arch. Allergy Appl. Immunol. 70*, 124 (1983).
35. A. W. Richter and E. Åkerblom, *Int. Arch. Allergy Appl. Immunol. 74*, 36 (1984).
36. I. N. Topchieva, *Russ. Chem. Rev. 49*, 494 (1980).
37. P. A. Harris, F. Tjerneld, A. A. Kozlowski, and J. M. Harris, unpublished results.
38. S.-S. Suh, M. E. Van Dam, G. E. Wuenschell, S. Plunkett, and F. H. Arnold, in: *Protein Purification* (M. R. Ladisch, R. C. Willson, C. C. Painton, and S. E. Builder, eds.), Chap. 10, Am. Chem. Soc. (1990).
39. R. Heusch, *Biotech-Forum 31* (1986).
40. W. Muller, *Eur. J. Biochem. 155*, 213 (1986).
41. J. S. Beckman, R. L. Minor, C. W. White, J. E. Repine, G. M. Rosen, and B. A. Freeman, *J. Biol. Chem. 263*, 6884 (1988).
42. A. Abuchowski, G. M. Kazo, C. R. Verhoest, Jr., T. Van Es, D. Kafkewitz, M. L. Nucci, A. T. Viau, and F. F. Davis, *Cancer Biochem. Biophys. 7*, 175 (1984).
43. M. S. Herschfield, R. H. Buckley, M. L. Greenberg, A. L. Melton, R. Schiff, C. Hatem, J. Kurtzberg, M. L. Markert, R. H. Kobayashi, A. L. Kobayashi, and A. Abuchowski, *N. Engl. J. Med. 316*, 589 (1987).
44. P. S. Norman, J. B. Alexander, T. P. King, P. S. Crettcos, A. K. Sobotka, and L. M. Lichtenstein, *J. Allergy Clin. Immunol. 69*, 99 (1982).
45. S. Zalipsky, C. Gilon, and A. Zilka, *Eur. Polym. J. 19*, 1177 (1983).
46. N. V. Katre, M. J. Knauf, and W. J. Laird, *Proc. Natl. Acad. Sci. U.S.A. 84*, 1487 (1987).
47. Y. Inada, A. Matsushima, Y. Kodera, and H. Nishimura, *J. Bioact. Compt. Polym. 5*, 343 (1990).
48. J. Senior, C. Delgado, D. Fisher, C. Tilcock, and G. Gregoriadis, *Biochim. Biophys. Acta 1062*, 77 (1991).
49. A. L. Klibanov, K. Maruyama, A. M. Beckerleg, V. P. Torchilin, and L. Huang, *Biochim. Biophys. Acta 1062*, 142 (1991).
50. G. M. Bonora, C. L. Scremin, F. P. Colonna, and A. Garbesi, *Nucleic Acids Res. 18*, 3155 (1990).
51. B. J. Herren, S. G. Shafer, J. M. Van Alstine, J. M. Harris, and R. S. Snyder, *J. Colloid Interface Sci. 51*, 46 (1987).

52. J. M. Van Alstine, J. M. Harris, S. Shafer, R. S. Snyder, and B. Herren, US Patent, 4,690,749 (1987).
53. G. J. M. Bruin, J. P. Chang, R. H. Kuhlman, K. Zegers, J. C. Kraak, and H. Poppe, *J. Chromatogr. 21*, 385 (1989).
54. M. Yamazaki and T. Ito, *Biochemistry 29*, 1309 (1990).
55. L. T. Boni, J. S. Hah, S. W. Hui, P. Mukherjee, J. T. Ho, and C. Y. Jung, *Biochem. Biophys. Acta 775*, 409 (1984).
56. P. Caliceti, O. Schiavon, A. Mocali, and F. M. Veronese, *Il Farmaco 44*, 711 (1989).
57. G. D. Bittner, M. L. Ballinger, and M. A. Raymond, *Brain Res. 367*, 351 (1986).
58. T. L. Krause and G. D. Bittner, *Proc. Natl. Acad. Sci. U.S.A. 87*, 1471 (1990).
59. N. Geron and H. Meiri, *Biochim. Biophys. Acta 819*, 258 (1985).
60. G. L. Clifton, B. G. Lyeth, L. W. Jenkins, W. C. Taft, R. J. DeLorenzo, and R. L. Hayes, *J. Neurotrauma 6*, 71 (1989).
61. M. Bundgaard and H. F. Cserr, *Brain Res. 206*, 71 (1981).
62. M. K. Spigelman, R. A. Zappulla, J. Johnson, S. J. Goldsmith, L. I. Malis, and J. F. Holland, *J. Neurosurg. 61*, 674 (1984).

2

Water Structure of PEG Solutions by Differential Scanning Calorimetry Measurements

KRIS P. ANTONSEN and ALLAN S. HOFFMAN

2.1. INTRODUCTION

As this volume attests, poly(ethylene glycol) or PEG is a material of growing importance in the biomedical world. It has been used in free solution as an agent for cell fusion[1] and protein precipitation.[2] It has also been conjugated to proteins and drugs to reduce immunological responses and control pharmacodynamics.[3] Finally, it has been used in biocompatible materials, either as a coating or incorporated into a hydrogel.[4] These surfaces are expected to be highly biocompatible because protein adsorption to them is low.[4,5] Both the amount of protein adsorption and the magnitude of other biochemical events, such as platelet adhesion, rapidly decline as the PEG molecular weight rises.[6,7] This decline is most marked at molecular weights up to 1000, after which the biointeractions tend to level out gradually.

All of the applications described above involve water. It is not surprising, then, that the interaction of PEG with its aqueous solvent has received intensive study. One of the primary goals of these studies has been to quantify the amount of water tenaciously bound to PEG in solution. This has been investigated using varying techniques, including NMR spectroscopy,[8] analysis of the water–PEG phase diagram,[9] and differential thermal analysis.[10–13] These studies have shown that between two and three water molecules are bound per PEG repeat unit, and have led to the development of structural models to account for the bound water.[14] None of these

KRIS P. ANTONSEN and ALLAN S. HOFFMAN • Center for Bioengineering, University of Washington, Seattle, Washington 98195. *Present address for K.P.A.*: Miles, Inc., Berkeley, California 94701.

Poly(Ethylene Glycol) Chemistry: Biotechnical and Biomedical Applications, edited by J. Milton Harris. Plenum Press, New York, 1992.

studies, however, has considered the effects of a wide range of PEG molecular weight on both the amount of bound water and the low-temperature phase behavior of aqueous PEG solutions. In this study, we explore these relationships more closely.

2.2. MATERIALS AND METHODS

2.2.1. Materials

Poly(ethylene glycol) samples came from the following sources: Aldrich Chemical (Milwaukee) supplied 99% triethylene glycol and tetraethylene glycol. Sigma Chemical (St. Louis) supplied PEGs of nominal molecular weights 1000, 1450, 3350, and 10,000, and methoxypoly(ethylene glycol)s (MPEGs) of 550, 750, and 5000. "PEG compound 20M," with nominal molecular weight 20,000, was supplied by Union Carbide (Danbury, CT), and poly(ethylene oxide) (PEO) 100,000 was obtained from Polysciences (Warrington, PA). All polymers were used as received. In this chapter, they are referred to by their nominal molecular weights.

The molecular weights of the polymers were measured using aqueous-phase gel permeation chromatography. The column was an Asahipak GS-510 (Asahi Chemical Industries, Yakoo, Japan), and the standards used were PEO. With two exceptions, the molecular weights corresponded to the nominal values. The exceptions were PEO 100,000, which had an actual molecular weight of 230,000, and the 20M compound, whose actual molecular weight was 4000 rather than the nominal value of 20,000.

2.2.2. DSC Measurements

Polymer solutions with PEG weight fractions between 0.2 and 0.4 were prepared by dissolving the polymer in distilled, deionized water. Approximately 50 μl of each sample was pipetted into a silver sample pan (capacity 70 μl) and hermetically sealed. An empty sample pan was used as a control for the DSC measurements. Samples were placed in the sample chamber of a Seiko (Torrance, CA) model DSC100 differential scanning calorimeter. Cooling experiments were performed by cooling from 30 °C at 2 °C min^{-1}, and terminated when the sample temperature reached −50 °C.

For heating experiments, samples were initially cooled from room temperature at approximately 2 or 5 °C min^{-1} to −60 °C. The sample temperature was then allowed to rise gradually (at ~1 °C min^{-1}) to −50 °C, where the measurements began. The samples were heated at a rate of 1 °C min^{-1} to 30 °C. In a few cases, samples were cooled to −80 °C, and the thermal scan was carried out from −75 °C. This was done in order to determine what, if any, phase transitions took place at temperatures below −50 °C.

2.2.3. Observation of Phase Separation

Liquid and solid phases in equilibrium were observed directly using the following procedure. A 30% (w/w) solution of PEG 3350 was rapidly cooled in a dry ice–acetone bath until it had solidified. It was then placed in a temperature-controlled centrifuge precooled to $-10\,^{\circ}C$ and spun at 800*g* for 30 min. A thermocouple was then inserted into the sample in order to verify that its temperature had reached $-10\,^{\circ}C$. No phase separation took place when the sample was cooled to $-10\,^{\circ}C$ from room temperature because the liquid merely became supercooled. Liquid–solid equilibrium was reached, however, when $-10\,^{\circ}C$ was approached from a lower temperature.

2.3. RESULTS AND DISCUSSION

2.3.1. Thermal Behavior

Heating scans of PEG solutions commonly yield thermograms of the form shown in Figure 1. There is one exothermic peak, at approximately $-50\,^{\circ}C$, and two endothermic peaks, both falling below the freezing point of pure water. The exotherm is associated with changes in ice structure.[15] The two endotherms, which are of the

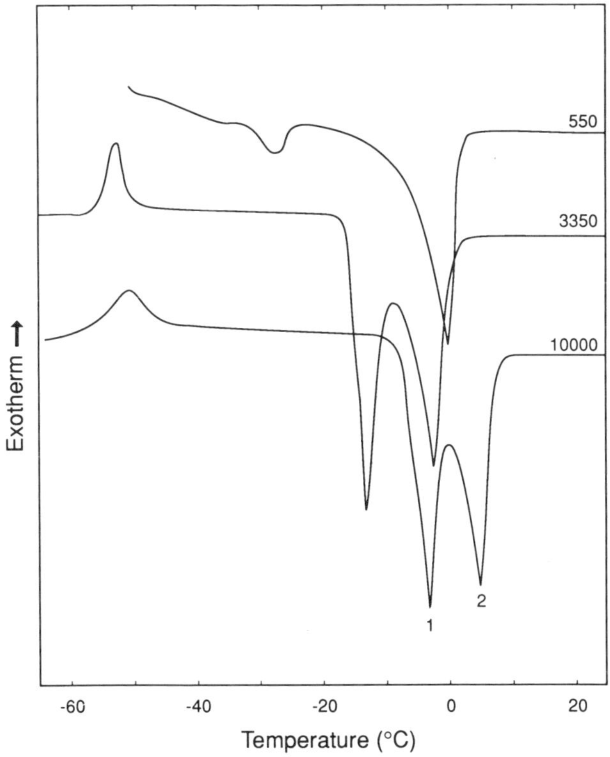

Figure 1. Thermograms during heating for 30% PEG solutions at three different molecular weights. The peak numbering system is shown for PEG 10,000.

most interest in this study, are identified as peaks 1 (lower) and 2 (upper), as shown in the figure. There are several lines of evidence that suggest that peak 1 is caused by the melting of a polymer hydrate of fixed composition, and that peak 2 is caused by the melting of ice. For example, in the latter case, the size of peak 2 increases as the water content of the sample rises.[9–11] The temperature of the peak 2 transition is at or just below the freezing point of water, which also supports the conclusion that the peak is due to the melting of ice. In the temperature range studied, PEG of molecular weights 400 and below yield only one, rather broad endothermic peak, which is presumably peak 2 and caused by the melting of ice.

With respect to peak 1, it has been shown that for PEG of a given molecular weight, the transition temperature is independent of the PEG concentration. On this basis, Bogdanov and Mihailov[10] and Hager and MacRury[9] argue that peak 1 is due to the melting of a polymer hydrate containing approximately two waters per polymer repeat unit. We have obtained direct visual confirmation that the two peaks correspond to two solid-to-liquid transitions. A 30% solution of PEG 3350, when cooled to approximately −40 °C, became a hard white solid. After it was warmed to −10 °C, the solution softened. Centrifugation at this temperature yielded a white solid phase, less dense than the underlying clear liquid. As Figure 1 shows, only one endothermic transition has occurred at this temperature. The high viscosity of the liquid phase at this temperature precluded phase separation for analysis. This suggests that at this temperature, a concentrated polymer solution is in equilibrium with ice.

As with most aqueous solutions, the enthalpy change associated with the melting of ice or the freezable water (peak 2) is less than the enthalpy change expected, based on the total water content of the sample.[16] Based on the evidence above, we used the area of peak 2 to estimate the amount of freezing (or "free") water in the sample, and took the remainder to be "bound" water. The results follow in the next section.

The thermograms obtained from cooling experiments were quite different, and only a single exotherm was seen for polymer molecular weights of 150, 750, 1450, and 3350 (only the cooling curve for PEG 3350 is shown in Figure 2). We found two exotherms only with the highest-molecular-weight sample studied, PEO 100,000. The area of the single exotherm in Figure 2 for PEG 3350 does not, however, correspond to the sum of the areas of peaks 1 and 2 for this PEG solution in Figure 1; rather, its area is close to (but not exactly the same as) the area of peak 2, corresponding to the melting of ice in Figure 1. The lack of similarity between the heating and cooling curves means that the single peak in the cooling curve must be interpreted separately. The area of the peak and its value when extrapolated to low polymer concentrations indirectly suggests that the single exotherm corresponds to the formation of only ice.[12] Since the expected two endotherms are seen upon rewarming a sample exhibiting a single exotherm during cooling, the polymer hydrate must form rather slowly. This may be caused by the high viscosity of the liquid phase, as noted above from visual observations, which would reduce the rate of reorientation of the polymer chains as the hydrate forms and lead to supercooling of the liquid phase. This supercooling is clearly seen in a comparison of the heating and cooling curves for PEG 3350 (Figure 3).

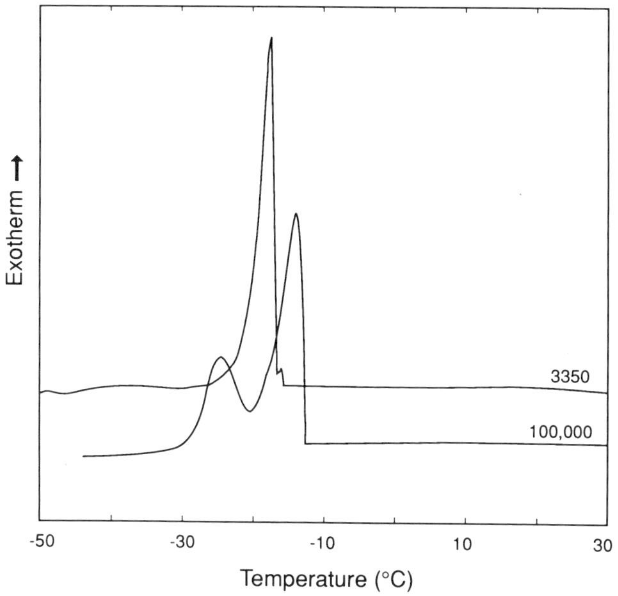

Figure 2. Thermograms during cooling for 30% PEG solutions at two different molecular weights.

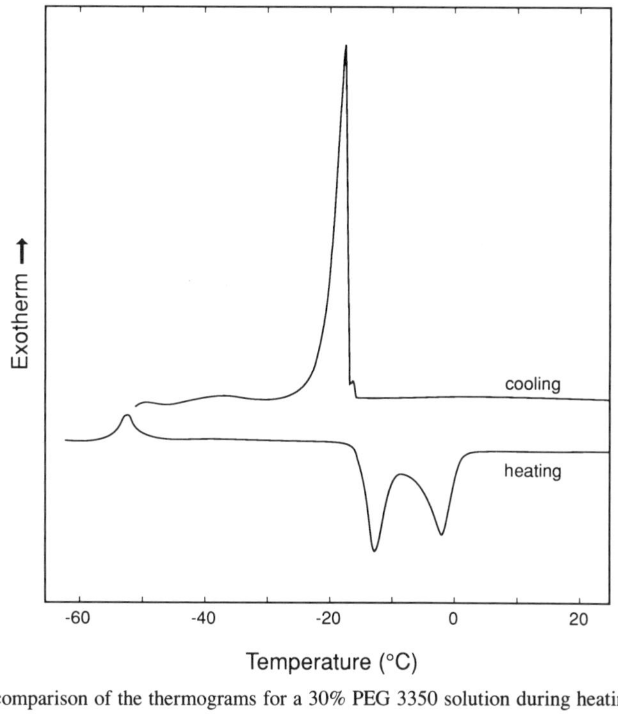

Figure 3. A comparison of the thermograms for a 30% PEG 3350 solution during heating and cooling.

2.3.2. Determination of Bound Water

Given that peak 2 of the thermograms from heating experiments is unambiguously caused by the melting of free water, its area, or the heat of fusion, ΔH_f, can be used to determine the fraction of freezable water in the sample. By difference, the amount of bound, or nonfreezing, water is obtained. This is expressed algebraically as[9,12]

$$\phi = \frac{1}{w_1}\left(w_2 - \frac{\Delta H_f}{\Delta H_w}\right)\frac{M_p}{M_w} \tag{1}$$

where ϕ is the number of bound water molecules per polymer repeat unit, M_p the molecular weight of polymer repeat unit, M_w the molecular weight of water, w_1 the weight fraction of polymer, w_2 the weight fraction of water, ΔH_f the heat of fusion from heating experiment, and ΔH_w the heat of fusion of pure water at the peak 2 transition temperature.

At low polymer molecular weights, there is significant freezing-point depression, so Eq. (1) must be modified to account for it. This correction is

$$\phi = \frac{1}{w_1}\left(w_2 - \frac{\Delta H_f}{\Delta H_w^0 + (C_{p,w} - C_{p,i})(T_f - T_0)}\right)\frac{M_p}{M_w} \tag{2}$$

where $C_{p,w}$ is the mean specific heat of liquid water over the temperature range T_f to T_0, $C_{p,i}$ the mean specific heat of ice, T_f the melting point of the polymer solution, T_0 the melting point of pure water (0 °C), and ΔH_w^0 the heat of fusion of pure water at 0 °C. The quantity $C_{p,w} - C_{p,i}$ can be taken as 2.3 J g^{-1}°C^{-1}.[12,17] A further correction for the heat of mixing is not needed because it is small compared to the heat of fusion.[12,18]

The heat of fusion of pure water was measured using a 50 μl sample under the same conditions as the other experiments. We found ΔH_w to be 307 J g^{-1}, a value that is somewhat below the literature value of 334 J g^{-1}. The measured value was used in Eq. (2).

Equation (2) was used to estimate the amount of bound water in 30% (w/w) PEG solutions from heating scans. This PEG concentration was chosen because peaks 1 and 2 in Figure 1 are approximately the same size over a wide molecular weight range. Thus, the melting processes of both the polymer hydrate and frozen free water could be monitored. The results are shown in Figure 4. No significant differences can be detected in the measurements of water bound to PEG or MPEG. Thus, a single terminal methoxy group, at least at molecular weights of 550 and higher, has no significant effect. More significantly, Figure 4 shows that the amount of "bound" water varies from 2.3 to 3.8 waters per repeat unit, with a break in the curve at a molecular weight of 1000. The amount bound at higher molecular weights is significantly higher than at low molecular weights.

Data obtained for low-molecular-weight PEGs by Tilcock and Fisher[13] are also shown in Figure 4. Their measurements were made by finding the polymer concentration at which peak 2 disappears, and then assuming that all of the water present was

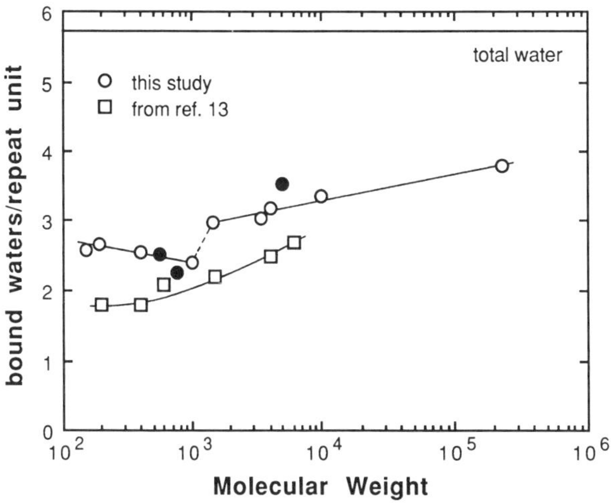

Figure 4. The amount of bound water, based on heating experiments, as a function of molecular weight. The circles represent our data for 30% PEG (open) and MPEG (closed) solutions, using Eq. (2). The squares represent data from ref. 13, obtained as discussed in text.

bound. Although their values are somewhat lower than ours, they are reasonably close, and in both cases the amount of bound water rises with molecular weight.

A number of methods can be used to assess the significance of the data shown in Figure 4. In Figures 5 and 6, ΔH_f is plotted as a function of polymer concentration for two polymer molecular weights. If the amount of water bound per polymer repeat unit

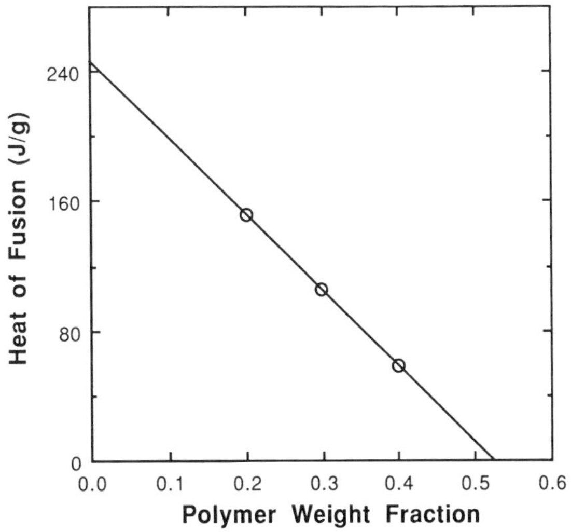

Figure 5. The heat of fusion, corrected for freezing point depression, as a function of PEG 400 concentration. The line represents the linear best fit to the data.

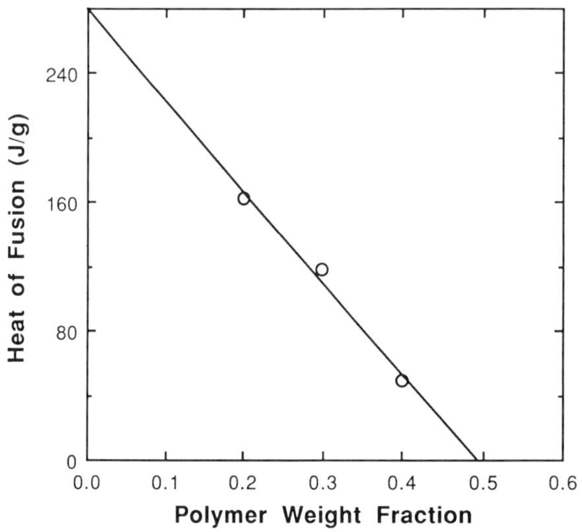

Figure 6. The corrected heat of fusion, based on peak 2, as a function of MPEG 750 concentration.

is independent of concentration, then this plot should be linear. The value of ΔH_f will be zero when all of the water present is bound; when extrapolated to pure water ($w_1 = 0$), ΔH_f should equal the pure-water value. With this method, the calculation of bound water does not depend on a single measurement, and the consistency of the best fit to the data with the known heat of fusion for pure water can be checked. For a large set of PEG samples, as used in this study, it would be impractical to make scans of solutions at multiple concentrations, so only two polymer molecular weights were chosen for this study.

Such plots were constructed for PEG solutions with polymer weight fractions between 0.2 and 0.4, shown in Figures 5 and 6. The ΔH_f values were corrected for freezing point depression by inverting Eq. (2). An initial estimate of the amount of bound water was used for the corrections, and then compared with the amount calculated using linear regression. The amount of bound water was then adjusted by trial and error until the calculated value was consistent with the assumed value.

We found that $\Delta H_f = 0$ for PEG 400 at a polymer weight fraction (w_1) of 0.53, corresponding to 2.2 bound water molecules per repeat unit. The corresponding value in Figure 4 was 2.5. Similarly, ΔH_f goes to zero for MPEG 750 at $w_1 = 0.49$, equivalent to 2.5 bound waters per repeat unit. The value in Figure 4 in this case was 2.3. Thus, the two methods for calculating the amount of bound water for these two molecular weights yielded comparable results.

The value of ΔH_f for pure water, $\Delta \hat{H}_f(0)$, obtained by extrapolating to $w_1 = 0$, was 280 J g^{-1} for MPEG 705—only 9% lower than the measured value (see above). However, such good agreement was not seen for PEG 400, for which $\Delta \hat{H}_f(0)$ was 246

J g^{-1}, almost 20% low. The origins of this discrepancy and its effect on our conclusions deserves particular comment.

Such low values of $\Delta\hat{H}_f(0)$ were also observed by de Vringer *et al.*[12] They attribute this to overlap between peaks 1 and 2 in the heating thermogram. The area assigned to peak 2 is then smaller than it ought to be. In addition, the assignment of peak area at low molecular weights is difficult because the peak is so broad. As the polymer molecular weight rises, the transition temperatures draw closer together, and peak overlap increases. This may account for some of the small increase in bound water with molecular weight seen in Figure 4.

There is no problem with broad or overlapping peaks when thermograms obtained during *cooling* are used, at least for PEG of molecular weight 3350 and below, as in Figure 2. On this basis, de Vringer *et al.*[12] advocate using cooling curves to calculate bound water. The validity of their approach was confirmed by both the linearity of plots of ΔH_f vs. w_1 and the accuracy of their values for $\Delta\hat{H}_f(0)$. We would argue that the use of cooling curves provides only a partial answer to the issues connected with the use of heating curves. First, multiple peaks may still be found in cooling curves of PEGs of sufficiently high molecular weight; peak overlap is still problematic in those cases. Second, we found that it was difficult experimentally to get reliable values for ΔH_f. For pure water, we obtained a value, corrected for supercooling, of 227 J g^{-1}, which is low by 26%.

Nevertheless, for the molecular weights studied here (from 150 to 100,000), the amounts of bound water calculated from experiments with 30% PEG were quite consistent, ranging from 2.3 to 2.8 (Figure 7). At low molecular weights, these values are the same as those obtained from heating curves (Figure 4). At higher molecular

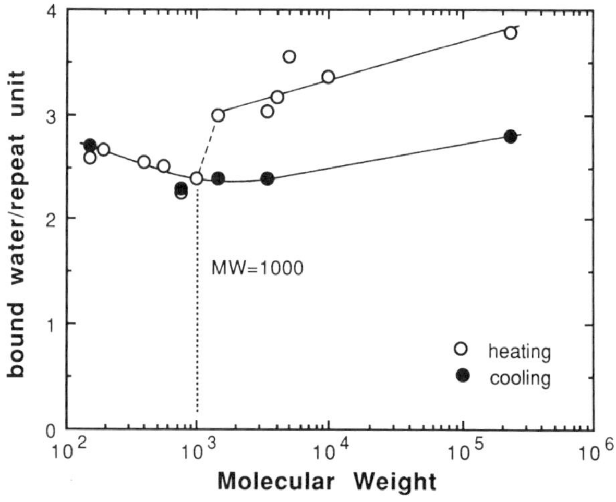

Figure 7. A comparison in the amount of bound water in 30% PEG solutions calculated using heating and cooling curves.

weights, however, these values remain lower. The reason for this difference will be explored in more detail below. In any case, these data confirm the general conclusion that between two and three water molecules are bound per repeat unit (consistent with earlier results),[8–13] and within this range there may be a gradual rise with molecular weight beginning between 800 and 1000.

2.3.3. Phase Transition Behavior

While the amount of bound water is only a weak function of molecular weight, the same cannot be said for the actual temperatures of phase transition. These were obtained from heating curves, in which supercooling is not present. The transition temperatures are plotted in Figure 8 for 30% solutions. The melting points, corresponding to peak 2, are lowest at low molecular weights, and rise toward the melting point of ice as the molecular weight rises. This can be attributed to the colligative effect, since at low molecular weights there are significantly more polymer molecules than at higher molecular weights. The magnitude of the freezing point depression is much greater than predicted by ideal-solution behavior, but is consistent with the highly nonideal character of PEG solutions, as shown by vapor-pressure[18] and osmotic-pressure[19] measurements. These studies demonstrated significant negative deviations from ideality.

The other temperature transition (peak 1) corresponds to the melting of a polymer hydrate whose composition is approximately 50% water.[10] The composition of this hydrate varies slightly with molecular weight, ranging from 50 to 55% water as the PEG molecular weight goes from 1000 to 1.8×10^6.[10] The hydrate at low PEG molecular weights could not be induced to form at temperatures as low as $-80\,°C$, but

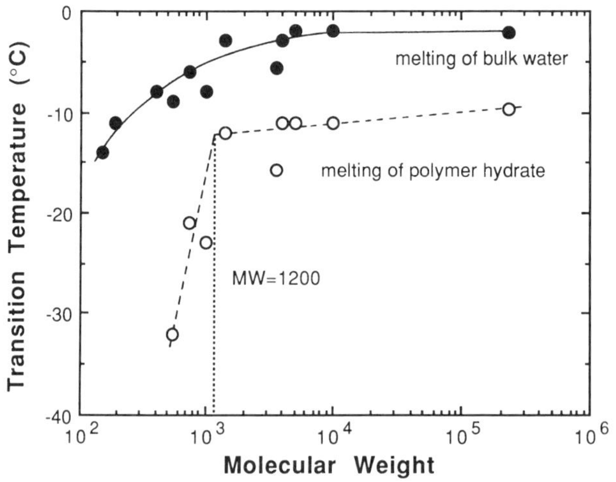

Figure 8. Transition temperatures for 30% PEG solutions as functions of molecular weight.

with molecular weights above 550, as shown in Figure 8, the melting points increased rapidly with molecular weight before leveling off at approximately −10 °C, at a molecular weight of 1200.

To make sense of the transition temperature data, it is helpful to examine the energetics of phase transition as well. Figure 9 shows the total enthalpy change required to go from solid to liquid (the sum of peaks 1 and 2). There is a break in the curve centered at a molecular weight of 550. Since the area of peak 2 is only a weak function of molecular weight, this break is primarily due to changes in the area of peak 1, or the heat of fusion of the polymer hydrate. The transition takes place rapidly over a molecular range between 400 and 1000. This is in the same range where the melting point of the hydrate rises rapidly.

2.3.4. Summary

The data presented here collectively show that a fundamental change in the structure of the hypothetical polymer hydrate takes place near a molecular weight of 1000. These data are summarized in Table 1 along with several other similar, molecular-weight-sensitive properties noted by others. The one exception is protein precipitation, which exhibits no break. The change in hydrate structure is consistent with the hypothesis that, as the polymer molecular weight rises and the polymer chains are more able to fold upon themselves (adopt a "secondary" structure), loosely-bound water molecules can be more easily shared between adjacent segments of a single chain, e.g., as in a helix. This accounts for the small rise with molecular weight in the amount of water contained in the polymer hydrate.[10] The data on the freezing of water are also consistent with this picture. The polymer hydrate in general

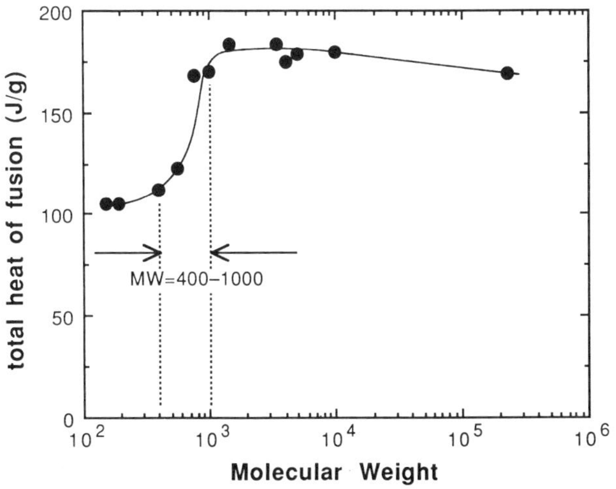

Figure 9. The total heat of fusion (i.e., based on the sum of peaks 1 and 2) for 30% PEG solutions.

Table 1. Critical Values of PEG Molecular Weight at Which "Changes" in Trends in Polymer Properties Are Noted

Property	Molecular weight	Source
Melting temperature of PEG-water	1200	this study
Amount of bound water (from heating scan)	1000	this study
Enthalpy of fusion of 30% aqueous solutions	400–1000	this study
Melting temperature (pure polymer)	2000	23
Platelet adhesion to PEG-containing hydrogels	660	7
Albumin solubility in aqueous PEG solutions	none	22
Polymer relaxation time (NMR)	3400	24
Onset of thermally-induced phase separation	2140–2180	25
IgG, albumin, and serum adsorption to PEG-coated surfaces	750–1900	20, 21

solidifies slowly.[12] Thus, during cooling only one exotherm is exhibited, corresponding to the freezing of free water. These cooling scans reveal that over a wide range of molecular weights, spanning three orders of magnitude, the amount of tightly bound water is constant at between two and three water molecules per repeat unit. This is in accord with measurements made using a broad array of techniques. At high PEG molecular weights, some of this free water is loosely trapped within the hydrate, becoming bound only as the hydrate forms. As the frozen ice–hydrate mixture is heated, this loosely bound water is released, causing the overlap in the two endotherms of Figure 1 and the higher values for bound water upon heating.

This study provides some insight into the effects of PEG molecular weight on the biocompatibility of PEG-coated surfaces. The amounts of both protein adsorption and platelet adhesion are strong functions of PEG molecular weight below 750–1900 daltons (depending on details such as the immobilization method and substrate), and thereafter level off.[7,20,21] The studies described here suggest that the underlying reason is that, at low molecular weights, the polymer chain is not long enough to fold into a hydrated coil; once it has done so, high-molecular-weight solutes are excluded from the PEG coil on the surface. Molecular weight effects level off at higher molecular weights for PEG on surfaces than for PEG in solution. This can be attributed to the relative lack of mobility due to the end of the polymer chain that is bonded to the surface.

2.4. CONCLUSIONS

The model of PEG solution behavior that has been developed here can be used to explain the unique behavior of PEG in conferring protein resistance to surfaces. We hypothesize that at low molecular weights, only tightly-bound water is associated with the PEG chain. As the molecular weight rises, the chain begins to fold in on itself, forming segment–segment interactions as it traps additional, more loosely bound water between the segments. This water functions to "bind together" the polymer coil, making it difficult for proteins to interact with it. Thus, chain coiling and water hydration interactions provide the common link connecting a number of molecular-weight-dependent properties of PEG.

ACKNOWLEDGMENTS

The authors are pleased to thank the Washington Technology Centers for financial support, Professor Paul Yager for the use of his DSC, and Dina Furin for assistance with the experiments. KPA is the recipient of a National Research Service Award through NIH grant HL07403.

REFERENCES

1. R. L. Davidson and P. S. Gerald, *Methods Cell Biol. 15*, 325 (1977).
2. M. Fried and P. W. Chun, *Methods Enzymol. 22*, 238 (1971).
3. A. Abuchowski, G. M. Kazo, C. R. Verhoest, T. van Es, D. Kafkewitz, M. L. Nucci, A. T. Viau, and F. F. Davis, *Cancer Biochem. Biophys. 7*, 175 (1984); J. S. Beckman, R. L. Minor, Jr., C. W. White, J. E. Repine, G. M. Rosen, and B. A. Freeman, *J. Biol. Chem. 263*, 6884 (1988); F. F. Davis, A. Abuchowski, T. van Es, N. C. Palczuk, R. Chen, K. Savoca, and K. Wieder, in: *Enzyme Engineering* (G. B. Brown, G. Nanecke, and L. B. Wingard, Jr., eds.), Vol. 4, p. 169, Plenum Press, New York (1978); N. V. Katre, M. J. Knauf, and W. J. Laird, *Proc. Natl. Acad. Sci. U.S.A. 84*, 1487 (1987).
4. E. W. Merrill and E. W. Salzman, *ASAIO J. 6*, 60 (1983).
5. S. Nagaoka, Y. Mori, H. Takiuchi, K. Yokota, H. Tanzawa, and S. Nishiumi, *Polym. Preprints 24*, 67 (1983).
6. S. Nagaoka, H. Takiuchi, K. Yokota, Y. Mori, H. Tanzawa, and T. Kikuchi, *Kobunshi Ronbunshu 39*, 165 (1982); S. Nagaoka, Y. Mori, H. Takiuchi, K. Yokota, H. Tanzawa, and S. Nishiumi, in: *Polymers as Biomaterials* (S. W. Shalaby, A. S. Hoffman, B. D. Ratner, and T. A. Horbett, eds.), p. 361, Plenum Press, New York (1984).
7. Y. Mori, S. Nagaoka, H. Takiuchi, T. Kikuchi, N. Noguchi, H. Tanzawa, and Y. Noishiki, *Trans. Am. Soc. Artif. Intern. Organs 28*, 459 (1982).
8. J. Breen, D. Huis, J. de Bleijser, and J. C. Leyte, *J. Chem. Soc., Faraday Trans. 1 84*, 293 (1988); G. N. Ling and R. C. Murphy, *Physiol. Chem. Phys. 14*, 209 (1982); V. D. Zinchenko, V. V. Mank, V. A. Moiseev, and F. D. Ovcharenko, *Kolloidn. Zh. 38*, 44 [*Chem. Abstr. 84*, 136208] (1976).
9. S. L. Hager and T. B. MacRury, *J. Appl. Polym. Sci. 25*, 1559 (1980).
10. B. Bogdanov and M. Mihailov, *J. Polym. Sci., Polym. Phys. Ed. 23*, 2149 (1985).
11. N. B. Graham, M. Zulfiqar, N. E. Nwachuku, and A. Rashid, *Polymer 30*, 528 (1989).
12. T. de Vringer, J. G. H. Joosten, and H. E. Junginger, *Colloid Polym. Sci. 264*, 623 (1986).
13. C. P. S. Tilcock and D. Fisher, *Biochim. Biophys. Acta 688*, 645 (1982).
14. R. Kjellander and E. Florin, *J. Chem. Soc., Faraday Trans. 1 77*, 2053 (1981).
15. Z. L. Zhang and G. N. Ling, *Physiol. Chem. Phys. Med. NMR 15*, 407 (1983).
16. F. Franks, in: *Water: A Comprehensive Treatise* (F. Franks, ed.), Vol. 7, p. 215, Plenum Press, New York (1982).
17. J. H. Awbery, in: *International Critical Tables* (E. W. Washburn, ed.), Vol. 5, p. 95, McGraw-Hill, New York (1929).
18. G. N. Malcolm and J. S. Rowlinson, *Tans. Faraday Soc. 53*, 921 (1957).
19. H. Vink, *Eur. Polym. J. 7*, 1411 (1971).

20. V. Hlady, R. A. Van Wagenen, and J. D. Andrade, in: *Surface and Interfacial Aspects of Biomedical Polymers* (J. D. Andrade, ed.), Vol. 2, p. 81, Plenum Press, New York (1985).
21. K. Bergstrom, K. Holmberg, A. Safranj, A. S. Hoffman, M. J. Edgell, B. A. Hovanes, and J. M. Harris, "Reduction of Fibrinogen Adsorption on PEG-Coated Polystyrene Surfaces," *Biomaterials*, in press.
22. D. H. Atha and K. C. Ingham, *J. Biol. Chem. 256*, 12108 (1981).
23. A. Altmeyer, V.-H. Karl, and K. Ueberreiter, *Makromol. Chem. 182*, 3311 (1981).
24. G. G. Hammes and P. B. Roberts, *J. Am. Chem. Soc. 90*, 7119 (1968).
25. S. Saeki, N. Kuwahara, M. Nakata, and M. Koneko, *Polymer 17*, 685 (1976).

3

Molecular Simulation of Protein–PEG Interaction

KAP LIM and JAMES N. HERRON

3.1. INTRODUCTION

Poly(ethylene glycol) (PEG) is a water-soluble polymer that exhibits properties such as protein resistance, low toxicity, and nonimmunogenicity.[1–5] These properties have been attributed to its segmental flexibility and its polar, but uncharged, chemical composition. This segmental flexibility produces a high degree of steric exclusion and entropy at PEG–water interfaces which in turn leads to protein resistance. Its exclusion property also enables the precipitation of proteins without denaturation.[6–9]

Although PEG exhibits a high degree of biocompatibility, it lacks the mechanical properties necessary to replace materials such as polyurethane and silica. However, grafted PEG chains on the surfaces of these relatively rigid materials can make them more biocompatible. One feature of grafted PEG surfaces is their low degree of protein adsorption. The key to this phenomenon is the mobility of the grafted chains. It is much harder for a protein to get a foothold on an uneven surface that constantly changes its topology than on a smooth, rigid surface. Moreover, since the flexible PEG chains can readily adapt their conformation to fit the surface topology of a protein, there is very little tendency for the protein to undergo significant conformational changes. In addition the electrical neutrality of PEG may diminish the differences in its interaction with the hydrophobic or hydrophilic parts of the protein surface.

In this chapter, we examine the use of computer simulation methods to study the interfacial behavior of proteins on grafted PEG surfaces. We used an atomistic approach; molecular mechanics and molecular dynamics techniques were used to study the energetics and dynamics of protein–PEG interactions. Historically, molecular modeling studies often employed molecular liquids because of their simplicity and

---

KAP LIM and JAMES N. HERRON • Center for Biopolymers at Interfaces, and Departments of Bioengineering and Pharmaceutics, University of Utah, Salt Lake City, Utah 84112.

Poly(Ethylene Glycol) Chemistry: Biotechnical and Biomedical Applications, edited by J. Milton Harris. Plenum Press, New York, 1992.

various applications in statistical mechanics.[10] Hard spheres were first used as simple models of liquid molecules in molecular dynamics simulations.[11,12] With advances in computer hardware, more complicated macromolecular simulations of proteins, nucleic acids, lipids, and synthetic polymers have become routine.

Molecular mechanics approaches are largely empirical in the sense that the nominal potential-energy parameters assigned for bond distances, bond angles, torsion angles, and nonbonded interactions are determined from experimental techniques such as infrared spectroscopy and X-ray diffraction. These parameters have also been determined from semiempirical and *ad initio* quantum mechanical calculations. Although empirical, molecular mechanics simulations are very useful in the study of systems such as biological macromolecules and synthetic polymers. True simulation of molecular interactions can be most closely approximated with quantum mechanics, which require many hours of today's fastest supercomputers to study only a few tens of atoms.

Molecular mechanics procedures include energy minimization techniques to study static structures, and molecular dynamics (MD) and Monte Carlo techniques to study the dynamic behavior of molecules. In Monte Carlo simulations, atoms are moved along a random trajectory, but only if the energy of the next step is lower than the previous one. If the energy is not lower, atomic positions are determined from the Boltzmann factor $\exp(-\Delta E/kT)$.[13,14] In molecular dynamics, Newton's equations of motion are solved for each atom of the system, which allows the time parameter to be recorded. Another technique called Brownian dynamics assigns frictional coefficients to molecules to overcome local energy barriers.[15,16] With the current capacity of supercomputers, a molecular system of several thousand atoms can be simulated over a period of a few hundred picoseconds. For a small peptide, a few nanoseconds of MD have been performed.[17]

In this chapter, we describe two different series of simulations. In the first series, grafted PEG surfaces with three different densities of PEG chains were simulated and their equilibrium properties were studied. In the second series of simulations, we examined the interactions of these surfaces with two different proteins—lysozyme and myoglobin. Each protein was rotated on top of a grafted PEG surface and the interaction energy of the protein with the surface was calculated. As a comparison, the interactions of these proteins with a crystalline polyethylene (PE) surface were also simulated. Finally, energy minimization and molecular dynamics simulations were used to provide a detailed molecular picture of the interfacial behavior of lysozyme on grafted PEG surfaces. Due to constraints in computation time, no solvent was included in simulations except for the MD of single polymer chains.

3.2. BACKGROUND

3.2.1. Theoretical Studies of Polymers

Several different theoretical approaches have been used to study the conformation and dynamics of polymers. These include statistical mechanics, molecular mechanics, Brownian dynamics, and Monte Carlo methods.[14,18–23] These approaches have also been used to model the adsorption of polymer chains on surfaces.[24–26] To date, most theoretical approaches to grafted polymeric surfaces have been based on scaling theory and self-consistent mean-field approximation theory.[27–31] The results of recent molecular dynamics simulations of a generalized grafted polymer brush were consistent with the latter theory.[32]

3.2.2. Modeling of Protein Adsorption

The focus of most computer simulations of biological macromolecules has been on the proteins themselves *in vacuo* or in solution. Also, molecular modeling studies of protein–ligand interactions have become more prevalent in the last few years, especially in the area of rational drug design.[33] However, only a few simulations of protein adsorption have been reported. An early computer simulation by Morrissey attempted to correlate the distance between lysozyme and a polymeric surface with the experimentally measured protein coverage.[34] More recent modeling studies include the characterization of protein–PEG surface interactions,[35,36] which will be further discussed in Sections 3.5.4.3 and 3.5.7, and the study of adsorption of an ionic homopolypeptide as a function of pH.[37]

Most recently, a modeling study of interactions between proteins and polymeric surfaces has been reported, but its methods and results were limited.[38] Our approach is to include all the atomic coordinates for both proteins and polymeric surfaces. This enables us to look directly at the conformation and dynamics of the protein–polymer interaction.

3.3. A Brief Summary of the Molecular Simulation Method

The methodology and application of molecular simulations of macromolecules are well established. There are several good reviews of this topic in the literature.[39–41] The basis of molecular simulations is the potential energy function

$$V = \underbrace{\tfrac{1}{2}\sum_b K_b(b - b_0)^2}_{(1)} + \underbrace{\tfrac{1}{2}\sum_\theta K_\theta(\theta - \theta_0)^2}_{(2)} + \underbrace{\tfrac{1}{2}\sum_\phi K_\phi[1 + s\cos(n\phi)]}_{(3)} + \underbrace{\sum\left[\frac{A}{r^{12}} - \frac{C}{r^6}\right]}_{(4)} + \underbrace{\sum\frac{q_i q_j}{Dr_{ij}}}_{(5)} \tag{1}$$

with terms for (1) bond length, (2) valence angle, (3) torsion angle, (4) van der Waals interaction (repulsion and attraction term, respectively), and (5) electrostatic interaction (D is the dielectric constant). The quadratic bond length term may be replaced by a Morse potential, and cross terms may be added in energy minimizations for accurate comparisons with the vibrational data.[42] The weakest part of the potential function is the electrostatic interaction term. The electrostatic term of Eq. (1) is only valid for charged particles with a uniform dielectric constant. Since the dielectric constant is a macroscopic parameter, it may not apply to an atomic level simulation. Attempts to describe the Coulombic effect have been made with varying degrees of success.[43]

Molecular dynamics calculations are made with the Newtonian equations of motion:

$$\mathbf{F} = m\mathbf{a}$$

$$-\frac{\partial V}{\partial \mathbf{r}_i} = \mathbf{F} = m_i \frac{d^2\mathbf{r}_i}{dt^2} \tag{2}$$

The force, or negative first derivative of the potential energy function, is solved analytically and the acceleration is calculated for each atom. The most common algorithm to solve the equations of motion is the leap-frog Verlet method:

$$\mathbf{r}(t + \Delta t) = \mathbf{r}(t) + \Delta t\, \mathbf{v}(t + \tfrac{1}{2}\Delta t) \tag{3a}$$

$$\mathbf{v}(t + \tfrac{1}{2}\Delta t) = \mathbf{v}(t - \tfrac{1}{2}\Delta t) + \Delta t\, \mathbf{a}(t) \tag{3b}$$

However, this method does not handle the velocities satisfactorily. The velocity Verlet method is an alternative that stores the positions, velocities, and accelerations all at the same time.[10] As a result, it minimizes roundoff errors:

$$\mathbf{r}(t + \Delta t) = \mathbf{r}(t) + \Delta t\, \mathbf{v}(t) + \tfrac{1}{2}\Delta t^2\, \mathbf{a}(t) \tag{4a}$$

$$\mathbf{v}(t + \tfrac{1}{2}\Delta t) = \mathbf{v}(t) + \tfrac{1}{2}\Delta t[\mathbf{a}(t) + \mathbf{a}(t + \Delta t)] \tag{4b}$$

Other methods, such as the Beeman method and the Gear method, have also been developed.[40]

3.4. Computational Method in PEG and Protein–PEG Interaction

3.4.1. PEG Simulation

Energy minimizations and molecular dynamics (MD) simulations of PEG and protein–PEG systems were performed with the DISCOVER program (BIOSYM Technologies, Inc.). MD simulations were performed at 300 K, unless otherwise noted. Periodic boundary conditions were used to allow a continuous field of the PEG chains. The initial system consisted of a molecular ensemble of twenty PEG chains (40 EO-units, 1762 MW) in a crystal conformation. Crystalline PEG is a helix that has torsion angles in a *trans–gauche*(+)–*trans* (TGT) conformation about the —O—C—C—O bonds.[44] Three sets of grafted PEG surfaces were constructed. In the first set, one end of each PEG chain was fixed and separated from the ends of the neighboring chains by 12 Å in a triangular pattern. In the second set, the separation distance was increased to 18 Å, and to 24 Å in the third set. The equivalent areas were 124.7, 280.6, and 498.8 $Å^2$ per chain, respectively. As will be shown later, these crystal PEG structures were very stable and remained extended throughout the MD runs. A second set of grafted PEG surfaces was constructed with the torsion angles in an all *gauche*(−) conformation (torsion angle of −60° for C—C bond and −105° for C—O bond). These surfaces collapsed into very compact structures during MD simulation. The different grafted PEG ensembles are summarized in Table 1.

Since no barrier surface was provided beneath the fixed ends of PEG chains, it was possible that the mobile parts of the chains would cross over the plane of the grafting surface. However, periodic boundary conditions and the nonbonded interactions [terms 4 and 5 of Eq. (1)] ensured that PEG chains would not approach the grafting surface too closely.

3.4.2. Protein–PEG Interaction

3.4.2.1. Simulation of Lysozyme and Myoglobin

Atomic coordinates for hen egg white lysozyme and sperm whale myoglobin were obtained from the Protein Data Bank.[45] Energy minimizations of both proteins were performed, and lysozyme was also equilibrated with molecular dynamics. Myoglobin's heme group was not included in the simulations due to the lack of parameters in the Consistent Valence Force Field employed by DISCOVER program.[46] However, since myoglobin was only used in rigid-body rotations and static calculations of interaction energies, the inclusion of heme was not essential.

3.4.2.2. Rotation of Lysozyme and Myoglobin over the Polymeric Surface

Lysozyme and myoglobin were "docked" with a PEG or PE surface with the ROT program written by one of us (K. L.). Potential energy calculations included a van der Waals term (the Lennard-Jones function) and a Coulombic energy term [terms 4 and 5 of Eq. (1), respectively]. The starting orientation of a protein on top of the polymeric surface was chosen arbitrarily: For lysozyme, the axis of the α-helical sequence from 86 to 101 was oriented parallel to the x-axis (Figure 1a), and this α-helix was on the positive y side relative to the rest of the lysozyme (Figure 1b). For myoglobin, the axis of the α-helical sequence from 125 to 148 was oriented parallel to the x-axis (Figure 2a), and this α-helix was on the positive y side relative to the rest of the myoglobin (Figure 2b). The polymer chains were aligned in the z-axis

Table 1. Identification of Different PEG Chain Systems

Initial conformation	Chain separation (Å)	Reference name
TGT	12	A1
TGT	18	A2
TGT	24	A3
All *gauche*(−)	12	B1
All *gauche*(−)	18	B2
All *gauche*(−)	24	B3

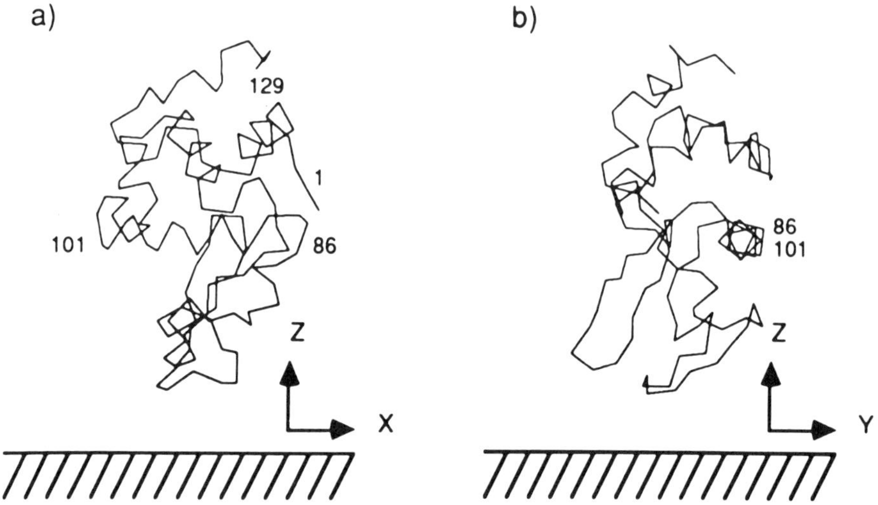

Figure 1. Starting orientation of lysozyme for rigid-body rotation on a polymeric surface: (a) view from *y*-axis; (b) view from *x*-axis.

direction. A cutoff criterion of 8.5 Å was used for all of the nonbonded calculations performed in this study. A switching function was employed to smooth the cutoff between 7.0 Å and 8.5 Å. For electrostatic calculations, a dielectric constant of 78 was used. The *z*-axis component of the center of mass of the protein was placed at different grid points over the PEG surface and the protein was rotated at each point. The protein was then rotated in 5° increments about the *x*- and *y*-axis for a total of 2592 rotations at each grid point.

For a homogeneous and symmetric surface such as PE, rotations around two axes can sample all of the rotational space for the protein. However, for a heterogeneous surface such as PEG, a *z*-axis rotation should also be made. Since this was computationally prohibitive, we elected to rotate the protein around the *x*- and *y*-axis at 25 different points on the PEG surface. Although this strategy did not fully compensate for the lack of *z*-axis rotation, more than enough data were obtained to show significant energetic differences between various regions of the protein.

The grid spacing of the 25 points is listed below for the different types of PEG surfaces:

PEG group	Chain separation distance (Å)	*x*-Axis grid spacing (Å)	*y*-Axis grid spacing (Å)
A1 or B1	12	2	1.5
A2 or B2	18	3	2
A3 or B3	24	4	3

Since periodic boundary conditions were not used in the rotation calculation, the grid spacings were confined to the center of the surface. In addition, grid spacings were

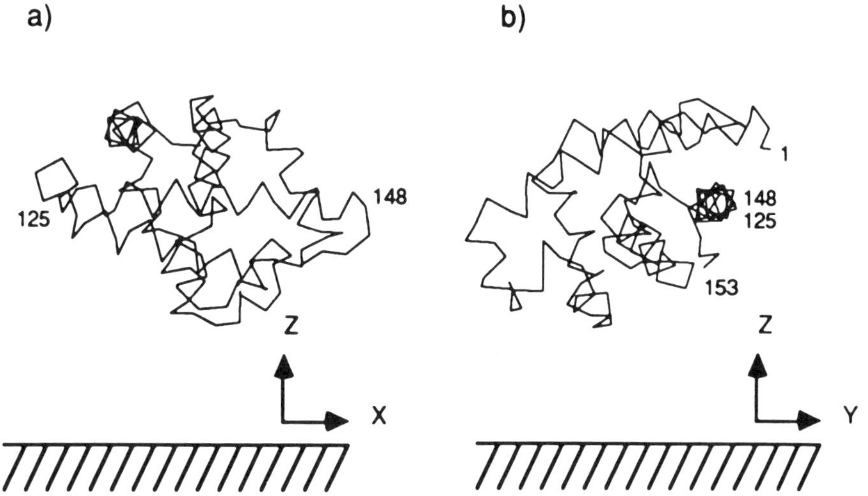

Figure 2. Starting orientation of myoglobin for rigid-body rotation on a polymeric surface: (a) view from *y*-axis; (b) view from *x*-axis.

varied in proportion to the density of each PEG group. For a protein on the flat PE surface, rotations were made only over the center of the polymeric surface. A 5° rotation was first made, and the protein was then placed above the polymeric surface in such a way that the distance between the lowest *z*-coordinate of the protein and the highest point of the polymeric surface was 8.5 Å. Then the protein was lowered along the *z*-axis toward the polymeric surface in steps of 0.5 Å until the interaction energy exceeded a somewhat arbitrary value of 10,000 kcal mol^{-1}, which signified an overlap between the protein and polymeric surface. The lowest interaction energy of the protein with the polymeric surface from these steps was recorded for each 5° rotation. At the point along the *z*-axis with the lowest interaction energy, all the exterior amino acid residues within 8.5 Å of the polymer chains were recorded. The identification of these residues described how the protein made contact with the polymeric surface.

The most time-consuming step of the protein rotation program was the generation of the neighbor list for nonbonded interactions. We used a cubic algorithm to generate neighbor lists in the ROT program. This algorithm was very similar to the one employed by the DISCOVER program, and eliminated the need for an exhaustive search of all possible atom pairs. The ROT program was vectorized and run on an IBM 3090 supercomputer, and DISCOVER was run on Silicon Graphics Iris 220/4S and 70D/GT computers.

3.4.2.3. Molecular Dynamics of the Lysozyme–Polymeric Surface System

Molecular dynamics simulations *in vacuo* were performed for lysozyme on three different polymeric surfaces—polyethylene, grafted PEG with 18 Å separation (A2 PEG), and grafted PEG with 24 Å separation (A3 PEG). For each of these surfaces, the most energetically favorable orientation of lysozyme was chosen as the starting model. The protein–polymer systems were first minimized for 5000 cycles and then MD runs were started. Initially, the temperature was gradually increased from 1 K to 300 K in increments of 30 K per 100 steps and then maintained at 300 K. The size of the step as 1 femtosecond (10^{-15} s). An MD run of 20,000 steps or 20 picoseconds was performed for each protein–polymer system. In lysozyme–PEG MD simulations, all the atoms were allowed to move except the fixed ends of PEG chains. In the lysozyme–PE MD simulation, only the protein's atoms were allowed to move. For comparison, an MD simulation of a free lysozyme molecule *in vacuo* was also performed.

3.5. RESULTS AND DISCUSSION

3.5.1. Analysis of Molecular Dynamics Trajectories of PEG

3.5.1.1. Molecular Dynamics of a Single Polymer Chain *in Vacuo* and in Water

Preliminary to the studies of grafted PEG chains, molecular dynamics simulations of single polymer chains were performed. MD simulations of an n-octadecane ($C_{18}H_{38}$) and PEG-600 (14 EO units) were performed *in vacuo* and in water. For octadecane in water, there were 478 water molecules in a 20×20×35 Å box with periodic boundary conditions. For PEG-600 in water, there were 579 water molecules in a 20×20×45 Å box. Radial distribution functions for these simulations are shown in Chapter 15 by Gölander *et al.*

A comparison of results for octadecane *in vacuo* and in water demonstrated the importance of the torsional barrier of the C—C bond. Although the energy barrier between the *trans* and *gauche* states is relatively high at 300 K (e.g., 4.6 kcal mol^{-1} for the middle C—C bond in octadecane), some conversions were observed. The extent of conversions of octadecane *in vacuo* seemed to depend upon the dielectric constant (Table 2). From the 50-picosecond MD, the average radius of gyration of octadecane (partial charge of H atom, $+0.1517e$) in water was 6.88 ± 0.08 Å. Also, only one torsional change was observed at one end of the octadecane chain. The dielectric constant was set to 1.0 when water was included in the simulation. These results showed that the solvent effect was minimal when the intramolecular energy barrier was a dominant factor for a long hydrocarbon molecule such as octadecane.

The simulations of octadecane provided an important test of the validity of the Consistent Valence Force Field used by the DISCOVER program. Of specific interest was the rate of conversion between the *trans* and *gauche* torsional states of a polymer chain. In MD simulations of liquid n-butane, a rate constant of 2.8×10^{11} s^{-1} at 292 K was predicted by transition state theory.[47] Results from Brownian dynamics showed that the rate varied with polymer length.[48] Also, studies by Helfand showed that the energy barrier between the *trans* and *gauche* states was frequently overcome by a cooperative transition of the two bonds separated by a *trans* bond.[49] The transition

Table 2. Characteristics of an Octadecane Molecule from MD *in Vacuo*[a,b]

A. Radius of gyration and end-to-end distance:

Partial charge of H atom (*e*)	Dielectric constant	Radius of gyration (Å)	End-to-end distance (Å)
+0.1000	1	6.40 ± 0.24	19.87 ± 1.76
+0.1000	78	6.58 ± 0.15	19.67 ± 1.14
0.0000	—	6.34 ± 0.27	19.18 ± 1.17
+0.1517	1	6.88 ± 0.14	22.49 ± 0.87
+0.1517	78	6.38 ± 0.16	19.67 ± 1.03

B. Total number of torsion angles:

Partial charge of H atom (*e*)	Dielectric constant	*trans*	*gauche*(+)	*gauche*(−)
+0.1000	1	12.7 ± 0.8	0.9 ± 0.4	1.4 ± 0.6
+0.1000	78	14.1 ± 0.8	0.3 ± 0.4	0.6 ± 0.6
0.0000	—	12.2 ± 0.9	1.6 ± 0.7	1.2 ± 0.5
+0.1517	1	14.9 ± 0.3	0.1 ± 0.3	0.0 ± 0.0
+0.1517	78	11.9 ± 0.7	2.0 ± 0.4	1.1 ± 0.7

[a]From 30–100 picoseconds; 700 coordinate files.
[b]All *trans* starting conformation.

was a localized one and aided by other bond rotations and contributions from bond stretching and bond-angle bending [terms 1 and 2 of Eq. (1), respectively]. A recent simulation of polypropylene showed conversions within 50 picoseconds of MD.[50] Given these results, our data were not unreasonable.

Simulations of a single PEG-600 chain showed that PEG exhibited a much greater segmental flexibility than octadecane. *In vacuo*, an extended helical PEG structure quickly assumed a compact, coiled conformation. Although this phenomenon was less pronounced when simulations were performed in water, PEG exhibited a relatively higher degree of segmental flexibility than octadecane both in water and *in vacuo*. The average radius of gyration for PEG-600 *in vacuo* was 5.34 ± 0.35 Å, while a value of 10.49 ± 0.24 Å was determined for the same molecule in water. These values were calculated from MD simulations over a period of 30–50 picoseconds (200 trajectory coordinates). Vacuum simulations approximated the polymer chain dynamics in a "poor" solvent, while results obtained in water reflected the dynamic behavior in a "good" solvent.

3.5.1.2. Molecular Dynamics of Grafted PEG Chains

Because simulations of multiple polymer chains in the presence of solvent molecules are computationally prohibitive, we used a different approach to simulate the conformation of grafted PEG surfaces in "poor" and "good" solvents. The thickness of PEG surfaces changes with different types of solvents. In simulations, the equilibrium structure of the PEG chains attained at the end of the simulation was influenced by their starting conformation. Specifically, an all-*gauche*(−) starting conformation was used to simulate grafted PEG surfaces in a "poor" solvent, and a *trans–gauche*(+)*–trans* (TGT) (*trans* for C—O torsion, *gauche*(+ for C—C torsion) starting conformation was used to simulate PEG surfaces in a "good" solvent.

The torsion angle energy plot of Figure 3a shows that there is essentially a downhill roll toward the *trans* conformation for the C—O bond of PEG. The all-*gauche*(−) starting conformation of PEG quickly collapsed into a relatively compact structure, compared to the TGT starting conformation. The initial torsion angle of the C—O bond in all-*gauche*(−) PEG was −105° (or 285°). Since this conformation was energetically unfavorable, the PEG structure changed rapidly into a mostly *trans* conformation. The equilibrium structure for the twenty-chain system was reached within 21 picoseconds. Table 3 lists averaged values of various parameters of interest. The density profiles of these PEG chains (Figure 4) show that they assumed relatively compact structures. The crystal conformation of TGT was used for the second set of PEG structures ("good" solvent). Because the initial TGT conformation was very stable, the overall appearance of the twenty-chain system did not change significantly during the course of the simulation. An average thickness of about 50 Å was observed for these PEG layers. Ellipsometry measurements were conducted in our laboratory to determine the thickness of PEG-2000 grafted to a silica surface.[51] These data were in close agreement with the results of the simulation. The thickness value of grafted PEG in air was comparable to that of equilibrated

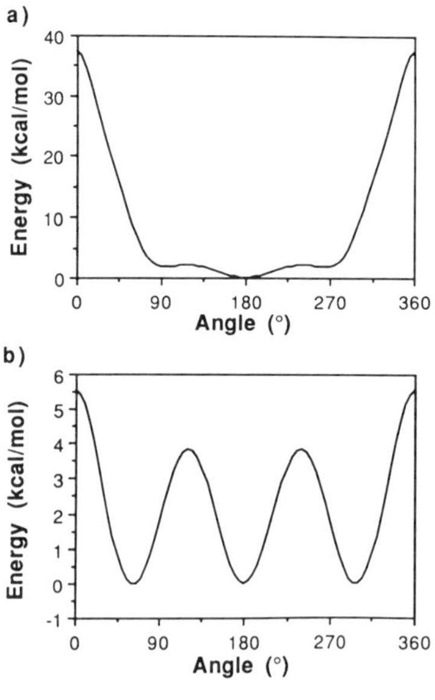

Figure 3. Torsional energy profile of PEG for the (a) C—O bond and (b) C—C bond.

Table 3. Characteristics of Twenty PEG Chains (40 EO Units) from MD Trajectories[a] for Different Fixed End Separations[b]

A. *Radius of gyration, z-component of radius of gyration (r.g.), end-to-end distance and average thickness*:

Separation distance (Å)	Radius of gyration (Å)	z-Component of r.g. (Å)	End-to-end distance (Å)	Thickness (Å)
12	11.74 ± 0.16	9.53 ± 0.25	31.20 ± 0.46	20.20
18	9.81 ± 0.08	6.82 ± 0.16	25.24 ± 0.44	15.63
24	9.34 ± 0.04	6.12 ± 0.19	20.83 ± 0.63	15.08

B. *Total number of C—O torsion angles*:

Separation distance (Å)	*trans*	*gauche*(+)	*gauche*(−)
12	1149.4 ± 13.7	83.5 ± 7.7	287.1 ± 18.5
18	1169.3 ± 14.1	76.7 ± 8.0	274.0 ± 18.3
24	1169.1 ± 12.9	68.7 ± 8.0	202.2 ± 16.5

C. *Total number of C—C torsion angles*:

Separation distance (Å)	*trans*	*gauche*(+)	*gauche*(−)
12	149.6 ± 6.8	218.4 ± 4.8	392.0 ± 9.4
18	106.7 ± 6.1	240.9 ± 10.0	274.0 ± 18.3
24	96.7 ± 7.7	205.5 ± 7.5	457.8 ± 14.8

[a]From 21–31 picoseconds; 100 coordinate files.
[b]All-*gauche*(−) starting conformation.

all-*gauche*(−) PEG, and the value of grafted PEG in water was close to that of TGT PEG.

One deficiency of these simulations was that the PEG chains could only interact with each other in the absence of solvents, which tended to overemphasize interchain interactions. Furthermore, in the case of the TGT PEG systems, this bias was localized near the free ends of the chains. If all parts of the PEG chain were allowed to move, they would seek a maximum number of contacts with each other, and the two

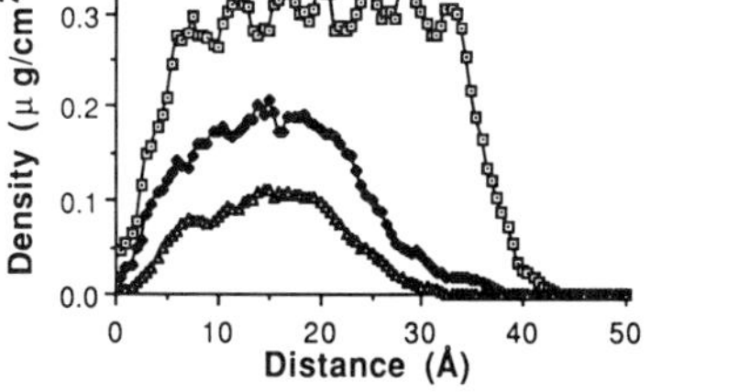

Figure 4. Density profiles of PEG systems of all-*gauche*(−) starting conformation for three different fixed end separation distances.

ends of the chain would be drawn close to each other. However, because only one end was allowed to move, the mobility of the chain was greatest at the free end and gradually decreased toward the fixed end. The TGT PEG systems were very stable over the duration of the MD runs, and very little compression of the chains occurred. Due to this relative incompressibility, the free ends were confined to their local positions and not allowed to move further down toward the fixed ends. Thus, the density distributions were skewed toward the free ends. However, the unfavorable energetic states of all-*gauche*(−) PEG systems caused the entire lengths of chains to convert their torsion angles rapidly, which produced a concerted movement toward collapsed structures.

In order to remove the density bias, high-temperature MD (simulated annealing) of the TGT PEG systems were performed. The temperature was increased from 300 K to 2000 K in steps of 0.5 K per iteration. After about 3 picoseconds of MD simulations at 2000 K, the temperature was reduced back to 300 K. An analysis of this simulated annealing process is listed in Table 4. It should be noted that the lowest possible energy *in vacuo* is achieved when the chains are in a completely collapsed state, or when the PEG atoms are within optimal distances of each other for nonbonded interactions. As a result of simulated annealing, the total number of torsion angles in *gauche*(+) and *gauche*(−) states were nearly equal.

Table 4. Characteristics of Twenty PEG Chains (40 EO Units) from MD Trajectories[a] for Different Fixed End Separations[b]

A. Radius of gyration, z-component of radius of gyration (r.g.), end-to-end distance and average thickness:

Separation distance (Å)	Radius of gyration (Å)	z-Component of r.g. (Å)	End-to-end distance (Å)	Thickness (Å)
12	28.23 ± 0.03	27.43 ± 0.04	89.29 ± 0.15	50.30
18	28.17 ± 0.07	26.85 ± 0.06	86.54 ± 0.22	50.44
24	28.40 ± 0.04	27.23 ± 0.02	87.69 ± 0.48	50.25

B. Total number of C—O torsion angles:

Separation distance (Å)	*trans*	*gauche*(+)	*gauche*(−)
12	1184.3 ± 9.5	180.0 ± 5.4	155.7 ± 7.9
18	1180.9 ± 8.6	160.4 ± 6.8	178.7 ± 4.7
24	1172.6 ± 10.2	164.4 ± 8.6	183.0 ± 5.9

C. Total number of C—C torsion angles:

Separation distance (Å)	*trans*	*gauche*(+)	*gauche*(−)
12	355.8 ± 1.0	210.2 ± 1.1	194.0 ± 1.2
18	386.4 ± 1.3	182.2 ± 0.9	191.4 ± 0.8
24	371.2 ± 1.4	188.9 ± 1.1	199.9 ± 1.0

[a]From the end of simulated annealing MD; 100 coordinate files.
[b]TGT starting conformation.

In summary, these PEG simulations have shown that a single chain *in vacuo* will assume a random coil conformation due to the lack of intermolecular constraints. However, when more than one chain is present in the molecular ensemble, the equilibrium conformation is a function of both the density of the chains and the type of solvent.

3.5.2. Protein Rotation on PEG and PE Surfaces

The objective of the protein rotation was to scan all possible orientations of the protein on top of a polymeric surface and determine which orientations were most energetically favorable. These orientations were then examined for amino acid residues that came into contact with the polymeric surface. The protein was rotated and placed at a distance of 8.5 Å from the polymeric surface. The protein was then brought down to the surface in steps of 0.5 Å, and three components of interaction energy (repulsion, attraction, and Coulombic) were computed (Figure 5). A value of 78 was used for the dielectric constant in calculations of Coulombic energy, which significantly reduced the contribution of electrostatic interactions to the overall energy. Consequently, the Lennard-Jones potential was the major component in the evaluation of interactions energies. A typical potential-energy diagram for the docking of a protein with a polymeric surface is shown in Figure 5.

For most of the rotations, the interaction energy became negative as the protein approached the surface and then increased very rapidly after the lowest energy had been reached. However, there were some cases where the interaction energy increased slightly before decreasing to a minimum value. This positive fluctuation indicated that a small overlap occurred as the protein approached the polymeric surface. The protein was able to squeeze through a narrow opening on the heterogeneous PEG surface and proceeded further down until an optimal contact could be made with the PEG chains.

There were two potential problems with our protein rotation simulations. First, the heterogeneity and infinity of the PEG surface quickly eliminated any possibility of a full sampling of all protein–polymer interaction. Second, since the protein can approach the surface from any orientation, all possible orientations needed to be sampled. A Monte Carlo approach of random selection of orientations would have reduced some of the calculations, but the results may not have been satisfactory. The

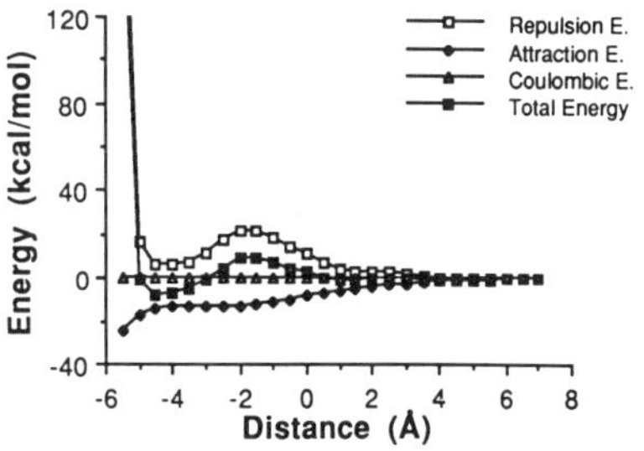

Figure 5. A nonbonded energy profile of lysozyme as it is brought down to the PEG surface. Distance zero is the highest point of PEG chains along the *z*-axis.

compromise was to rotate the protein at several different points on the randomized PEG surface in order to obtain a statistical average of protein–polymer interactions. Finally, MD simulations were used to explore the conformational space around favorable orientations.

3.5.3. Contact Residues of Lowest Interaction Energy with PEG and PE Surfaces

With the rotation and docking of lysozyme and myoglobin on six different sets of PEG surfaces and one PE surface, a very large amount of data was generated and had to be distilled into a few pertinent facts. These data were processed in several steps in order to identify those surface residues of the proteins that participated in the lowest overall energetic interactions with PEG and PE. First, the protein was rotated in 5° increments and docked with the surface. For each 5° increment, the lowest interaction energy was identified. After all rotations and dockings of the protein with a given surface were performed, a distribution of lowest interaction energies was formed. A typical distribution is shown in Figure 6. These distributions had an average of about -1.2 kcal mol^{-1} and a standard deviation of 1.3 kcal mol^{-1}. We decided to examine only those rotations with energies less than -5 kcal mol^{-1} (three standard deviation units below the mean). For each of these selected rotations, the amino acid residues of the protein that came into contact with the polymeric surface were identified.

This selection process was repeated for lysozyme and myoglobin on six sets of PEG surfaces and one PE surface. For each protein–polymeric surface set, the number of contacts per residue was tabulated and divided by the mean number of contacts for all residues of the protein. Then, only those residues with normalized frequency values greater than two standard deviation units above the mean were chosen. Finally, residues with a high frequency of contacts were compared across different protein–PEG sets. Residues that appeared two or more times in different sets were selected as regions of lowest interaction energy. For rotations on the homogeneous and symmetrical PE surface, only one rotation data set was obtained for each protein and the final comparison step was not necessary.

In summary, orientations that gave interaction energies $\leqslant -5$ kcal mol^{-1} were selected and then all the amino acid residues that came into contact with the polymeric

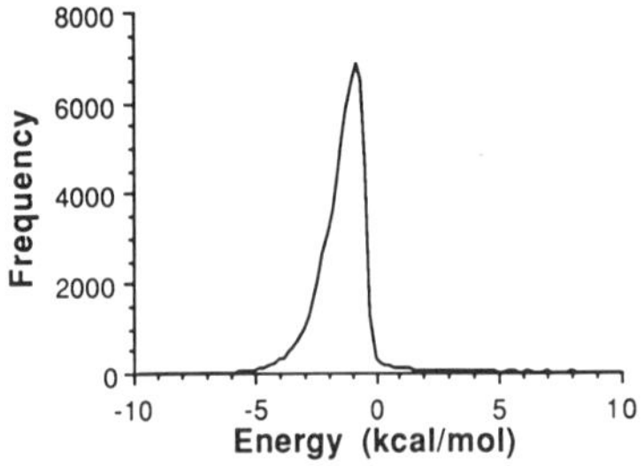

Figure 6. An interaction energy distribution from lysozyme rotation on the PEG surface.

surface were tabulated. This procedure was used to identify specific regions of the protein that participated in the low-energy interactions with the surface. Alternatively, we could have tabulated only those residues with the most favorable interaction energies, but this approach would not have taken account of the effect of all the residues which participated in the interaction with the surface.

3.5.4. Protein Surface Regions of Low Interaction Energy

3.5.4.1. Lysozyme

Three different classes of contact residues were identified for lysozyme. The first class consisted of eleven residues which only interacted with PEG. The second class consisted of twenty-eight residues which only interacted with PE. The third class consisted of twenty-one residues which interacted with both PEG and PE. The location of these residues in the amino acid sequence of lysozyme is shown in Figure 7. Their locations on the surface of lysozyme are shown in Figures 8 and 9. As a

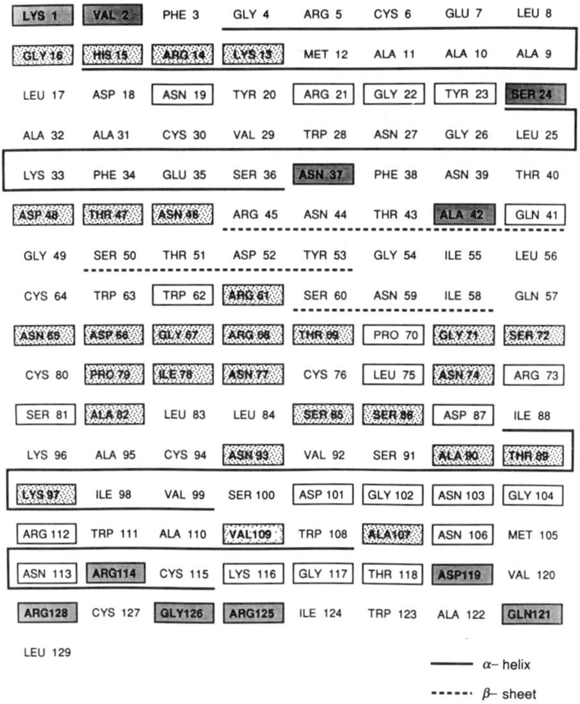

Figure 7. Amino acid sequence of lysozyme. Residues in heavily shaded rectangles represent contact residues of lysozyme orientations with lowest interaction energy on the PEG surface; residues in dotted rectangles represent those on the PE surface; and residues in plain rectangles represent those on both PEG and PE surfaces.

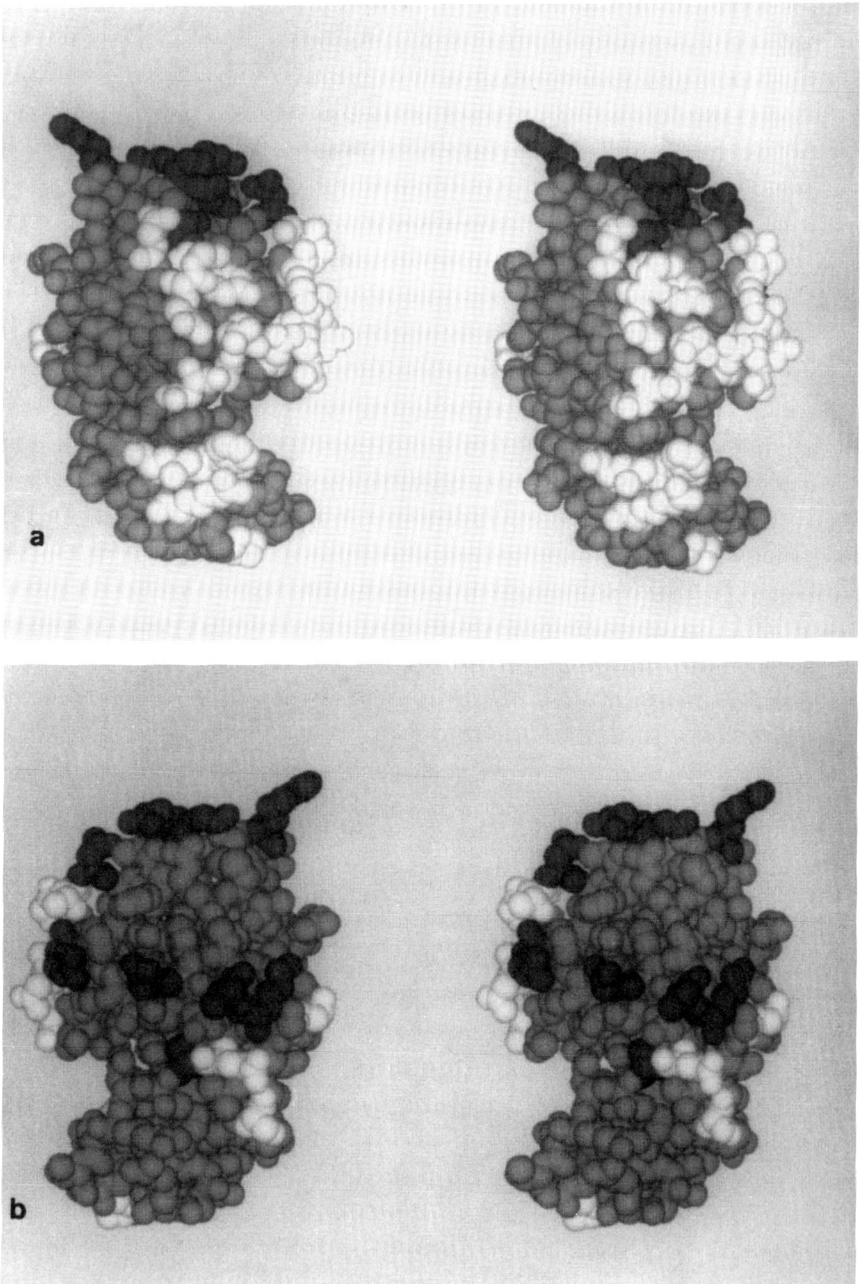

Figure 8. Stereo views of lysozyme. Residues of low interaction energy on PEG surfaces only are in dark shade and residues of low interaction energy on both PEG and PE surfaces are in white shade. Parts a and b represent two sides of lysozyme rotated 180° about the vertical axis.

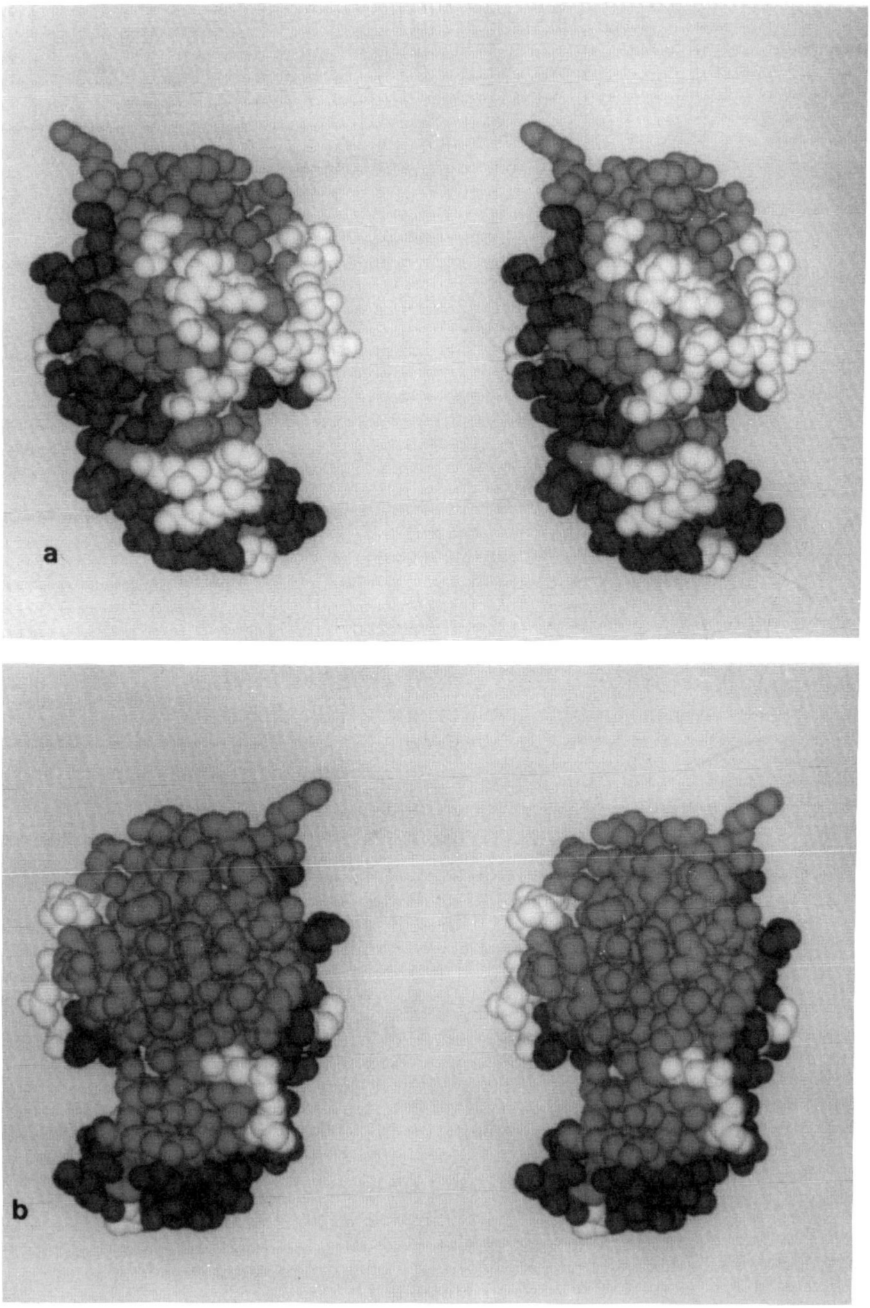

Figure 9. Stereo views of lysozyme. Residues of low interaction energy on the PE surface only are in dark shade and residues of low interaction energy on both PEG and PE surfaces are in white shade. Parts a and b represent two sides of lysozyme rotated 180° about the vertical axis.

general observation, most of the contact residues were found in regions which did not contain much secondary structure, such as the loops and turns at the ends of α-helices and β-sheets. For example, many residues of the second class were located in a long sequence between ARG 61 to SER 86. They are shown as heavily shaded residues in Figure 9a, and their locations are in contrast with the locations of residues of the first class shown in Figure 8a. Many of the residues of the second and third classes were located around the active site cleft of lysozyme.

The total solvent accessible surface area of lysozyme was 5835.6 $Å^2$ (calculated with Connolly's surface calculation program.[52] All atoms were included in the calculation. Only 17 of 129 lysozyme residues did not have any solvent accessible areas. The residues of the first class accounted for 15.2% of the total surface area, those of the second class accounted for 40.1%, and those of the third class account for 26.1%.

3.5.4.2. Myoglobin

Unlike lysozyme, myoglobin only had two classes of residues—those which only interacted with PEG (55 residues) and those which interacted with both PEG and PE (9 residues). The locations of these residues in the amino acid sequence of myoglobin are shown in Figure 10 and on the surface of the protein in Figure 11. Although over 70% of myoglobin sequence is α-helical, the 9 residues which interacted with both PEG and PE are not part of an α-helical sequence and are concentrated only on one part of the protein's surface. Unlike lysozyme, most of the 55 residues of myoglobin which interacted only with PEG are parts of α-helices. Many of them are near the heme pocket, while the opposite side of the protein that consists of α-helix sequences from GLY 23 to LEU 40 ad from LYS 96 to ILE 111 is devoid of contact residues. The total solvent accessible surface area of myoglobin was 7181.6 $Å^2$, and the 64 contact residues made up 58.4% of the total area.

3.5.4.3. Differences in Protein Contact Regions

Our rigid-body rotation data were consistent with a dynamical view of protein adsorption. Polar but neutral PEG chains can interact favorably with many different proteins. However, the fact that so many possible interaction sites for PEG were observed on the surface of the protein suggests that no single conformation will be preferred. Consequentially, this system has a high configurational entropy. This same entropic argument could not be employed in the case of the myoglobin–PE simulation, where only one interaction site was found.

Adsorption of a protein occurs after many collisions with the polymeric surface. If the differences in the interaction energies between various orientations of the protein on the polymeric surface are small, the protein will adsorb weakly and may never find a permanent foothold. However, when only a few orientations with favorable interaction energies are available, the protein will adsorb more strongly and then probably denature in order to further reduce the interaction energy. The rotation

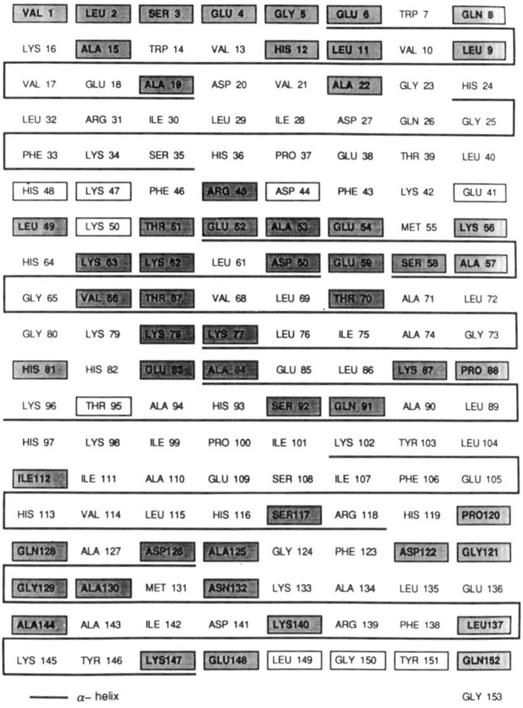

Figure 10. Amino acid sequence of myoglobin. Residues in heavily shaded rectangles represent contact residues of myoglobin orientations with lowest interaction energy on the PEG surface and residues in plain rectangles represent those on both PEG and PE surfaces.

data showed a significant difference between lysozyme and myoglobin. For myoglobin, there was only one small surface patch that was energetically favorable for interactions with the PE surface. However, lysozyme exhibited quite a few more interaction sites for polyethylene, and should be less susceptible to surface-induced denaturation. This view is consistent with the experimental results by Wei *et al.*,[53] which showed that myoglobin is more susceptible to denaturation at the air/water interface than lysozyme. The air/water interface has some relationship to crystalline polyethylene in that both are smooth and hydrophobic. In addition, the mobility of the polymeric surface is an important factor. When a protein adsorbs to a mobile polymeric interface, it will move with the polymer chains and constantly change its orientation with respect to the surface. This issue will be addressed further in our discussion of molecular dynamics simulations of lysozyme on various polymeric surfaces (see Section 3.5.6 below).

One main point made by Jeon *et al.* in their paper on protein–PEG interactions was that the van der Waals attraction was minimal when proteins interact with a PEG surface that had a low tendency for hydrophobic attraction.[35] Our rotation data showed that while most of the energy distribution for proteins on the PE surface lay in the negative region, the energy distribution for PEG was narrower and contained a

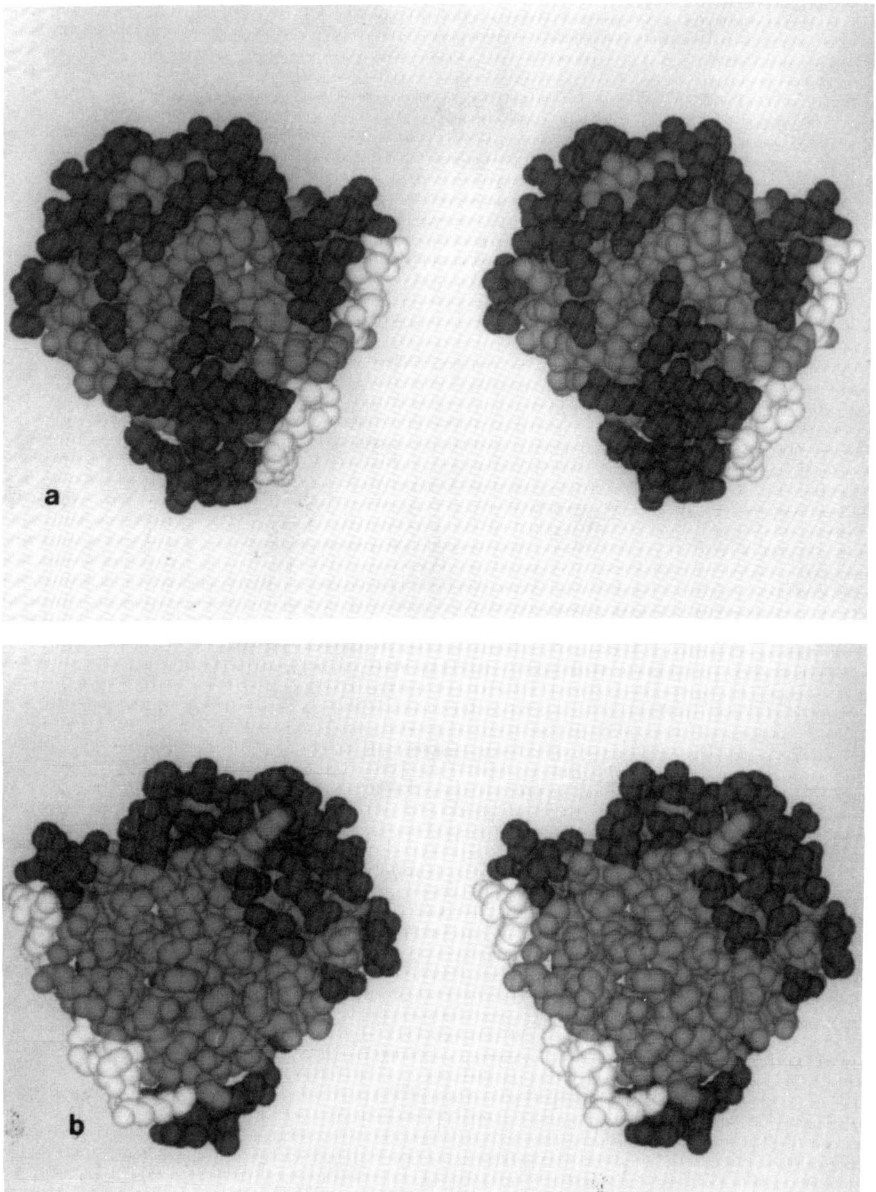

Figure 11. Stereo views of myoglobin. Residues of low interaction energy on both PEG and PE surfaces are in white shade and residues of low interaction energy on PEG surface only are in dark shade. Parts a and b represent two sides of myoglobin rotated 180° about the vertical axis. The heme pocket is viewed from part a.

positive component. This indicated that the repulsion of the protein was seen more often for PEG than for crystalline PE.

3.5.5. Initial Structures of the Lysozyme–Polymer System for Molecular Dynamics

Although our rotation studies identified potential interaction sites on the surfaces of lysozyme and myoglobin, they did not provide any information about what happened after the protein adsorbed to the surface. To address this issue, we ran molecular dynamics simulations of lysozyme on PE and PEG surfaces. Two different TGT PEG surfaces were selected for this study—one with the fixed ends of the PEG chains separated by 18 Å (A2 PEG) and the other with a separation distance of 24 Å (A3 PEG). The most energetically favorable orientation of lysozyme on each surface was chosen as the starting model.

The interaction energy of the most favorable orientation of lysozyme on the PE surface was -11.5 kcal mol^{-1}. For this particular orientation, there were 22 contact residues which were located at about 90 degrees left of the active site cleft as viewed directly in Figure 9a. This interaction site was probably the flattest part of the protein and could make the maximum number of contacts with the flat PE surface. The interaction site was formed from two regions of the protein—a large loop which connects two α-helices (LYS 13 to SER 24) and an α-helical region (THR 89 to GLY 102).

The rigid-body rotations of lysozyme on PE did not identify any residues that were in the active site cleft; the active site was formed by an invagination of the protein and was inaccessible to the flat PE surface. However, 16 contact residues were identified in and around the active site cleft in the most favorable orientation of lysozyme with A2 PEG. In fact, one PEG chain inserted itself directly into the active site. The Interaction energy of this orientation was -7.5 kcal mol^{-1}.

The most favorable orientation of lysozyme on A3 PEG (24 Å separation) revealed another contrast with lysozyme on PE. In this orientation, the most globular part of lysozyme's surface came into contact with several PEG chains. The interaction energy was -9.2 kcal mol^{-1}. The interaction site contained 16 contact residues which were located mostly in β-sheet strands and turn regions. Comparison of the A2 PEG and A3 PEG surfaces showed that lysozyme penetrated more deeply into the less dense A3 PEG matrix and interacted with PEG chains on multiple faces of the protein surface.

3.5.6. Molecular Dynamics of the Lysozyme–Polymer System

There were four sets of MD simulations of lysozyme: (a) free lysozyme, (b) lysozyme on rigid PE, (c) lysozyme on A2 PEG (18 Å separation), and (d) lysozyme on A3 PEG (24 Å separation). All simulations were performed *in vacuo* for 20 picoseconds and periodic boundary conditions were used in the lysozyme–PEG simulations. Data analyses included calculations of radius of gyration (Table 5) and

Table 5. Averages of Radius of Gyration (Å) of Lysozyme from MD Simulations

Time period (ps)	Free lysozyme	Lysozyme on PE	Lysozyme on A2 PEG	Lysozyme on A3 PEG
0–5	14.47 ± 0.16	14.41 ± 0.12	14.43 ± 0.78	14.47 ± 0.19
5–10	14.77 ± 0.31	14.63 ± 0.30	14.74 ± 0.30	14.66 ± 0.30
10–15	14.72 ± 0.32	14.48 ± 0.34	14.71 ± 0.31	14.62 ± 0.33
15–20	14.86 ± 0.32	14.44 ± 0.31	14.56 ± 0.31	14.59 ± 0.31

root-mean-square (RMS) deviations of atomic coordinates (Table 6). Both radius of gyration and RMS deviation values were accumulated over 5-picosecond intervals and then averaged. Values for RMS deviations of atomic coordinates were computed relative to the minimized structure of free lysozyme.

The average values of radius of gyration of lysozyme on PE and PEG surfaces did not deviate significantly from that of the free protein. These data suggested that lysozyme maintained its globular structure in different environments. Root-mean-square deviations of lysozyme on polymeric surfaces were somewhat larger than comparable values for free lysozyme. This indicated that lysozyme modified its conformation to accommodate the surface. Also, the standard deviations of the RMS measurements indicated how much lysozyme's structure fluctuated during a given time period.

The RMS deviations of individual residues were also examined in order to determine which parts of the protein were most mobile and to give clues about how environmental factors influenced mobility. These results are shown in Figure 12. Positive RMS values mean that the conformation of a given residue of lysozyme on a polymeric surface changed more during 20 picoseconds of MD than the corresponding residue of free lysozyme, and the opposite for negative RMS values. Many of the residues with high RMS values (greater than one standard deviation from the average) were also identified as contact residues in the rotation studies. In addition, smaller solvent accessible areas were observed for residues with low RMS values (less than one standard deviation from the average) than for residues with high RMS values.

Table 6. Averages of RMS Coordinate Deviations (Å) of Lysozyme from MD Runs Compared to the Minimized Structure of Free Lysozyme

Time period (ps)	Free lysozyme	Lysozyme on PE	Lysozyme on A2 PEG	Lysozyme on A3 PEG
0–5	1.20 ± 0.49	1.21 ± 0.46	1.22 ± 0.41	1.19 ± 0.37
5–10	2.00 ± 0.12	2.00 ± 0.21	2.07 ± 0.26	1.99 ± 0.20
10–15	2.14 ± 0.17	2.56 ± 0.26	2.57 ± 0.10	2.36 ± 0.13
15–20	2.58 ± 0.18	2.75 ± 0.12	3.04 ± 0.18	2.81 ± 0.18

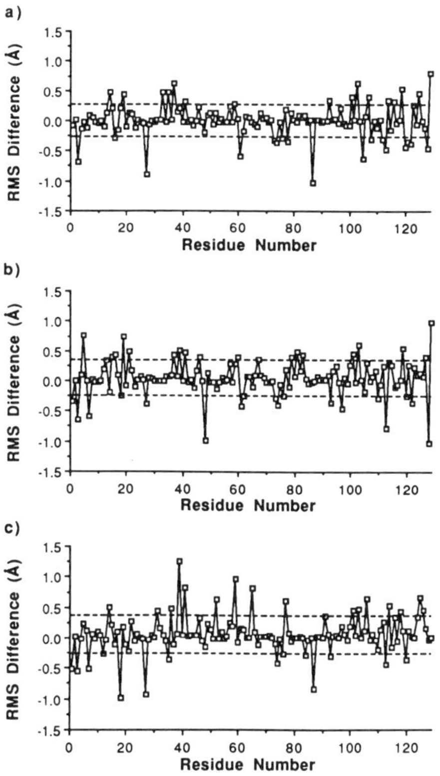

Figure 12. RMS coordinate differences between lysozyme on a polymeric surface and free lysozyme: (a) lysozyme on PE, (b) lysozyme on A2 PEG (18-Å separation), and (c) lysozyme on A3 PEG (24-Å separation). Dashed lines indicate the ± 1 standard deviation from the average.

This is a sign that they were constrained by factors such as hydrogen bonds and close association with neighboring residues.

While the RMS deviations of individual residues may show how the protein changed internally, they do not indicate if the protein moved with respect to the surface. A current hypothesis is that proteins adsorb and denature on rigid surfaces, but are often repelled by mobile ones. Although our 20-picosecond MD simulations were too short to capture the denaturation process, they did reveal significant differences in the dynamic behavior of lysozyme on rigid and mobile surfaces. Figure 13 shows the movement of lysozyme away from the surface (along the *z*-axis) during the course of the MD simulations. The results were striking in that lysozyme remained attached to the rigid PE surface, while it gradually diffused away from the mobile PEG surfaces. In fact, as the protein moved away from the PEG surface, two of the PEG chains moved with it until they became almost fully extended (Figure 14a).

It should be remembered that the protein and PEG chains interacted with each

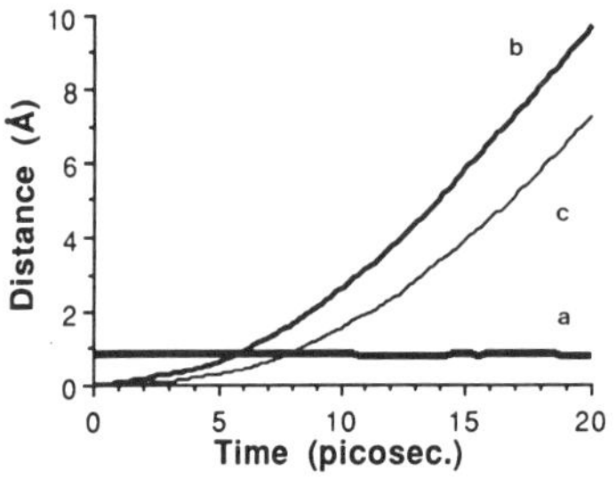

Figure 13. Deviation of the z-component of the center of mass of lysozyme during the 20-picosecond MD: (a) lysozyme on PE; (b) lysozyme on A2 PEG (18-Å separation); (c) lysozyme on A3 PEG (24-Å separation).

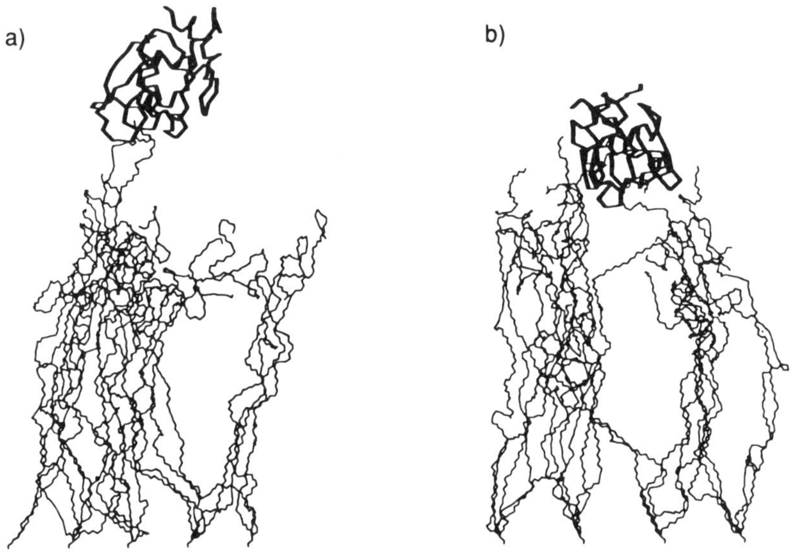

Figure 14. Lysozyme on grafted PEG chains at the end of 20-picosecond MD: (a) lysozyme on A2 PEG (18-Å separation); (b) lysozyme on A3 PEG (24-Å separation).

other more strongly *in vacuo* than when solvent molecules were present. Although this was an artifact due to the absence of water in the molecular ensemble, there was a realistic component to it. In the lysozyme–PE system, nonbonded interactions between lysozyme and the immobile PE surface significantly lowered the total energy of the system. During the course of the MD simulation, lysozyme had to remain within the sphere of influence of the PE surface in order to maintain this favorable interaction. This was also true for lysozyme on the PEG surface. However, the PEG chains were mobile which allowed them to move in concert with the lysozyme molecule.

The effects of PEG chain density on the polymer–protein interaction was clearly demonstrated in Figure 14. As mentioned previously, lysozyme penetrated further into the A3 PEG matrix than it penetrated into the more dense A2 PEG matrix. Initially, there were 16 lysozyme residues that were in contact with the A3 PEG matrix. This number increased to 46 after the 20-picosecond MD simulation. The potential energy of lysozyme on A3 PEG was lower than either the free lysozyme or lysozyme on the A2 PEG set, but this was due to the fact that this lysozyme had more nonbonded interactions with PEG chains.

3.5.7. The Excluded Volume and Effect of Solvents

PEG's ability to precipitate proteins has been attributed to its excluded volume. Proteins are sterically excluded from the hydrodynamic volume of PEG. This effect concentrates proteins until their solubility is exceeded and precipitation occurs. Experimental data showed a correlation between protein precipitation and PEG size, and between protein solubility and PEG concentration.[8] A theoretical study by Jeon and Andrade suggested that there was an optimal combination of protein size, polymer size, and polymer density for protein resistance.[36] According to their calculations, PEG chains with a separation distance of 13–17 Å will optimally resist proteins of 60 to 80 Å in radius. PEG chains with separation distances of 18 Å (A2 PEG) and 24 Å (A3 PEG) were used in our MD studies. Since the radius of gyration of lysozyme was about 15 Å, it was probably too small for optimum protein resistance. The A1 PEG system with 12-Å separation would be a more suitable surface for lysozyme. However, the lack of solvent molecules must be taken into account for the relationship between protein size and polymer density and length in our studies.

Molecular dynamics simulations of a single PEG-600 chain showed that water molecules greatly reduced the number of intramolecular contacts within the PEG chain. Furthermore, the TGT conformation of PEG enabled it to insert itself into the hydrogen-bonded network of bulk water.[54,55] However, if the density of PEG chains is too high, the hydrogen bond network will be disrupted. Water is then excluded from the PEG matrix which results in a dense, rigid surface. In contrast, if the density is too low, proteins penetrate too far into the PEG matrix and are trapped. Thus, an optimum packing density of the PEG chains is needed for protein resistance and polymer fluidity.

The simulation techniques described in this chapter were used to identify sites on the surface of proteins which could interact with polymeric surfaces. However, in order to rank these orientations on an absolute energy scale, solvent will have to be included in the simulations. Preliminary energy minimization data of the lysozyme–PEG–water system showed that the total energy was lower when lysozyme was placed beyond the nonbonded energy cutoff distance for the PEG surface than near it. The same effect was observed when a hydrocarbon molecule and a PEG-600 chain were solvated, or when two PEG chains were solvated together. However, in the lysozyme–PE–water system, the total energy was lower when lysozyme was placed close to the PE surface. Furthermore, when two octadecane molecules were solvated, the total energy was lower when they were closely associated than when they were separated. The electrostatic energy for the PEG–water interaction becomes dominant in the lysozyme–PEG–water system, while the van der Waals energy for the lysozyme–PE interaction becomes dominant in the lysozyme–PE–water system. These data seemed to indicate that the structural arrangement of water molecules influences the total energy. It should be noted that these energy values are equivalent to enthalpies. Free-energy calculation involves either a direct calculation of entropy or special thermodynamic integration methods.[56–58] At present, methods to calculate free energy are still under development and computationally very expensive.[59] To date, we have not performed any free-energy calculations with our protein–PEG–water systems, but once the methodology is validated, such calculations will enable direct comparisons between experiments and simulations.

3.6. SUMMARY AND CONCLUSION

Molecular simulations were performed to study grafted-PEG surfaces and their interactions with proteins. The equilibrium conformations of grafted-PEG surfaces were determined largely by the choice of starting models. The TGT conformation of PEG was very stable and kept the PEG chains in an extended state. This conformation was used in further studies with proteins. A crystalline PE surface was also used for comparison.

Protein molecules (lysozyme or myoglobin) were rotated on top of a polymeric surface and interaction energies for the protein were calculated. The surface amino acid residues of the proteins were used to identify regions of low interaction energy. For lysozyme, most of the contact residues were located at the ends of the secondary structural elements (α-helix and β-sheet) or in the loops and turns. For myoglobin, most of the contact residues were located around the heme pocket. Rotation studies showed that both proteins had more potential interaction sites for PEG surfaces than for crystalline polyethylene. This result is consistent with the entropic view of PEG repulsion of proteins.

Molecular dynamics simulations were performed *in vacuo* with lysozyme adsorbed to three different types of surfaces: crystalline polyethylene, PEG with the chains separated by 18 Å, and PEG with the chains separated by 24 Å. While lysozyme remained virtually stationary on the rigid PE surface, it was swept along by the mobile PEG chains and drifted away from the polymeric surface. Significant differences were also observed in the interfacial behavior of lysozyme on the PEG surface with 18-Å chain separation versus the PEG surface with 24-Å chain separation. From the rotation studies, lysozyme was initially placed deeper in the matrix of the 24-Å-separation PEG than in the matrix of 18-Å separation PEG. In MD simulations, the protein moved away much faster from the more dense surface (18-Å separation) than it did from the 24-Å-separation surface. Hence, molecular dynamics simulations showed that the mobility and density of the grafted polymeric surface can determine the extent of protein–polymer interaction. Solvent molecules were not explicitly included in these simulations but only approximated, due to limits in computation time. Their effect should be fully taken into account for direct comparisons between experiments and simulations. Nevertheless, these studies demonstrated the potential of computer simulations for screening different types of polymeric systems and their interactions with proteins.

ACKNOWLEDGMENTS

The authors wish to thank Silicon Graphics, Inc. for their generous donation of computer graphics workstations, BIOSYM Technologies, Inc. for the DISCOVER and INSIGHT programs, and the University of Utah Supercomputing Institute for computing time, and Professor Joseph D. Andrade for helpful comments on this manuscript. K. L. also wishes to thank the University of Utah Center for Biopolymers at Interfaces for financial support.

REFERENCES

1. A. Kondo, M. Kishimura, S. Katoh, and E. Sada, *Biotech. Bioeng. 34*, 532 (1989).
2. T. Suzuki, K. Ikeda, and T. Tomono, *J. Biomater. Sci. Polymer Ed. 1*, 75 (1989).
3. S. Zalipsky, C. Gilon, and A. Zilkha, *Eur. Polym. J. 19*, 1177 (1983).
4. I. N. Topchieva, *Russ. Chem. Rev. 49*, 260 (1980).
5. A. Abuchowski, T. van Es, N. C. Palczuk, and F. F. Davis, *J. Biol. Chem. 252*, 3578 (1977).
6. K. C. Ingham, *Methods Enzymol. 104*, 351 (1984).
7. D. D. Knoll and J. Hermans, *Biopolymers 20*, 1747 (1981).
8. D. H. Atha and K. C. Ingham, *J. Biol. Chem. 256*, 12108 (1981).
9. J. Wilf and A. P. Minton, *Biochim. Biophys. Acta 670*, 316 (1981).
10. M. P. Allen and D. J. Tildesley, *Computer Simulation of Liquids*, Oxford University Press, New York (1987).
11. B. J. Alder and T. E. Wainwright, *J. Chem. Phys. 27*, 1208 (1957).
12. B. J. Alder and T. E. Wainwright, *J. Chem. Phys. 31*, 459 (1959).
13. N. Metropolis and S. Ulam, *J. Am. Stat. Assoc. 44*, 335 (1949).
14. R. H. Boyd, *Macromolecules 22*, 2477 (1989).
15. D. L. Ermak and J. A. McCammon, *J. Chem. Phys. 69*, 1352 (1978).
16. E. Helfand, Z. R. Wasserman, and T. A. Weber, *Macromolecules 13*, 526 (1980).
17. R. H. Reid, C. A. Hooper, and B. R. Brooks, *Biopolymers 28*, 525 (1989).

18. K. A. Dill, J. Naghizadeh, and J. A. Marqusee, *Annu. Rev. Phys. Chem. 39*, 425 (1988).
19. A. Baumgärtner, *Annu. Rev. Phys. Chem. 35*, 419 (1984).
20. K. Kremer, G. S. Grest, and I. Carmesin, *Phys. Rev. Lett. 61*, 566 (1988).
21. T. A. Weber and E. Helfand, *J. Phys. Chem. 87*, 2881 (1983).
22. D. Rigby and R.-J. Roe, *J. Chem. Phys. 89*, 5280 (1988).
23. D. N. Theodorou and U. W. Suter, *Macromolecules 18*, 1467 (1985).
24. D. R. Fitzgibbon and R. L. McCullough, *J. Polym. Sci., Part B 27*, 655 (1989).
25. G. Forgacs, V. Privman, and H. L. Frisch, *J. Chem. Phys. 90*, 3339 (1989).
26. A. C. Balaze and S. Lewandowski, *Macromolecules 23*, 839 (1990).
27. P. G. de Gennes, *Scaling Concepts in Polymer Physics*, Cornell University Press, Ithaca, New York (1985).
28. P. G. de Gennes, *Macromolecules 13*, 1069 (1980).
29. S. Alexander, *J. Phys. (Paris) 38*, 983 (1977).
30. M. Muthukumar and J.-S. Ho, *Macromolecules 22*, 965 (1989).
31. S. T. Milner, T. A. Witten, and M. E. Cates, *Macromolecules 21*, 2610 (1988).
32. M. Murat and G. S. Grest, *Macromolecules 22*, 4054 (1989).
33. N. Tomioka, A. Itai, and Y. Iitaka, *J. Comput.-Aided Mol. Design 1*, 197 (1987).
34. B. W. Morrissey and R. R. Stromberg, *J. Colloid Interface Sci. 46*, 152 (1974).
35. S. I. Jeon, J. H. Lee, J. D. Andrade, and P. G. de Gennes, *J. Colloid Interface Sci. 142*, 149 (1991).
36. S. I. Jeon and J. D. Andrade, *J. Colloid Interface Sci. 142*, 159 (1991).
37. J. Vila and J. L. Alessandrini, *J. Theor. Biol. 134*, 445 (1988).
38. D. R. Lu and K. Park, *J. Biomat. Sci. Polym. Ed. 1*, 243 (1990).
39. C. L. Brooks, M. Karplus, and B. M. Pettitt, *Adv. Chem. Phys. 71* (1988).
40. J. A. McCammon and S. C. Harvey, *Dynamics of Proteins and Nucleic Acids*, Cambridge University Press, Cambridge (1987).
41. A. T. Hagler, *The Peptides 7*, 213 (1985).
42. A. T. Hagler, J. R. Maple, T. S. Thacher, G. B. Fitzgerald, and U. Dinur, in: *Computer Simulation of Biomolecular Systems* (W. F. van Gunsteren and P. K. Weiner, eds.), p. 149, ESCOM, Leiden, Netherlands (1989).
43. S. C. Harvey, *Proteins 5*, 78 (1989).
44. Y. Takahashi and H. Tadokoro, *Macromolecules 6*, 672 (1973).
45. Protein Data Bank Newsletter, No. 46, Brookhaven National Laboratory, Upton, NY 11973 (1988).
46. P. Dauber-Osguthorpe, V. A. Roberts, D. J. Osguthorpe, J. Wolff, M. Genest, and A. T. Hagler, *Proteins 4*, 31 (1988).
47. R. Edberg, D. J. Evans, and G. P. Morriss, *J. Chem. Phys. 87*, 5700 (1987).
48. J. Skolnick and E. Helfand, *J. Chem. Phys. 72*, 5489 (1980).
49. E. Helfand, *Science 226*, 647 (1984).
50. A. J. Hopfinger and D. C. Doherty, *Polym. Prepr. 30*, 5 (1989).
51. J. Wang, S.-C. Huang, J. Lin, and J. D. Andrade, submitted.
52. M. L. Connolly, *J. Appl. Crystallogr. 16*, 548 (1983).
53. A.-P. Wei, J. N. Herron, and J. D. Andrade, in: *From Clone to Clinic* (D. J. A. Crommelin, ed.), p. 305, Kluwer Academic Publishers, Netherlands (1990).
54. C. Tanford, *The Hydrophobic Effect*, Chap. 5, Wiley–Interscience, New York (1980).
55. R. Kjellander and E. Florin, *J. Chem. Soc., Faraday Trans. 1, 77*, 2053 (1981).
56. R. M. Levy, M. Karplus, J. Kushick, and D. Perahia, *Macromolecules 17*, 1370 (1984).
57. M. Mezei, S. Swaminathan, and D. L. Beveridge, *J. Am. Chem. Soc. 100*, 3255 (1978).
58. P. H. Berens, D. H. J. Mackay, G. M. White, and K. R. Wilson, *J. Chem. Phys. 79*, 2375 (1983).
59. W. L. Jorgensen, *Acc. Chem. Res. 22*, 184 (1989).

4

PEG-Derivatized Ligands with Hydrophobic and Immunological Specificity

Applications in Cell Separation

DONALD E. BROOKS, JAMES M. VAN ALSTINE, KIM A. SHARP, and S. JILL STOCKS

4.1. INTRODUCTION

In the last few years the use of poly(ethylene glycol) (or PEG) as an agent with which to modify the properties of macromolecules and surfaces has greatly increased, as witnessed by the contributions in this volume. In most instances, PEG is used because it exhibits the interesting property of being highly compatible with water (i.e., highly water soluble) while exhibiting strong incompatibility with a wide variety of other water-soluble substances. Incompatibility means that an unfavorable free-energy change occurs when a second species interacts with a solvated PEG molecule, resulting in a statistical tendency for the second species to be excluded from the region within or near the PEG chain. Such excluded volume effects are manifested in a variety of ways, including phase separation in mixtures with a second

DONALD E. BROOKS, JAMES M. VAN ALSTINE, KIM A. SHARP, and S. JILL STOCKS • Departments of Pathology and Chemistry, University of British Columbia, Vancouver, British Columbia V6T 2B5, Canada. *Present address of J.M.V.A.*: Department of Biological Sciences, The University of Alabama in Huntsville, Huntsville, Alabama 35899. *Present address of K.A.S.*: Department of Biochemistry and Molecular Biophysics, College of Physicians and Surgeons of Columbia University, New York, New York 10032. *Present address of S.J.S.*: Biotechnology Process Services Ltd., Durham DL1 6YL, England.

Poly(Ethylene Glycol) Chemistry: Biotechnical and Biomedical Applications, edited by J. Milton Harris. Plenum Press, New York, 1992.

water-soluble polymer of salt, enhanced exclusion of PEG from chromatographic gel beads relative to other polymers of similar molecular weight, protein precipitation and reduced binding of external proteins to surfaces or molecules derivatized with PEG.[1,2] Many of these kinds of interactions are described in this book.

The focus of the current chapter is on the use of PEG derivatization to produce reagents which will affect the partitioning of biological or model cells in the liquid two-phase systems which form when incompatible materials are added to aqueous solutions of PEG. The work we will review has all been done in two-phase mixtures in which the incompatible partner is dextran, a neutral bacterial polysaccharide.[1] Combinations of these two polymers in aqueous solution at concentrations above a few percent in both result in the formation of two phases, each enriched in one of the polymers. When PEG is bound in some way to a third molecular species the distribution of this material between the two phases in general will be changed in favor of the top, PEG-rich phase since the PEG derivatized to the species of interest will also be excluded to some extent from the dextran-rich phase. If the PEG-derivatized species has an affinity for a group or structure present in a mixture from which the target material is to be purified, it can be used as an affinity ligand to selectively enhance the partition of the target into the PEG-rich phase. Selective extractions of this kind are discussed by Johansson (Chapter 5) and Tjerneld (Chapter 6) in this volume for isolation of molecular species. The present contribution focuses on the behavior of cells and related particles in the presence of PEG-derivatized molecules of various kinds.

Affinity-based isolations offer high selectivity at the cost of significant effort to develop the ligands. A recognized drawback of the approach is that one has to understand enough about the target substance to select appropriate affinity ligands. However, as will be seen, in the area of hydrophobic affinity partitioning of cells this is not always the case. This approach utilizes the advantages of a liquid-phase support for the affinity ligand which eliminates nonspecific irreversible losses associated with the use of solid affinity ligand supports. Combined with the generality of multiple-step separations, analagous to chromatography, hydrophobic partitioning is shown to be capable of subtle, unexpected enrichments of cell subpopulations.

4.2. HYDROPHOBIC AFFINITY PARTITIONING OF CELLS AND MEMBRANES

One of the most interesting and useful classes of PEG-based affinity ligands of use in cell separations are the PEG esters of fatty acids. These molecules were first used for affinity partitioning by Shanbag and Johansson[3] who demonstrated an alteration in human serum albumin partition upon exposure to the palmitate (C-16:0) ester of PEG. Applications to red cells,[4,5] cell membranes,[6,7] and organelles[8] also were soon described. Membrane models in the form of lipid bilayer vesicles were utilized by Eriksson and Albertsson, who showed that the increase in partition into the top phase induced by PEG fatty acid esters was sensitive to the lipid composition used.[9,10] A report also appeared showing that hydrophobic cell partition detected erythrocyte alterations in blood from rats bearing a subcutaneous Leydig cell tumour.[11]

4.2.1. Solution Properties of PEG-Alkyl Esters

Our own work began with the development of an easy method to purify gram quantities of PEG-fatty acid esters, and an assay to estimate the molar concentrations of the ligands.[12] The resulting chromatographically pure materials were used in a series of investigations aimed both at better understanding the detailed mechanism by which the esters acted and at applying them to interesting separation problems. It was first demonstrated that the micromolar quantities of PEG-fatty acid esters required to shift the partition of most cells into the PEG-rich phase had no significant effect on interfacial tension, phase viscosity, or the electrostatic potential between the bulk phases,[13,14] rather their effects had to be due to a direct interaction with the cells.

The solution properties of the esters were further defined by measurements of the critical micelle concentration (CMC) of a series of saturated and unsaturated PEG-fatty acid esters (Table 1, Figure 1).[15] As expected, the CMC decreased with increasing acyl chain length and degree of saturation, reflecting the increased hydrophobicity of the tails. However, the CMCs were found to be two orders of magnitude lower than predicted by simple extrapolations from the literature based on low molecular weight ethylene oxide oligomeric head groups. The low CMC values may indicate relatively high values for the average number of molecules per micelle.[16] The result also suggests that once the head groups become sufficiently large, they are able to distort or interpenetrate to accommodate the hydrophobic association of the alkyl chains. A related type of accommodation presumably is also required if PEG-fatty acid esters are to gain access to the lipid bilayer of a cell membrane in the presence of the glycocalyx. The glycocalyx is the name given to the glycoproteins and glycolipids anchored in the lipid bilayer that occupy the interfacial region

Table 1. Estimates of the CMC Values of the PEG 8000-Fatty Acid Esters Studied

Acyl tail[a]	CMC (μM) (means ± S.D.)
Dodecanoate (12:0)	100 ± 20
Tetradecanoate (14:0)	70 ± 10
Hexadecanoate (16:0)	42 ± 4
Octadecanoate (18:0)	1.1 ± 0.4
cis-9-Octadecenoate (18:1)	12 ± 3
cis-9,*cis*-12-Octadecadienoate (18:2)	40 ± 5

[a]Fatty acyl tail designation (M:N), where M = number of carbon atoms in acyl tail, N = number of double bonds.

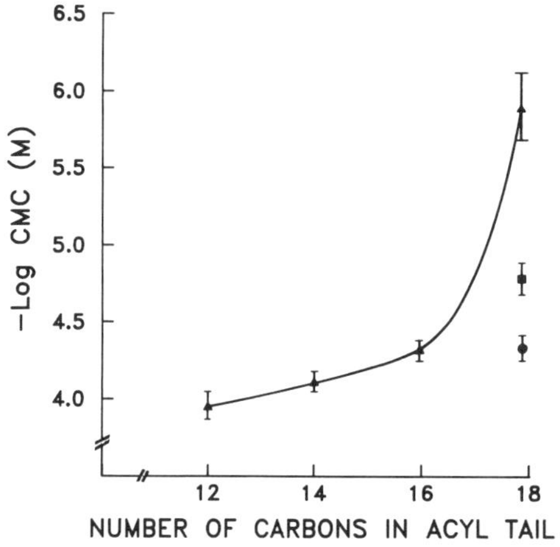

Figure 1. Log CMC versus acyl chain length for the PEG 8000-fatty acid esters in Table 1: (▲) saturated tail; (■) mono-unsaturated tail; (●) di-unsaturated tail.

between the bulk suspending medium and the lipid head groups. If the PEG was totally incompatible with the structures in the glycocalyx, interaction of PEG-alkyl esters with the lipid bilayer might be inhibited and the esters ineffective as hydrophobic affinity ligands.

4.2.2. Liposome Hydrophobic Affinity Partitioning

Both cells and liposomes behave similarly in hydrophobic affinity partition. This suggests that the effects of the ligands on cells may reflect primarily membrane lipid composition, since Eriksson and Albertsson reported that the nature of the lipid head group had a significant effect on such partition.[9,10] They suggested that the primary interaction between liposomes and fatty acid esters was a hydrophobic insertion of the acyl tail into the lipid bilayer. Longer ligand acyl tails increased partition at equal concentrations, but the nature of the acyl region of the lipid bilayer had little effect. The chemical nature of the lipid head group was a significant determinant of the efficiency of the ligand, however.

We investigated another aspect of this issue by attempting to detect by hydrophobic affinity partitioning the thermotropic gel-to-liquid-crystalline phase transition at approximately 21.5 °C in small unilamellar liposomes of dimyristoyl phosphatidylcholine (DMPC).[13] By carefully controlling the experimental conditions we were able to detect a shift in partition of DMPC liposomes over the temperature range 18 to 23 °C. This shift was opposite to that expected from any increase in liposome volume accompanying the phase transition.[17] The results, summarized in Table 2 and Figure 2, demonstrate that the phase transition can be detected by this technique and that PEG unsaturated fatty acid ester preparations, when purified of free fatty acid contaminants, are more discriminating than saturated tails in detecting such phenomena (see below). The mechanism most likely responsible for the observed partition

Table 2. Effect of Temperature and Temperature Alteration on the Partition of DMPC Small Unilamellar Liposomes in a (5,4) V Ester-Containing Phase System

	Upper phase percent partition	
Incubation	PEG 6000-18:2 0.5 μM	PEG 6000-18:0 0.5 μM
18 °C	65.2[a] ± 2.3	85.4 ± 3.2
18 °C transferred to 23 °C	88.6 ± 2.6	92.1 ± 1.7
23 °C	92.8 ± 1.8	94.8 ± 1.2
23 °C transferred to 18 °C	91.2 ± 0.8	94.1 ± 2.4

[a]Each value represents the mean ± S.D. of four individual partitions. The phase composition for all two-phase systems, unless otherwise noted, was 5% w/w Dextran 500, 4% w/w PEG 8000, 130 mM NaCl, and 10 mM phosphate buffer.

increase is the increased area per lipid molecule above the phase transition[18] associated with the liquid state allowing more ligand binding per unit area. It is interesting to note that the partition increase was not reversed when the temperature was decreased below the phase transition temperature in the presence of equilibrated ligand (Table 2). This may indicate that the bound ligand is not released from the expanded surface once the temperature is dropped. In cellular systems, however,

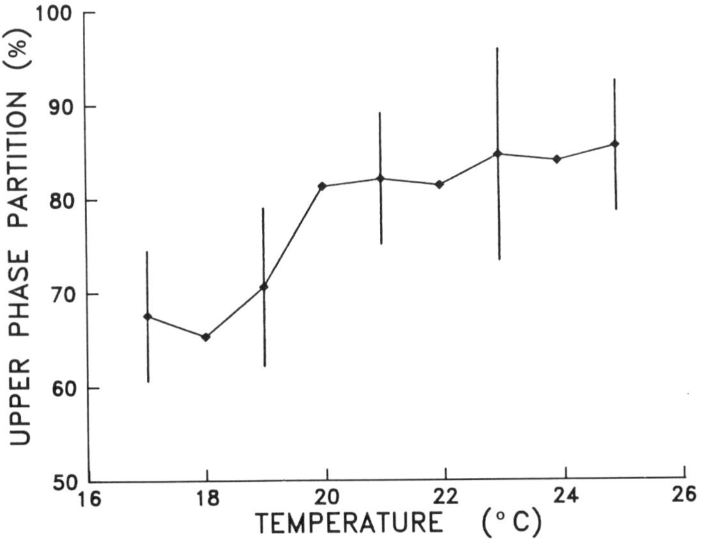

Figure 2. Upper phase partition versus temperatures for DMPC small unilamellar liposomes. The two-phase system was supplemented with 0.5 μM PEG 6000-18:2 ester.

PEG-fatty acid ester binding is largely reversible upon dilution (*vide infra*) so the observed irreversibility in the liposome system is surprising.

4.2.3. Hydrophobic Affinity Partitioning of Cells

4.2.3.1. Binding of PEG and PEG-Alkyl Esters to Cells

The simple picture of affinity partitioning presented above is that the PEG-ligand binds to the target and, by virtue of its tendency to partition into the PEG-rich phase, selectively extracts the target material into the top phase. The binding event can be more complicated than might first be supposed, however. This is particularly true for binding of PEG-alkyl esters to cell surfaces since the ligands must reach equilibrium in the presence of the glycocalyx which, as mentioned earlier, potentially affects the binding. Such effects have, in fact, been observed.[14] Here we summarize our studies on the adsorption of PEG and PEG-palmitate to erythrocytes and the effects of this binding on the partition behavior that results. They are discussed in more detail elsewhere.[1,14]

As a general rule, adsorption to any site or surface is increased from poorer solvents for the binding solute, because free energy is reduced when the solute reduces its area of contact with the solvent by contact with the binding site or surface. When the binding of PEG-palmitate is considered, it is the difference in compatibility of the polymers in the two phases with the PEG head group which is relevant in this regard since the fatty tail is almost equally compatible with both phases. Obviously, the PEG ester will be less compatible with the dextran-rich lower phase than with the PEG-rich upper phase. Hence, one might expect stronger binding from the lower than from the upper phase, an effect which is in the *opposite direction* from that required to produce an affinity partitioning effect. Countering this tendency, however, is the lower concentration of PEG ester in the lower phase due to its partition coefficient. If the ligand were able to completely remove itself from contact with the phase polymers upon binding, these two effects would just balance out and no affinity partition would be observed.[1,14] On the other hand, if there were no difference in the exposure of the PEG head group to the phase polymers in the bound and unbound states, the binding would be expected to be identical in both phases and affinity partitioning would be observed due to the difference in concentration of ligand between the phases. Interestingly, both these effects are seen experimentally.

The amount of PEG-^{14}C-palmitate bound at equilibrium to human erythrocytes suspended in either the top or bottom phase of a system comprising 5% dextran 500, 4% PEG 8000, 10 mM phosphate buffer, pH 7.16, and 130 mM NaCl has been measured as a function of the equilibrium ester concentration.[14] Saturating binding isotherms were found but binding from the bottom, dextran-rich phase was significantly stronger than from the top phase at all subsaturating concentrations. Linearization of this data by Scatchard analysis led to the values for numbers of sites and dissociation constants shown in Table 3. It is seen that the binding constants differ

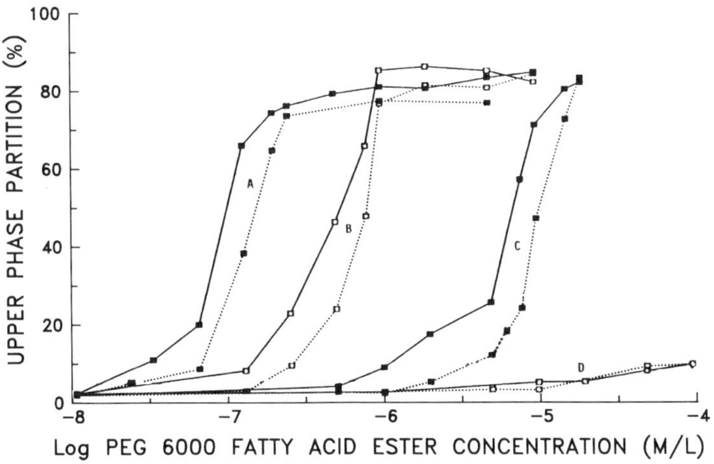

Figure 3. Partition of fresh washed (——) and 1% glutaldehyde fixed (· · ·) erythrocytes in a two-[phase system with the following esters: A, stearate (18:0); B, palmitate (16:0); C, myristate (14:0); and D, laurate (12:0).

significantly but the number of binding sites is indistinguishable when estimated from the two phases, consistent with expectation.

The reversibility of binding was studied by washing the cells in buffer, top phase containing no ester or ester-free bottom phase and measuring the amount left on the cells and the ester concentration in solution. The cells equilibrated with radiolabeled ester were spun down, resuspended in ester-free medium, equilibrated, sampled, and the washing process continued to progressively reduce the free and bound ester concentration. When the cells were washed in buffer or bottom phase, very little ester was removed, but when the data for top-phase washes were analyzed and the amount of ester bound plotted as a function of its solution concentration, the result was very similar to the binding isotherm shown in Figure 3.[14] The slow desorption in buffer is thought to be due to multiple weak attachments sites between the PEG head group and membrane components, secondary to the acyl tail insertion in determining binding energy. The head group behavior dominates desorption kinetics, however, because of the low probability of all attachments spontaneously releasing simultaneously. In the presence of competing PEG from the top phase, on the other hand, head groups

Table 3. Ester Binding Data from Scatchard Plots

Phase	Binding sites (10^6 per cell)	Dissociation constant (μM)	Binding energy (*kT* per molecule)	Regression coefficient
Top	8.48 ± 0.3	3.3 ± 0.3	−16.6 ± 0.2	0.966
Bottom	8.68 ± 0.6	0.85 ± 0.2	−18.0 ± 0.4	0.886

can be released by the "unzipping" of occupied sites and replacement of each in turn with competing polymer. The bottom phase is poor at desorbing bound ester because of incompatibility of the washing medium for PEG headgroups.

The adsorption of ^{14}C-PEG to erythrocytes from a variety of media was also studied.[14] The binding is much weaker than that of PEG-alkyl esters, being measurable at concentrations four orders of magnitude higher than those associated with PEG-palmitate effects. Moreover, in marked contrast to the results with the fatty acyl ester, adsorption of PEG itself is indistinguishable whether occurring from buffer, top phase or bottom phase. By the above arguments, this implies that the polymer is fully exposed to the phase polymers in the bound state. The simplest explanation of this difference is that, by intercalating into the lipid bilayer, the palmitate tail of the ester anchors the PEG headgroup at the plane of the bilayer, among the other anchored molecules which make up the glycocalyx. Evidently, not all of the PEG is exposed to the phase polymers in this location. In the absence of the anchoring acyl tail, the PEG molecule interacts with regions of the glycocalyx which are more exposed to the exterior solution, perhaps nearer the outside of the surface-attached layer.

The effects of the bound PEG and PEG esters on partitioning depend both on the numbers of molecules bound per unit area of cell surface and on the binding energies characteristic of the interaction.[1] In a phase system in which the partition into the top phase is essentially zero in the absence of PEG-palmitate and 50% into the PEG-rich phase in the presence of ester there is more PEG adsorbed to the cell than ester, yet the ester is necessary for the partition to occur. Therefore the presence of bound PEG alone is not sufficient to produce partition into the top phase. The theory of such

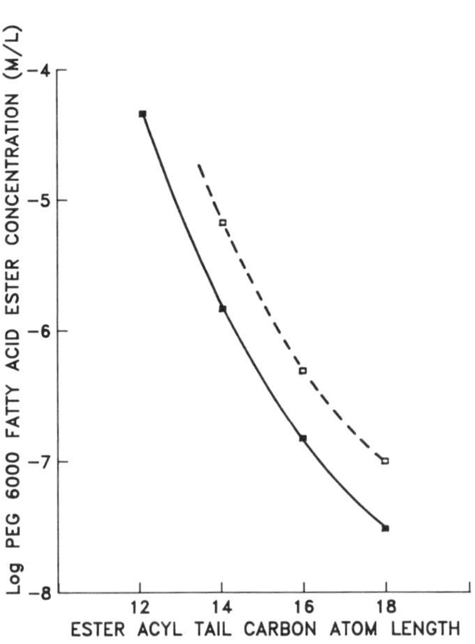

Figure 4. Log PEG 6000-saturated fatty acid ester concentration yielding 10% (——) or 50% (- - -) erythrocyte partition in a two-phase system versus ester acyl tail length.

effects[1] shows that the binding energy is primarily responsible for the change in surface free energy which causes the partition, consistent with the effects observed and with the much higher affinity constant of the PEG-palmitate compared to the PEG itself (the latter has not been measured but is estimated to be about 10^4 times smaller).

4.2.3.2. Applications of PEG-Alkyl Esters in Cell Partitioning

As part of a study in which hydrophobic affinity partitioning was used to examine potential differences between erythrocytes obtained from multiple sclerosis patients and normal controls, Van Alstine examined some of the factors which had to be controlled to obtain reproducible results from healthy donors' blood samples.[13,19] The most important factors appeared to be strict reproducibility in the handling of the cell sample and purity of the affinity ligand preparation, particularly with respect to contamination by free fatty acids or other amphipaths which could compete for hydrophobic binding sites.[12,20] Insufficient care in handling the cell sample could denature proteins or rupture cells to expose hydrophobic sites and hydrophobic affinity partitioning is sensitive to changes in protein structure.[21] Hydrophobic affinity compounds synthesized using ether as opposed to ester linkages to the PEG head group[12] had no advantage with respect to sensitivity of the partition coefficient, but the ether bond is much more stable to biodegradation.[22] The ligands are more difficult to synthesize, however, and are not widely available.

The ability of a PEG-alkyl ester to increase cell partition into the PEG-rich phase is related to its chain length and degree of saturation, as expected. Figure 3 demonstrates the influence of chain length on the hydrophobic affinity partition of fresh and glutaraldehyde fixed erythrocytes using a series of PEG 8000-saturated fatty acid esters. Similar results were obtained for unsaturated fatty acid esters of the C-18 and C-16 class.[13] Fatty acid esters can completely shift cell partition into the upper phase when added at micromolar concentrations. It can be seen that crosslinking of cell membrane proteins with glutaraldehyde only slightly decreases the ability of the affinity ligands to affect partition. The steric hindrance provided by membrane proteins is apparently little affected by the reaction.

If the mode of action of hydrophobic ligands is as postulated in Section 4.2.2, it would be expected that intercalation of the alkyl group into the lipid bilayer would behave like partition into an oil phase with respect to dependence on chain length.[23] Hence, a monotonic change in K would be expected as a function of the number of carbons in the alkyl tail. Such an effect is demonstrated in Figure 4 in which the logarithm of the total ester concentration required to achieve the indicated cell partition is plotted as a function of chain length. Similar results were found for unsaturated esters and for fixed cells.[13]

The above correlation, and the similarity of results obtained with liposomes and cells, suggest that lipid intercalation is responsible for PEG-alkyl ester binding to mammalian cells and hence for the partition behavior observed. Other hydrophobic structures on cell surfaces ought to be targets for binding of these ligands as well, however. We investigated a bacterial system which illustrates this principle by comparing the hydrophobic partition of gram negative *Aeromonas salmonicida* strains which differed in their ability to produce a hydrophobic surface protein array known as the A layer and in their ability to produce smooth lipopolysaccharide (O-antigen) on the bacterial surface. Both of these components show significant correlations with pathogenicity of the organism. PEG 8000-linoleate was found to be the most discriminating ligand.[24] Figure 5 demonstrates the ability of this ester to differentially partition *A. salmonicida* cells of surface phenotypes possessing A-layer and O-antigen; a plot of erythrocyte partition is also provided for comparison. It is clear that in these bacteria, nonbilayer lipid surface components play a role in determining interaction with hydrophobic affinity partition ligands.

An example of the sensitivity of hydrophobic affinity partitioning applied to mammalian cells is found in the study of metastatic sublines derived from primary tumors. Metastasis, the spread of secondary tumors from a primary tumor, results from the ability of a small fraction of the cells which are shed from a primary site to evade the host's defence systems and grow into an independent lesion. The cells responsible for this process are believed to be successful due to undefined differences in surface properties. Miner *et al.*[25] reported the ability to differentiate between murine lymphosarcoma sublines of low and high liver colonizing potential when partitioning in phase systems sensitive to cell surface charge. In an effort to expand these results we undertook to study the partition characteristics of B16 melanoma cells from sublines with differing metastatic potential.[26] Cells with lower (B16-F1) and higher (B16-F10) lung colonizing capability were partitioned in systems containing various PEG-fatty acid esters. Again, unsaturated fatty acid esters were the most discriminatory.

Figure 6 illustrates the countercurrent distribution (CCD) (multiple sequential

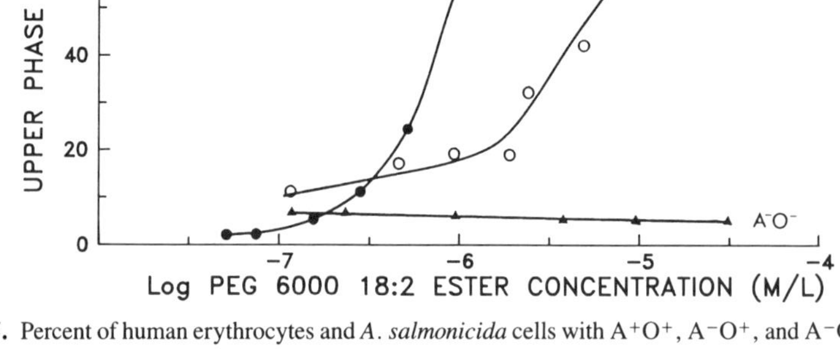

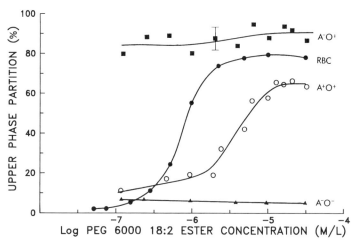

Figure 5. Percent of human erythrocytes and *A. salmonicida* cells with A^+O^+, A^-O^+, and A^-O^- surface phenotypes partitioning into the upper phase versus PEG 6000-lineolate (18:2) concentration in a two-phase system.

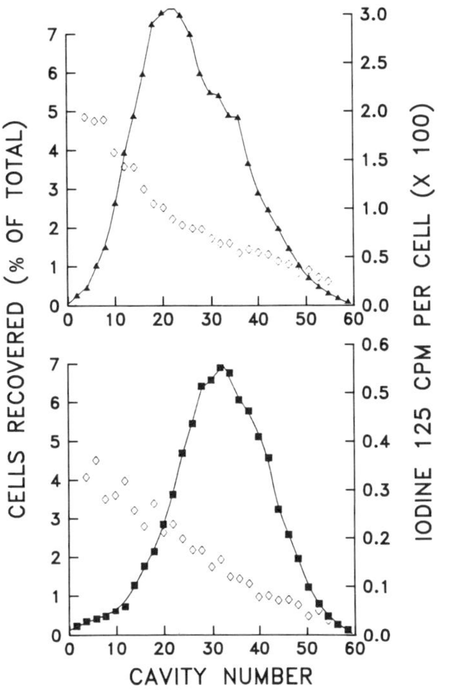

Figure 6. Countercurrent distribution of B16-F10 (▲) and B16-F1 (■) at 4 °C in a two-phase system containing 1 μM PEG-linoleate: [^{125}I] UdR specific cell activity (□). The first 3 cavities were loaded with cell suspension, the others with phase system. 57 transfers were made with 7.5 min settling time after 30 s mixing.

partition steps[27]) of the two B16 cell populations in a phase system containing PEG 8000-linoleate ester. Cells were labelled with ^{51}Cr and with ^{125}I-labelled deoxyuridine (IUdR). The chromium label allowed cell counts to be made independent of any aggregation while IUdR was used to indicate nucleic acid synthesis rates. Both populations were extremely heterogeneous in their surface properties, evident from the widths of the distributions.[27] The difference in peak location for the two sublines induced by the presence of PEG-linoleate indicates differences in surface properties between the most prevalent cell types in the two populations. Such discrimination was not seen in the absence of ester. Overall, the B16-F10 cells had greater IUdR uptake than the B16-F1 cells. In both cases specific activity of IUdR per cell decreased from left to right in the CCD curve, implying that the cells with the strongest interaction with PEG-linoleate, and hence the highest partition coefficient, were the least active in taking up IUdR and synthesizing nucleic acids. In this instance, therefore, a variety of subtle characteristics were revealed by hydrophobic affinity partitioning.

4.3. IMMUNOAFFINITY PARTITIONING

Immunoaffinity partitioning involves the use of antibodies, antibody fragments, or related immunological reagents, such as protein A or G, to carry out affinity partitioning. Usually, the immunoactive molecule is chemically modified to enhance its partition into the desired phase, typically via derivatization with PEG or some other substance, such as dye[28] with a relatively high affinity for one phase. The first use of immune specificity to alter partition behavior was carried out without modified reagents, however. Philipsson and Albertsson used anti-virus antibodies to produce alterations in viral partitioning via antibody binding.[29,30] The native properties of the unmodified proteins were sufficiently different from the viral surface properties to produce partition changes, but the effects were not as large as are found in systems containing derivatized proteins. Later work of this type was carried out with unmodified antibodies directed against bacteria[31] and peptide hormones.[20]

In 1986 two groups reported on the use of PEG-derivatized antibodies (Abs) for the immunoaffinity partition of cells.[32,33] We utilized cyanuric chloride coupling of PEG to polyclonal IgG from rabbits, directed against human erythrocytes. Both the ability of the PEG-Ab to change red cell partition and the agglutinating activity of the PEG-Ab were measured as a function of PEG molecular weight and derivatization density. The latter assays the effect of coupled polymer on the ability of the Ab to bind antigen (Ag) at both its combining sites, since both sites must be available if the Ab is to agglutinate cells by binding to antigenic sites on both surfaces simultaneously. Affinity partition effects, on the other hand, should result if only one of the Ag binding sites on the Ab molecule is active, since all that is required is that the modified Ab bind to the cell immunospecifically.

As would be expected, it was found that agglutination activity was sensitive to derivatization, activity being lost as either the molecular weight or the average number of lysines derivatized per Ab molecule increased.[32] The molecular weight effects were of particular interest, agglutination being almost unaffected by PEG 200 at all PEG:lysine ratios but being completely eliminated when less than half of the lysines were reacted with PEG 1900. This indicated that the derivatization chemistry had no deleterious effects on Ab activity up to at least PEG:lysine ratios of 5:1 in the reaction mixture. Instead, it seemed that the size of the bound polymer was the most important characteristic, presumably because molecules with larger radii of gyration would interfere with Ag binding over greater distances on the IgG surface.

Since bound PEG increased the Ab partition coefficient, the effect increasing with molecular weight and amount bound, the conditions chosen for application to cell immunoaffinity partition represented a balance between the loss of binding activity and the increase in partition coefficient. PEG 1900, reacted at a PEG:lysine ratio of 3:1, resulted in 31 moles PEG bound per mole IgG and a 17× increase in partition coefficient of the derivatized molecule.[32] This material, when added to a mixture of human and rabbit erythrocytes, immunospecifically separated the population as shown in the CCD in Figure 7. The modified Ab had to be present to provide the effect, however, as the modified Ab bound too weakly to allow the cells to be washed and run in the absence of unbound PEG-Ab.

Similar cell partition results were reported by Karr *et al.*[33] utilizing PEG 5000- and PEG 1900-derivatized Abs. Extensive single-step cell partition studies as a function of PEG-Ab concentration demonstrated an inverse dependence of cell partition on degree of substitution. Again, a separation by CCD between Ag-bearing

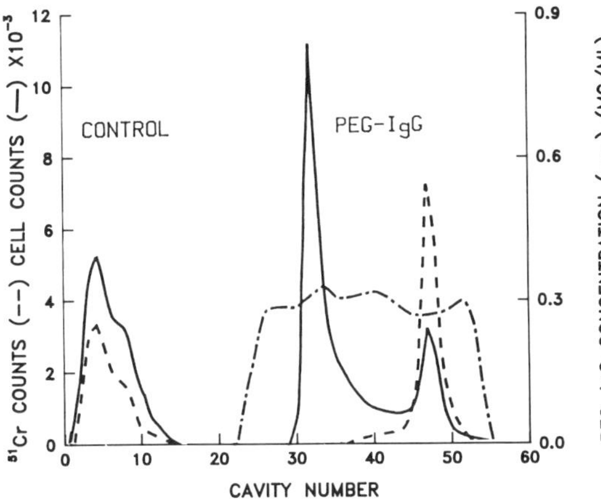

Figure 7. Countercurrent distribution of erythrocytes. A mixture of rabbit and ^{51}Cr-labeled human cells was loaded in cavities 1, 2, 31, and 32. A phase system containing 0.3 mg per ml [^{125}I] PEG-IgG was loaded in cavities 7–37. Other cavities were loaded with a control phase system containing buffer. The run consisted of 20 transfers with 30 s mixing and 8 min settling before each transfer.

and Ag-negative cell types was demonstrated with PEG-derivatized polyclonal IgG. In this case the Ab preparations bound sufficiently strongly that it was not necessary to include the PEG-Ab in the two-phase system while CCD was carried out, a 15-min prior incubation being sufficient exposure to produce the desired effect. The dependence of agglutination inhibition on PEG concentration and molecular weight differed somewhat from those summarized above however, presumably due to differences in the preparations used.

Since these initial reports, significant extensions have been made in the technique. Derivatization chemistry felt to be milder than the cyanuric chloride coupling has been successfully applied to cell immunoaffinity isolation.[34] More general reagents, which do not require the tedious derivatization optimization procedures required for each new separation problem, have been developed based on a sandwich technique.[28,34,35] These methods utilize a reagent such as PEG-derivatized protein A which is reacted with cells exposed to, then washed from, the Ab on which the separation is to be based. The derivatized protein A binds to the antibodies coating the target cell, causing a specific change in its partition coefficient.[35] Provided the primary Ab is of those subclasses which bind to protein A, immunospecific separation results. In another approach we utilized a monoclonal Ab directed against the Fc fragment of IgG and derivatized it with PEG to again vary the partition of cells bearing any primary Ab.[28,35] Both these reagents allow many commercially available monoclonal and polyclonal Ab preparations to be used as the basis of the isolation.

4.4. SUMMARY

PEG derivatization of molecules which exhibit an affinity for various features of the cell surface allows isolations or enrichments of populations bearing the feature of interest via partition in two polymer aqueous phase systems. The basic work which has been accomplished to date has demonstrated the strength and generality of this approach. The results of isolations based on hydrophobic or antigenic properties of the cell surface provide the impetus for a much wider application of these techniques in the future.

ACKNOWLEDGMENTS

The work on which this chapter is based was supported by grant MT-5759 of the Medical Research Council of Canada.

REFERENCES

1. D. E. Brooks, K. A. Sharp, and D. Fisher, in: *Partitioning in Aqueous Two-phase Systems Theory, Methods, Uses and Applications to Biotechnology* (H. Walter, D. E. Brooks, and D. Fisher, eds.), p. 11, Academic Press, Orlando (1985).
2. K. A. Sharp, M. Yalpani, S. J. Howard, and D. E. Brooks, *Anal. Biochem. 154*, 110 (1986).
3. V. P. Shanbhag and G. Johansson, *Biochem. Biophys. Res. Commun. 61*, 1141 (1974).
4. E. Eriksson, P.-Å. Albertsson and G. Johansson, *Mol. Cell. Biochem. 10*, 123 (1976).
5. H. Walter, E. J. Krob and R. Tung, *Exp. Cell Res. 102*, 14 (1976).
6. G. Johansson, *Biochim. Biophys. Acta 451*, 517 (1976).
7. H. Walter and E. J. Krob, *FEBS Lett. 61*, 290 (1976).
8. H. Westrin, P.-Å. Albertsson, and G. Johansson, *Biochim. Biophys. Acta 436*, 696 (1976).
9. E. Eriksson and P.-Å. Albertsson, *Biochim. Biophys. Acta 507*, 425 (1978).
10. E. Eriksson, *J. Chromatogr. 205*, 189 (1981).
11. P. S. Gascoine, C. J. Dix, and D. Fisher, *Exp. Cell Biol. 51*, 322 (1983).
12. J. M. Harris, E. C. Struck, M. G. Case, M. S. Paley, J. M. Van Alstine, M. Yalpani, and D. E. Brooks, *J. Polymer Sci. Polymer Chem. Ed. 22*, 341 (1984).
13. J. M. Van Alstine, Ph.D. Thesis, University of British Columbia, Vancouver (1984).
14. K. A. Sharp, Ph.D. Thesis, University of British Columbia, Vancouver (1985).
15. J. M. Van Alstine, K. A. Sharp, and D. E. Brooks, *Colloids and Surfaces 17*, 115 (1986).
16. M. G. Styring, H. H. Teo, C. Price, and C. Booth, *J. Chromatogr. 388*, 421 (1987).
17. C. P. S. Tilcock, T. Dempsey, P. R. Cullis, and D. Fisher, in: *Separations Using Aqueous Phase Systems: Applications in Cell Biology and Biotechnology* (D. Fisher and I. A. Sutherland, eds.), p. 179, Plenum Press, New York (1989).
18. S. A. Simon, W. L. Stone, and P. B. Bennet, *Biochim. Biophys. Acta 550*, 38 (1979).
19. J. M. Van Alstine and D. E. Brooks, *Clin. Chem. 30*, 441 (1984).
20. B. Desbuquois and G. D. Aurbach, *Biochem. J. 12*, 717 (1972).
21. C. G. Axelsson, *Biochim. Biophys. Acta 533*, 34 (1978).
22. J. M. Harris, M. G. Case, B. Hovanes, J. M. Van Alstine, and D. E. Brooks, *Ind. Eng. Chem. Prod. Res. Dev. 23*, 86 (1984).
23. C. Tanford, *The Hydrophobic Effect*, 2nd ed., Wiley, New York (1980).
24. J. M. Van Alstine, T. J. Trust, and D. E. Brooks, *Appl. Environ. Microbiol. 51*, 1309 (1986).
25. K. M. Miner, H. Walter, and G. L. Nicholson, *Biochemistry 20*, 6244 (1981).
26. J. M. Van Alstine, P. Sorensen, T. J. Webber, R. Greig, G. Poste, and D. E. Brooks, *Exp. Cell Res. 164*, 366 (1986).
27. T. E. Treffry, P. T. Sharpe, H. Walter, and D. E. Brooks, in: *Partitioning in Aqueous Two-phase Systems Theory, Methods, Uses and Applications to Biotechnology* (H. Walter, D. E. Brooks, and D. Fisher, eds.), p. 132, Academic Press, Orlando (1985).

28. S. J. Stocks and D. E. Brooks, in: *Separations Using Aqueous Phase Systems: Applications in Cell Biology and Biotechnology* (D. Fisher and I. A. Sutherland, eds.), p. 183, Plenum Press, New York (1989).
29. P.-Å. Albertsson and L. Philipsson, *Nature (London) 185*, 38 (1960).
30. L. Philipsson and H. Bennich, *Virology 29*, 330 (1966).
31. O. Stendahl, C. Tagesson, and L. Edebo, *Infect. Immun. 10*, 316 (1974).
32. D. E. Brooks, K. A. Sharp, and S. J. Stocks, *Makromol. Chem. Macromol. Symp. 17*, 387 (1988).
33. L. J. Karr, S. G. Shafer, J. M. Harris, J. M. Van Alstine, and R. S. Snyder, *J. Chromatogr. 354*, 269 (1986).
34. L. J. Karr, J. M. Van Alstine, R. S. Snyder, S. Shafer, and J. M. Harris, *J. Chromatogr. 442*, 219 (1988).
35. S. J. Stocks and D. E. Brooks, *Anal. Biochem. 173*, 86 (1988).

5

Affinity Partitioning in PEG-Containing Two-Phase Systems

GÖTE JOHANSSON

5.1. INTRODUCTION

Water and other liquids can be divided into two fluid compartments or phases (in direct contact with each other) by using them as solvents for poly(ethylene glycol) (PEG) together with another polymeric substance.[1] The incompatibility of polymers in solution that gives rise to this phase separation also has the consequence that the two polymers are accumulated in opposite phases. The difference in polymer structures and polymer concentrations in the phases may cause significant divergences in the solvating properties for high molecular weight substances added in low concentration. When water is used as the solvent, added salts partition more or less evenly between the phases. Proteins, on the other hand, partition more unequally. The actual partition of a substance is described by the partition coefficient, K, which is defined as the ratio of the concentration of partitioned substance between the upper and lower phase. The most popular aqueous two-phase systems for partitioning of biological substances have been the ones containing PEG and dextran.[2–4] The top phases of these systems contain, besides water, mainly PEG (5–15%) while the bottom phases contain dextran (10–25%) and some PEG (0.2–2%). Three PEG–dextran systems which have identical phase compositions are shown in Figure 1.

The unequal distribution in the two phases of the polymers makes them useful as carriers for various kinds of chemical substances, such as affinity ligands. Phase-restricted groups influence the partitioning of substances via electrostatic interaction or by direct binding to the target substance (Figure 2). PEG is by far the most used

---

GÖTE JOHANSSON • Department of Biochemistry, Chemical Center, University of Lund, S-22100 Lund, Sweden.

Poly(Ethylene Glycol) Chemistry: Biotechnical and Biomedical Applications, edited by J. Milton Harris. Plenum Press, New York, 1992.

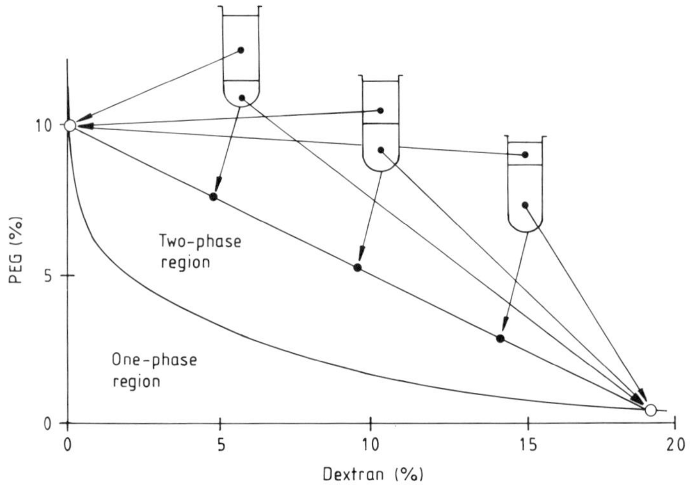

Figure 1. Phase diagram for the system water–PEG 8000–dextran 500 at 20 °C. Compositions above the curved line (the binodal curve) give two phases. Three systems, with their total polymer concentrations belonging to the same tie line, are shown. The composition of the top phase and the bottom phase are found on the binodal curve. Systems with compositions above or below the tie line shown here belong to other tie lines approximately parallel with the first one. The phase diagram is based on data presented by Albertsson.[2] From Johansson.[5]

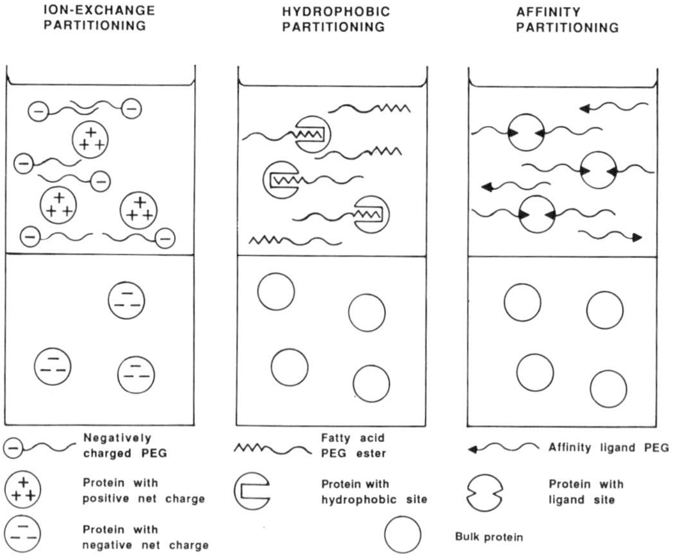

Figure 2. Three ways of affecting the partition of proteins by using PEG-bound groups. From Johansson.[5]

ligand carrier. Affinity ligands can be used in this way for specific extraction of special proteins, nucleic acids or cell particles containing binding sites for the ligand. This process is called affinity partitioning.[6]

In practice, such extractions (Figure 3) can be carried out in the following way: A complex mixture (the sample) is added to a system which is chosen so that the target substance, after equilibration, is collected in one phase. The equilibration is usually a very rapid process which is completed within 10–30 s by careful but effective mixing. A much longer time, 10–60 min, is needed for complete settling of the phases due to the similarity in density and extremely low interfacial tension of the phases. The complementary phase (with low concentration of the target substance) is removed together with extracted "impurities" and it is replaced by a new phase containing polymer-bound affinity ligand. After re-equilibration, the target substance is now transferred into the ligand-containing phase while the remaining substances do not change their partition but largely remain in the original phase. To complete the separation, the ligand/product-containing phase may be "washed" by equilibration with pure opposed phase. The effectiveness of the extraction depends on a number of factors, which are summarized in the following section.

5.2. SYSTEM PARAMETERS

5.2.1. Type and Concentration of Polymer

A number of water-soluble polymers are known to give rise to two-phase systems when paired with PEG.[7] A general rule of thumb is that the more the two polymers differ in chemical structure and the higher their molecular weights, the lower are the polymer concentrations necessary for formation of two phases. By choosing increasing concentrations of polymers, the one-sidedness of the polymer distribution between the phases can be accentuated as can be seen in Table 1. For PEG

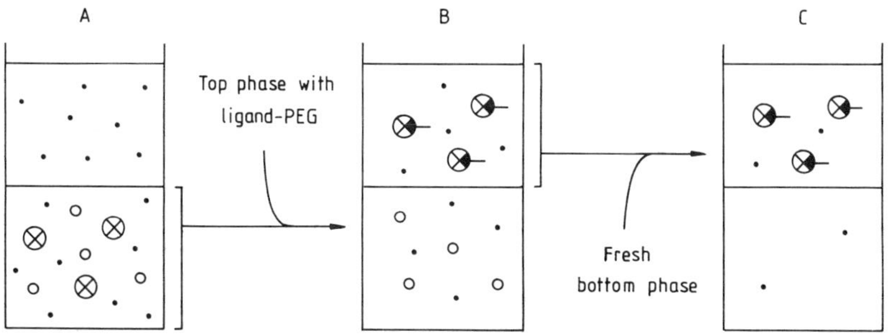

Figure 3. Purification of a target substance, ⊗ (such as an enzyme) from a complex mixture by partitioning in three steps: (a) pre-extraction; (b) affinity extraction with ligand-PEG, ◄—; (c) "washing" step.

Table 1. The Distribution of PEG 8000 and Dextran 500 between the Phases in Two-Phase Systems at 20 °C[a]

System composition		Distribution of PEG[b]		Distribution of dextran[b]	
% dextran	% PEG	K'	$\lvert\log K'\rvert$	K'	$\lvert\log K'\rvert$
5.00	3.50	1.93	0.29	0.245	0.61
6.20	4.40	6.70	0.83	0.023	1.64
7.00	5.00	12.2	1.09	0.0088	2.06
9.80	7.00	51	1.71	0.0022	2.66

[a]Calculated from data in Ref. 2.
[b]Quantity K' is defined as the ratio between the concentration (in % w/w) of the polymer in the top and bottom phases.

the partition coefficient of the polymer itself can be more than 50, while for dextran it can be as extreme as 0.002. A measure of the effectiveness of the polymers as ligand carriers is the absolute value of their partition coefficients, $\lvert\log K_{polymer}\rvert$, which are also given in Table 1. The introduction of an affinity ligand on (normally a minor part of) the polymer may, however, influence the $\lvert\log K_{polymer}\rvert$ value of the ligand–polymer derivative.[8] The partition of an enzyme in PEG–dextran systems of various polymer concentrations, both with and without ligand–PEG, is shown in Figure 4.

5.2.2. Type and Concentration of Ligand

The choice of ligand is guided by the kinds of binding sites present on the target substance as well as on contaminating substances. Attachment of ligand to the

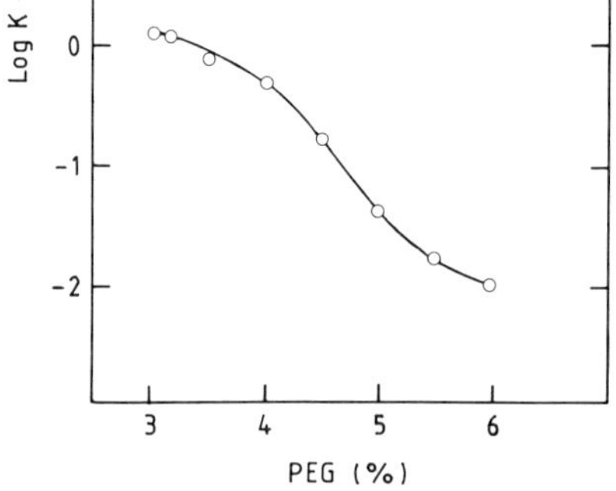

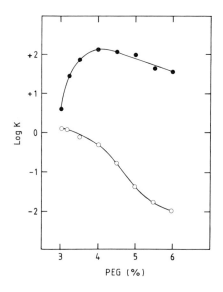

Figure 4. Partition of phosphofructokinase (from yeast), 4 nkat ml^{-1} without (○) and with (●) ligand–PEG (Cibacron blue F3G-A PEG) in systems with various concentrations of dextran is 1.5 times the concentration of PEG and the concentration of Cibacron blue PEG is 3% of total PEG. The system also contains 50 mM sodium phosphate buffer, pH 7.0. Temperature: 0 °C. From Johansson *et al.*[8]

polymer may involve considerable effort as well as tedious purification of the ligand–polymer product to remove free ligand and other impurities. Covalent binding of affinity ligand to the polymer may also influence the binding properties toward the target substance.[9] A number of the ligands which have been used for affinity partitioning are found in Table 2.

The effect of the concentration of ligand–polymer, keeping the total amount of polymer constant, is illustrated in Figure 5. The enzyme phosphofructokinase is partitioned in a system containing increasing concentrations of PEG-bound Cibacron blue F3G-A, which binds strongly to the ATP sites of this enzyme.[20]

For the common laboratory-scale separations, usually only a few percent of the PEG carries one, or in some cases two, mole of ligand per mole polymer. In the case of fully monosubstituted PEG with a molecular weight of 8000 dalton, in a system containing 80 g PEG per liter, the overall ligand concentration will be 10 mM. If we further assume that the ligand–polymer strongly binds to a protein with two binding sites for the ligand and with a molecular weight of 100,000 dalton, the limiting extraction capacity of the system would be 500 g target protein per liter. Since in the equilibrium process only one out of ten ligand–polymer molecules may at a certain moment be bound to the target protein, the capacity may in reality be reduced to 50 g protein per liter. This is still a large capacity, thus making affinity partition interesting for biotechnical use, especially since the method works well even with very crude extracts.

5.2.3. Type and Concentration of Salt

The kind of salt that is included in a two-phase system is of great importance for success of the affinity partitioning procedure. Usually a buffer is added to keep a favorable pH value. If necessary, additional salt is used to obtain partitioning of the main contaminating material to the phase poor in ligand. The effect of salts on the partition is due to the electric charge of the partitioned material.[21] This is, for membrane particles and nucleic acids, normally a negative net charge at the commonly used pH values of 5–9. Some proteins, on the other hand, are positively charged in part of this pH region. As a rule, negatively charged material is excluded from the PEG phase (in a PEG–dextran system) by using potassium chloride and even

Table 2. Selected Works in Affinity Partitioning of Proteins

Protein	Ligand	Reference
Albumin	Fatty acids	10
Colipase	Lecithin	11
Dehydrogenases, kinases	Textile dyes	8, 12–15
S-23 Myeloma protein	Dinitrophenol	6
3-Oxosteroid isomerase	Estradiol	16
Prealbumin	Remazol yellow GGL	17, 18
Trypsin	*p*-Aminobenzamidine	19

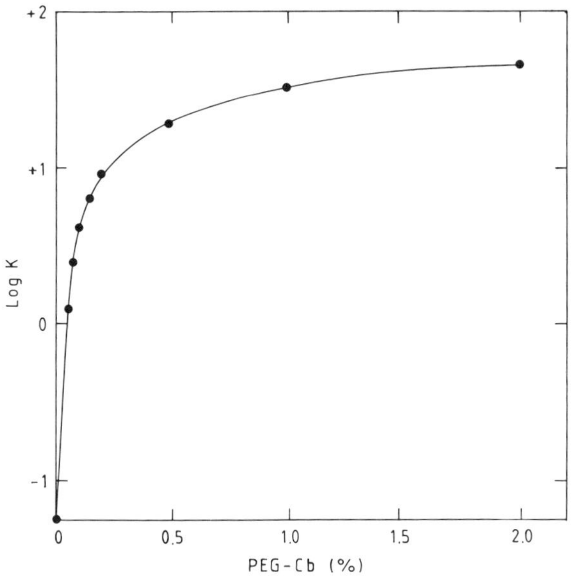

Figure 5. The partition coefficient of yeast phosphofructokinase as a function of the concentration of Cibacron blue F3G-A PEG (PEG-Cb), given as percent of total PEG. System composition: 7% dextran 500, 5% PEG 8000, 50 mM sodium phosphate buffer, pH 7.0, and 4 nkat ml^{-1} of phosphofructokinase. Temperature: 0 °C. From Johansson *et al.*[8]

more by adding sodium perchlorate. Higher partition coefficients are obtained for the same material by using lithium phosphate or tetrabutylammonium phosphate buffers. However, these salts may also influence binding between the target substance and the ligand. Salts enhance the binding if this is of a pure hydrophobic character, but more specific bindings, which may include electrostatic elements, can be considerably weakened. In the case of the commonly used ligand Cibacron blue F3G-A, phosphate or acetate buffer, up to 0.1 M, can be included without adverse effects on affinity partitioning, but chlorides and perchlorates should be avoided.[8,22]

5.2.4. Temperature

Temperature influences both composition of the phases (i.e., the two polymers will partition less extremely) and binding of ligand to the target molecules. The latter effect depends on the balance between hydrophobic and electrostatic factors in the interaction between ligand and target substance.

5.2.5. pH Value

By changing the pH value the target–ligand interaction may be affected and the charge of the target material may be altered, causing changes in its partition coefficient. If the main salt of the system is a buffer, its electrostatic directing will be modified.

5.2.6. Ligand Density on Carrier Polymer

The normal linear PEG cannot carry more than two ligand molecules attached to its end groups. With other types of polymers, such as dextran, the number of ligands (the degree of substitution) can be chosen to be very high. At the same time, the original character of the polymer is increasingly changed and the partition of ligand–polymer may deviate increasingly. Also, the introduction of charges (on the ligand) may make the partition of the ligand–polymer more sensitive to the kind of salt present in the system.[23] This salt-dependence gives, however, the interesting possibility of using the same ligand–polymer for extraction either into the lower or into the upper phase, depending on the kind of salt used.

5.2.7. Sample Concentration

The effect of sample concentration has not been studied in detail. Affinity extraction has, however, been found to work very well with crude concentrated protein extracts up to 150 g per kg system.[24]

5.2.8. Free Ligand

The presence of free ligands (e.g., the natural ligand present in cell extracts) may have considerable influence on affinity partitioning, especially when free ligand binds to the target substance more strongly than does the polymer-bound ligand. Free ligands can, on the other hand, be used for selective "stripping" of the ligand–polymer after the extraction step to return the target material to a fresh opposite phase.[25]

5.3. MODELING

5.3.1. Thermodynamic Model

A simple but useful model for affinity partitioning was formulated by Flanagan and Barondes.[6] This model was obtained by merging the relations for the partition coefficients for all possible molecular arrangements (such as ligand–PEG, target substance, and ligand–PEG and target substance complexes), the association equilibria in upper and lower phases, and the mass balance. With an excess of mono-ligand–PEG (i.e., when the binding sites are saturated with ligand–polymer in both lower and upper phase) the partition coefficient K_{max} will, according to the model, be given by equation 1:

(1) $$K_{max} = K_0\, k_t/k_b\, K_L^n$$

where K_0 is the partition of the target substance in a system without ligand–PEG, k_t and k_b are the total association constants for the complex in the top and bottom phases, respectively, K_L is the partition coefficient for the ligand–PEG, and n is the maximal number of bound ligand–PEG molecules per target molecule (or particle). By applying this model on partition data for enzymes, assuming $k_t = k_b$, the n values were found to be 1–3 depending on the enzyme.[8,26] This was the case even when a much higher number of binding sites had been established for free ligands, as in the case of phosphofructokinase.[8] The attachment of two PEG molecules or one dextran molecule to a protein of medium size (molecular weight of 100,000 dalton) will form a polymer "atmosphere" around the protein molecule (Figure 6). The molecular principle of affinity partitioning can be visualized as the inclusion of target molecules in volume elements of the ligand containing phase.

5.3.2. Molecular Model

A model based on entropy calculations of the random flight of the polymer chain attached to the target molecule depending on the phase occupied has been presented by Baskir *et al.*[28]

5.4. APPLICATIONS

5.4.1. Purification of Proteins

5.4.1.1. General Strategy

The first step, when applying the two-phase affinity partition technique for a new separation problem, is the synthesis of ligand–polymer. A number of well-established methods for attaching ligands to PEG have been presented, and they have been summarized by Harris.[29] The importance of the additional purification of the ligand–polymer to remove contaminating free ligands or ligands with small attached polymer fragments cannot be overestimated. The study of the influence of various parameters (Section 5.2) on the partition that follows can often, compared with the synthetic work, be considered as a pleasurable event.

Two-phase systems can be used in a number of ways for separation purposes.

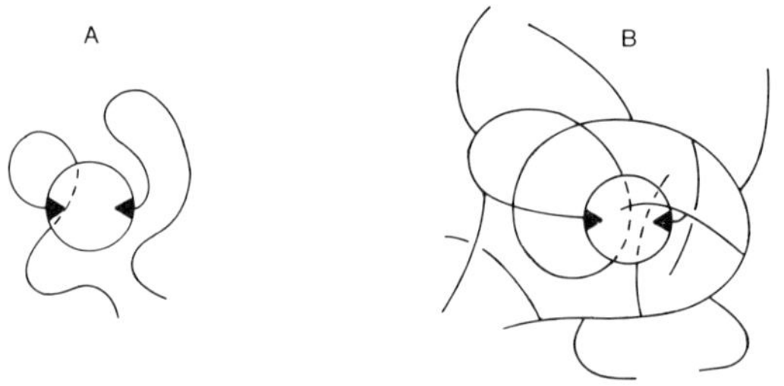

Figure 6. Scale models of complexes between a globular protein molecule (MW 100,000) and ligand polymers: (A) mono-ligand PEG, MW 8000; (B) di-ligand–dextran, MW 70,000. From Johansson.[27]

The purity obtained after a single extraction step can often be considerably improved by applying "washing" steps with fresh opposite phase. Changing phases also opens the possibility of using two ligands, one in each phase, or successively introducing several ligands and stepwise extracting a number of components of a mixture.

A multipartition method named, countercurrent distribution (according to Craig),[30] is an excellent tool for estimating the homogeneity of the target substance to be extracted. It may in many cases be found that what was assumed to be a single substance actually splits up into several discrete fractions, as is the case with isoenzymes. This knowledge is extremely useful for the formulation of batch extraction procedures.

Aqueous two-phase systems can further be used for column chromatography as developed by Müller.[31,32] In this approach the dextran phase is retained by a specially designed adsorption bed of beads covered by polyacrylamide, while the PEG phase acts as the mobile phase.

5.4.1.2. Use of Dye Ligands

A number of studies have been conducted on the possibility of using textile dyes (such as dyes on a triazine basis) for affinity partitioning of proteins (Table 1). The reason for this is that dyes are available in a great number of structures that show high affinity and selectivity to a number of proteins. Furthermore, the dyes are extremely cheap and easy to attach to polymers containing hydroxy or primary amino groups. Several enzymes in the kinase and dehydrogenase groups have been purified by affinity partitioning with PEG-bound dye ligands. Also, some serum proteins have been isolated by this method. Isolations in a large scale of formate dehydrogenase from *Candida boidinii*[33] and lactate dehydrogenase from pig muscle[34] have been presented, and the favorable economy of the process was shown due to excellent recovery of the ligand–PEG (96% per extraction cycle).

5.4.1.3. Hydrophobic Ligands

The most commonly used hydrophobic ligands have been long-chained fatty acids bound to PEG. Some examples are found in Table 1. Combination of an aliphatic chain with a charged end group has been shown to either decrease or increase the affinity toward target proteins.[35] Systems with hydrophobic ligands are also used for determination of the hydrophobicity of proteins and particles.[36]

5.4.1.4. Specific Ligands

Examples of more specific ligands for affinity partitioning are still few. One very promising class of ligands is the metal chelates which were earlier successfully applied for affinity chromatography. A commonly used ligand was iminodiacetic acid (IDA), which can be bound to PEG and "loaded" by metal ions (e.g., Ni^{2+} or Cu^{2+}). This ligand has been used for affinity partitioning of proteins obtained by gene manipulation.[37]

The use of immunoligands would make it possible to combine very high specificity with high binding strength. However, the introduction of an increasing number of PEG chains on the antibody molecule reduces its binding strength.[38] An interesting way to solve this problem has been presented by Karr *et al*. They suggested the use of protein-A bind to PEG to localize the antibody to the upper phase.[39]

5.4.2. Purification of Nucleic Acid

Müller *et al.* have used PEG-bound triphenylmethyl dyes, such as malachite green, for selective extraction of DNA fragments. When these two-phase systems were used for column chromatography (see above), fragments of DNA could be effectively fractionated according to their size. By the choice of dye the fragments could be eluted also according to their relative content of cytosine–guanine and adenine–thymine base pairs.[31,32]

5.4.3. Purification of Membranes and Cells

Affinity partitioning has also been applied for membrane fragments, cell organelles, and cells such as erythrocytes. In a number of works PEG-bound hydrophobic ligands have been used for studies of cells (see Chapter 4).

Membranes containing neurotransmitter receptors have been selectively extracted by using PEG-bound compounds with high affinity for the receptor. Cholinergic membranes from Torpedo electroplax have been purified by affinity partitioning with trimethylamino–PEG, hexamethonium–PEG, and bis-*O*, *O'*-(triethylaminoethyl)-resorcinolamino–PEG.[40,41] Also, membranes containing opiate binding sites from brain cortex have been extracted by affinity partitioning using Naloxone–PEG.[9]

When the green organelles of the leaves, the chloroplasts, are prepared from plant material, a considerable number lose their surrounding envelope, and the inner membrane system, the grana, is expanded. By partitioning in a PEG–dextran system it is possible to separate intact (class I) chloroplasts from those which have lost the envelop (class II). To achieve this extraction it is necessary to use PEG-bound fatty acids (PEG esters) of certain critical lengths. Class I chloroplasts are extracted by PEG-esterified capric acid (C_{10}), but class II chloroplasts require lauric acid (C_{12}) to be transferred into the upper phase.[42]

By attaching PEG to antibodies, a specific extraction of red cells from a certain species has been demonstrated.[38] Use of liquid–liquid systems, in combination with countercurrent distribution, is an elegant way to avoid the problem of irreversible adsorption to matrices, which is a problem for column chromatography of cells.

5.4.4. Studies of Molecular Interactions

Affinity partitioning has also been used to study interactions of ligands and proteins.[43] If one ligand is bound to PEG and the protein extracted into the upper phase, the competing binding or allosteric influence of an additional (free) ligand can be studied by following the change in the partition coefficient of the protein. Such studies have been made on yeast phosphofructokinase and α-2-macroglobulin from blood plasma.[43]

5.5. CONCLUSIONS

Affinity partitioning using PEG-bound affinity ligands offers a broad spectrum of applications from small-scale separations of cells and cell constituents in the laboratory to the biotechnical large-scale extractions. A great deal of research has been done with this technique and affinity partitioning can now be regarded as a mature science, which is ripe for a wide range of laboratory and industrial separation.

REFERENCES

1. G. Johansson and M. Joelsson, *J. Chromatogr. 464*, 49 (1989).
2. P.-Å. Albertsson, *Partition of Cell Particles and Macromolecules*, 3rd ed., Wiley, New York (1986).
3. H. Walter, D. E. Brooks, and D. Fisher (eds.), *Partitioning in Aqueous Two-Phase Systems: Theory, Methods, Uses, and Applications to Biotechnology*, Academic Press, Orlando (1985).
4. H. Walter and G. Johansson, *Anal. Biochem. 155*, 215 (1986).
5. G. Johansson, *Biofutur 74*, Supplement No. 24, 7 (1988).
6. S. D. Flanagan and S. H. Barondes, *J. Biol. Chem. 250*, 1484 (1975).
7. F. Tjerneld and G. Johansson, *J. Bioseparation 1*, 255 (1990).
8. G. Johansson, G. Kopperschläger, and P.-Å. Albertsson, *Eur. J. Biochem. 131*, 589 (1983).
9. B. Olde and G. Johansson, *Neuroscience 15*, 1247 (1985).
10. G. Johansson and V. P. Shanbhag, *J. Chromatogr. 284*, 63 (1984).
11. C. Erlanson-Albertsson, *FEBS Lett. 117*, 295 (1980).
12. G. Kopperschläger, G. Lorenz, and E. Usbeck, *J. Chromatogr. 259*, 97 (1983).
13. G. Johansson and M. Joelsson, *Enzyme Microbiol. Technol. 7*, 629 (1985).
14. H. K. Kroner, A. Cordes, A. Schelper, M. Morr, A. F. Bückmann, and M.-R. Kula, in: *Affinity Chromatography and Related Techniques* (T. C. J. Gribnau, J. Visser, and R. J. F. Nivard, eds.), p. 491, Elsevier, Amsterdam (1982).
15. M. Joelsson and G. Johansson, *Enzyme Microbiol. Technol. 9*, 233 (1987).
16. P. Hubert, E. Dellacherie, J. Neel, and E.-E. Baulieu, *FEBS Lett. 65*, 169 (1976).
17. G. Birkenmeier, G. Kopperschläger, and G. Johansson, *Biomed. Chromatogr. 1*, 64 (1986).
18. G. Birkenmeier, B. Tschechonien, and G. Kopperschläger, *FEBS Lett. 174*, 162 (1984).
19. G. Takerkart, E. Segard, and M. Monsigny, *FEBS Lett. 42*, 218 (1974).
20. E. Hofmann and G. Kopperschläger, *Methods Enzymol. 90*, 49 (1982).
21. G. Johansson, *Mol. Cell. Biochem. 4*, 169 (1974).
22. G. Johansson and M. Andersson, *J. Chromatogr. 303*, 39 (1984).
23. G. Johansson and M. Joelsson, *J. Chromatogr. 393*, 195 (1987).
24. A. Szöke, M. Joelsson, and G. Johansson, unpublished results.
25. G. Johansson and M. Joelsson, *Enzyme Microbiol. Technol. 7*, 629 (1985).
26. G. Johansson and M. Joelsson, *J. Chromatogr. 537*, 219 (1991).
27. G. Johansson, in: *Protein–Dye Interaction: Developments and Applications* (M. A. Vijayalakshmi and O. Bertrand, eds.), p. 165, Elsevier, London (1989).
28. J. N. Baskir, T. A. Hatton, and U. W. Suter, *J. Phys. Chem. 93*, 969 (1989).

29. J. M. Harris, *J. Macromol. Sci. C-25*, 325 (1985).
30. L. C. Craig, in: *Comprehensive Biochemistry* (M. Florkin and E. H. Stotz, eds.), Vol. 4, p. 1, Elsevier, Amsterdam (1962).
31. W. Müller, in: *Partitioning in Aqueous Two-Phase Systems: Theory, Methods, Uses, and Applications to Biotechnology* (H. Walter, D. E. Brooks, and D. Fisher, eds.), p. 227, Academic Press, Orlando (1985).
32. W. Müller, *J. Bioseparation 1*, 265 (1990).
33. A. Cordes and M.-R. Kula, *J. Chromatogr. 376*, 375 (1986).
34. F. Tjerneld, G. Johansson, and M. Joelsson, *Biotechnol. Bioeng. 30*, 809 (1987).
35. V. P. Shanbhag and G. Johansson, *Eur. J. Biochem. 93*, 363 (1979).
36. V. P. Shanbhag and L. Backman, in: *Separations Using Aqueous Phase Systems* (D. Fisher and I. A. Sutherland, eds.), p. 25, Plenum Press, New York (1989).
37. R. Todd, M. Van Dam, D. Casimiro, B. L. Haymore, and F. H. Arnold, *Proteins 10*, 156 (1991).
38. K. A. Sharp, M. Yalpani, S. J. Howard, and D. E. Brooks, *Anal. Biochem. 154*, 110 (1986).
39. L. J. Karr, J. M. van Alstine, R. S. Snyder, S. G. Shafer, and J. M. Harris, in: *Separations Using Aqueous Phase Systems* (D. Fisher and I. A. Sutherland, eds.), p. 193, Plenum Press, New York (1989).
40. S. D. Flanagan, S. H. Barondes, and P. Taylor, *J. Biol. Chem. 251*, 858 (1976).
41. G. Johansson, R. Gysin, and S. D. Flanagan, *J. Biol. Chem. 256*, 9126 (1981).
42. G. Johansson and H. Westrin, *Plant Sci. Lett. 13*, 201 (1978).
43. G. Kopperschläger and G. Birkenmeier, *J. Bioseparation 1*, 235 (1990).

6

Aqueous Two-Phase Partitioning on an Industrial Scale

FOLKE TJERNELD

6.1. INTRODUCTION

Aqueous polymer two-phase systems are increasingly being used in biochemistry and cell biology for the separation of macromolecules, membranes, cell organelles, and cells.[1,2] The great interest in aqueous phase partitioning is due to the unique separation properties of the systems and the mild conditions during the separation process. The unique properties of the systems make them also very interesting for large-scale industrial applications. In the biotechnical industry this technique is starting to be used for large-scale enzyme extractions.[3,4] Many applications of aqueous polymer two-phase systems in biotechnology are currently being explored, both for separations of biomolecules, cell organelles, and cells, and for bioconversions.

Poly(ethylene glycol) (or PEG) is used in almost all applications of aqueous phase partitioning. For biochemical separations on the laboratory scale the most commonly used aqueous phase system is composed of PEG and dextran. Separations of biological macromolecules, membranes, cell particles, and cells have been performed with this system.[1,2] For large-scale separations in the biotechnical industry PEG/salt systems have been the most widely used, primarily for large-scale enzyme extractions.[3,4] The application of the PEG/salt systems for large-scale extractions is to a large extent due to economic reasons because of the inexpensive chemicals that are used to create these systems. For large-scale applications of the PEG/dextran systems,

FOLKE TJERNELD • Department of Biochemistry, Chemical Center, University of Lund, S-22100 Lund, Sweden.

Poly(Ethylene Glycol) Chemistry: Biotechnical and Biomedical Applications, edited by J. Milton Harris. Plenum Press, New York, 1992.

dextran is a very expensive polymer. This has been shown by economic calculations of these systems.[5]

6.2. AQUEOUS POLYMER TWO-PHASE SYSTEMS

6.2.1. General Principles

Aqueous polymer two-phase systems are obtained in water solutions of two polymers with differing chemical structure. Aqueous two-phase systems containing only one polymer are obtained in water solutions of PEG with high concentrations of inorganic salts. The physical chemical basis for phase separation in water solutions of polymers has recently been studied.[6–11] Phase separation in systems of two polymers in water is due to polymer incompatibility, and can be explained with the use of the Flory–Huggins theory for polymer solutions.[6–10,12] The mechanistic description of polymer incompatibility is as follows[7]: The entropy increase upon mixing of polymer molecules will favor mixing and thus the formation of a single phase. However, for long polymer chains the entropy of mixing is a relatively small term in the total free energy of mixing for the polymer solution. Counteracting the entropy term is the interaction between the monomer segments of each polymer. Generally "like prefer like," which leads to an attractive interaction between monomer segments from the same polymer. This will lead to a net repulsive interaction between the different polymers. Due to this repulsive interaction and the small entropy of mixing per monomer unit in long polymers, the solution will separate into two phases. The degree of polymerization, i.e., the polymer molecular weight, has a strong influence on the phase separation. As the degree of polymerization increases the entropy of mixing is further reduced. Thus, by increasing the molecular weight of the polymers the phase separation will occur at lower polymer concentrations.[7]

6.2.2. PEG/Dextran Systems

The most used aqueous phase system for biochemical separations is composed of PEG, dextran, and water.[1,2] The PEG is enriched in the top phase and dextran in the bottom phase. Both phases have a water content of 80–90%. A phase diagram (Figure 1) shows the binodial curve separating the one- and two-phase regions and the tie lines indicating the composition of the phases in equilibrium. The temperature dependence of the phase diagram for the PEG/dextran system has been explained by the Flory–Huggins theory combined with a new model for PEG in solution developed by Gunnar Karlström.[13] This model for PEG solutions is based on earlier models proposed by Goldstein[68] and Kjellander.[69] The new model for PEG is derived from quantum chemical calculations of PEG conformations in water.[70] At room temperature polar conformations of PEG are predominant, but when the temperature is increased the nonpolar conformations become dominant. The polymer thus becomes more hydrophobic due to the temperature increase which ultimately leads to phase

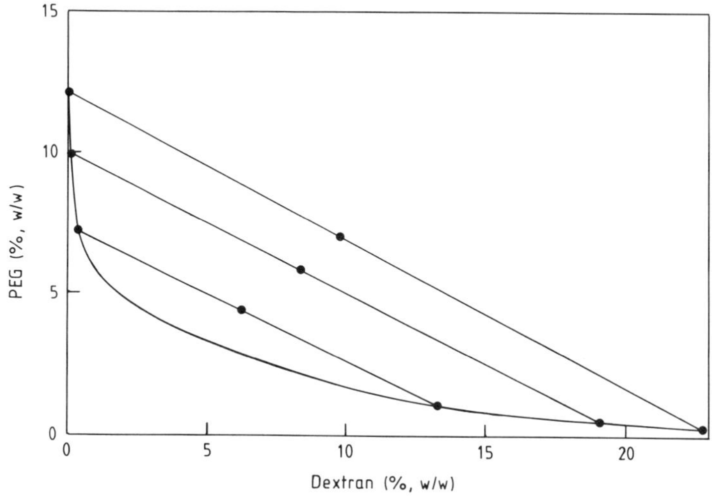

Figure 1. Phase diagram for the system composed of PEG 8000, Dextran T-500 (MW 500,000) and water at 20 °C. From Albertsson.[1]

separation. Experimentally, PEG phase separates in water solutions at temperatures above 100 °C.[14] With the application of the new PEG model it was possible to explain the appearance of a three-phase area in the phase diagram of the PEG/dextran/water system at temperatures around 110 °C.[15]

The PEG/dextran systems are the prime ones used in laboratory-scale separations with aqueous phase systems. This is due to the multitude of ways to adjust the partition which are possible in these systems. The drawback for industrial-scale use has been the high cost for fractionated dextran.[5] Crude dextran has been evaluated as an alternative to fractionated dextran for large-scale enzyme extractions,[16] and has been used with PEG in phase systems for bioconversions.[17,18]

6.2.3. PEG/Salt Systems

The phase separation of PEG in water solutions at high temperatures, described in Section 6.2.2. can be made to occur at lower temperatures by the addition of salts. Phosphate and sulfate salts are the most effective. Aqueous two-phase systems are formed at room temperature in solutions of PEG with high concentrations of phosphate or sulfate salts. A phase diagram for the PEG/phosphate system is shown in Figure 2. The phase diagram shows that PEG is almost exclusively found in the top phase and the salt is enriched in the bottom phase. However, the salt concentration is high in both phases. The PEG/salt systems are used for enzyme extractions in industrial scale[3,4] (see Section 6.4).

The drawback of the PEG/salt systems is the high salt concentration in both

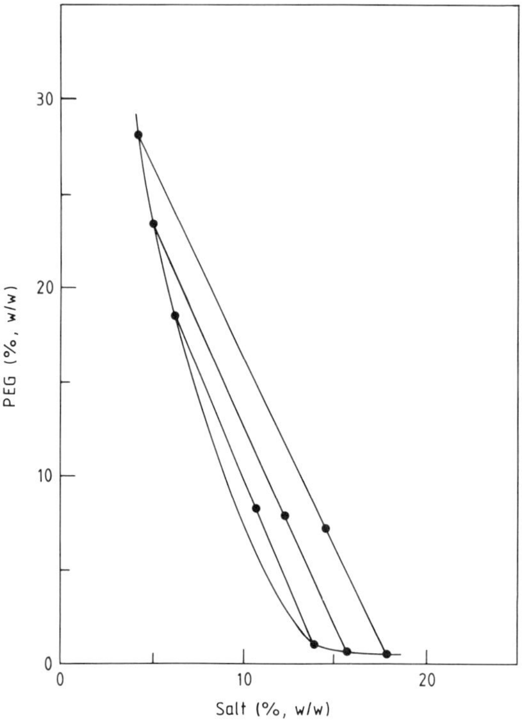

Figure 2. Phase diagram for the system composed of PEG 8000, potassium phosphate, and water at 20 °C. The potassium phosphate was a mixture with the ratio 306.9 g K_2HPO_4 to 168.6 g KH_2PO_4. From Albertsson.[1]

phases. This can lead to denaturation of salt-sensitive biological structures and makes it impossible to use affinity ligands where the binding is based on electrostatic interactions. When aqueous phase systems based on two polymers are used (such as PEG and dextran) no salt is necessary for phase formation and the systems can be made with the desired buffers.

6.2.4. PEG/Starch Systems

New polymer phase systems suitable for large-scale processes have been studied in recent years. An important aim has been the development of an inexpensive polymer as a substitute for dextran in the PEG/dextran systems. A series of starch derivatives were studied and it was found that hydroxypropyl starch (HPS) was optimal for aqueous phase systems with PEG.[19] An especially designed HPS polymer has been developed for biotechnical use. The derivatization of the starch molecule with hydroxypropyl ether groups was performed to overcome the strong gel-forming properties of starch solutions. Figure 3 shows the phase diagram of HPS of molecular weight 35,000 (trade name Aquaphase PPT from Perstorp Biolytica, Lund, Sweden) with PEG 8000. Another HPS polymer for aqueous phase partitioning has been

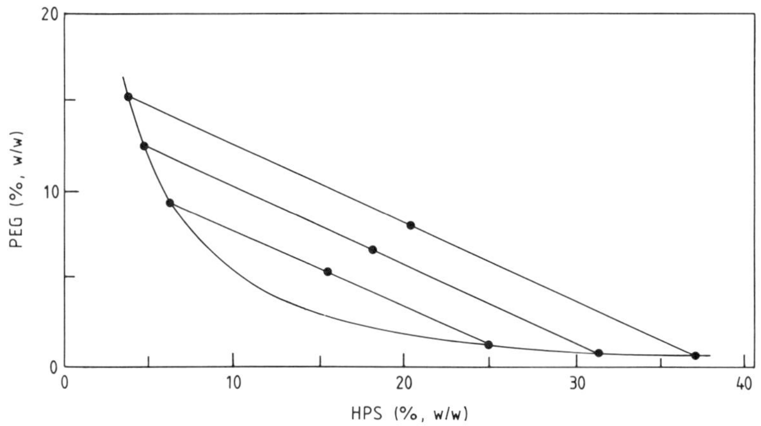

Figure 3. Phase diagram for the system composed of PEG 8000, hydroxypropyl starch (Aquaphase PPT, MW 35,000), and water at 20 °C. From Sturesson *et al.*[21]

introduced on the market (Reppal PES from Reppe Glykos, Växjö, Sweden). It has similar partitioning properties to Aquaphase PPT in phase systems containing PEG.[20]

The partitioning behavior of macromolecules, membranes, cell organelles, and cells in PEG/HPS (Aquaphase PPT) systems have been compared with PEG/dextran systems.[21] The properties of the systems are in many ways comparable. A higher solubility of serum proteins was observed in the PEG/HPS system. The Reppal polymer has been used in bioconversions of steroids in aqueous phase systems.[22]

Another alternative to obtain nongelating solutions of starch polymers is to reduce the molecular weight, e.g., by using dextrins. A low-cost phase system based on maltodextrin and PEG has been employed for enzyme extraction.[23]

6.2.5. Phase Systems with Cellulose Derivatives

Important economic advantages can be obtained by lowering the polymer concentrations when the phase systems are used for industrial extractions. Cellulose derivatives have been shown to form aqueous phase systems at low polymer concentrations. The system methylcellulose/dextran was tested in the early development of the aqueous two-phase partitioning technique.[1] Ethylhydroxyethyl cellulose (EHEC) has been studied recently as a phase-forming polymer.[24] EHEC forms aqueous phase systems with dextran or HPS at low polymer concentrations. Figure 4 shows the phase diagram using EHEC of three different molecular weights together with Dextran 500. Phase separation takes place at 1–2% total polymer concentration, which is a reduction by a factor of 10 relative to the PEG/dextran systems. With hydroxypropyl starch the phase system with EHEC is formed at 5–6% total polymer concentration, a reduction by a factor of 4 relative to the PEG/HPS system.[21]

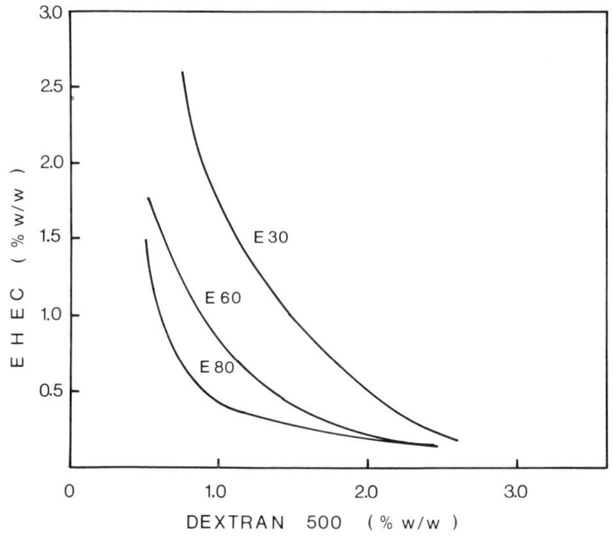

Figure 4. Binodial curves for the system composed of EHEC, Dextran T-500 (MW 500,000) and water at 20 °C. The molecular weights of EHEC 30, 60, and 80 are 30,000, 60,000, and 80,000, respectively. From Tjerneld.[24]

6.2.6. PEG/Pullulan Systems

The microbial polysaccharide pullulan is commercially available at a low price. Phase systems based on PEG and pullulan have been investigated with the object of using the latter polymer as a replacement for dextran in biotechnical applications.[25] The PEG/pullulan systems were used for extractive bioconversions with enzymes or microorganisms retained in the pullulan-rich bottom phase.

6.2.7. PEG/PVA Systems

Polymer phase systems can also be formed with only synthetic polymers, as in the system PEG/polyvinyl alcohol (PVA). Best stability in solution was obtained with PVA that was 90% hydrolyzed, i.e., it had 10% residual acetate groups. Enzymes were strongly partitioned to the top PEG-rich phase in the PEG/PVA system.[24] The partitioning of yeast cells, cell debris, and proteins has also been investigated in the PEG/PVA system.[26]

6.2.8. Phase Systems with Polyacrylates

Acrylate co-polymers have been shown to form two-phase systems with PVA.[27] The polymers are ampholytic as they contain both anionic and cationic co-monomers. The anionic monomers are acrylic acid and the cationic are aminoethylmethacrylic and dimethylaminoethylmethacrylic acid. The polyampholytic acrylic co-polymers have been investigated for the purification of proteins from crude extracts.[27] The acrylates could be removed from the solution by isoelectric precipitation.

6.3. PARTITIONING OF SUBSTANCES

6.3.1. General Principles

Partitioning in aqueous two-phase systems depends on the size and surface properties of the substance. Several different mechanisms are in operation and can be used to influence the partitioning.[1,2,10] The ionic composition determines the electrical potential between the phases (below). The charge of the partitioned substance will then have a strong influence on the partition coefficient. Direct interactions between the partitioned substance and the phase-forming polymers also contribute to the partitioning. These interactions, and thus the partition coefficient, can be influenced by the polymer concentration and the polymer molecular weights. The strongest effects on the partitioning are obtained by including charged polymers in the phase system or by employing biospecific ligands attached to one of the phase-forming polymers.

6.3.2. Ionic Composition

Some salts have an unequal distribution between the two phases in PEG/dextran systems. Phosphate salts partition more to the bottom phase while lithium salts partition more to the top phase. The partition of NaCl is almost equal between the two phases.[28] An unequal distribution of ions in the phase system will give rise to an electrical potential difference between the phases.[1,2] The partition of phosphate ions toward the bottom phase creates an electrical potential with a negative charge in this phase which forces negatively charged macromolecules or cells to partition to the top phase. By increasing the NaCl concentration the effect of the unequal distribution of the phosphate is either reduced or eliminated because NaCl partitions equally between the two phases. Thus by changing the ionic composition the partition coefficient can be strongly influenced. If the charge of the substance depends on the pH of the solution, as for proteins and cells, a pH change will thus also affect the partitioning.[1,2]

6.3.3. Polymer Concentration

The polymer concentrations can also be used to influence the partition coefficient (K). At polymer concentrations close to the critical point in the phase diagram the K-value for macromolecules is close to one. When the polymer concentration is increased the macromolecules are partitioned unequally between the phases, that is, K will either increase above 1 or below 1. For macromolecules there is a linear relationship between log K and the polymer concentration.[24] Cells and cell particles are partitioned to one of the phases at low polymer concentrations and to the interface at high concentrations of polymers when the interfacial tension is increased.[1,2]

6.3.4. Polymer Molecular Weight

Polymer molecular weight has a strong influence on the partitioning of substances. Generally, if the bottom-phase polymer molecular weight is increased, the substance is partitioned more to the top phase, and by increasing the top-phase polymer molecular weight the substance is partitioned more to the bottom phase. For proteins, the higher the molecular weight of the protein the greater its partitioning is affected by a change in the molecular weight of either polymer. This effect is strong only for proteins of molecular weights above 50,000.[29]

6.3.5. Charged Polymers

The partition of charged molecules or cells can be strongly affected if polymers with charged groups are included in the phase system. For example, trimethyl-amino-PEG and carboxymethyl-PEG have been used to steer the partition of proteins.[30]

6.3.6. Affinity Ligands

By using biospecific affinity interactions very large changes in the partition coefficient can be obtained.[31] The most commonly used method is to covalently bind an affinity ligand to one of the phase-forming polymers. A substance can then be extracted selectively into the phase containing the polymer-bound ligand. This is described in Chapter 4 of this book.

6.4. LARGE-SCALE PROTEIN EXTRACTIONS

6.4.1. Process Design

Protein extractions can be performed on an industrial scale by use of PEG/salt systems in methods developed by M.-R. Kula and co-workers.[3,4,32,33] A general scheme for this process is shown in Figure 5. In the purification of intracellular proteins, the debris produced by cell homogenization can be effectively removed by partition in a PEG/salt system. The debris partitions to the lower (salt) phase while the main part of the target proteins can be recovered in the upper phase by the choice of a suitable phase composition. Low partition coefficients of target proteins can still give good yield in the upper phase if a system is chosen with a large volume of this phase relative to the lower one. The protein mixture in the upper (PEG) phase can be further fractionated by equilibrating it with a new salt phase. The proteins in the upper phase are extracted into the new salt phase by changing pH, salt composition, or molecular

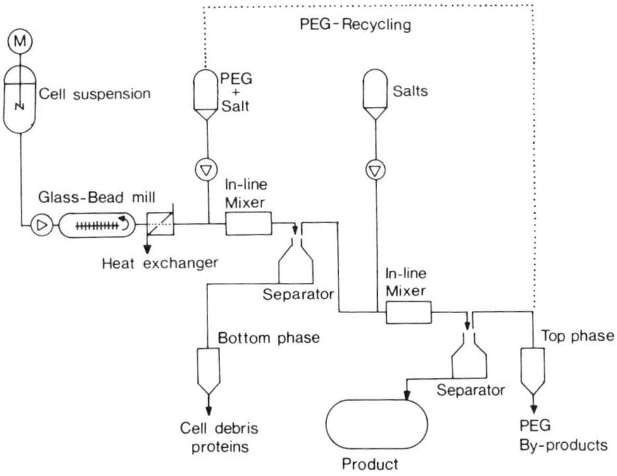

Figure 5. Scheme of protein purification by liquid–liquid extraction. The cells are disrupted by wet milling and, after passing through a heat exchanger, PEG and salts are added into the process stream of broken cells. After mixing and obtaining of equilibrium the phase system is separated, the outflowing bottom phase going to waste. The product-containing PEG-rich top phase goes to a second mixer after addition of more salt to the process stream. The product is recovered in the resulting bottom phase while the concentrated PEG solution (upper phase) goes to waste or is recycled. From Hustedt *et al.*[33]

weight of PEG. In this way the main part of PEG can be directly recycled.[4,33] Examples of large-scale (25 to 500 L) protein purification with the use of PEG–salt systems are given in Table 1. In industrial scale mostly intracellular enzymes have been purified with this technique.

6.4.2. Continuous Extractions

Extractive enzyme purifications have mostly been performed as batchwise separations, but continuous processes have also been developed. Very rapid continuous extractions can be carried out by combining the use of static mixers (for mixing of phase components) and centrifugal separators (for phase separation).[3,4,33–36] A fully automated process for continuous cross-current extraction of enzymes has been developed.[37] By integrating the cell disruption and two-phase extraction steps it is possible to perform rapid purifications of intracellular proteins. This is shown in a process for continuous extraction of β-galactosidase from *E. coli* cells[36] (see Figure 6). The average residence time was 6.3 minutes from the disintegrator inlet to the separator outlet with the PEG phase containing the enzyme. Rapid extraction protects the target protein from proteolytic degradation. In the β-galactosidase extraction only 3% of the protease in the disrupted cells was extracted together with the enzyme.[36]

Table 1. Examples of Proteins Purified by Subsequent Extraction Steps in the Enzyme Technology Department of GBF, Braunschweig[a]

Enzyme	Organism	Number of extraction steps[b]	Enrichment factor	Final yield	Remarks	Ref.
Aspartase	*E. coli*	3	18	82	Removal of interfering fumarase	32
Formate dehydrogenase	*Candida boidinii*	3	4.4	78	Removal of polysaccharides and nucleic acids (enzyme purity > 70%)	38
Fumarase	*Brevibacterium ammoniagenes*	2	22	75	Removal of polysaccharides	41
Penicillin acylase	*E. coli*	2	10	78		41
L-2-Hydroxyisocaproate dehydrogenase	*Lactobacillus confusus*	2	21	90		42
Interferon	Human fibroblasts	1	>350	75		43

[a]From Hustedt *et al.*[33]
[b]The first extraction step is designed to remove all debris, with the exception of the interferon purification.

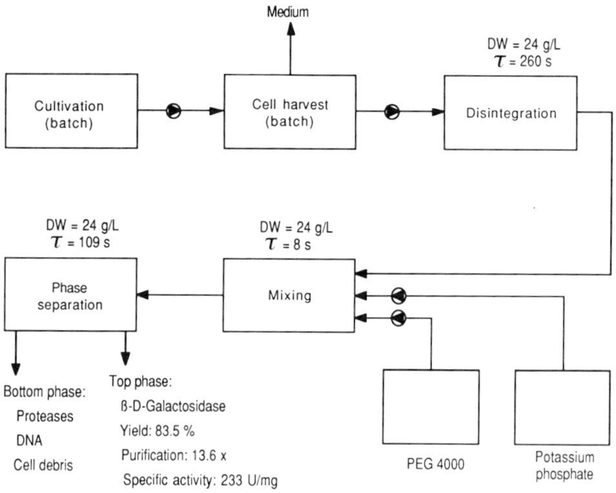

Figure 6. Continuous extraction of β-galactosidase in a PEG/salt aqueous two-phase system. DW, dry cell weight; τ, residence time. From Veide *et al.*[36]

6.4.3. Industrial Aspects

A key feature of two-phase extractions is that they can be scaled up from laboratory scale to industrial scale with performance data retained. This has been shown for the extractive purification of formate dehydrogenase.[38] The enzyme yield in 10 mL scale and in process scale (250 L) was 74 and 70%, respectively.

The use of PEG/salt systems in industrial-scale partitioning is very much dictated by economic considerations. Inexpensive chemicals are used to create the phase systems which makes them attractive compared to systems based on fractionated dextran. However, for protein purification the aqueous phase systems have to compete with alternative methods. An economic evaluation of large-scale enzyme extractions in PEG/salt systems has been made.[5] Comparisons were made with other techniques for enzyme extraction from cell debris: centrifugation, centrifugation plus precipitation, filtration by drum filter, and cross-flow filtration. The values for the capacity, throughput, space time yield, and recovery yield were highest for the aqueous phase system. The costs for labor, energy, and investment were lowest for the aqueous phase systems. The consumption of chemicals is higher, compared to other techniques, but this can be reduced by recycling of the phase components.[39] The economic analysis shows that the phase systems give high space time yield together with low total cost.[5]

High phosphate or sulfate concentrations in the effluent streams are obtained when the PEG/salt systems are used industrially. This can cause environmental problems. To provide a solution a process scheme, which includes both PEG and salt recycling, has been developed.[39] Recently a PEG/citrate system has been introduced for large-scale enzyme extractions.[40] Citrate is biodegradable and nontoxic and therefore should cause fewer waste problems compared to inorganic salts.

6.5. LARGE-SCALE AFFINITY EXTRACTIONS

6.5.1. Process Design

The use of the PEG/salt systems for affinity partitioning is limited by high salt concentration. Only affinity interactions which are insensitive to or strengthened by high salt concentrations can be employed. Affinity partitioning in laboratory scale has mostly been performed in PEG/dextran systems (see Chapter 4 of this book), but this technique has not been scaled up due to the high cost for fractionated dextran. The recent availability of inexpensive phase-forming polymers has changed the situation. The new phase systems composed of PEG and hydroxypropyl starch have been used for affinity partitioning on a larger scale. The enzyme lactate dehydrogenase was extracted from swine muscle in up to 50-liter scale[44,45] in a process scheme shown in Figure 7. The dye Procion yellow HE-3G bound to PEG in the top phase was used for the extractions. To speed up separation of the phases, a centrifugal separator was used. Separation of ligand–PEG from the enzyme in the recovered top phase was

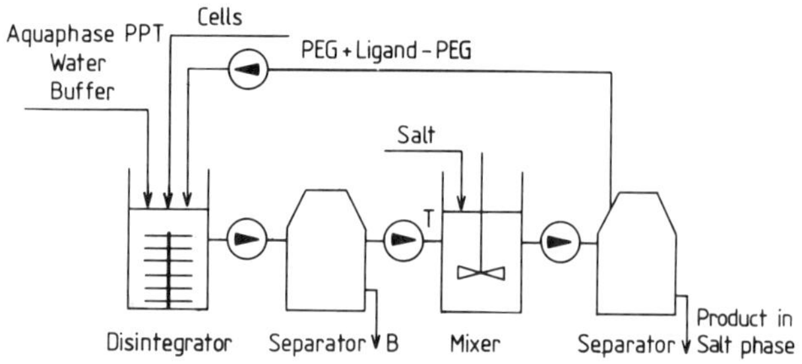

Figure 7. Scheme for continuous extraction of enzyme by affinity partition. The cells are disintegrated and mixed with phase-forming components (hydroxypropyl starch and PEG) and ligand–PEG. The bottom phase (B) after phase separation contains the cell debris. Enzyme and PEG are separated by addition of salt to the top phase (T). PEG and ligand–PEG are recycled. From Tjerneld *et al.*[44]

achieved by addition of a salt solution (sodium or potassium phosphate) at high concentration. In the resulting PEG/salt system the enzyme was found in the salt-rich bottom phase in practically quantitative yield. The ligand–PEG was recovered in the PEG-containing top phase. Both PEG and ligand–PEG could be recycled to more than 95% in each extraction cycle. Two process alternatives were compared: direct homogenization of muscle in the phase system or use of a muscle extract where the debris was removed by centrifugation. Direct homogenization in the presence of ligand–PEG simplified the process, but more ligand–PEG was consumed and the capacity of muscle per liter of phase system was lower. Accordingly, a better recovery of LDH was obtained with the muscle extract, which gives significantly lower total costs calculated per kU of enzyme when compared with direct homogenization.[44,45]

A process for large-scale affinity extraction of formate dehydrogenase from homogenized *Candida boidinii* cells has also been developed.[46] A PEG/crude dextran phase system with PEG-bound Procion red HE-3G affinity ligand was used. The composition of the phase system was 9% PEG 10,000 and 1% crude dextran. The very low concentration of crude dextran was possible because of the high cell concentration (20% wet weight) and the partition of the cell debris to the bottom phase where it substitutes for the crude dextran in the phase formation. This process was scaled up from 5 g to 220 kg phase system with retained enzyme yield (70%) and a slight reduction of specific activity (from 5.9 to 3.5 U per mg). Economic calculations of different process alternatives for formate dehydrogenase purification show that the costs for affinity extraction are comparable to other methods.[46]

6.5.2. Selective Extractions Using Fused Proteins

A new method for directing the partitioning of proteins in aqueous phase systems has been introduced with the use of recombinant DNA techniques. A fused protein of *S. aureus* protein A and *E. coli* β-galactosidase was constructed and produced in *E. coli*.[47] The pure β-galactosidase has a strong partitioning to the top phase in PEG/salt systems. The protein A-β-galactosidase fused protein is also partitioned to this phase.[47,48] The β-galactosidase part of the fused protein dominates in the partitioning and directs the protein to the top phase. Thus, the selective extraction of fused proteins from cell homogenates is possible in PEG/salt systems where the cell debris can be partitioned to the salt phase and the fused protein to the PEG phase.[49,50] Depending on the final use of the product, the fused protein can be split in the next step in order to recover the native protein. Alternatively, the fused protein is recovered for use in a technical application.

6.6. BIOCONVERSIONS IN AQUEOUS TWO-PHASE SYSTEMS

6.6.1. General Principles

Aqueous two-phase systems can be used for bioconversions.[51–53] The biocatalyst, which can be an enzyme, a cell or a microorganism, is partitioned to one of the phases together with the substrate (Figure 8). During mixing of the phases the bioconversion of substrate into product can take place. Because of the low interfacial tension[1] there is a rapid equilibrium of the formed product across the interface. The low interfacial tension also leads to the formation of very small phase droplets during mixing of the phases. This minimizes migration distances and facilitates mass transfer. After phase separation the product-containing phase can be removed. The biocatalyst is retained inside the reactor in the second phase and can be supplied with new substrate. The advantages with bioconversions in aqueous phase systems are: (1) the product is continuously removed, (2) the polymers stabilize the biocatalyst,

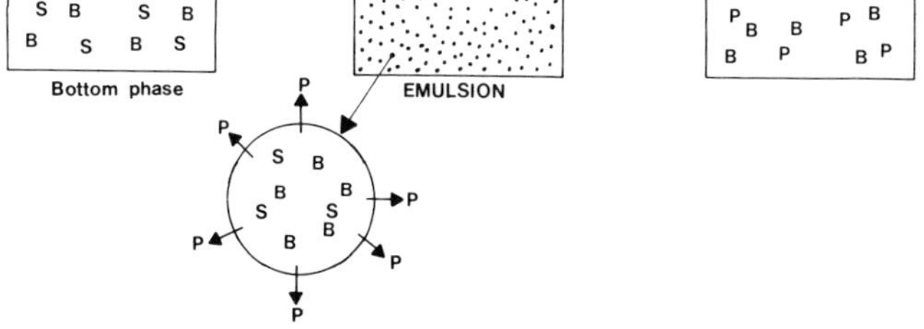

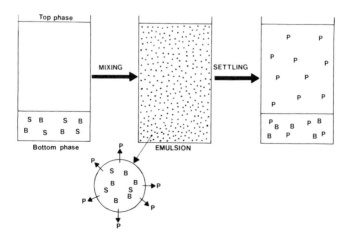

Figure 8. A common arrangement when aqueous two-phase systems are applied in bioconversion processes. S, substrate; B, biocatalyst (such as microorganism, cell, or enzyme); P, product. From Andersson and Hahn-Hägerdal.[52]

(3) the biocatalyst is recycled, and (4) macromolecular and particulate substrates can be used without the steric hindrance encountered when biocatalysts are immobilized in a gel matrix.

There is as yet no industrial process which uses aqueous phase systems for bioconversions. However, there are a number of potential applications in the biotechnical industry. A wide range of bioconversion processes using aqueous phase systems have been performed on laboratory scale. Enzymatic conversions have been: starch[53,54] and cellulose[17,18,55] hydrolysis (both in PEG/crude dextran systems), penicillin conversion[56] (in PEG/phosphate), and lactose hydrolysis[25] (in PEG/pullulan). Bioconversions using microorganisms in aqueous phase systems have been: ethanol production[58,59] (in PEG/dextran), butanol and acetone production[60] (in PEG/dextran), α-amylase production[61,62] (in PEG/dextran), cellulase production[63–65] (in PEG/dextran), β-glucosidase production[66] (in PEG/dextran and PEG/PVA), steroid conversion (in PEG/dextran[67] and PEG/HPS[22]), and fumarate conversion[57] (in PEG/phosphate).

There have been problems associated with the use of aqueous phase systems for bioconversions which have limited the industrial applications. One problem has been loss of the biocatalysts. When cells or microorganisms are used, they can be completely partitioned to one phase and thus retained inside the bioreactor. Enzymes, on the other hand, are difficult to partition 100% to one phase which means that there will be a loss of enzymes in the product-containing phase. One way of solving this problem has been to combine the aqueous phase system with an ultrafiltration unit, as described in Section 6.6.2.[55] The product is often obtained in a polymer solution and must then be separated from the polymer. This can be done with low-molecular-weight products by letting the product-containing phase pass over a UF membrane with a cutoff that lets the product go through but retains the phase- forming polymers (see Section 6.6.2).[54,55] For proteins extracted with the PEG phase the product–polymer separation can be performed by salt addition to the PEG phase creating a PEG/salt system where the product can be obtained in the salt phase. Two examples of bioconversions will be discussed in more detail, one using enzymes and one using a microorganism.

6.6.2. Enzymatic Cellulose Hydrolysis

Hydrolysis of cellulose by enzymes has been carried out in aqueous phase systems.[17,18,55] The cellulose particles were partitioned to the bottom phase together with the cellulose degrading enzymes and the sugar products were removed with the top phase. The affinity of the enzymes for the cellulose substrate was the most important factor in the retention of enzymes in the bottom phase. Also, use of a low-molecular-weight bottom-phase polymer and a high-molecular-weight top-phase polymer contributed to obtaining low partition coefficients for the enzymes.[17] In order to be able to scale up the process, the inexpensive PEG/crude dextran system was used, although the molecular weight of crude dextran was too high to be optimal. The reactor could be run for 300 hours.[18] Because of the incomplete partitioning of enzymes to the bottom phase, there was a loss of enzyme activity in the withdrawn top phase.

An improvement of the bioreactor was made by the addition of an ultrafiltration unit.[55] The sugar-containing top phase was run through the UF unit (see Figure 9). With a membrane cutoff of 10,000 MW the top-phase polymer (PEG 20,000) was retained as well as the small amount of enzyme that was lost with the top phase in the previous reactor construction. The produced sugars could pass through the membrane. This reactor could be run for 1200 hours on the enzymes added at the start of the reaction. A concentration of sugars (glucose, cellobiose, and xylose) of 75 gL^{-1} in the effluent of the reactor was maintained by semicontinuous addition of cellulose. The enzyme consumption in this reactor was only 25% of the consumption for a batch hydrolysis in a stirred tank reactor.[55]

6.6.3. Enzyme Production

Cells or microorganisms can be cleanly partitioned 100% to one of the phases in an aqueous phase system. The cells can then be supplied with a substrate and the products can be extracted with the other phase, which is free of cells. An example of this process follows.

Production of cellulolytic enzymes with *Trichoderma reesei* Rutgers C30 in aqueous phase systems was studied in shake flask cultivations.[63,64] The phases were separated every 24 h, the fungus was partitioned to the bottom phase, and the enzyme-containing top phase was withdrawn. Fresh top phase, cellulose substrate, and media components were added. Cultivations in shake flasks with different polymers and polymer molecular weights were made in order to find the optimal phase system, both in respect to extraction of enzymes with the top phase (a high partition

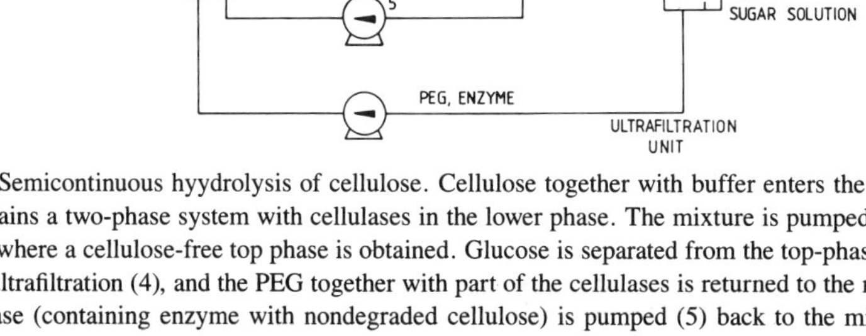

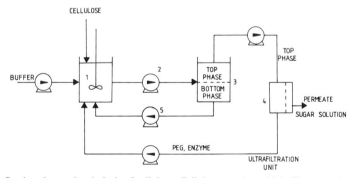

Figure 9. Semicontinuous hyydrolysis of cellulose. Cellulose together with buffer enters the mixer (1), which contains a two-phase system with cellulases in the lower phase. The mixture is pumped (2) to the settler (3), where a cellulose-free top phase is obtained. Glucose is separated from the top-phase polymer (PEG) by ultrafiltration (4), and the PEG together with part of the cellulases is returned to the mixer. The bottom phase (containing enzyme with nondegraded cellulose) is pumped (5) back to the mixer. From Tjerneld *et al.*[55]

coefficient is needed) and for enzyme production (yield and productivity).[64] The phase system composed of PEG 8000–Dextran T500 was chosen for extractive enzyme production.

Fermentor cultivations with *T. reesei* were made in a 7L fermentor with a working volume of 4L in a PEG/dextran system.[65] The fermentor was fitted with a steam sterilizable valve for exchange of top phase under sterile conditions. In the aqueous two-phase system an extractive fed-batch cultivation was maintained for 360 hours. The enzyme-containing top phase was withdrawn after phase separation. New cellulose substrate and nutrients were added with the new top phase. The enzyme extraction was started after 120 h and repeated every 72 h. The cellulolytic enzymes were obtained in the cell-free top phase.[65]

6.6.4. Integration of Production and Purification

With the use of the aqueous phase systems for protein production, as described in Section 6.6.3 it is possible to integrate the production and purification steps in a biotechnical process. During the fermentation the produced proteins can be continuously extracted into a cell-free top phase containing PEG, and the proteins can then be further purified by addition of a concentrated salt solution to this phase. In the formed PEG/salt system the proteins can be partitioned to the bottom phase. By desalting in an ultrafiltration unit a purified protein solution is obtained.

6.7. CONCLUDING REMARKS

Partitioning in aqueous two-phase systems is well established in biochemistry and cell biology as a powerful separation technique. In the biotechnical industry the technique is so far almost exclusively used for enzyme extractions. Expansion to a large number of industrial applications depends on the development of purification processes and bioconversions, where the properties of the aqueous phase systems are used to the greatest advantage. Separations with aqueous phase systems have been demonstrated in small and large scale for many processes. Phase systems suitable for industrial use have been developed. The industrial engineers now have the possibility to use the available knowledge for the application of phase partitioning to technical processes. We expect to see the same dramatic increase in applications of two-phase partitioning in the biotechnical industry as has occurred in biochemistry and cell biology during the last decade. For biotechnical companies who wish to be at the frontier of developments it will be necessary to master the technique of aqueous two-phase partitioning.

ACKNOWLEDGMENTS

The author wishes to thank the National Swedish Board for Technical Development (STU) and the National Energy Administration of Sweden for financial support.

REFERENCES

1. P.-Å. Albertsson, *Partition of Cell Particles and Macromolecules*, 3rd ed., Wiley, New York (1986).
2. H. Walter, D. E. Brooks, and D. Fisher (eds.), *Partitioning in Aqueous Two-Phase Systems: Theory, Methods, Uses, and Applications to Biotechnology*, Academic Press, New York (1985).
3. M.-R. Kula, K. H. Kroner, and H. Hustedt, in: *Advances in Biochemical Engineering* (A. Fiechter, ed.), Vol. 24, p. 73, Springer, Berlin (1982).
4. H. Hustedt, K. H. Kroner, and M.-R. Kula, in: *Partitioning in Aqueous Two-Phase Systems: Theory, Methods, Uses, and Applications to Biotechnology* (H. Walter, D. E. Brooks, and D. Fisher, eds.), p. 529, Academic Press, New York (1985).
5. K. H. Kroner, H. Hustedt, and M.-R. Kula, *Process Biochem. 19*, 170 (1984).
6. D. E. Brooks, K. A. Sharp, and D. Fisher, in: *Partitioning in Aqueous Two-Phase Systems: Theory, Methods, Uses, and Applications to Biotechnology* (H. Walter, D E. Brooks, and D. Fisher, eds.), p. 11, Academic Press, New York (1985).
7. Å. Gustafsson, H. Wennerström, and F. Tjerneld, *Polymer 27*, 1768 (1986).
8. Å. Sjöberg, G. Karlström, and F. Tjerneld, *Macromolecules 22*, 4512 (1989).
9. J. N. Baskir, T. A. Hatton, and U. W. Suter, *J. Phys. Chem. 93*, 2111 (1989).
10. J. N. Baskir, T. A. Hatton, and U. W. Suter, *Biotechnol. Bioeng. 34*, 541 (1989).
11. C. A. Haynes, R. A. Beynon, R. S. King, H. W. Blanch, and J. M. Prausnitz, *J. Phys. Chem. 93*, 5612 (1989).
12. P. J. Flory, *Principles of Polymer Chemistry*, Cornell University Press, Ithaca, NY (1953).
13. G. Karlström, *J. Phys. Chem. 89*, 4962 (1985).
14. S. Saeki, N. Kuwahara, M. Nakata, and M. Kaneko, *Polymer 17*, 685 (1976).
15. Å. Sjöberg and G. Karlström, *Macromolecules 22*, 1325 (1989).
16. K. H. Kroner, H. Hustedt, and M.-R. Kula, *Biotechnol. Bioeng. 24*, 1015 (1982).
17. F. Tjerneld, I. Persson, P.-Å. Albertsson, and B. Hahn-Hägerdal, *Biotechnol. Bioeng. 27*, 1036 (1985).
18. F. Tjerneld, I. Persson, P.-Å. Albertsson, and B. Hahn-Hägerdal, *Biotechnol. Bioeng. 27*, 1044 (1985).
19. F. Tjerneld, S. Berner, A. Cajarville, and G. Johansson, *Enzyme Microbiol. Technol. 8*, 417 (1986).
20. H. Walter and E. J. Krob, *J. Chromatogr. 441*, 261 (1988).
21. S. Sturesson, F. Tjerneld, and G. Johansson, *Appl. Biochem. Biotechnol. 26*, 281 (1990).
22. R. Kaul and B. Mattiasson, *Appl. Microbiol. Biotechnol. 24*, 259 (1986).
23. D. C. Szlag and K. A. Giuliano, *Biotechnol. Techniques 2*, 277 (1988).
24. F. Tjerneld, in: *Separations Using Aqueous Phase Systems* (D. Fisher and I. A. Sutherland, eds.), p. 429, Plenum Press, New York (1989).
25. A. L. Nguyen, S. Grothe, and J. Luong, *Appl. Microbiol. Biotechnol. 27*, 341 (1988).
26. A. Kokkoris, J. B. Blair, and J. A. Shaeiwitz, *Biochim. Biophys. Acta 966*, 176 (1988).
27. P. Hughes and C. R. Lowe, *Enzyme Microbiol. Technol. 10*, 115 (1988).
28. G. Johansson, *Biochim. Biophys. Acta 221*, 387 (1970).
29. P.-Å. Albertsson, A. Cajarville, D. E. Brooks, and F. Tjerneld, *Biochim. Biophys. Acta 926*, 87 (1987).
30. G. Johansson, A. Hartman, and P.-Å. Albertsson, *Eur. J. Biochem. 33*, 379 (1973).
31. G. Johansson, in: *Methods in Enzymology* (W. B. Jakoby, ed.), Vol. 104, p. 356, Academic Press, New York (1984).
32. H. Hustedt, K. H. Kroner, H. Schütte, and M.-R. Kula, in: *Enzyme Technology, 3rd Rothenburger Fermentation Symposium* (R. M. Lafferty, ed.), p. 135, Springer, Berlin (1983).
33. H. Hustedt, K. H. Kroner, U. Menge, and M.-R. Kula, *Trends Biotechnol. 3*, 139 (1985).
34. K. H. Kroner, H. Hustedt, S. Granda, and M.-R. Kula, *Biotechnol. Bioeng. 20*, 1967 (1978).
35. H. Hustedt, K. H. Kroner, and M.-R. Kula, in: *Proc. 3rd Eur. Congr. Biotechnol.*, Vol. 1, p. 597, Verlag Chemie, Weinheim (1984).
36. A. Veide, T Lindbäck, and S.-O. Enfors, *Enzyme Microbiol. Technol. 6*, 325 (1984).
37. H. Hustedt, B. Börner, K. H. Kroner, and N. Papamichael, *Biotechnol. Techniques 1*, 49 (1987).

38. K. H. Kroner, H. Schütte, W. Stach, and M.-R. Kula, *J. Chem. Technol. Biotechnol. 32*, 130 (1982).
39. H. Hustedt, *Biotechnol. Lett. 8*, 791 (1986).
40. J. Vernau and M.-R. Kula, *Biotechnol. Appl. Biochem. 12*, 397 (1990).
41. H. Hustedt, unpublished results.
42. W. Hummel, H. Schütte, and M.-R. Kula, in: *Enzyme Engineering VII* (A. J. Laskin, G. T. Tsao, and L. B. Wingard, eds.), *Ann. N.Y. Acad. Sci. 434*, 194 (1984).
43. U. Menge, M. Morr, U. Mayr, and M.-R. Kula, *J. Appl. Biochem. 5*, 75 (1983).
44. F. Tjerneld, G. Johansson, and M. Joelsson, *Biotechnol. Bioeng. 30*, 809 (1987).
45. G. Johansson and F. Tjerneld, *J. Biotechnol. 11*, 135 (1989).
46. A. Cordes and M.-R. Kula, *J. Chromatogr. 376*, 375 (1986).
47. A. Veide, L. Strandberg, and S.-O. Enfors, *Enzyme Microbiol. Technol. 9*, 730 (1987).
48. K. Köhler, L. von Bonsdorff-Lindeberg, and S.-O. Enfors, *Enzyme Microbiol. Technol. 11*, 730 (1989).
49. K. Köhler, A. Veide, and S.-O. Enfors, *Enzyme Microbiol. Technol. 13*, 204 (1991).
50. S.-O. Enfors, K. Köhler, and A. Veide, *Bioseparation 1*, 305 (1990).
51. B. Mattiasson and B. Hahn-Hägerdal, in: *Immobilized Cells and Organelles* (B. Mattiasson, ed.), CRC-Press, Boca Raton, FL (1983).
52. E. Andersson and B. Hahn-Hägerdal, *Enzyme Microbiol. Technol. 12*, 242 (1990).
53. R. Wennersten, F. Tjerneld, M. Larsson, and B. Mattiasson, in: *Proc. Int. Solvent Extraction Conf. ISEC'83*, Denver, p. 506 (1983).
54. M. Larsson, V. Arasaratnam, and B. Mattiasson, *Biotechnol. Bioeng. 33*, 758 (1989).
55. F. Tjerneld, I. Persson, P.-Å. Albertsson, and B. Hahn-Hägerdal, *Biotechnol. Bioeng. Symp. 15*, 419 (1985).
56. E. Andersson, B. Mattiasson, and B. Hahn-Hägerdal, *Enzyme Microbiol. Technol. 6*, 301 (1984).
57. Y. L. Yang, H. Hustedt, and M.-R. Kula, *Biotechnol. Appl. Biochem. 10*, 173 (1988).
58. I. Kühn, *Biotechnol. Bioeng. 22*, 2393 (1980).
59. B. Hahn-Hägerdal, B. Mattiasson, and P.-Å. Albertsson, *Biotechnol. Lett. 3*, 53 (1981).
60. B. Mattiasson, M. Souminen, E. Andersson, L. Häggström, P.-Å. Albertsson, and B. Hahn-Hägerdal, in: *Enzyme Engineering 6* (I. Chibata, S. Fukui, and L. B. Wingard, eds.), p. 153, Plenum Press, New York (1982).
61. E. Andersson, A.-C. Johansson, and B. Hahn-Hägerdal, *Enzyme Microbiol. Technol. 7*, 333 (1985).
62. E. Andersson and B. Hahn-Hägerdal, *Appl. Microbiol. Biotechnol. 29*, 329 (1988).
63. I. Persson, F. Tjerneld, and B. Hahn-Hägerdal, *Enzyme Microbiol. Technol. 6*, 415 (1984).
64. I. Persson, F. Tjerneld, and B. Hahn-Hägerdal, *Appl. Biochem. Biotechnol. 27*, 9 (1991).
65. I. Persson, H. Stålbrand, F. Tjerneld, and B. Hahn-Hägerdal, *Appl. Biochem. Biotechnol. 27*, 27 (1991).
66. I. Persson, F. Tjerneld, and B. Hahn-Hägerdal, *Biotechnol. Techniques 3*, 265 (1989).
67. S. Flygare and P.-O. Larsson, *Enzyme Microbiol. Technol. 11*, 752 (1989).
68. R. E. Goldstein, *J. Chem. Phys. 80*, 5340 (1984).
69. R. Kjellander and E. Florin-Robertsson, *J. Chem. Soc., Faraday Trans. 77*, 2053 (1981).
70. M. Andersson and G. Karlström, *J. Phys. Chem. 89*, 4957 (1985).

7

PEG-Modified Protein Hybrid Catalyst

KOHJI YOSHINAGA, HITOSHI ISHIDA, TAKASHI SAGAWA, and KATSUTOSHI OHKUBO

7.1. INTRODUCTION

Chemical modification of proteins by attachment of poly(ethylene glycol) (or PEG) is of much interest in medical applications,[1,2] enzymatic organic synthesis,[3] and affinity separations.[4] Some of the desired properties obtained by PEG modification are reduced antigenicity, control of partitioning of affinity ligands, and solubility in organic solvents.

The concept we wish to explore here is construction of hybrid, semisynthetic enzymes consisting of three parts: (1) a catalytically inactive protein, (2) a catalytically active, nonprotein metal complex, and (3) PEG. Generally, proteins have hydrophobic domains capable of entrapping organic compounds, so introduction of appropriate functional groups, which have catalytic activity for a specific reaction, gives a protein hybrid that can potentially both bind reactants and catalyze a desired reaction. Attachment of PEG results in several advantages, including in particular control of solubility in organic solvents.

We have been interested for some time in the role of metalloporphyrins and transition-metal complexes in tryptophan dioxygenase reaction.[5–7] In the course of this work we have examined the catalytic activity of metalloporphyrin attached to bovine serum albumin (BSA), which is chosen because it is known to entrap hydrophobic compounds.[8] Examples of similar studies come from the work of Whitesides and coworker,[9] who have reported that a rhodium(I)–phosphine–avidin complex catalyzes asymmetric hydrogenation of prochiral enamines to produce the

---

KOHJI YOSHINAGA • Faculty of Engineering, Kyushu Institute of Technology, Sensui, Tobata-ku, Kitakyushu 804, Japan. HITOSHI ISHIDA, TAKASHI SAGAWA, and KATSUTOSHI OHKUBO • Faculty of Engineering, Kumamoto University, Kurokami, Kumamoto 860, Japan.

Poly(Ethylene Glycol) Chemistry: Biotechnical and Biomedical Applications, edited by J. Milton Harris. Plenum Press, New York, 1992.

amino acid derivatives with 33–44% enantiomeric excess (ee), and from the work of Kubota *et al.*,[10] who have found that OsO_4 coupled to BSA has asymmetric catalytic activity for the hydroxylation of alkenes with up to 68% ee.

Here we wish to describe our recent work on the catalytic properties of manganese–tetraphenylporphyrin complexes, attached to both PEG-modified and unmodified BSA, for stereoselective oxygenation of L- or D-tryptophan ester and for asymmetric oxidation of olefins and sulfides.

7.2. THE Mn(III)–PORPHYRIN–BSA CATALYTIC SYSTEM

7.2.1. Preparation

Covalent linkage of the Mn–porphyrin complex to BSA was carried out as shown in Scheme 1. Coupling the succinimidyl ester of 5,10,15,20-tetra(*p*-carboxyphenyl)

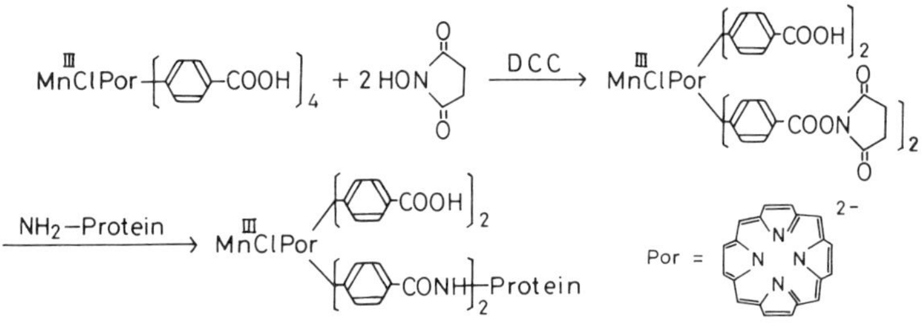

Scheme 1

porphyrin manganese(III) chloride [designated Mn(III)ClTCPP][11] to BSA was conducted in pH 9.3 borate buffer. The mixture was gently stirred for four days at room temperature, and free porphyrin complex was removed by gel filtration chromatography (Sephadex G-50) and ultrafiltration (cutoff MW 50,000) with the borate buffer containing sodium azide. In this way, a BSA conjugate containing one molecule of Mn porphyrin complex could be isolated. The binding number between the Mn–porphyrin complex and BSA can be changed by employing the mono, di, and tetra(succinimidyl ester) of Mn(III)ClTCPP.

Absorption spectra of the Mn–porphyrin complex and the Mn–porphyrin–BSA conjugate [designated Mn(III)ClTCPP–BSA] are shown in Figure 1. Attachment of Mn(III)ClTCPP to BSA did not shift the Soret absorption bands, but Q bands (assigned to a d–π^* transition) were red-shifted. This suggests that protein electron-donating groups (such as imidazole, indole, or amine groups) in the porphyrin–BSA conjugate are interacting electronically with Mn(III)ClTCPP to lower the d–π^* transition energy.

Attachment of Mn(III)ClTCPP, with four carboxyl groups, to BSA requires a surprisingly long reaction time of four days, presumably because the hydrophilic carboxyl groups must react with hydrophobic regions of the protein. On the other hand, Mn(III)Cl–tetraphenylporphyrin was adsorbed by BSA, in buffer solution, so

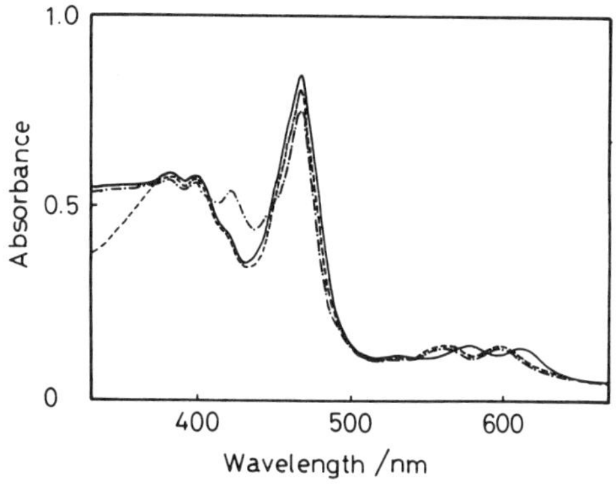

Figure 1. Absorption spectra of Mn(III)ClTCPP (- - -), Mn(III)ClTCPP–BSA (———), and *in situ* prepared Mn(III)ClTCPP/BSA (–·–) in the borate buffer solution (pH 9.3); concentration of the complexes, 1.72×10^{-5} mol dm^{-3}.

fast that changes in the absorption spectrum were observed a few minutes after mixing. We also attempted covalent coupling of Mn(III)Cl–tetra(*p*-aminophenyl)-porphyrin to BSA, in the presence of dicyclohexylcarbodiimide, but the protein denatured and precipitated. Similarly, coupling of this amino–porphyrin complex using glyoxal gave protein precipitation.

In related work, preparations of OsO_4/BSA[10] and chloroiron(III) porphyrin/human serum albumin[12] composites were achieved simply by mixing. Wilson and Whitesides[9] coupled chlororhodium(I) diphosphine complex, covalently bound to biotin, to avidin by utilizing the strong coordination affinity of biotin and avidin.

7.2.2. Catalytic Activity and Stereoselectivity

In our previous work, we found that the tryptophan dioxygenase model reaction, addition of molecular oxygen to 3-methylindole to give ring cleavage, was catalyzed by Fe(II), Mn(II), Co(II), and Cu(III) phthalocyanine complexes,[5] $FeCl_2$ or $FeCl_3$/bipyridine/pyridine complexes,[6] and Mn(II), Mn(III), Fe(II), Fe(III), Co(II), Co(III) complexes with tetraphenylporphyrin (TPP). Among these complexes, Mn(II) and Mn(III)–TPP complexes, having the highest activity for reaction of 3-methylindole in THF, were also observed to exhibit effective activity at 25°C for oxygenation of methyl *N*-acetyl-L(or D)-tryptophanate (abbreviated as L-Trp-OCH_3 or D-Trp-OCH_3).

The porphyrin–BSA conjugate prepared in Scheme 1, Section 7.2.1. [i.e., Mn(III)ClTCPP–BSA] catalyzed the oxygenation of L-Trp-OCH_3 and D-Trp-OCH_3. Pseudo-first-order rate plots were linear, and the enantiomeric rate ratio was unchanged with reaction time (Figure 2). These results indicate that the enantioselectivity of the catalyst can be characterized by the enantiomeric rate ratio, k(L)/k(D).

We have compared catalytic activities of the BSA conjugates with Mn(III)ClTPP, Mn(II)TPP, and Mn(III)ClTCPP in 18 vol% aqueous THF (Table 1).[13] The catalytic systems prepared *in situ* (i.e., nonisolated catalysts made by mixing the Mn complex with BSA) exhibited higher activities but significantly lower stereoselectivities (1.13–1.17 vs. 1.63) for the oxygenation reaction than the isolated Mn(III)ClTCPP–BSA

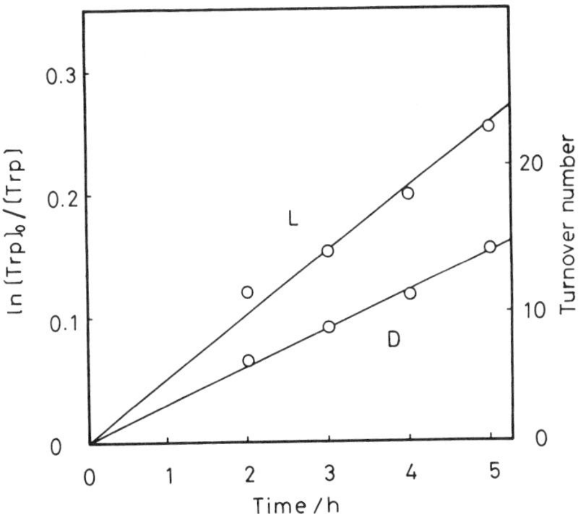

Figure 2. The pseudo-first-order plots in the oxygenation of L-Trp-OCH_3 and D-Trp-OCH_3 (2.5×10^{-2} mol dm^{-3}) by Mn(III)ClCTPP–BSA (2.5×10^{-4} mol dm^{-3}) under O_2 atmosphere at 298 K.

conjugate. Although the reasons for the difference in catalytic behavior between the *in situ* and isolated conjugates is unclear, we suspect that orientation, mobility, or location of the porphyrin complex in the BSA may be a contributor.

The number of covalent bonds (or binding number, *n*) between Mn(III)ClTCPP and BSA also affects catalytic properties (Table 2). Binding via one covalent bond resulted in almost similar catalytic activity and stereoselectivity as that via two covalent bonds, but increasing the binding number to 4 suppressed selectivity. Apparently, enantiomeric molecular recognition requires flexibility around the catalytically active species. Moreover, estimates of the distance between the Mn complex and tryptophan residues of BSA (see Section 7.3.3.) indicate that binding through four bonds deforms or expands the protein tertiary structure. This deformation of BSA could also be unfavorable for enantiomeric molecular recognition. For these reasons we have used the system with a binding number of 2 for the PEG-modified system.

7.3. PEG-MODIFIED Mn(III)–PORPHYRIN–BSA CATALYTIC SYSTEM

7.3.1. PEG Attachment

The monomethyl ether of PEG-5000 was activated by preparing the *N*-succinimidyl succinate according to the method of Abuchowski *et al.*[14] Coupling the activated PEG to the BSA–porphyrin conjugate was carried out in pH 9.3 borate buffer. We estimate that PEG was bound to 33% of the lysine amine groups, based on Habeeb analysis of PEG–BSA prepared under the same conditions. The PEG–porphyrin–BSA UV-Vis spectrum contains a large absorbance in the required region

Table 1. Kinetic Parameters for the Stereoselective Oxygenation of L- or D-Trp-OCH_3 by PEG-Modified and -Unmodified MnTPP or MnTCPP–BSA and Mn Chiral Porphyrin Complexes[a]

Catalyst	THF cont. (vol %)	$10^5 \times k$ (s^{-1}) L	$10^5 \times k$ (s^{-1}) D	k(L)/k(D)
Mn(III)ClTCPP	30	6.25		
Mn(III)ClTCPP–BSA	10	2.97	1.96	1.52
	18	1.37	0.835	1.63
Mn(III)ClTCPP–BSA–PEG	10	2.09	1.42	1.47
Mn(III)ClTCPP/BSA[b]	18	2.44	1.93	1.13
Mn(II)TPP/BSA[b]	18	1.68	1.44	1.17
Mn(III)ClTPP/BSA[b]	18	1.89	1.674	1.15
1[c]	100	0.218	0.238	1/1.09
2[c]	100	0.163	0.107	1.52
3[c]	100	0.0284	0.0428	1/1.51

[a]In 0.05 mol dm^{-3} $Na_2B_4O_7$ buffer (pH 9.3) and under O_2 atmosphere at 298K.
[b]*In situ* prepared from the Mn complex (2.5×10^{-4} mol dm^{-3}) and BSA (2.5×10^{-4} mol dm^{-3}).
[c]The following complex was used (1.0×10^{-5} mol dm^{-3}):

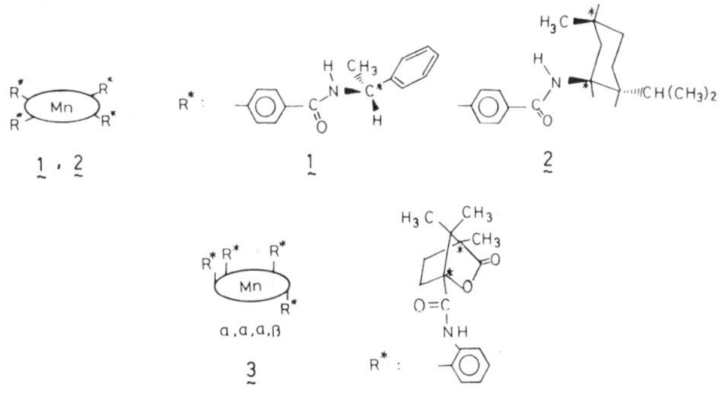

Table 2. Effects of Binding Number (*n*) between Mn(III)ClTCPP and BSA[a]

n	$10^5 \times k$ (s^{-1}) L	$10^5 \times k$ (s^{-1}) D	k(L)/k(D)	R^b (nm)
1	2.22	1.36	1.63	2.39
2	2.97	1.96	1.52	2.48
4	2.64	2.24	1.18	4.69

[a]The reaction conditions are the same as in Table 1.
[b]*R* denotes the distance between the Mn complex and tryptophan residue in BSA (cf in Section 7.3.3).

for Habeeb analysis.[15] The succinimidyl ester was chosen for this reaction because of the previous demonstration of little loss of catalytic activity for PEG–alkaline phosphatase prepared with this particular derivatives.[16,17]

7.3.2. PEG Modification Effects on Catalytic Activity and Stereoselectivity

PEG modification of Mn(III)TCPP–BSA lowered the catalytic activity slightly, but had little effect on stereoselectivity (Table 1). Possibly PEG interferes with diffusion of the substrate to the reactive site and thus lowers activity; once the substrate is bound, however, there is no PEG effect on stereoselectivity. It is also noteworthy that the PEG modified and unmodified porphyrin–BSA conjugates showed higher activities than optically active Mn(III)ClTCPP complexes **1**, **2**, and **3**, though their stereoselectivities were almost the same (Table 1).

PEG modification of the hybrid catalysts dramatically affected the catalytic behavior in THF–water solvent (Figure 3). The activity of the unmodified catalyst increased and the selectivity decreased with increasing THF content. The activity of the PEG-free catalyst was still less than Mn(III)ClTCPP itself, even at 30% THF. On

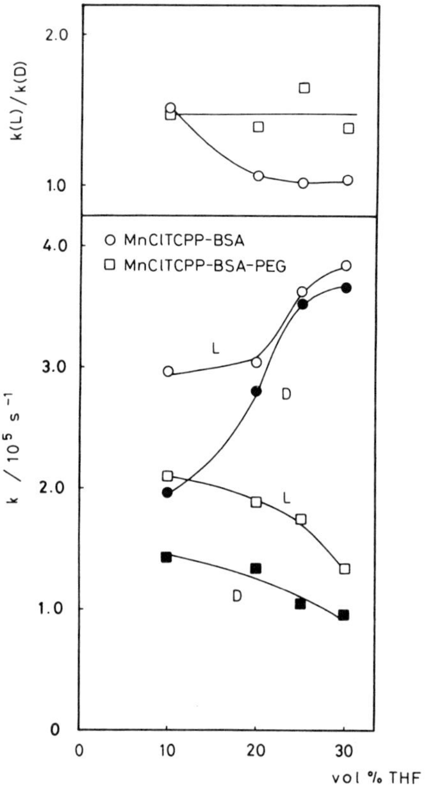

Figure 3. Effects of THF contents in aqueous solvent on catalytic activity and stereoselectivity of PEG-modified and unmodified Mn(III)ClTCPP–BSA (5.0×10^{-7} mol dm^{-3}) at 298 K. L-Trp-OCH_3 and D-Trp-OCH_3: 5.0×10^{-5} mol dm^{-3}.

the other hand, the PEG catalyst lost activity with an increase in THF content, while stereoselectivity was essentially invariant. There is further discussion of this point below. Higher percents of THF could not be studied because of limited solubility of the catalysts in these solvent mixtures.

Michaelis–Menten parameters for the oxidation by PEG-modified and unmodified hybrid catalysts are given in Table 3. Since the experimental conditions were different in each oxidation, absolute values of K_m and k_2 cannot be directly compared to each other. The Michaelis constant K_m(L) for PEG hybrid is higher than the K_m(D) value, while for the unmodified hybrid the reverse is true. These results indicate that PEG attachment on the hybrid alters the preference of substrate binding. However, overall stereoselectivity is not altered since overall reaction rate is controlled by the reaction after the substrate is bound to the catalyst, with the result that the PEG-modified system has similar stereoselectivity to the unmodified system.

The catalysts were essentially insoluble and inactive in aqueous dioxane (Table 4). The catalysts were soluble in aqueous acetonitrile and, interestingly, they exhibit a reversed selectivity in this solvent, relative to that found in THF (Table 4). These phenomena are related to deformation of BSA in higher polar solvents, as reflected in the CD spectra (next section).

7.3.3. Effects of PEG Attachment on BSA Conformation in Hybrid Catalysts

To elucidate the effects of PEG attachment on the porphyrin–BSA hybrid catalysts, we have obtained CD spectra of the catalysts in THF–water. In 30 vol% THF, the intensity of the CD spectrum decreased steadily over a period of days (Figure 4). Increasing the THF content to 60 vol% produced an even more rapid decay. In contrast, the CD spectrum of PEG-modified Mn(III)ClTCPP–BSA remained unchanged after several days in 60 vol% THF. These results indicate that PEG modification retards deformation of BSA and keeps the protein stable in an aqueous solvent having a high organic content.

It is noteworthy that the peak at 221 nm for the unmodified catalyst is transformed into a shoulder in the spectrum for the PEG-modified catalyst (Figure 4). This

Table 3. Michaelis–Menten Parameters for Trp-OCH_3 Oxygenation

Catalyst	Sub.	$k_{overall}$ (k_2/K_m)	K_m (mol dm^{-3})	k_2 (s^{-1})
Mn(III)C1TCPP–BSA[a]	L	2.53×10^{-2}	8.58×10^{-4}	2.17×10^{-1}
	D	1.32×10^{-2}	1.30×10^{-3}	1.72×10^{-1}
	(L/D)	(1.91)	(1/1.52)	(1.26)
Mn(III)C1TCPP–BSA–PEG[b]	L	6.03×10^{-3}	1.79×10^{-2}	1.08×10^{-4}
	D	4.58×10^{-3}	1.38×10^{-2}	0.632×10^{-4}
	(L/D)	(1.32)	(1.30)	(1.71)

[a]Catalyst, 2.5×10^{-5} mol dm^{-3} in THF (10 vol%)/H_2O (borate buffer) (pH 9.3) at 298K.
[b]In THF (30 vol%)/H_2O (borate buffer).

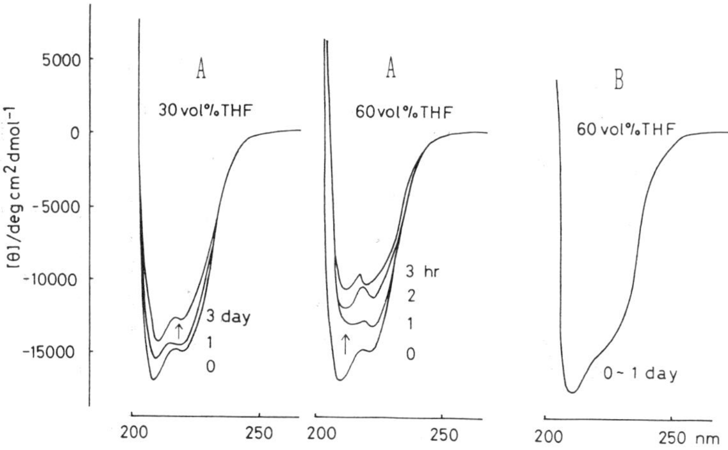

Figure 4. CD spectra of Mn(III)ClTCPP–BSA (A) and Mn(III)ClTCPP–BSA–PEG (B) in THF–borate buffer (pH 9.3).

can be explained by assuming that PEG modification brings about a conformational change in the secondary structure of the protein, possibly partial deformation of the α-helix or the β-sheet structure.

Conformational change can also be followed by calculating the distance between bound Mn(III)ClTCPP and BSA tryptophan residues. According to Foster's theory, this distance can be determined from fluorescence energy transfer from the tryptophan chromophore to the quenching group of the Mn complex.[18] We have used the method of Honore and Pedersen, in which the distance is calculated from the fluorescence emission of tryptophan at 340 nm resulting from irradiation at 290 nm.[19] There are two tryptophan residues in BSA,[20] so R in Table 5 indicates the average distance between tryptophan group and Mn(III)ClTCPP. As can be seen from Table 5, increasing THF content from zero to 30 to 60 vol% gives an increase in separation

Table 4. Solvent Effects on Catalytic Activities and Stereoselectivities of Mn(III)ClTCPP–BSA and Mn(III)ClTCPP–BSA–PEG[a]

Solvent	Cont. (vol%)	Catalyst	$k \times 10^5$ (s^{-1}) L	$k \times 10^5$ (s^{-1}) D	k(L)/k(D)
Dioxane	30	MnClTCPP–BSA	0	0	
		MnClTCPP–BSA–PEG	0	0	
THF	30	MnClTCPP–BSA	3.83	3.67	1.04
		MnClTCPP–BSA–PEG	1.34	0.97	1.38
Acetonitrile	30	MnClTCPP–BSA	2.49	3.94	0.63
		MnClTCPP–BSA–PEG	2.77	3.24	0.85

[a]The reaction conditions are the same as in Figure 3.

Table 5. Distance between Tryptophan Residue and Bound Mn Porphyrin in BSA

Catalyst	THF Cont. (vol%)	*R* (nm)
Mn(III)ClTCPP–BSA	0	2.48
	30	2.63
	60	3.55
Mn(III)ClTCPP–BSA–PEG	0	2.73
	30	2.76
	60	2.91

distance *R* by a factor of 1.5 for the unmodified catalyst, while *R* for the PEG-modified catalyst is essentially unchanged. These results therefore provide direct conformation for the hypothesis presented above regarding the effects of solvent organic content on conformational stability of both modified and unmodified catalysts. Also, the value of *R* in borate buffer is slightly larger for the unmodified catalyst, again supporting our contention that PEG modification brings about slight conformational change.

These results aid in interpreting the catalytic activities presented earlier as well. Thus it appears that solution of unmodified Mn(III)ClTCPP–BSA in THF–water leads to deformation of the complex, so that the active species becomes more accessible to substrate and the catalyst becomes more active. This same change results in a loss of stereoselectivity. The PEG-modified catalyst, however, resists deformation, substrate entry to the hydrophobic binding site is restricted, catalytic activity is relatively low, and catalyst stereoselectivity is relatively high.

7.4. ASYMMETRIC OXIDATION BY THE Mn(III)–PORPHYRIN–BSA CATALYST[21]

7.4.1. Epoxidation of Olefins

Catalytic oxidation or epoxidation of unsaturated compounds with metalloporphyrin is usually carried out in the presence of a quarternary ammonium phase transfer agent.[22] When a mixture of styrene and Mn(III)ClTPP in methylene chloride or chloroform was stirred together with the phosphate buffer (pH 7.2) containing BSA and NaOCl, optically active styrene oxide of 4.8% ee was produced in yield of 80–85% with benzaldehyde formation in the range of only 2–3% (Table 6). Thus, BSA can act both as a phase transfer agent and an asymmetric catalyst. This latter approach did, however, proceed more slowly. It has been reported that the phase transfer process involves transfer of an ammonium-hypochlorite ion pair from aqueous to organic phase, followed by production of the Mn(=O)ClTPP complex.[22] In the BSA system, BSA could serve to trap Mn(III)ClTCPP and substrate at the organic–water interface, where reaction could occur with aqueous hypochlorite.

Table 6. Asymmetric Epoxidation of Styrene by *in Situ* Prepared Mn(III)ClTPP/BSA, Mn(III)ClTCPP–BSA, and Mn(III)ClTCPP–BSA–PEG[a]

Catalyst	Org. sol. (vol%)		Time (day)	Yield (%)	$[\alpha]_D$ (deg)	ee[b] (%)
MnClTPP/BSA[c]	$CHCl_3$	(33)	4	85	1.2	4.8
MnClTPP/BSA[c]	CH_2Cl_2	(33)	4	79	0.9	3.6
MnClTPP/BSA[d]	CH_2Cl_2	(33)	2	81	1.2	4.8
MnClTPP/BSA[c]	THF	(10)	1	85	2.4	9.6
MnClTPP/BSA[c]	THF	(17)	1	54	2.2	8.8
MnClTCPP–BSA[e]	$CHCl_3$	(33)	4	87	1.2	4.8
MnClTCPP–BSA[e]	THF	(10)	4	92	2.9	11.7
MnClTCP–BSA–PEG[e]	$CHCl_3$	(33)	4	92	1.0	4.0

[a]Styrene: 1.5 g; NaOCl solution: 15 cm^3 at pH 7.2 and at 298K.
[b]Calculated from $[\alpha]_D$ −24.89° (*ca* 2.86 in $CHCl_3$)[23] for (*R*)-styrene oxide.
[c]From 30 mg of Mn(III)ClTPP and 30 mg of BSA.
[d]From 30 mg of Mn(III)ClTPP and 30 mg of BSA–PEG.
[e]Consisted of 6 mg of Mn(III)ClTCPP and 0.1 g of BSA.

With regard to the reaction catalyzed by the PEG-modified hybrid catalyst, an increase of reaction rate with PEG modification was observed, but enantioselectivity was not affected (Table 6). This result suggests that PEG attached on BSA contributes to the phase transfer process but not to catalytic asymmetric induction. Possibly the low enantioselectivity comes from less rigid coordination of styrene in hydrophobic binding sites.

Interestingly, the THF–water solvent system gave slightly higher ee (8–12%) of styrene oxide than the methylene chloride–water or chloroform–water systems. In the heterogeneous system, the BSA hybrid catalyst exists in aqueous phase, so that the conformation of BSA is possibly retained. Thus, two explanations can be considered for the solvent effects on catalyst enantioselectivity: (1) the hydrophilic environment of the THF–water system drives more rigid coordination of styrene to the BSA binding site, or (2) the BSA conformation in aqueous solvent is favorable for the enantiomeric coordination of styrene. We consider the first explanation to be most plausible.

Olefinic compounds related to styrene, such as *trans*-stilbene, α-methylstyrene, and ethyl cinnamate, also gave the corresponding oxides, but inseparable by-products were formed as well.

7.4.2. Oxidation of Sulfides

PEG-modified and unmodified Mn(III)ClTCPP-hybrid catalysts also facilitate oxidation of sulfides using hydrogen peroxide. Unfunctionalized phenyl methyl sulfide and 1-(phenylthio)acetic acid esters give slight asymmetric induction, as shown by low optical rotations of the products (Table 7). Oxidation of 1-(*p*-tolylthio)-

Table 7. Asymmetric Oxidation of Sulfides (R^1 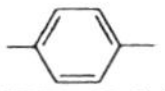 S-CH_2COOR^2) by Mn(III)ClTCPP–BSA and Mn(III)ClTCPP–BSA–PEG[a]

Sulfide R^1	Sulfide R^2	Catalyst	Org. sol. (vol%)	Time (day)	Yield (%)	$[\alpha]_D$ (deg)	ee (%)
H	Me	MnClTCPP–BSA	(0)	1	68	−0.3	
H	Bu	MnClTCPP–BSA	(0)	2	37	1.5	
Me	Me	MnClTCPP–BSA	(0)	1	92	3.7	<5[b]
		MnClTCPP–BSA	THF (17)	1	94	5.2	7[b]
		MnClTCPP–BSA	CH_3CN (17)	1	96	0.9	
		MnClTCPP–BSA–PEG	THF (17)	1	81	4.1	<5[b]
		MnClTCPP–BSA–PEG	$CHCl_3$ (17)	1	30	3.1	<5[b]
Me	Ph	MnClTCPP–BSA	(0)	3	71	9.7	5.4[c]
		MnClTCPP–BSA	THF (17)	3	15	10	5.5[c]
		MnClTCPP–BSA	$CHCl_3$ (33)	2	89	9.1	5.0[c]
		MnClTCPP–BSA–PEG	THF (17)	2	25	9.4	5.2[c]
		MnClTCPP–BSA–PEG	$CHCl_3$ (33)	2	79	13	7.2[c]

[a]Sulfide: 0.8 g; H_2O_2 solution (31%): 6 cm^3 at pH 7.2 and at 298K.
[b]Determined by ^1H NMR with $Eu(hfc)_3$ as a shift reagent.
[c]Calculated from $[\alpha]_D$ + 180.9° (*ca* 1 in $CHCl_3$)[24] for (*R*)-(+)-phenyl 1-(*p*-tolylsulfinyl)acetate.

acetic acid esters gave optically active sulfoxide with 5–7% ee. The polar solvent acetonitrile–water influenced the reaction rate, but suppressed enantioselectivity. The bulky phenyl ester group retarded reaction, as compared with the methyl group, but did not bring about an increase in the ee value for the sulfoxide. It is notable that reaction in the heterogeneous solvent system of chloroform–water proceeded faster than that in homogeneous THF–water, since the reverse was observed for epoxidation of styrene. In this case, the retained conformation of BSA in the heterogeneous system may be favorable for inclusion of the phenyl ester sulfide.

The effectiveness of PEG modification of BSA was realized in the reaction in 17 vol% THF–water. Here, yield increased from 15% after three days reaction to 25% after two days reaction. However, it was observed that the oxidation catalyzed by the unmodified hybrid involved a spontaneous (i.e., noncatalyzed) reaction in considerable extent (about 50% conversion under the condition shown in Table 7). Probably the ee values of sulfoxides produced by net catalytic reaction would be approximately two times higher than values shown in Table 7. Inhibition of the spontaneous reaction is needed to improve the ee values of the products.

7.5. SUMMARY

In this work we have reviewed the catalytic activities of PEG-modified manganese(III)–tetraphenylporphyrin hybrid complexes as artificial enzymes for a variety of chemical processes. In the case of stereoselective oxygenation of L- or D-tryptophan ester, it was observed that the hybrid catalyst was able to distinguish between enantiomers with enantiomeric rate ratios $k(\mathrm{L})/k(\mathrm{D})$ of 1.1–1.6. PEG modification of the catalyst prevented protein deformation and loss of activity even in high-THF solvent systems. The hybrid catalysts also exhibited asymmetric induction ability in epoxidation of olefins and oxidation of sulfides to sulfoxides.

In the present cases, PEG modification of the protein-hybrid catalysts imparted tolerance of organic solvents, but it did not greatly alter catalytic activities and stereoselectivities. Further improvement in these properties appears difficult in the present case because of difficulty in identifying the location of the catalytically active Mn(III) complex in BSA. From the general viewpoint of the design of a more effective protein hybrid catalyst, we feel it will be necessary to employ proteins having known secondary structure and to precisely control the site of binding of the catalytically active species to this secondary structure.

REFERENCES

1. A. Abuchowski, T. van Es, N. C. Palczuk, and F. F. Davis, *J. Biol. Chem. 252*, 3582 (1977).
2. Y. Inada, K. Takahashi, T. Yoshimoto, A. Ajima, A. Matsushima, and Y. Saito, *Trends Biotechnol. 4*, 190 (1986).
3. Y. Inada, K. Takahashi, T. Yoshimoto, Y. Kodera, A. Matsushima, and Y. Saito, *Trends Biotechnol. 6*, 131 (1988).
4. J. M. Harris and Y. Yalpani, in: *Partitioning in Aqueous Two-Phase Systems* (H. Walter, D. E. Brooks, and D. Fisher, eds.), Academic Press, New York (1985).
5. K. Ohkubo and K. Takano, *J. Coord. Chem. 14*, 169 (1985).
6. K. Ohkubo, M. Iwabuchi, and K. Takano, *J. Mol. Catal. 32*, 285 (1985).
7. K. Ohkubo, T. Sagawa, M. Kuwata, T. Hata, and H. Ishida, *J. Chem. Soc., Chem. Commun.*, 352 (1989).
8. T. Peter, Jr., *Adv. Protein Chem. 37*, 161 (1985).
9. M. E. Wilson and G. W. Whitesides, *J. Am. Chem. Soc. 100*, 306 (1978).
10. T. Kokubo, T. Sugimoto, T. Uchida, S. Tanimoto, and M. Okano, *J. Chem. Soc., Chem. Commun.*, 769 (1983).
11. A. Harriman and G. Porter, *J. Chem. Soc., Faraday Trans. 2 75*, 1532 (1979).
12. N. Datta-Cupta, D. Malakar, and J. Dozier, *Res. Commun. Chem. Pathol. Pharm. 63*, 289 (1989).
13. K. Ohkubo, H. Ishida, and T. Sagawa, *J. Mol. Catal. 53*, L5 (1989).
14. A. Abuchowski, G. M. Kazo, C. R. Verhoest, T. Van Es, D. Kafkewitz, M. L. Nucci, A. T. Viau, and F. F. Davis, *Cancer Biochem. Biophys. 7*, 175 (1984).
15. A. F. S. A. Habeeb, *Anal. Biochem. 14*, 328 (1966).
16. K. Yoshinaga, S. G. Shafer, and J. M. Harris, *J. Bioact. Compatible Polym. 2*, 49 (1987).
17. K. Yoshinaga and J. M. Harris, *J. Bioact. Compatible Polym. 4*, 17 (1989).
18. T. Foster, *Ann. Phys. (Leipzig) 2*, 55 (1948).
19. B. Honore and A. O. Pedersen, *Biochem. J. 258*, 199 (1989).
20. J. M. Brown, *Federation Proc. 35*, 2141 (1976).
21. K. Yoshinaga, N. Itoh, and T. Kito, *Polymer J. 23*, 65 (1991).
22. R. H. Holm, *Chem. Rev. 87*, 1401 (1987).
23. J. Biggs, N. B. Chapman, and V. Wray, *J. Chem. Soc. (B)* 71 (1971).
24. R. Annuziata, M. Cinquini, and F. Cozzi, *J. Chem. Soc., Perkin Trans. 1*, 1687 (1979).

8

PEG-Coupled Semisynthetic Oxidases

TETSUYA YOMO, ITARU URABE, and HIROSUKE OKADA

8.1. INTRODUCTION

Enzymes are naturally designed catalysts with high efficiency and specificity. To know how to design such enzymes is a major goal for enzymologists, and they have tried to elucidate the structural basis of enzyme catalysis. Information on structure–function relationships has made it possible to design or redesign enzymes by protein engineering and also to design artificial enzymes.[1–3]

We have prepared a malate dehydrogenase–PEG–NAD conjugate[4] (MDH–PEG–NAD) and a glucose dehydrogenase–PEG–NAD conjugate[5] (GlcDH–PEG–NAD) by covalently linking NAD to malate and glucose dehydrogenases, respectively, via a long spacer of PEG (M_4 3000). These two enzyme–NAD conjugates show quite different effects of the covalent linking of NAD on the rate of the reaction between the enzyme and the NAD moieties: for GlcDH–PEG–NAD the reaction rate between GlcDH and PEG–NAD is much enhanced by the covalent linking, but not for MDH–PEG–NAD. These findings on the kinetic properties of these conjugates have shown us the importance of kinetic information for designing enzymes and enzyme-like catalysts.

There are three main characters in an enzyme reaction: a substrate-binding site, a catalytic group, and substrate. To understand the rate-acceleration mechanisms used in enzyme reactions, we planned to prepare these characters separately, make several types of enzymatic catalysts by linking two or three of them, and investigate their kinetic properties.

TETSUYA YOMO, ITARU URABE, and HIROSUKE OKADA • Department of Fermentation Technology, Faculty of Engineering, Osaka University, 2-1 Yamada-oka, Suita-shi, Osaka 565, Japan.

Poly(Ethylene Glycol) Chemistry: Biotechnical and Biomedical Applications, edited by J. Milton Harris. Plenum Press, New York, 1992.

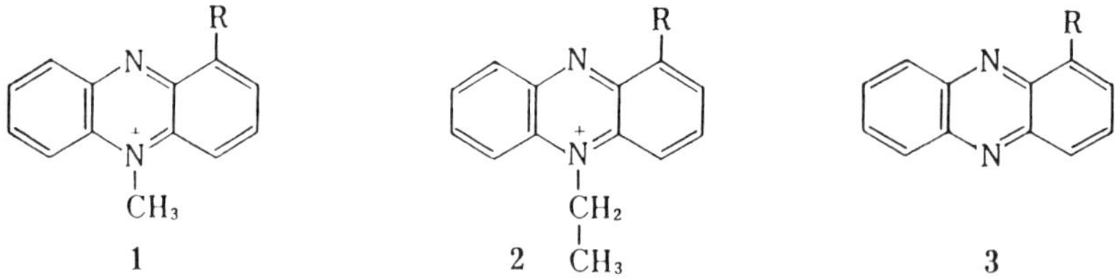

Figure 1. Phenazine derivatives. **A**: R = H. **B**: R = $O(CH_2)_3CO_2H$. **C**: R = $O(CH_2)_3CO_2CH_2CH_3$. **D**: R = $O(CH_2)_3CONH(CH_2)_2NH_2$. **E**: R = OH. **F**: R = $O(CH_2)_3CONH(CH_2CH_3O)_n(CH_2)_2NH_3^+$.

Recently we have prepared 1-(3-carboxypropyloxy)-5-ethylphenazine (Figure 1, **2B**), a stable derivative of 5-alkylphenazine with a carboxyl group at the end of the side chain at position 1.[6] This ethylphenazine (EP) derivative is an electron mediator that accepts two electrons from NAD(P)H and is reoxidized with oxygen or other electron acceptors, such as 3-(4′,5′-dimethylthiazole-2-yl)-2,5-diphenyltetrazolium bromide (MTT). Therefore, the EP derivative works as a catalyst that oxidizes NAD(P)H with oxygen or MTT by the reaction cycle as shown in Figure 2.

In this work, we have used NAD(H), the coenzyme-binding site of a dehydrogenase, and the EP group for the three main characters in an enzyme reaction: the substrate, the binding site, and the catalytic group, respectively. We have prepared the following conjugates using PEG as a long, flexible, and hydrophilic linker: 5-ethylphenazine–poly(ethylene glycol)–NAD conjugate (EP–PEG–NAD), 5-ethylphenazine–poly(ethylene glycol)–glutamate dehydrogenase conjugate (EP–PEG–GltDH), and 5-ethylphenazine–glucose dehydrogenase–NAD conjugate (EP–GlcDH–NAD). These conjugates shown new catalytic activities, and the EP moiety works as an artificial catalytic group for the oxidation of NADH (or the NADH moiety) with oxygen or MTT in the new catalytic reactions. In addition, the PEG linker enables

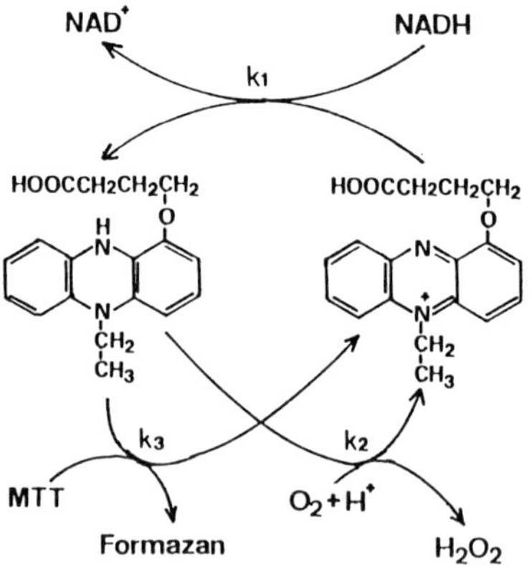

Figure 2. NADH oxidation catalyzed by an ethylphenazine derivative.

NAD(H), its binding site, and EP on a conjugate to interact with each other, and makes the reactions between them into an intramolecular type. These conjugates are unique enzyme-like catalysts or semisynthetic oxidases, and provide us with a kinetic basis for artificially designing enzymes.

8.2. PHENAZINE DERIVATIVES

5-Methylphenazine (Figure 1, **1A**) is a unique electron mediator as it can accept electrons from NAD(P)H and is widely used in various redox reactions. However, its application is still limited, mainly because of its instability to light and to alkaline solution. 5-Ethylphenazine (Figure 1, **2A**) has been recommended as a more stable electron mediator in alkaline solutions,[7,8] and light-insensitive derivatives have also been prepared and used in the assay of dehydrogenases,[9–12] but the mechanism of these instabilities are still unclear.

We intended to use a phenazine derivatives as an artificial catalytic group for semisynthetic oxidases. For this purpose, it is necessary to prepare a stable phenazine derivative with a functional group suitable for binding to dehydrogenases. We prepared new phenazine derivatives carrying a carboxyl group (Figure 1, **1B** and **2B**) and an amino group (Figure 1, **2D** and **2F**).[6] Figure 3 shows the scheme for the preparation of derivative **2B**. Details of the procedure were reported previously.[6]

As shown in Table 1, the 5-methylphenazine derivatives (**1A** and **1B**) are unstable above pH 5, while 5-ethylphnazine derivatives (**2A**, **2B**) and **2D**) are much more stable in neutral and alkaline solutions. In addition, **2B** was stable for one month at pH 9 under scattered light. The alkali decomposition of **1** seems to be due to

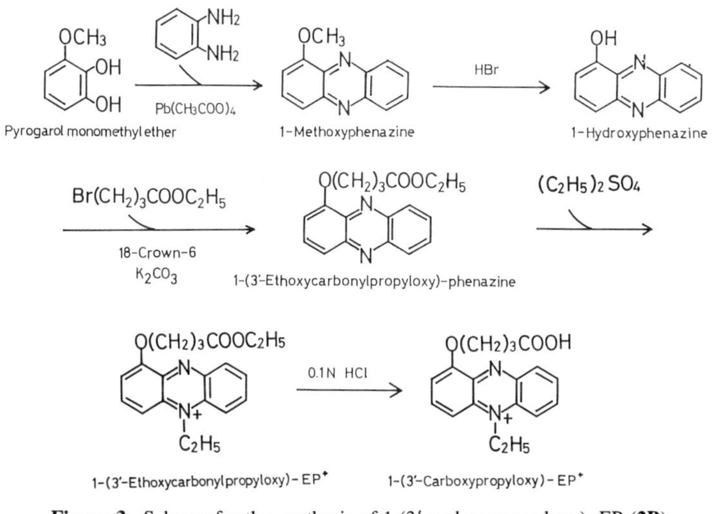

Figure 3. Scheme for the synthesis of 1-(3′-carboxypropyloxy)–EP (**2B**).

Table 1. Stability of Phenazine Derivatives

Compound	Stability under light	at pH 5	at pH 8
1A	no	yes	no
1B	yes	yes	no
2A	no	yes	yes
2B	yes	yes	no
2D	yes	yes	yes

elimination of the 5-methyl group, and **1E** is more resistant to this elimination than **1A** and **1B**. The ethyl group of **2** is not eliminated, and the alkaline decomposition of **2** seems to be due to a ring-opening reaction.[13,14] As the ring-opening reaction occurs under more alkaline conditions than the elimination of the 5-methyl group of **1**, **2** is more stable than **1**. On the other hand, the photodecomposition of 5-alkylphenazines is associated with hydroxylation at position 1, as has been reported for **1A**,[15–18] and 1-substitution makes them photostable, probably by preventing the formation of an adduct of 5-alkylphenazine with water.[17]

When these phenazine derivatives are used as electron mediators in enzyme reactions, they are reduced with NAD(P)H, and the reduced derivatives are reoxidized with oxygen or other electron acceptors, such as MTT. We assume the scheme shown in Figure 2 for these coupled reactions, where k_1, k_2, and k_3 are the second-order rate constants of the indicated reactions and the value of k_2 depends on the proton concentration, which was kept constant throughout our study. We determined these kinetic constants by a steady-state method described previously.[6] Table 2 shows the k_1 values thus obtained. The k_2 and k_3 values for **2B** were also determined to be 1.21 $mM^{-1}s^{-1}$ and 91 $mM^{-1}s^{-1}$, respectively.[6] These results indicate that the k_1 value depends on the substituents at position 1 and 5. It has been demonstrated that the second-order rate constants for outer-sphere electron-transfer reactions are related to the difference in the standard oxidation–reduction potential (E'_0) between the reactants.[19] The difference in the k_1 values between the derivatives with the 5-methyl and 5-ethyl groups seems to be due to the difference in the E'_0 values, because the E'_0 values for **1A** and **2A** are 80 mM and 55 mV, respectively[20]; and the ratios of the k_1

Table 2. Second-Order Rate Constants of NADH Oxidation by Phenazine Derivatives

Compound	k_1 ($mM^{-1}s^{-1}$)	Compound	k_1 ($mM^{-1}s^{-1}$)
1A	1.83	**2C**	1.68
1B	3.33	**2D**	2.03
2A	0.75	**2F**	2.78
2B	1.42		

values for **1A** and **2A** and the ratios of those for **1B** and **2B** are almost the same. The effects of the substituents at position 1 on the k_1 values seem to include potential and charge effects, and these effects were also analyzed.[21]

The phenazine derivative, **1B**, **2B**, **2D**, and **2F**, have a carboxyl or amino groups at position 1. These functional groups can be used for linking these derivatives to other molecules. In the following work, we covalently linked **2F** to dehydrogenases as an artificial catalytic group, and converted the dehydrogenases into oxidases, i.e., semisynthetic oxidases.

8.3. ETHYLPHENAZINE–NAD CONJUGATE

Now we have three main characters in an enzyme reaction: the coenzyme-binding site of a dehydrogenase for a substrate-binding site, EP for a catalytic group, and NAD(H) as a substrate. There are several combinations of these characters and, at first, we prepared EP–PEG-NAD (Figure 4), a covalently linked 5-ethylphenazine–poly(ethylene glycol)–NAD conjugate.[22] This corresponds to the conjugate of a catalytic group and a substrate. Using this conjugate, the rate-accelerating effect of linking two reactants by PEG can be estimated as follows. The first-order rate constant (k_1) of the intramolecular reaction between the EP and the NAD moieties of EP–PEG–NAD was measured to be 1.1 s^{-1}, and the second-order rate constant (k_2) of the intermolecular reaction between EP and PEG–NADH to be 2.8 $mM^{-1}s^{-1}$.[22] The ratio of k_1/k_2 is known as the effective concentration,[23] and the value is 0.4 mM. This means that the two moieties on a single molecule of EP–PEG–NAD feel the concentration of the other moiety to be 0.4 mM, irrespective of the actual concentration of each moiety. The ratio of the effective and the actual concentrations corresponds to the ratio of the intramolecular and intermolecular rates at the same actual concentration of the reactants; when the actual concentration is 0.4 μM, for example,

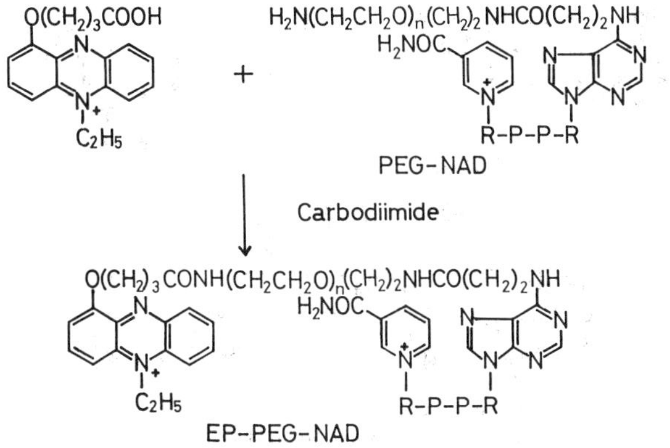

Figure 4. Scheme for the linking of PEG–NAD and 1-(3′-carboxypropyloxy)–EP (**2B**).

the reaction rate increases 1000 times by the linking of the reactants, and the lower the actual concentration, the larger the rate acceleration.

EP–PEG–NAD has a structure in which EP and NAD are linked with a linear, flexible, hydrophilic chain of PEG. The effective concentration of 0.4 mM corresponds to the concentration where one molecule is present in a sphere of radius 10 nm and this value is reasonable as the polymer is about 25 nm long when stretched (Figure 5). In enzymic reactions, substrates are bound at the active sites and the chemical steps at the active sites are accelerated by keeping the substrates near the catalytic groups. Thus, EP–PEG–NAD has the enzyme-like rate-acceleration mechanism of keeping the two reactants a short distance apart.

8.4. SEMISYNTHETIC NADH OXIDASE

Next, we planned to link a catalytic group to a substrate-binding site; that is, linking the EP group to a dehydrogenase. The conjugate of this type is expected to function as an NADH oxidase, if the bound NADH reacts well with the EP moiety. Figure 6 shows all the possible reactions catalyzed by EP–PEG–dehydrogenase.

The reactivity of the bound NADH may be different depending on the structure of the binding site, and we must select an appropriate dehydrogenase for this purpose. The reactivity of NADH bound in the binding site can be examined by comparing the k_1 and k_2 values in Figure 7 in the presence of different concentration of the binding site.[24] After several experiments, we selected glutamate dehydrogenase (GltDH) as the substrate-binding site; this enzyme catalyzes the interconversion of L-glutamate and oxaloacetate using NAD(H) as a coenzyme.

EP–PEG–GltDH was prepared according to the procedure described for the preparation of dehydrogenase–PEG–NAD conjugates.[4,5] The terminal amino group

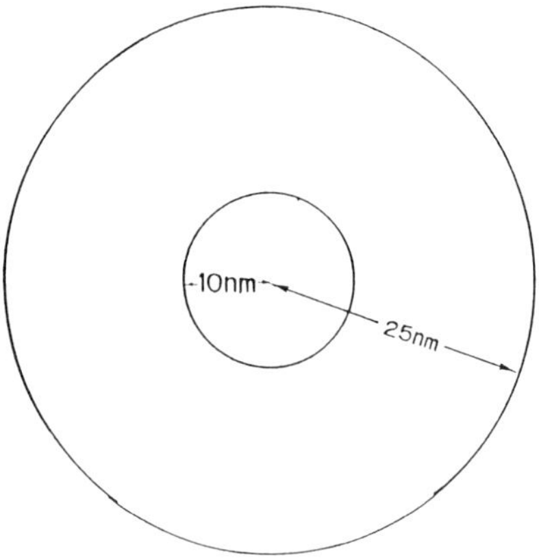

Figure 5. Effective radius for EP–PEG–NAD.

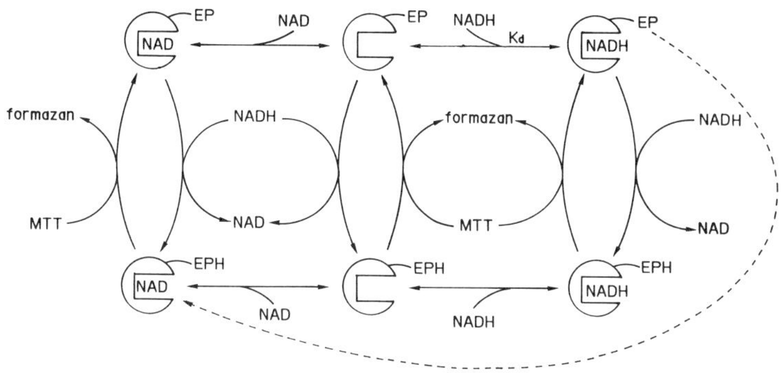

Figure 6. Reactions catalyzed by EP–PEG–dehydrogenase.

of poly(ethylene glycol)-bound ethylphenazine (EP–PEG) was activated with a bifunctional reagent [3,3′-(1,6-dioxo-1,6-hexanediyl)bis-2-thiazolidinethione], and the activated EP–PEG was linked to GltDH (Figure 8).[24]

EP–PEG–GltDH thus prepared worked as an NADH oxidase; that is, GltDH was converted to NADH oxidase just by the linking with EP–PEG. The oxidase activity increased with the increase in the NADH concentration. The activity is higher than the intermolecular-reaction rate between EP and NADH, and the difference between them corresponds to the rate acceleration by the presence of the substrate-binding site near the catalytic group, because the observed activity includes the intramolecular-reaction rate between the EP moiety and bound NADH in addition to the intermolecular-reaction rate between the EP moiety and free NADH; the intramolecular rate constant was 0.38 s^{-1} and the intermolecular one was 11.6 $mM^{-1} s^{-1}$. This rate-acceleration effect can be explained as the increase in the effective concentration of NADH around the catalytic group due to the presence of the bound NADH. The maximum value of this extra effective concentration of NADH was estimated to be 0.33 mM.[24]

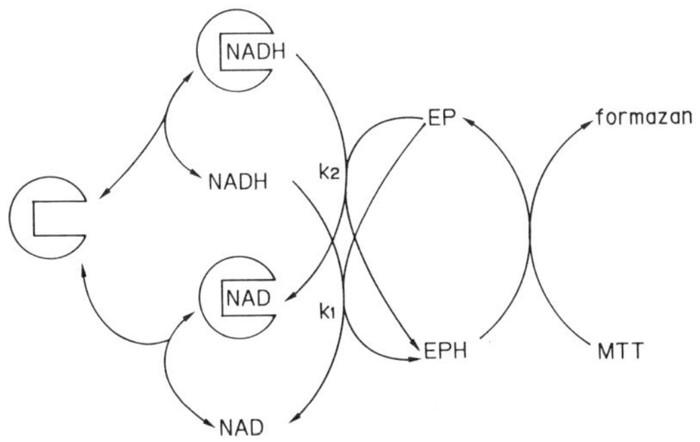

Figure 7. Reaction system for determination of reactivity of NADH bound to dehydrogenase.

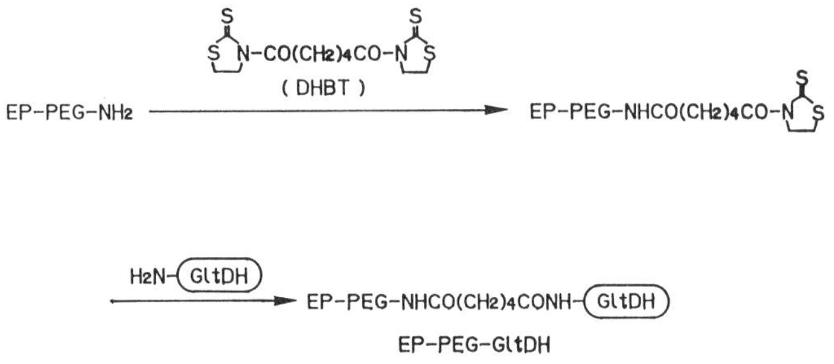

Figure 8. Scheme for the preparation of EP–EPG–GltDH.

8.5. SEMISYNTHETIC GLUCOSE OXIDASE[25]

Glucose dehydrogenase (GlcDH) catalyzes the oxidation of glucose to gluconolactone using NAD or NADP as a coenzyme (Figure 9). In our previous work, we found that this enzyme shows much lower activity for NAD derivatives, such as PEG–NAD, than for NAD, but when PEG–NAD is covalently linked to the enzyme, the conjugate shows good activity due to the increase in the effective concentration of the NAD moiety.[5] When the GlcDH reaction is coupled with the reaction cycle of EP, as shown in Figure 9, the overall reaction becomes the oxidation of glucose with oxygen; i.e., the dehydrogenase becomes glucose oxidase. Therefore, we linked both EP and NAD to GlcDH using PEG as a spacer by the methods used for the preparation of EP–PEG–GltDH.

EP–GlcDH–NAD thus prepared worked as a glucose oxidase by the catalytic cycle shown in Figure 10. The K_m value for oxygen was 1.6 mM; that is, GlcDH was converted to glucose oxidase. In the reaction cycle of EP–GlcDH–NAD, the following two catalytic reactions are coupled intramolecularly: the reduction of the NAD moiety by the active site of GlcDH and the reoxidation of the NADH moiety by the EP moiety. Figure 11 shows the importance of the intramolecular coupling of these two

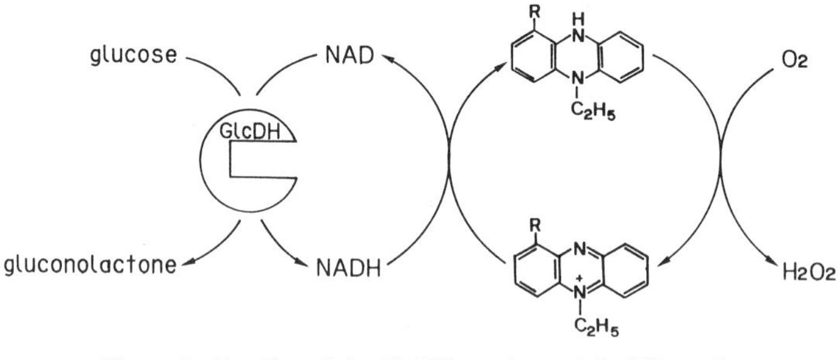

Figure 9. Coupling of the GlcDH reaction and the EP reaction.

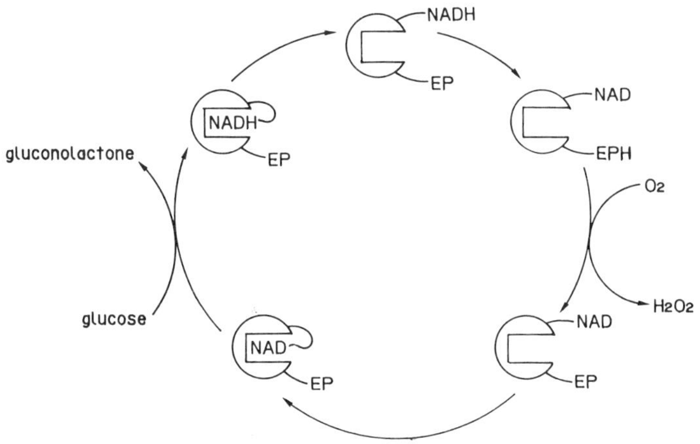

Figure 10. Reaction scheme of semisynthetic glucose oxidase.

reactions. When the EP moiety is separated from the conjugate, the V_{max} value decreases to 0.14% due to the decrease in the rate of the second catalytic reaction. When the first reaction is made into an intermolecular type, keeping the second reaction intramolecular type, the V_{max} value also decreases to 0.13%. Therefore, this intramolecular coupling of the two successive reactions is essential to the glucose oxidase activity of EP–GlcDH–NAD.

8.6. STRATEGY FOR DESIGNING ENZYME-LIKE CATALYSTS[26]

We have changed dehydrogenases into oxidases. This method seems to be generally applicable to preparing new oxidases. On the basis of the above results, the following strategies for designing enzyme-like catalysts are presented. For simple catalysts having one substrate-binding site and one catalytic group, use a binding site with higher affinity for the substrate, and make the ratio of the intramolecular and intermolecular rate constants as large as possible. To increase the intramolecular rate

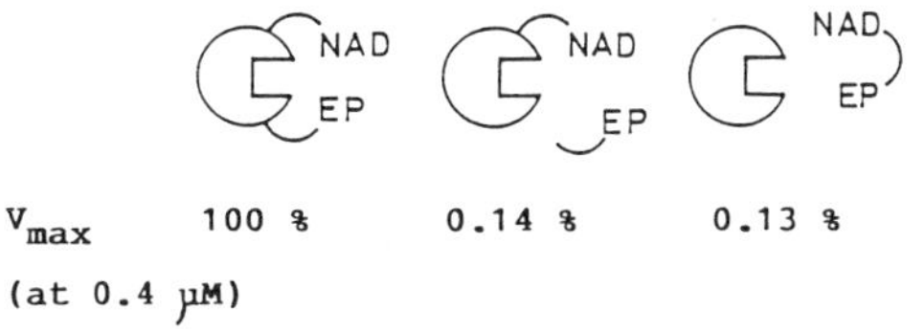

Figure 11. Effects of the presence and absence of linkers between GlcDH, NAD, and EP on the oxidase activity of EP–GlcDH–NAD.

constant, select a good binding site and use a flexible linker of an appropriate length. A catalyst with a subunit structure is also effective at increasing the specific activity, because the specific activity of a catalyst is proportional to the number of the catalytic groups and also to the number of the binding sites. For example, when a monomer has one catalytic site and one substrate-binding site, the dimer prepared from the monomer has four times as high activity as the monomer. A long and flexible linker is important for this rate acceleration.

For a more complex catalyst having one substrate-binding site and two kinds of catalytic groups, it is important to couple the two catalytic reactions intramolecularly. For a reaction system in which two reactions are coupled by the recycling of an intermediate, covalent linking of the intermediate is the best way. If the intermediates are covalently linked, the catalyst does not need to have the binding sites for the intermediates.

We have used PEG as a flexible linker of an appropriate length for preparing enzyme-like catalysts. In fact, the enzyme-like conjugates prepared so far have good catalytic activity. This may also reflect that the modification of enzymes with PEG does not cause large change in their structures and activities. In addition, PEG increase the solubility of NAD and EP in organic solvents such as CH_2Cl_2, when PEG is covalently linked to them. This enables us to use a wider variety of conditions for preparing the conjugates. Thus, PEG is one of the best linkers for preparing enzyme-like catalysts.

REFERENCES

1. A. Wiseman, in: *Topics in Enzyme and Fermentation Biotechnology* (A. Wiseman, ed.), Vol. 9, p. 202, Ellis Horwood, Chichester (1984).
2. R. Breslow, in: *Cold Spring Harbor Symposia on Quantitative Biology*, Vol. 52, p. 75 (1987).
3. A. Pluckthun, R. Glockshuber, I. Pfitzinger, A. Skerra, and J. Stadlmuller, in: *Cold Spring Harbor Symposium on Quantitative Biology*, Vol. 52, p. 105 (1987).
4. T. Eguchi, T. Iizuka, T. Kagotani, J. H. Lee, I. Urabe, and H. Okada, *Eur. J. Biochem. 155*, 415 (1986).
5. A. Nakamura, I. Urabe, and H. Okada, *J. Biol. Chem. 261*, 16792 (1986).
6. T. Yomo, H. Sawai, I. Urabe, Y. Yamada, and H. Okada, *Eur. J. Biochem. 179*, 293 (1989).
7. C. Bernofsky and M. Swan, *Anal. Biochem. 53*, 452 (1973).
8. R. Ghosh and J. R. Quayl, *Anal. Biochem. 99*, 112 (1979).
9. R. Hisada and T. Yagi, *J. Biochem. 82*, 1469 (1977).
10. S. Nakamura, K. Arimura, K. Ogawa, and T. Yagi, *Clin. Chim. Acta 101*, 321 (1980).
11. J. L. Orsonneau, K. Meflah, P. Lustenberger, G. Cornu, and S. Bernard, *Clin. Chim. Acta 125*, 177 (1982).
12. J. L. Orsonneau, K. Meflah, P. Lustenberger, G. Cornu, and S. Bernard, *Clin. Chim. Acta 125*, 185 (1982).
13. S. L. Johnson and K. A. Rumon, *Biochemistry 9*, 847 (1970).
14. S. L. Johnson and D. L. Morrison, *Biochemistry 9*, 1460 (1970).
15. H. McIlwain, *J. Chem. Soc.*, 1704 (1937).
16. V. S. F. Chew and J. R. Bolton, *J. Phys. Chem. 84*, 1903 (1980).

17. V. S. F. Chew, J. R. Bolton, R. G. Brown, and G. Porter, *J. Phys. Chem. 84*, 1909 (1980).
18. F. G. Halaka, G. T. Babcock, and J. L. Dye, *J. Biol. Chem. 257*, 1458 (1982).
19. R. A. Marcus, *Annu. Rev. Phys. Chem. 15*, 155 (1964).
20. H. A. Sober, *Handbook of Biochemistry*, 2nd ed., J-36, The Chemical Rubber Co., Cleveland (1970).
21. T. Yomo, I. Urabe, and H. Okada, *Biochim. Biophys. Acta 1017*, 139 (1990).
22. T. Yomo, H. Sawai, I. Urabe, and H. Okada, *Eur. J. Biochem. 179*, 299 (1989).
23. M. I. Page, *Chem. Soc. Rev. 2*, 295 (1973).
24. T. Yomo, I. Urabe, and H. Okada, *Eur. J. Biochem. 196*, 343 (1991).
25. T. Yomo, I. Urabe, and H. Okada, *Eur. J. Biochem. 203*, 533 (1992).
26. T. Yomo, I. Urabe, and H. Okada, *Eur. J. Biochem. 203*, 543 (1992).

9

Preparation and Properties of Monomethoxypoly(Ethylene Glycol)-Modified Enzymes for Therapeutic Applications

F. M. VERONESE, P. CALICETI, O. SCHIAVON, and L. SARTORE

9.1. INTRODUCTION

The therapeutic value of enzymes as drugs would be considerably increased if drawbacks such as immunogenicity and antigenicity, rapid clearance from circulation, difficulty in targeting, instability, and inadequate supply were overcome. Genetic engineering seems to be promising in obtaining large amounts of useful enzymes, although doubts exist concerning the correct folding of expressed proteins.[1] Nevertheless, the disadvantages of limited tissue distribution and rapid clearance from circulation of enzymes remain a major problem. Moreover, genetic methods cannot be used for producing enzymes carrying post-transcriptional modifications, but with this aim the so-called transgenic animal technology appears to be quite promising. For these reasons, alternative strategies are being actively investigated in several laboratories. These strategies include surface modification of the enzymes by chemical modification or compartmentalization of the enzyme onto complex structures which isolate it from body cells, tissues, and proteolytic enzymes.[2,3]

Natural or artificial polymers are being used for surface enzyme modification. These include mainly monomethoxypoly(ethylene glycol) (MPEG), but also dex-

F. M. VERONESE, P. CALICETI, O. SCHIAVON, and L. SARTORE • Department of Pharmaceutical Sciences (Centro di Studio di Chimica del Farmaco e dei Prodotti Biologicamente Attivi del CNR), University of Padua, 35100 Padua, Italy.

Poly(Ethylene Glycol) Chemistry: Biotechnical and Biomedical Applications, edited by J. Milton Harris. Plenum Press, New York, 1992.

trans, heparin, homologous albumin, polyvinylpyrrolidone, poly(ethylene-co-maleic acid anhydride), or unusual copolymers such as poly(styrene-co-maleic acid anhydride).[4–9] In this context, MPEG is the polymer receiving the greatest attention, due to its lack of toxicity, commercial availability, and relatively simple chemistry required for its covalent biding to proteins. The main advantages of MPEG modification of enzymes involve the enhanced half-life *in vivo* circulation and the decrease of antigenicity and immunogenicity.[10–16] Additional examples of these effects will be reported here, as well as in other chapters of this volume.

In our laboratory surface protein modification by MPEG has been investigated, using a few enzymes as protein models. In addition, studies are being carried out with soluble polymers of different structure such as *N*-acryloylmorpholine-*N*-acryloxysuccinimide copolymers or polyvinylpyrrolidone.[17,18]

In this chapter we wish to summarize the outcome of recent experiments dealing with methods of MPEG activation and coupling to proteins, structural aspects of MPEG-modified proteins, as well as biological and pharmacological properties.

9.2. ENZYME MODIFICATION BY MPEG

Several methods of MPEG activation for coupling to proteins are reported in the literature as well as in patents, and some of these are described in other chapters of this volume. Most of these methods involve the preparation of an active MPEG intermediate with a functional group possessing reactivity toward amino groups in proteins. The reaction conditions must be mild enough to maintain the protein in its native conformation and, with this aim, aqueous buffer and pH close to neutrality should be employed. Known methods for MPEG binding to proteins were recently compared in J. M. Harris's laboratory using alkaline phosphatases as models. The condition of preparation of the reactive MPEG, the modification yield, and the degree of enzyme inactivation were taken into consideration.[19,20]

In this context we note that easy preparation of activated MPEG and low enzyme inactivation are obtained with a method proposed in our laboratory.[21] This method takes advantage of the rapid reaction of MPEG (**1**) with phenylchloroformates (**2**) to give the active compound (**3**) which is characterized by significant stability in water and good reactivity with proteins. Interestingly, the same urethane linkage between MPEG and proteins (**4**) is obtained by carbonyldiimidazole activation of the polymers.[11] We now report here a method recently developed in our laboratory that may present specific advantages.

$$\underset{\mathbf{1}}{CH_3O(C_2H_4O)_nH} + \underset{\mathbf{2}}{ClCOOR} \rightarrow \underset{\mathbf{3}}{CH_3O(C_2H_4O)_nCOOR} + H_2NProtein$$

$$\downarrow$$

$$\underset{\mathbf{4}}{CH_3O(C_2H_4)_nCONHProtein}$$

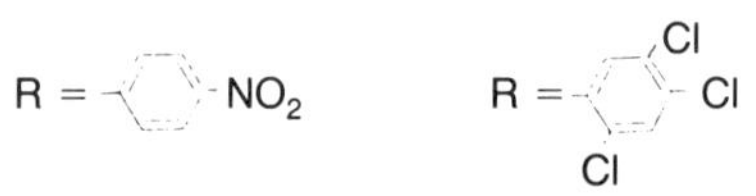

9.2.1. MPEG Activation and Linking to Proteins through an Amino Acid as Spacer Arm

Compound **3** was employed in a method suitable for the introduction of an amino acid (**5**) between oligomer and protein.[22] This kind of spacer arm has the advantage

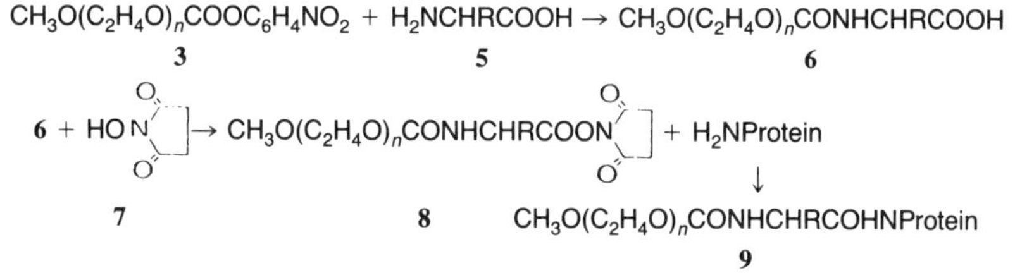

R = C_α residue of Trp, Gly, Nle, Phe

of a wide variety of opportunities offered by the different characteristics of the amino acids. For example, evaluation of the extent of modification may be performed by using norleucine as spacer (see below). Advantage also comes from the use of tryptophan as a spacer. Tryptophan is seldom present in proteins, and that and its strong fluorescence allow study of the microenvironment of the water–protein interface in close proximity to the polymer. Moreover, specifically labeled products useful for pharmacokinetic experiments may be obtained using available radioactive amino acids as spacers. This method of labeling disturbs the structure and conformation of the protein less than the commonly used iodination procedure.

Last, a bifunctional amino acid such as lysine allows introduction at a double amount of MPEG chains at the protein surface, resulting in more hindrance to the approach of proteolytic enzymes or antibodies.

With the exception of lysine, the methodology of MPEG linking to an amino acid arm is based on the reaction of compound **3**, usually as 4-nitrophenyl derivative, in aqueous solution, or water/acetone 1:1, at pH 8.3 with excess of amino acid. The MPEG–amino acid (**6**), extracted with chloroform at acidic pH and crystallized from ethanol, is activated at the carboxylic group of the amino acid as the *N*-succinimidyl ester.

The aqueous reaction conditions allowing the direct coupling of unprotected amino acid is the advantage of this strategy. Furthermore, some activated MPEG9, which does not react with the spacer amino acid, decomposes to the starting MPEG–OH by standing in the aqueous reaction solution. This method of differential activation of MPEG and MPEG–amino acid spacer avoids any linkage to the protein of MPEG with unbound arm. Any unreacted MPEG–OH is removed from the modified protein (**9**) by gel filtration in the final step of protein purification.[22]

9.2.2. Evaluation of Extent of Protein Modification

The extent of protein modification by MPEG is commonly determined by colorimetric evaluation of amino groups before and after modification. TNBS (2,4,6-

trinitrobenzenesulfonic acid) was the reagent originally employed by Abuchowski *et al.*[10] and, for its simplicity, it has been used with few exceptions in all studies dealing with MPEG proteins. However, colorimetric assays applied to native proteins are generally subject to large errors, because of differing accessibility of the reagent to the hindered region of the protein or to changes in the molar extinction coefficient of the chromophore in the protein environment. Furthermore, this methodology requires a separate estimation of protein content, which is usually determined by the biuret method since the modification may change the extinction coefficient of the protein itself.

To overcome such limitations we are now applying MPEG with norleucine as a spacer arm in protein modification; in this case the number of linked MPEG chains per protein molecule is calculated from the norleucine content.[23] This method exploits the stability of this unnatural amino acid to acid hydrolysis and its easy separation and evaluation in any standard amino acid analyzer. By amino acid analysis both number of bound polymer chains and protein concentrations are contemporarily obtained in the same sample and analysis.

This direct norleucine method and a colorimetric one were applied to different proteins extensively modified by MPEG with the norleucine arm, and also to samples of proteins with lower degrees of modification obtained changing the reagent to protein molar ratio; the results are reported in Table 1. The values of colorimetric estimation in the table were obtained by the Snyder TNBS procedure.[24]

The results reported in Table 1 show that the percent of amino group modification evaluated by the norleucine method is, in most cases, lower than that calculated using the colorimetric one. Indeed, values that indicated a lower degree of modification were in some cases found also when the TNBS method was performed according to the Habeeb procedure.[25] This may be explained by an increased accessibility to some buried amino group in the modified protein due to the different reaction conditions employed in this procedure.

In any case we propose that one should better rely on the norleucine method for the above-reported considerations.

The results of Table 1 are very interesting, since they demonstrate that the protein amino groups with linked polymer are less than generally thought. A lower degree of MPEG derivatization is a new key to the interpretation of the heterogeneity of MPEG-proteins, as often verified by gel filtration or electrophoresis. In fact, the less the amount of modified amino groups in the protein, the higher the possibility of modification at different sites of the protein molecules: in theory, a single species-modified protein is obtained only when MPEG chains are linked to all the available protein amino groups.

In this context, it is of interest to recall a recent study on MPEG–superoxide dismutase (MPEG–SOD), in which heterogeneity was demonstrated in detail using various techniques.[26]

In the case of MPEG–SOD, we found also the presence of high molecular weight modified SOD species that were related to cross-linking products arising from nonmethoxylated MPEG 5000 polymer present as impurities in the MPEG samples.[27]

Table 1. Degree of Protein Modification with MPEG Evaluated by Two Methods

Proteins[a]	Modified —NH_2 (%) (by TNBS analysis)[b]	Modified —NH_2 (%) (by Nle analysis)[c]
Cytocrome C	45.8	46.7
	78.7	71.8
Superoxide dismutase	26.6	20.0
	44.0	26.6
	64.6	45.0
	80.0	61.3
Bovine serum albumine	16.0	11.3
	28.6	14.7
	48.2	25.9
	50.3	39.2
Ribonuclease A	43.7	45.6
	80.5	71.3
Thermolysine	48.5	46.5

[a]The proteins were reacted with MPEG with Nle as amino acid spacer. With the exception of thermolysine, samples at different extent of modification were obtained on changing the ratio of reagent to protein.
[b]The degree of MPEG binding was evaluated by TNBS and the value expressed as percent of polymer-modified amino groups with respect to the total present.
[c]The degrees of MPEG binding were evaluated from the amount of Nle after amino acid hydrolysis and expressed as in the latter note *b*.

The evidence of bifunctional nonmethoxylate PEG comes from the end-group titration of carboxylate MPEG derivatives and this is in agreement with previous findings of other authors.[28]

Furthermore, in commercial MPEG 5000 samples, by gel filtration with BIO GEL P 100 Fine column, we demonstrated the presence of small amounts of a higher molecular weight oligomer that may also contribute to the final heterogeneity of the modified protein.

These results indicate that there is a need for careful control and characterization of samples of commercial MPEG, as well as of MPEG-proteins when prepared for biomedical studies.

9.3. STRUCTURAL INVESTIGATIONS ON MPEG-ENZYMES

The covalent binding of MPEG to enzymes is often accompanied by decreased enzyme activity. This event is generally of limited extent if the substrate is a small molecular weight compound, and steric hindrance, due to MPEG chains, cannot be advocated. To investigate the reasons for this reduced enzyme activity, structural characterizations have recently been carried out in two enzyme models, superoxide dismutase (SOD) and ribonuclease-A (RNase).

9.3.1. MPEG–Superoxide Dismutase

Copper–zinc superoxide dismutase (Cu/Zn SOD), an enzyme used in the native form in many countries for its antiinflammatory activity, was extensively modified by using a carboxylate MPEG activated as succinimidyl ester. The enzyme with about 18 MPEG chains bound to its surface still retains 80% of its activity toward the substrate $O_2^{\overline{\cdot}}$.

Circular dichroism investigation in the near or far UV as well as differential UV studies in the visible region of the Cu/Zn chromophore does not reveal significant conformational differences between native and modified enzyme. A more subtle investigation of the active site region has been carried out on a MPEG–Cu/Co SOD.[29] Substitution of Zn^{2+} by Co^{2+} drastically changes the spectroscopic properties of SOD, turning the copper ion into a ^{1}H-NMR probe without altering its enzyme activity.[30,31] This change gives rise to well-resolved ^{1}H-NMR signals of all the protons of the histidines coordinated to both the cobalt and copper ions and therefore offers a detailed picture of the active center of the enzyme. Since the ^{1}H-NMR spectrum of Cu/Co SOD and all the bands assigned have already been reported in the literature, it could be compared with the MPEG–Cu/Co SOD spectrum.[30] In this case, signals were found to be broader than those of the unmodified enzyme, due to the effect of molecular weight on linewidth. Interestingly, histidine hydrogens involved in metal binding were the same as in the native enzyme.

The decreased activity of the MPEG-modified SOD could be interpreted by calculating the affinity constants of a specific negative SOD inhibitor N_3^-, which binds the active enzyme site similarly to $O_2^{\overline{\cdot}}$. By NMR titration, the N_3 affinity constants were found to be $K = 154\ M^{-1}$ and $75\ M^{-1}$ for the native and polymer-modified SOD, respectively.[29] A reduction in affinity constant was similarly found by UV titration at 370 nm of the N_3/native Cu/Zn SOD and the N_3^-/MPEG–Cu/Zn SOD complexes.[27] This evidence allows us to conclude that decreased enzyme activity following MPEG modification is not related to structural changes in the overall conformation of the enzyme or in variations in active site geometry, but to decreased channeling of the negative $O_2^{\overline{\cdot}}$ ion toward the enzyme active site. In fact, MPEG binding to lysine residues is accompanied by a reduction of surface positive charges, which have been demonstrated by electronic field calculation to play an important role in directing the superoxide ion inside the active site cavity.[31]

9.3.2. MPEG–Ribonuclease-A

Ribonuclease-A, an enzyme already in use for its antiviral activity, is now under investigation in our laboratory to verify the influence of MPEG modification on its structural conformation and stability.[32] Samples of the enzyme with about 4, 9, and 10 molecules of bound MPEG chains per protein molecule were prepared. Reduced activity toward the synthetic substrate cytidine-2′,3′-cyclic monophosphate occurs during modification to reach a minimum of 66% in the most extensively modified sample.[33]

Spectroscopic analysis showed that modification is not accompanied by a significant variation in the UV spectrum. In particular, the ratio of A280/A260 nm and the molar extinction coefficient are unchanged in the various samples. Similarly, the CD spectra of the large band centered at 275 nm (due to tyrosine) as well as that centered at 245 nm (due to tyrosine and disulfide bridges) are unchanged in the various RNase species.

The maximum wavelength of fluorescence emission and quantum yield do not change after protein modification, indicating that the overall conformation of the native protein is maintained.

The thermal stability of the modified enzyme differs from that of the native one, as assessed by continuous recording of the circular dichroism signal in the far-ultraviolet region upon increasing temperature. The melting temperature of 67 °C of the native enzyme decreases to 64, 59, and 57 °C for protein samples containing 4, 9, and 10 MPEG bound molecules, respectively.[33] Decreased stability of RNase in samples with increasing degrees of modification also appeared from denaturation studies performed by circular dichroism measurements in the presence of denaturating agents such as guanidinium chloride, urea, and detergents.[34]

The decrease in structural stability following MPEG modification of RNase is at variance with other observations regarding the immobilization to polysaccharides, as well as to other multifunctional matrices.[35] This apparent contradiction may be explained by the multipoint attachment that takes place between enzymes and these matrices, which results in freezing and stabilization of the protein native structure.[36] This situation does not seem to occur in the one-point attachment of MPEG to the protein surface.

9.4. BIOLOGICAL BEHAVIOR OF MPEG ENZYMES

One relevant aspect of the use of MPEG in protein modification is its influence on the contact of the polymer-modified protein with interacting molecules or microstructures. This is due to the unusual properties of MPEG, which is both hydrophilic and hydrophobic. The hydrated polymer cloud created by MPEG at the protein surface is in fact responsible for various biological and pharmacological effects.

We report here recent studies on antibody recognition following MPEG binding, data on interaction with biological membranes, and the pharmacokinetic and pharmacological properties of superoxide dismutase modified to various extents.

9.4.1. Antibody Recognition of MPEG-Proteins

The reduction of antigenicity and immunogenicity of proteins is one of the most important results of MPEG surface modification, and is the topic of a special chapter of this volume. However, the study of antigen/antibody (Ag/Ab) recognition, where the antigen is covalently linked to a hydrophilic polymer such as MPEG, may raise new problems. In fact, increased solubility due to the polymer chains may prevent or reduce precipitation of the immuno complex. To overcome this risk and to have direct and quantitative estimation of Ag/Ab complex formation, we verified the applicability of analytical affinity chromatography to a MPEG-protein.[37] This method, previously employed to quantify the binding of inhibitors to enzymes, was recently demonstrated by one of us to allow also estimation of the Ag/Ab interaction while the antigen remains in solution and the antibody is properly bound to a column chromatography matrix.[38] If binding occurs, it results in delayed elution time of the antigen from the column.

Dissociation constants were calculated by mathematical treatment of the results obtained on a series of experiments in which either the column is loaded with decreasing amounts of antigen and the peak elution volume is recorded (zonal elution method), or in which the column is eluted with decreasing concentrations of antigen and the antigen elution volume is recorded (frontal elution method).[37] The enzyme model system chosen for this study was native RNase and RNase samples containing 4 or 9 molecules of MPEG chains per enzyme molecule.

Purified IgG antibodies anti-native RNase were covalently linked to a Protein A derivatized resin. Protein A was used to immobilize the antibodies in an oriented manner to improve the functional capacity of the resin. The immuno-affinity matrix, packed in a chromatographic column, was then used to quantify immuno-recognition of native and MPEG-derivatized RNase by frontal and zonal elutions in native conditions while the protein elution profiles were monitored by an "on line" UV detector.[33]

Both zonal and frontal elution experiments allowed the evaluation of the dissociation constants for native RNase, kd $= 7.6 \times 10^{-8}$ M, RNase with 4 MPEG bound molecules, kd $= 1.3 \times 10^{-7}$ M, and with 9 MPEG bound molecules, kd $= 1.2 \times 10^{-6}$ M.[33]

These results demonstrated the suitability of the quantitative affinity chromatography procedure to calculate the affinity constants for MPEG enzymes. The method may be further improved to evaluate samples with higher kd, if a suitable method of protein detection in the eluate is employed, as, for instance, if MPEG with radioactive amino acid as spacer is used.

9.4.2. Biological Membrane Interaction of MPEG-Enzymes

The surface-active properties of MPEG are well known, as demonstrated by the induction of cell fusion. It is therefore expected that MPEG proteins may show different behavior toward cell membranes with respect to native proteins. Evidence of this came during a study designed to set up a new method of SOD evaluation in blood. Unmodified and MPEG-modified SOD were added to whole blood and enzyme activity was evaluated in plasma after purification of the proteins of interfering substances by gel filtration and cationic exchange chromatography. It was found that, unlike the underivatized enzyme, MPEG–SOD was only partially recovered in plasma and also that decreased recovery was higher as the number of blood cells in whole blood increased.[39] It was also found that the binding of MPEG–SOD to blood cells was reversed by the addition of free MPEG to the blood. Actually, displacement was not complete, since not all of the added MPEG–SOD was recovered in plasma even at high MPEG concentrations. This suggests either strong irreversible binding to the membrane or penetration into the cells.

Similar phenomena of strong binding and possible penetration of MPEG–SOD in endothelial cells have also been demonstrated by other authors.[40] In this context, it is worth recalling an interesting phenomenon observed in a study of MPEG–SOD activity in carrageenan-induced pleurisy in rats.[41] It was found that after plasma injection the exudate to plasma concentration ratio of MPEG–SOD increased from 0.16 to 0.66 and 1.06 at the 3rd, 6th, and 24th hours after IV injection. This behavior may be explained by oriented membrane diffusibility of the MPEG-protein.

9.5. PHARMACOKINETIC AND PHARMACOLOGICAL STUDIES ON MPEG–SUPEROXIDE DISMUTASE

The potential therapeutic applications of SOD are based on its scavenger action on the toxic superoxide ion that may occur in different pathological states such as inflammation, burns, or reperfusion injury associated with heart ischemia or kidney transplantation.[15,42] The major drawback of the use of SOD in therapy is its rapid clearance, which is of the order of a few minutes. In this context, we were interested in studying the influence of MPEG modification of SOD on the enzyme pharmacokinetics in rats, as well as its possible use in different inflammation models.

A study on rats following intravenously injected MPEG–SOD showed that the number and also molecular weight of the polymer bound to the enzyme is directly reflected in clearance time. While half-time clearance of native SOD from the blood is about 6 min, that of the modified form is 1.5 and 25 hours in samples with 3 molecules of MPEG 1900 or 18 molecules of MPEG 5000 bound per enzyme molecule, respectively. Intermediate clearance values were found for differently modified forms.[43]

The administration route also plays an important role in clearance from blood. In fact, when extensively modified enzyme was administered IP, IM, or SC half-life values of 80, 100, and 90 hours were found, respectively.[27] These and other data not reported here demonstrate not only the very high systemic bioavailability of MPEG adducts by any administration route but also the stability and diffusibility in the body of these high molecular weight derivatives.

The therapeutic activity of MPEG–SOD modified to different extent was evaluated in rats in two models of acute inflammation. MPEG samples with about 3 or 18 MPEG 5000 bound chains per enzyme molecule were tested in a pharmacological model based on foot oedema induced by carrageenan and in experimental pleurisy. In both models it was found that the degree of polymer derivatization deeply influenced antiinflammatory activity, the long-lasting compound being by far the most active.[41,42]

To assess the activity of MPEG–SOD in chronic inflammation, the enzyme was recently tested in adjuvant arthritis in rats. In this model MPEG–SOD was administered IM to rats on alternative days according to different treatment schedules.[44]

In prophylactic treatment, MPEG–SOD was administered from days 3 to 21, after *Mycobacterium butyrricum* injection into rat tail, while in therapeutic treatment the drug was administered from days 14 to 28. On days 14, 21, and 28 inflammation was assessed through an arthritic score based on primary and secondary lesions. It was shown that MPEG–SOD derivatives reduced arthritic lesions significantly when administered according to the prophylactic scheme. Moreover, when administered according to the therapeutic protocol, MPEG–SOD did not show antiinflammatory activity, indicating that the enzyme probably interferes with the active evolutive phase of arthritis which develops between days 4 and 13.[44]

9.6. CONCLUSIONS

Since the first important publication, which appeared in 1977, a large body of information regarding the use of MPEG to improve the enzymatic, pharmacological, and immunological properties of enzymes as drugs has been obtained.[10] Several methods of MPEG activation for protein binding are now available and these have proved to be suitable for modification of different enzymes. Moreover, studies on the structure-function of MPEG-modified proteins revealed some interesting aspects, such as their interaction with the immunological system and cell membranes.

However, in order to develop MPEG-enzymes into useful therapeutic agents, more work must be done for a better understanding of their toxicological aspects, with special regard to a precise definition of their fate in the body as well as of their degradation products. A particular effort is necessary also for a better characterization of both starting MPEG and protein adducts, which must be well defined in their structural properties in order to respond to specifications now required for drugs. This will involve the development of new methods of analysis, since those so far employed in protein chemistry or polymer chemistry are often insufficient for characterization of the novel complex structures of MPEG-protein adducts. Nevertheless, the area of MPEG-protein modification is very promising indeed, as already demonstrated by the positive clinical trials with MPEG-enzymes, employed to cure some diseases.[15,45]

ACKNOWLEDGMENT

This research was partially supported by the contribution of the Finalized Project "Biotechnologie e Biostrumentazione" of CNR.

REFERENCES

1. M. O. Thorner, J. Reschke, J. Chitwood, A. D. Rogol, R. Furlanetto, J. River, W. Vale, and R. M. Blizzard, *N. Engl. J. Med. 312*, 994 (1985).
2. A. K. Larsen, R. J. Linhardt, D. Topper, M. Klein, and R. Langer, *Artif. Organs 8*, 198 (1984).
3. L. Bourget and T. M. S. Chang, *FEBS Lett. 180*, 5 (1985).
4. R. G. Melton, C. N. Wiblin, A. Baskerville, R. L. Foster, and R. F. Sherwood, *Biochem. Pharmacol. 36*, 113 (1987).

5. A. V. Maksimenko and V. P. Torchilin, *Thrombosis Res. 38*, 289 (1985).
6. K. Wong, L. G. Cleland, and M. J. Poznanski, *Agents Actions 10*, 231 (1980).
7. B. Geiger, B. U. von Spect, and R. Arnon, *Eur. J. Biochem. 73*, 141 (1977).
8. K. G. Raghavan and U. Tarachand, *FEBS Lett. 195*, 101 (1986).
9. H. Maeda, T. Matsumoto, T. Konno, K. Iwai, and M. Veda, *J. Prot. Chem. 3*, 181 (1984).
10. A. Abuchowski, J. R. Mc Coy, N. C. Palczuk, T. Van Es, and F. F. Davis, *J. Biol. Chem. 252*, 3582 (1977).
11. C. O. Beauchamp, S. L. Gonias, D. P. Menapace, and S. V. Pizzo, *Anal. Biochem. 131*, 25 (1983).
12. Y. Kamisaki, H. Wada, T. Yagura, H. Nishimura, A. Matsushima, and Y. Inada, *Gann 73*, 470 (1982).
13. M. Leonard and E. Dellacherie, *Biochim. Biophys. Acta 791*, 219 (1984).
14. A. Abuchowski and F. F. Davis, in: *Enzymes as Drugs* (J. S. Holcenberg and J. Roberts, eds.), p. 367, Wiley, New York (1981).
15. F. Fuertges and A. Abuchowski, *J. Controlled Release 11*, 139 (1990).
16. Y. Inada, K. Takahashi, T. Yoshimoto, Y. Kodera, A. Matsushima, and Y. Saito, *Trends Biotechnol. 6*, 131 (1988).
17. F. M. Veronese, R. Largajolli, C. Visco, P. Ferruti, and A. Miucci, *Appl. Biochem. Biotechnol. 11*, 269 (1985).
18. F. M. Veronese, L. Sartore, P. Caliceti, and O. Schiavon, *J. Bioactive Compat. Polymers 5*, 167 (1990).
19. K. Yoshinga and J. M. Harris, *J. Bioactive Compat. Polymers 4*, 17 (1989).
20. K. Yoshinaga, S. G. Shafer, and J. M. Harris, *J. Bioactive Compat. Polymers 2*, 49 (1987).
21. F. M. Veronese, R. Largajolli, E. Boccu, C. A. Benassi, and O. Schiavon, *Appl. Biochem. Biotechnol. 11*, 141 (1985).
22. L. Sartore, P. Caliceti, O. Schiavon, and F. M. Veronese, *Appl. Biochem. Biotechnol. 27*, 45 (1991).
23. L. Sartore, unpublished results.
24. S. L. Snyder and P. Z. Sobocinsky, *Anal. Biochem. 64*, 248 (1975).
25. A. F. S. A. Habeeb, *Anal. Biochem. 14*, 328 (1966).
26. P. McGoff, A. C. Baziotis, and R. Maskiewics, *Chem. Pharm. Bull. 36*, 3079 (1988).
27. F. M. Veronese, P. Caliceti, A. Pastorino, O. Schiavon, L. Sartore, L. Banci, and L. Monsu Scolaro, *J. Controlled Release 10*, 145 (1989).
28. J. M. Harris, *J. Macromol. Sci., Rev. Polym. Phys. Chem. C 25*, 325 (1985).
29. L. Banci, I. Bertini, P. Caliceti, L. Monsu Scolaro, O. Schiavon, and F. M. Veronese, *J. Inorganic Biochem. 39*, 149 (1990).
30. L. Banci, I. Bertini, C. Luchinat, and A. Scozzafava, *J. Am. Chem. Soc. 109*, 2328 (1987).
31. D. Cocco, L. Rossi, D. Barra, F. Bossa, and G. Rotilio, *FEBS Lett. 150*, 303 (1982).
32. B. N. Glukhof, A. P. Jerusalimsky, V. M. Canter, and R. I. Salganik, *Arch. Neurol. 33*, 598 (1990).
33. P. Caliceti, O. Schiavon, F. M. Veronese, and I. M. Chaiken, *J. Molecular Recognition 3*, 89 (1990).
34. C. Grandi, unpublished results.
35. F. M. Veronese, L. Sartore, O. Schiavon, and P. Caliceti, *Ann. N.Y. Acad. Sci. 613*, 468 (1990).
36. A. M. Klibanov, *Anal. Biochem. 93*, 1 (1979).
37. H. E. Swaisgood and I. M. Chaiken, in: *Analytical Affinity Chromatography* (I. M. Chaiken, ed.), p. 65, CRC Press, Boca Raton (1987).
38. P. Caliceti, F. Fassina, and I. M. Chaiken, *Appl. Biochem. Biotechnol. 16*, 119 (1987).
39. P. Caliceti, O. Schiavon, A. Mocali, and F. M. Veronese, *Il Farmaco 44*, 711 (1989).
40. J. S. Beckman, R. L. Minor Jr., C. W. White, J. E. Repine, G. M. Rosen, and B. A. Freeman, *J. Biol. Chem. 263*, 6884 (1988).
41. A. Conforti, L. Franco, R. Milanino, G. P. Velo, E. Boccu, R. Largajolli, O. Schiavon, and F. M. Veronese, *Pharmacol. Res. Commun. 19*, 287 (1987).
42. F. M. Veronese, A. Conforti, and G. P. Velo, in: *New Developments in Antirheumatic Therapy* (K. D. Rainsford and G. P. Velo, eds.), Vol. 3, p. 305, Kluwer Academic Publishers, Dordrecht (1989).
43. E. Boccu, G. P. Velo, and F. M. Veronese, *Pharmacol. Res. Commun. 14*, 113 (1982).
44. A. Conforti, P. Caliceti, L. Sartore, O. Schiavon, F. M. Veronese, and G. P. Velo, *Pharmacol. Res. 23*, 51 (1990).
45. T. Yoshimoto, H. Nishimura, Y. Saito, K. Sakurai, Y. Kamisaki, H. Wada, M. Sako, G. Tsijino, and Y. Inada, *Jpn. J. Cancer Res. (Gann) 77*, 1264 (1986).

10

Suppression of Antibody Responses by Conjugates of Antigens and Monomethoxypoly(Ethylene Glycol)

ALEC H. SEHON

10.1. INTRODUCTION

The possibility of *selectively* downregulating the host's immune response to a given antigen represents one of the most formidable challenges of modern immunology in relation to the development of new therapies for IgE-mediated allergies, autoimmune diseases, and the prevention of immune rejection of organ transplants. Similar considerations apply to an increasing number of promising therapeutic modalities for a broad spectrum of diseases, which would involve the use of foreign biologically active agents potentially capable of modulating the immune response, provided they were not also immunogenic. Among these agents, one may cite (1) xenogeneic monoclonal or polyclonal antibodies (collectively referred to here as xIg) against different epitopes of the patients' $CD4^+$ cells,[1,2] administered alone or in combination with immunosuppressive drugs for treatment of rheumatoid arthritis and other autoimmune diseases, or for the suppression of graft versus host reactions and

---

ALEC H. SEHON • MRC Group for Allergy Research, Department of Immunology, Faculty of Medicine, The University of Manitoba, Winnipeg, Manitoba R3E 0W3, Canada. This chapter represents a reprint of the Carl Prausnitz Memorial Lecture delivered by the author at the XVIII Symposium of the Collegium Internationale Allergologicum in Funchal, Madeira on September 25, 1990. The original article was published in *Symposium Proceedings in International Archives of Allergy and Applied Immunology*, Vol. 94, p. 11, Karger, Basel (1991).

Poly(Ethylene Glycol) Chemistry: Biotechnical and Biomedical Applications, edited by J. Milton Harris. Plenum Press, New York, 1992.

of the immune rejection of organ transplants,[1,2] and (2) "magic bullets" for the destruction of tumor cells,[3–5] which consist of anti-tumor xIg to which are coupled toxins (Tx), or radionuclides, or chemotherapeutic drugs. However, in most cases the patients produce antibodies to the injected xIg and to the even more immunogenic immunotoxins (xIg–Tx); consequently, the therapeutic effectiveness of these immunological strategies is undermined by the patients' antibodies which prevent these "bullets" from reaching their target cells. In addition, the repeated administration of these agents may result in serious complications, namely, serum sickness, anaphylactic symptoms (i.e., bronchospasm, dyspnea, and hypotension), and/or the deposition in the liver of toxic immune complexes leading frequently to hepatotoxicity.[6,7] Similar limitations apply to the use of hormones or other regulatory factors, such as lymphokines and growth factors, synthesized by recombinant DNA technology, which are often immunogenic[8] probably because of small differences in their conformational characteristics or in their glycosidic constituents in relation to their natural, human counterparts.

The broad area of immunological tolerance[9] and of the regulation of the immune response[10] is beyond the scope of this presentation, which is limited primarily to a brief review of work done in the author's laboratory within the last 15 years in relation to the *chemical conversion of antigens to tolerogenic derivatives* capable of inducing long-term specific suppression of the antibody response. Because of space limitation no reference will be made here to our methods developed for the suppression of the immune response to low molecular weight hapten-like epitopes, as was achieved for allergenic drugs or industrial chemicals by their conjugates with polyvinyl alcohol, and which may also be applicable to epitopes of common allergens. A description of all these studies prior to 1982 and the names of my many coworkers who contributed assiduously to their realization may be found in a more comprehensive review[11] and in more recent articles referred to here.

10.2. THE USE OF MODIFIED ALLERGENS FOR IMMUNOTHERAPY OF ALLERGIC PATIENTS

Since the traditional hyposensitization therapy consisting of injections of allergenic extracts is associated with the risk of systemic anaphylactic reactions, the doses of the allergens (AL) administered are by necessity extremely small. Therefore, generally, the effectiveness of the course of classical therapeutic regimens cannot be achieved unless extended over a number of years; even then its effectiveness is frequently only marginal. To circumvent these disadvantages, many attempts have been made to decrease the allergenicity of the preparations used therapeutically by chemical modification of the AL, such as urea denaturation,[12] reaction with formaldehyde,[13] and polymerization with glutaraldehyde.[14] An original type of chemically modified AL was developed in our laboratory by the coupling of an optimum number (n) of monomethoxypoly(ethylene glycol) (MPEG) molecules (MW 3000 or 6000) onto a given protein AL. The resulting co-polymeric conjugates, $AL(MPEG)_n$, were not only essentially nonallergenic, but also immunosuppressive.[11] It is to be noted that in our initial studies[15] the conjugates were prepared with PEG which contains two terminal hydroxyl groups and which, therefore, may give rise to intra- and intermolecularly crosslinked conjugates; to avoid this possible complication we have used MPEG in subsequent experiments.[11,16]

The inspiration for developing tolerogenic MPEG conjugates of diverse antigens stemmed from a number of facts: (1) the demonstration that covalent coupling of an enzyme to PEG resulted in a conjugate of substantially lower immunogenicity and clearance rate than those of the unmodified enzyme,[17] and (2) the long-established observations that administration of some antigens (e.g., xenogeneic immunoglobulins) in a deaggregated form resulted in specific tolerance[18] and that the clearance rate of deaggregated antigens was significantly lower than that of the original cruder form of the antigens.[19] Hence, the author hypothesized that PEG and MPEG conjugates of various antigens would be not only nonimmunogenic, but also tolerogenic. Indeed, as documented by experimental results obtained in his[11] and other laboratories,[20–22] this hypothesis was confirmed.

By way of illustration, reference shall be made to the original data[15] given in Table 1. For these experiments, Chester Beatty rats in the control group were sensitized i.p. with 1 μg of dinitrophenylated ovalbumin, i.e., OVA(DNP)$_3$, in the presence of 1 mg of Al(OH)$_3$ and 10^{10} *Bordetella pertussis*, and the rats in the test group received additionally an injection of 2.5 mg of OVA–PEG conjugate 4 hours prior to sensitization. Both groups were bled 14 days later for determination of their anti-OVA and anti-DNP IgE titers by passive cutaneous anaphylaxis (PCA) in

Table 1. Induction of Tolerance in Rats by OVA–PEG Conjugates

Groups of rats	Treatment[a]	Rat number	PCA titers on day 14 after sensitization: anti-OVA	PCA titers on day 14 after sensitization: anti-DNP	Systemic reaction
Control group	PBS	1	80	90	Death in 24 min
	+				
	OVA (DNP)$_3$	2	100	260	Death in 33 min
	+				
	Al (OH)$_3$	3	100	260	Death in 36 min
	+				
	B. pertussis	4	80	270	Death in 28 min
Test group	OVA–PEG	5	<10[b]	<10	No symptoms
	+				
	OVA (DNP)$_3$	6	20	20	No symptoms
	+				
	Al (OH)$_3$	7	<10	<10	No symptoms
	+				
	B. pertussis	8	<10	<10	No symptoms

[a]The control and test rats were pretreated with PBS and 2.5 mg of OVA–PEG, respectively; 4 h later all rats received an i.p. bolus of OVA(DNP)$_3$, Al(OH)$_3$, and *B. pertussis*.
[b]The lowest dilution of the sera used for PCA was 1:10.

random-bred Long–Evans hooded rats. Thereafter, on the same day, all animals received an i.v. injection of 2 mg of OVA in 1 ml of phosphate buffered saline (PBS). As is obvious from the results listed in Table 1, the IgE antibody response of the rats treated with OVA–PEG conjugates was essentially completely suppressed by comparison to IgE levels of control rats; most importantly, while on i.v. challenge with OVA all the animals in the test group survived, all the animals in the control group died within 40 minutes. Thus, these earliest experiments demonstrated that administration of PEG conjugates of a highly immunogenic antigen to rodents resulted in the suppression of their capacity to mount IgE antibody responses to the unmodified antigen and these animals were also systemically anergic to the injection of the antigen.

However, to our surprise, in clinical trials designed to validate the possible efficacy of immunotherapy with MPEG conjugates of diverse AL in close to 300 atopic patients, *it was established that while this treatment did not affect markedly the pre-existing IgE levels, the patients' AL-specific IgG levels were significantly and rapidly increased*. The results of these clinical trials, which were conducted under the aegis of Pharmacia AB of Uppsala in a number of medical centers in different countries, are documented in a recent comprehensive review article.[23] The main conclusions of these clinical studies are: (1) in comparison with conventional immunotherapeutic preparations containing unmodified pollen AL, substantially higher doses of the essentially nonallergenic $AL(MPEG)_n$ conjugates could be tolerated by allergic patients and did not lead to untoward physiological effects, (2) the beneficial effects of a short course of injections with $AL(MPEG)_n$ were similar to those obtained with conventional allergenic extracts over long periods of years, i.e., most of these patients showed fewer symptoms of rhinoconjunctivitis and required less medication during the pollination season, and (3) similar conclusions were derived from the results of the treatment of individuals allergic to honeybee venom with the corresponding MPEG conjugates.[23] The above-noted loss in allergenicity is attributable to the masking of some of the epitopes of the allergenic molecules by the long MPEG chains grafted onto the AL molecule, or to the conformational distortion of these epitopes as a result of this coupling reaction.

In spite of the fact that treatment of allergic patients over a relatively short period of time with $AL(MPEG)_n$ did not result in the anticipated drop in IgE antibodies, but led to an increase in AL-specific IgG antibodies, these MPEG conjugates may still be considered useful agents for immunotherapy of IgE-mediated allergies since the increase in IgG antibodies is still regarded as one of the principal goals of a successful hyposensitization regimen and, in particular, since these conjugates were essentially nonallergenic.[24] It may also be appropriate to point out that way back, in the 1950s and in the early 1960s, it had been shown[25] that *sera of allergic patients contained even prior to initiation of immunotherapy not only "reaginic" but also "blocking" antibodies, which correspond in modern terms to IgE and IgG antibodies, respectively*. Therefore, it may be suggested that the discrepancy between the enhancement of IgG responses in allergic patients and the induction of tolerance in mice, brought about by the administration of MPEG conjugates of the same AL, may be attributable to (1) the much higher dose (per kg) of the conjugates administered to mice, and (2) the difference in the immunological status of these two types of recipients, i.e., while the mice had not been stimulated by AL prior to administration of the tolerogenic AL(MPEG)$_n$, the immunological system of the patients had been stimulated "spontaneously" by repeated exposures to the AL prior to beginning their immunotherapy. In essence, therefore, the results of these clinical trials confirmed the fundamental rule of immune responsiveness, namely, that it is much more difficult to inactivate or eliminate long-lived memory cells of an established ongoing immune response, which have been expanded in response to repeated exposures to a given antigen, than the precursor cells prior to induction of the immune response. Moreover, these results suggested that MPEG conjugates may upregulate or downregulate the immune response depending on the immunological status of their recipient. The plausibility of this interpretation was supported by the results of experiments in mice which had been sensitized to produce anti-OVA IgE antibodies prior to receiving injections of the tolerogenic OVA(MPEG)$_{10}$ conjugates; indeed, although the IgE responses of these mice were markedly reduced, their IgG responses were enhanced after secondary immunization with OVA.[26] *However, the mechanism underlying the intriguing, diverging effects of tolerogenic MPEG conjugates on different isotypes (i.e., IgE versus IgG), depending on the immunological status of the recipient (i.e., naive versus presensitized) and on the doses of the conjugates remains to be elucidated.*

It is important to point out that the different allergenic preparations available at the time, when the exploratory clinical trials with AL(MPEG)$_n$ were being performed, contained multiple molecular components and, therefore, the resulting conjugates were complex mixtures of chemically undefined MPEG derivatives. However, in recent years, as a result of the application of hybridoma and recombinant DNA technologies, the principal AL responsible for some of the common IgE-mediated allergies which afflict masses of people around the world have been isolated, characterized, and synthesized as chemically well-defined and pure molecules, and even their B and T cell epitopes are being delineated in molecular terms.[27–29] Hence, it is obvious that there is no justification of continuing the "time honored" practice of immunotherapy with commercial preparations which consist of heterogeneous extracts of allergenic pollens, mites, or other complex allergens and which may contain constituents bearing no relation to the few allergenic constituents to which a given patient is actually allergic and which may be present in minute amounts in the extracts used. In this connection, it is important to note that, by the application of recombinant DNA technology, the existence of previously unrecognized major mite[30] and grass pollen[31] AL was recently demonstrated which had been missed by the use of standard immunochemical techniques. This discovery underscores the inadequacy of chemically complex extracts of AL or of their MPEG conjugates for use in immunotherapy, as was the case for the above-mentioned and many other clinical trials in the past. It is therefore not surprising that the long saga of many older clinical studies of the efficacy of hyposensitization therapies is filled with inconsistencies and speculative interpretations of the effects of immunotherapy.

However, in recent times, the production of well-defined AL on an industrial scale has become feasible by the application of recombinant DNA technology.[27–29] One may therefore anticipate that the availability of pure (i.e., standardizable and consistently identical) AL, and of their B and T cell epitopes, would revolutionize not only the diagnosis of IgE-mediated allergies, but also their therapeutic modalities. Thus, one may visualize that molecularly designed, specific, therapeutic immunosuppressants, capable of downregulating IgE antibodies to unique AL or haptens, would be synthesized in the near future. Similarly, attempts are being made to design immunosuppressive analogues of endogenous epitopes which are implicated in the production of antibodies responsible for devastating autoimmune diseases.[32] It is therefore regrettable that, in the present era of rapid advances of cellular and molecular biology which hold promise for the development of novel immunotherapeutic procedures based on solid scientific principles, the anachronistic recommendations of the U.K. Committee on Safety of Medicines formulated in 1986 threaten to lead to the discontinuation of immunotherapy of IgE-mediated allergies in the U.K.[33–36] and, possibly by imitation, also in other countries of the European Community, if not even globally.

10.3. MECHANISM OF INDUCTION OF SPECIFIC SUPPRESSION BY ANTIGEN–MPEG CONJUGATES

In earlier experiments[11] it had been shown that transfer of spleen cells from mice tolerized by OVA(MPEG)$_n$ conjugates into naive syngeneic recipients led to significant suppression of anti-OVA antibody responses in the latter to subsequent injections of OVA. It was therefore concluded that the downregulation of the primary antibody responses was due, at least in part, to the activation of antigen-specific Ts cells which were present among the spleen cells of actively immunosuppressed animals. In more recent experiments[37] injection of spleen cells from mice tolerized with MPEG conjugates of human monoclonal (myeloma) IgG, referred to as HIgG, induced HIgG-specific tolerance in syngeneic recipients. The extracts of the spleen cells of these immunosuppressed animals, which were obtained by freezing and thawing of the cells and which are referred to as F/T extracts, were also capable of downregulating the specific immune response to HIgG; hence, this extract was deemed to contain the suppressor T cell factor (TsF). The characteristics of the Ts cells induced *in vivo* to OVA and HIgG were studied with the aid of cell culture systems designed for the induction of antibody responses *in vitro*.[38,39] Thus, we determined that injection of OVA(MPEG)$_{13}$ into mice induced splenic Ts cells which could suppress *in vitro* secondary IgG responses to OVA(DNP)$_3$.[38] Similarly, spleen cells from mice tolerized with HIgG(MPEG)$_n$ were shown to possess HIgG-specific Ts cells and to lack HIgG-specific helper T (Th) cell activity (unpublished data).

In very recent experiments we succeeded in generating *nonhybridized HIgG-specific Ts clones* from spleen cells of BDF1 mice tolerized with HIgG(MPEG)$_{26}$.[39] One of these clones (clone 23.32), possessing the highest immunosuppressive activity for *in vitro* antibody formation, was shown to consist of cells which were Thy1.2$^+$, CD3$^+$, CD4$^-$, CD5$^-$, CD8$^+$ and expressed the $\alpha\beta$ heterodimer of TCR; moreover, these cells and their F/T extracts suppressed antibody responses in an HIgG-specific and MHC class I (H-2K^d) restricted manner. Similarly, the TsF of clone 23.32 was able to inactivate *in vitro* the function of HIgG-primed Th cells via a mechanism requiring the participation of accessory cells of the H-2^d haplotype; from these results it was inferred that the MHC restriction of the Ts cells was operative at the level of accessory cells (unpublished data). On the basis of immunochemical results, it was also concluded that the TsF produced by clone 23.32 had a determinant which was serologically related to that of the α chain of TCR. However, because the TsF did not react with two monoclonal antibodies to the TCR β chain, it was not possible to conclude that the TsF of this clone was identical to TCR. In very recent experiments, OVA-specific Ts clones with similar phenotypic and functional characteristics were generated from spleen cells of mice tolerized with OVA(MPEG)$_{12}$ and these Ts cells were shown to be devoid of cytotoxic activity; moreover, the TsF of these cloned Ts cells shared the epitopes of both the α and β chains of TCR. All these results taken together support the view that the specific immunosuppression induced by tolerogenic MPEG conjugates involves the activation of the corresponding antigen-specific Ts cells. Experiments designed to clarify the precise relationship between the receptors of these Ts cells and their suppressor factors at the molecular genetic level are in progress.

Furthermore, we have recently discovered (unpublished data) the phenomenon of "linked immunological suppression" which may be described by the statement that pretolerization of an animal to a given antigen, Ag_A, by the respective Ag_A(MPEG) conjugates results in abrogation of the antibody response also to a second antigen, Ag_B, *only* if Ag_B is injected into this immunosuppressed animal in the form of a stable covalent Ag_A–Ag_B adduct, but not as a mixture of Ag_A and Ag_B.

10.4. THE POSSIBLE THERAPEUTIC APPLICATIONS OF TOLEROGENIC MPEG CONJUGATES OF ANTIGENS TO DISEASES OTHER THAN IMMEDIATE HYPERSENSITIVITY

Although MPEG derivatives of AL appear to be useful for hyposensitization immunotherapy,[23] it may be envisaged that tolerogenic MPEG conjugates of some biologically active, immunogenic agents may have even a greater potential for the development of novel therapeutic strategies in relation to some of the devastating diseases referred to below. However, as documented earlier, tolerogenic MPEG conjugates are not effective in abrogating an established antibody response, i.e., an ongoing, immune response, as would be the case in autoimmunity. Therefore, MPEG conjugates would not be useful for patients producing—already prior to initiation of immunosuppressive therapy—antibodies to the antigens, or to crossreacting antigens, incorporated in the corresponding MPEG derivatives. *Nevertheless, this hurdle may possibly be circumventable by pretreating such patients for a reasonable period with a cocktail of nonspecific immunosuppressants in an attempt to wipe out their memory B and T cells, and administering thereafter the tolerogenic MPEG conjugates of the appropriate antigen(s).*

As alluded to in Section 10.1, xIg are being increasingly used for diverse *in vivo* diagnostic and therapeutic purposes, alone or in the form of immunoconjugates with Tx, cytotoxic drugs, or radionuclides.[3–6,40] These include *primarily murine monoclonal antibodies to* (1) tumor-specific antigens used for therapy of cancer of the colon, ovary, breast, and so on, or for diagnostic imaging of tumors,[41–44] (2) CD3 for treatment of renal allograft rejection,[45–47] (3) CD5 for treatment of acute graft-versus-host disease (GVHD) following bone marrow transplantation,[48,49] (4) CD4 for treatment of autoimmune diseases [1], and (5) IL-2 receptor for treatment of GVHD, allograft rejection, and autoimmune diseases.[2,50,51] Although some encouraging clinical results have been obtained by the use of the above xIg or xIg–Tx, one of the main complications which remains to be overcome is the inherent immunogenicity of xIg and Tx, which leads to the production by the patient of antibody responses (1) to epitopes of Tx, as well as (2) to both murine isotypic and idiotypic determinants, the predominant response being anti-idiotypic.[52] This complication cannot be overemphasized since multiple injections of the xIg or of their immunoconjugates are required for achieving the desired clinical effects.[41] Moreover, the patients' antibody responses lead to changes in the pharmacokinetics of the immunoconjugates,[53] e.g., xIg–Tx are rapidly cleared in the form of immune complexes from the circulation. Attempts to render the xIg less immunogenic have been made by different methods, e.g., by preparing chimeric antibodies consisting of the variable region of the mouse mAb coupled to the constant region of a human IgG molecule[54,55] or by constructing xIg–Tx by genetic engineering consisting of the variable region of a mouse mAb coupled to a toxin.[56,57] It is expected, however, that patients would still produce antibodies to the idiotypic determinants of these molecules, particularly when these determinants are attached to the patients' cells.[41–44]

In the belief that these complications could be overcome by tolerizing the patient to xIg (or to xIg–Tx) by administration of the tolerogenic MPEG conjugate of xIg (or of a mixture of tolerogenic MPEG conjugates of xIg and Tx) prior to beginning the therapeutic treatment with the corresponding nonpegylated agents, we have recently used the following *in vivo* model system. In the absence of access to nonhuman primates, the experimental system selected for testing the validity of this premise consisted of mice which were injected with HIgG as a source of xIg in conjunction with its tolerogenic MPEG derivatives; heat aggregated HIgG (haHIgG) served as the immunizing antigen[37] to increase its immunogenicity without recourse to adjuvants.

A diversity of protocols differing in the number of doses of the tolerogenic conjugate HIgG(MPEG)$_n$ and of haHIgG, which were injected over various lengths of time and at various intervals between injections, were used for assessing the capacity of the conjugate to induce long-term suppression.[58] In general, for screening the tolerogenic capacity of a given HIgG(MPEG)$_n$ conjugate (with n values in the range of 20–35), the protocol consisted of the i.v. injection of 100 μg of the conjugate into a group of 3 mice each, followed 7 days later by an i.p. immunization of 20 μg of haHIgG; the control groups received PBS in lieu of the conjugate. As illustrated in Figure 1, even the administration of a single i.v. injection of 100 μg of $HIgG(MPEG)_{25}$ on day −7 suppressed the murine anti-HIgG IgG1 response for at least 304 days in animals which received during this period six i.p. immunizing doses of 20 μg of HIgG at intervals of 40–90 days. (The reason for confining routinely the antibody assays to the determination of IgG1 antibodies is that this is the main class of immunoglobulins produced by B6D2F1 mice used in these experiments on repeated immunizations.)

As mentioned earlier, coupling of MPEG molecules to an antigen or allergen resulted, respectively, in a substantial decrease in antigenicity or allergenicity, which was attributable to conformational changes and/or masking of the epitopes caused by the coupling reaction. By analogy, the antigen-binding capacity of xIg was shown to be markedly reduced, if not totally impaired, as a result of conjugation with MPEG (unpublished data). Therefore, to ensure the efficacy of the annihilation of the target cells by "magic bullets," it is suggested that the clinical regimen include two phases. *The first, immunosuppressive phase* would consist of a series of injections of (1) tolerogenic MPEG conjugates of xIg, or (2) a mixture of tolerogenic MPEG conju-

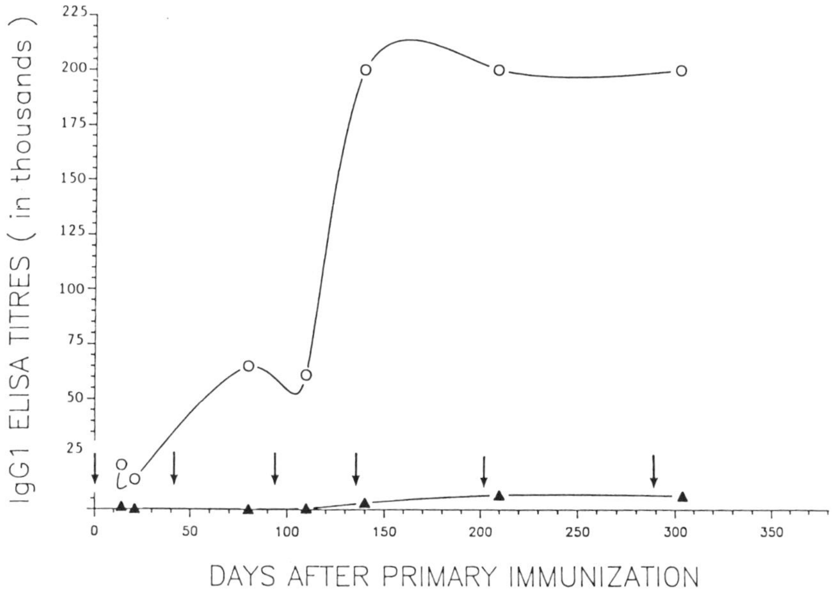

Figure 1. Induction of long-lasting suppression by a single injection of $HIgG(MPEG)_{25}$. Test mice (▲) were injected (i.v.) with 100 μg of $HIgG(MPEG)_{25}$ conjugate on day −7 and the control group (○) received PBS in lieu of conjugate. Both groups were immunized (i.p.) with 20 μg of haHIgG on days 0, 42, 95, 134, 200, and 290, as indicated by arrows. They were bled on days 14, 21, 80, 110, 140, 210, and 304, and their sera were tested for anti-HIgG IgG1 antibodies.

gates of xIg and Tx for induction of a state of immunological unresponsiveness, respectively, to the epitopes of (1) xIg, or (2) both xIgG and Tx. *The second, effector phase* would consist correspondingly of multiple injections of the respective "non-pegylated" xIg or xIg–Tx. The envisioned requirement for administration of a *mixture of tolerogenic MPEG conjugates of xIg and Tx*, rather than of the MPEG conjugate of the covalent xIg–Tx, is predicated by the following two considerations: (a) because of the possible difference in chemical reactivities of xIg and Tx with the activated MPEG intermediate,[16] it is likely that either of these two constituents of xIg– Tx may not be optimally pegylated in the presence of the other, i.e., either constituent may not be tolerogenic by itself; (b) because the xIg–Tx conjugate may dissociate *in vivo* into its constituents prior to reaching its target, the suboptimally pegylated constituent would elicit an antibody response and this would vitiate the proposed strategy.

In 1988 I also proposed that the two-step immunotargeting strategy suggested above for annihilation of malignant cells could be adapted for *in vivo* inactivation of HIV and destruction of HIV-infected cells in HIV-infected, clinically asymptomatic individuals by the use of conjugates of Tx with antibodies to appropriate, conserved epitopes of HIV.[59] This proposal rests on the hypothesis that conserved HIV epitopes[60] present on HIV-infected cells, e.g., on $CD4^+$ cells, may be accessible to anti-HIV xIg *in vivo*, as has been shown to be the case *in vitro*.[61,62] The obvious reservation, that the presence of the HIV genome in a cryptic, latent state in cells other than $CD4^+$ cells would invalidate the proposed strategy, may be countered by the argument that the anti-HIV xIg–Tx present in circulation would bind to the virus, released from or appearing on these cells, and thus inactivate it and/or destroy the corresponding cells. Clearly, the injection of the protective anti-HIV immunotoxins (and possibly also of its corresponding MPEG conjugate or of the mixture of the corresponding MPEG conjugates of xIg and Tx) may have to be continued at regular intervals for as long as the HIV has not been totally eradicated, or the patient has not developed effective immunity for the inactivation of all residual virus. It is obvious that to ensure the efficacy of this regimen, it ought to be initiated as soon as possible after infection. Even if all the cells bearing the virus could not be eliminated by this strategy, it is reasonable to anticipate that (1) under the protective umbrella of the passive immunization with anti-HIV xIg–Tx, *the proposed regimen would result at least in the reduction of the viral load*, and (2) consequently, *this strategy would slow down, if not arrest altogether, the progression from the asymptomatic phase to full-blown AIDS. This regimen may thus extend the critical period prior to the development of AIDS*, which might be sufficient (1) for the induction of a strong autologous anti-HIV response in patients by suitable vaccines which are being actively designed in many laboratories, and/or (2) for initiation of the treatment of the patients with effective anti-viral agents if, and when, these become available.

In spite of the promising results in experimental animal model systems suggesting that MPEG conjugates of immunogenic agents may prove useful for clinical applications, it still remains to be demonstrated that MPEG derivatives of xIg suppress also the immune response to the idiotypic determinants of cell bound xIg, since these determinants have been shown to be more immunogenic than isotypic epitopes.[63] Nevertheless, it is important to emphasize that administration of MPEG conjugates of immunogenic proteins, such as recombinant lymphokines,[8] allergens,[23] and xenogeneic enzymes,[64,65] has proven to be a safe procedure in man; hence, one may anticipate an increase in the clinical use of MPEG conjugates of diverse therapeutic agents.

ACKNOWLEDGMENTS

The author should like to acknowledge the unstinting efforts and the sound scientific insights of Dr. Weng Y. Lee who participated in all aspects of these studies prior to 1980. He should also like to acknowledge the more recent invaluable and multifarious contributions made by some of his former and present colleagues, i.e., Drs. Brian G. Carter, James L. Charlton, Youhai Chen, Valerie Holford-Strevens, Chung-Ja C. Jackson, Danuta Kierek-Jaszczuk, Glen M. Lang, Pradip K. Maiti, Edward S. Rector, Masaru Takata, and Ian Wilkinson, and his colleagues at Pharmacia in Uppsala, in particular Drs. Eva Åkerblom and Sten Dreborg. The skillful secretarial assistance of Mrs. Yvonne Hein and Ms. Kathy Risk is also greatly appreciated. These studies were supported by generous grants from the Medical Research Council of Canada, the National Institute of Allergy and Infectious Diseases of the U.S. National Institutes of Health, and by Pharmacia AB, Uppsala.

REFERENCES

1. J. M. Cruse and R. E. Lewis Jr. (eds.), *Therapy of Autoimmune Diseases*, Karger, Basel (1989).
2. T. Diamantstein and H. Osawa, in: *Immunol. Rev.* (G. Möller, ed.), Vol. 92, pp. 5–27, Munksgaard, Copenhagen (1986).
3. C-W. Vogel (ed.), *Immunoconjugates: Antibody Conjugates in Radioimaging and Therapy of Cancer*, Oxford University Press, Oxford (1987).
4. A. E. Frankel (ed.), *Immunotoxins*, Kluwer Academic Publishers, Boston (1988).
5. H. H. Sedlacek, G. Schulz, A. Steinstraesser, L. Kuhlmann, A. Schwarz, L. Seidel, G. Seemann, H. P. Kraemer, and K. Bosslet, *Monoclonal Antibodies in Tumor Therapy—Present Stage, Chances and Limitations*, Karger, Basel (1988).
6. R. O. Dillman, in: *CRC Crit. Rev. Oncol. Hematol.* (S. Davis, ed.), Vol. 1, pp. 357–385 (1985).
7. A. M. Zimmer, R. E. Goldman-Lieken, and J. M. Kazikiew (Abstract No. 193), *J. Nucl. Med. 28*, 603 (1987).
8. N. V. Katre, *J. Immunol. 144*, 209 (1990).
9. G. J. V. Nossal, *Science 245*, 147 (1989).
10. E. Sercarz (ed.), *Antigenic Determinants and Immune Regulation*, Karger, Basel (1989).
11. A. H. Sehon, in: *Regulation of the IgE Antibody Response* (K. Ishizaka, ed.), Vol. 32, pp. 161–202, *Prog. Allergy*, Karger, Basel (1982).
12. K. Ishizaka, H. Okudaira, and T. P. King, *J. Immunol. 114*, 110 (1975).
13. D. G. Marsh, *Int. Arch. Allergy Appl. Immunol. 41*, 199 (1971).
14. L. C. Grammer, M. A. Shaughnessy, and R. Patterson, *J. Allergy Clin. Immunol. 76*, 397 (1985).
15. W. Y. Lee and A. H. Sehon, *Nature 267*, 618 (1977).

16. C-J. C. Jackson, J. L. Charlton, K. Kuzminski, G. M. Lang, and A. H. Sehon, *Anal. Biochem. 165*, 114 (1987).
17. A. Abuchowski, J. R. McCoy, N. C. Palczuk, T. van Es, and F. F. Davis, *J. Biol. Chem. 252*, 3582 (1977).
18. D. W. Dresser, *Immunology 4*, 13 (1961).
19. W. Weigle (ed.), in: *Adv. Immunol.* (F. J. Dixon and H. G. Kunkel, eds.), Vol. 16, pp. 61–122 (1973).
20. T. P. King, L. Kochoumian, and N. Chiorazzi, *J. Exp. Med. 149*, 424 (1979).
21. K. V. Savoca, F. F. Davis, and N. C. Palczuk, *Int. Arch. Allergy Appl. Immunol. 75*, 58 (1984).
22. K. Kawamura, T. Igarashi, T. Fujii, T. Kamisaki, H. Wada, and S. Kishimoto, *Int. Arch. Allergy Appl. Immunol. 76*, 324 (1985).
23. S. Dreborg and E. Åkerblom, in: *CRC Crit. Rev. Therapeutic Drug Carrier Systems* (S. D. Bruck, ed.), Vol. 6, pp.315–365 (1990).
24. A. H. Sehon, *Allergol. et Immunopath. I* (suppl.) 19 (1974).
25. A. H. Sehon and L. Gyenes, in: *Sensitivity Chest Diseases* (M. C. Harris and N. Shure, eds.), pp. 1–35, Davis Co., Philadelphia (1964).
26. A. H. Sehon and G. M. Lang, in: *Mediators of Immune Regulation and Immunotherapy* (S. K. Singhal and T. L. Delovitch, eds.), pp. 190–203, Elsevier, New York (1986).
27. A. S. El Shami and T. G. Merrett (eds.), *Allergy and Molecular Biology*, Pergamon Press, Oxford (1989).
28. A. H. Sehon, D. Kraft, and G. Kunkel (eds.), *Epitopes of Atopic Allergens*, Proc. Workshop, XIV Congr. European Acad. Allergy and Clin. Immunol. (Berlin), pp. 1–2, UCB Institute of Allergy Monograph, Brussels (1989).
29. B. A. Baldo (ed.), *Molecular Approaches to the Study of Allergens*, Karger, Basel (1990).
30. W. R. Thomas, K. Y. Chua, W. K. Greene, and A. G. Stuart, in: *Epitopes of Atopic Allergens* (A. H. Sehon, D. Kraft, and G. Kunkel, eds.), Proc. Workshop, XIV Congr. European Acad. Allergy and Clin. Immunol. (Berlin), pp. 77–81, UCB Institute of Allergy Monograph, Brussels (1989).
31. S. S. Mohapatra, R. Hill, J. Astwood, A. K. M. Ekramoddoullah, E. Olsen, A. Silvanovitch, T. Hatton, F. T. Kisil, and A. H. Sehon, *Int. Arch. Allergy Appl. Immunol. 91*, 362 (1990).
32. A. A. Sinha, M. T. Lopez, and H. O. McDevitt, *Science 248*, 1380 (1990).
33. CMS Update: Desensitizing vaccines, *Br. Med. J. 293*, 948 (1986).
34. Forum on Immunotherapy and the Practice of Allergy: An enquiry in the United Kingdom, *Allergy and Clin. Immunol. News 1*, 147 (1989).
35. J. Charpin, *Allergy and Clin. Immunol. News 1*, 131 (1989).
36. J. Bousquet and F. B. Michel, *Allergy and Clin. Immunol. News 1*, 7 (1989).
37. I. Wilkinson, C-J. C. Jackson, G. M. Lang, V. Holford-Strevens, and A. H. Sehon, *J. Immunol. 139*, 326 (1987).
38. S. Mokashi, V. Holford-Strevens, G. Sterrantino, C-J. C. Jackson, and A. H. Sehon, *Immunol. Lett. 23*, 95 (1989).
39. M. Takata, P. K. Maiti, R. T. Kubo, Y. Chen, V. Holford-Strevens, E. Rector, and A. H. Sehon, *J. Immunol. 145*, 2846 (1990).
40. H. Waldmann (ed.), *Monoclonal Antibody Therapy*, in *Prog. Allergy*, Karger, Basel (1989).
41. M. V. Pimm, in: *CRC Crit. Rev. Ther. Drug Carrier Syst.* (S. D. Bruck, ed.), Vol. 5, pp. 189–227 (1988).
42. V. S. Byers and R. W. Baldwin, *Immunology 65*, 329 (1988).
43. E. S. Vitetta, R. J. Fulton, R. D. May, M. Till, and J. W. Uhr, *Science 238*, 1098 (1987).
44. R. W. Baldwin and V. S. Byers, *Curr. Opin. Immunol. 1*, 891 (1989).
45. H. Waldmann, *Annu. Rev. Immunol. 7*, 407 (1989).
46. Ortho Multicenter Transplant Study Group: A randomized clinical trial of OKT3 monoclonal antibody for acute rejection of cadaveric renal transplants, *N. Engl. J. Med. 313*, 337 (1985).
47. J. T. Mayes, J. R. Thistlethwaite, J. K. Stuart, M. R. Buckingham, and F. P. Stuart, *Transplantation 45*, 349 (1988).
48. N. A. Kernan, V. S. Byers, P. J. Scannon, R. P. Mischak, J. Brochstein, N. Flomenberg, B. Dupont, and R. J. O'Reilly, *J. Am. Med. Assoc. 259*, 3154 (1988).

49. V. Byers, P. Henslee, N. Kernan, B. Blazar, R. Gingrich, G. Phillips, J. Antin, R. Mischak, R. O'Reilly, and P. Scannon (Abstract No. 1071), *Blood 70* (suppl), 304a (1987).
50. D. Cantarovich, B. LeMauff, M. Hourmant, P. Peyronnet, Y. Jacques, F. Boeffard, M. Hirn, and J. P. Soulillou, *Am. J. Kidney Disease 11*, 101 (1988).
51. E. L. Ramos, E. L. Milford, R. L. Kirkman, N. L. Tilney, T. B. Strom, M. E. Shapiro, T. A. Waldmann, I. G. Wood, M. R. Rollins, and C. B. Carpenter, *Transplantation 48*, 415 (1989).
52. L. Chatenoud, M. F. Baudrihaye, N. Chkoff, H. Kreis, G. Goldstein, and J. F. Bach, *J. Immunol. 138*, 830 (1986).
53. M. B. Khazaeli, M. N. Saleh, R. H. Wheeler, W. J. Huster, H. Holden, R. Carrano, and A. F. LoBuglio, *J. Natl. Cancer Inst. 80*, 937 (1988).
54. L. Riechmann, M. Clark, H. Waldmann, and G. Winter, *Nature 332*, 323 (1988).
55. D. R. Shaw, M. B. Khazaeli, L. K. Sun, J. Ghrayeb, P. E. Daddona, S. McKinney, and A. F. LoBuglio, *J. Immunol. 138*, 4534 (1987).
56. V. K. Chaudhary, C. Queen, R. P. Junghans, T. A. Waldmann, D. J. FitzGerald, and I. Pastan, *Nature 339*, 394 (1987).
57. V. K. Chaudhary, J. K. Batra, M. G. Gallo, M. C. Willingham, D. J. FitzGerald, and I. Pastan (Abstract), *Second Int. Symp., Immunotoxins*, Lake Buena Vista, FL (1990).
58. P. K. Maiti, G. M. Lang, and A. H. Sehon, *Int. J. Cancer* (suppl.)*3*, 17 (1988).
59. A. H. Sehon, in: *Progress in Allergy and Clinical Immunology* (W. J. Pichler, B. M. Stadler, C. Dahinden, A. R. Pecoud, P. C. Frei, C. Schneider, and A. L. de Weck, eds.), pp. 570–574, Hogrefe & Huber Publishers, Toronto (1989).
60. B. R. Starcich, B. H. Hahn, G. M. Shaw, P. D. McNeely, S. Modrow, H. Wolf, E. S. Parks, W. P. Parks, S. T. Joseph, R. C. Gallo, and F. Wong-Staal, *Cell 45*, 637 (1986).
61. S. H. Pincus, K. Wehrly, and B. Chesebro, *J. Immunol. 142*, 3070 (1989).
62. M. A. Till, V. Ghetie, T. Gregory, E. J. Patzer, J. P. Porter, J. W. Uhr, D. J. Capon, and E. S. Vitetta, *Science 242*, 1166 (1988).
63. R. J. Benjamin, S. P. Cobbold, M. R. Clark, and H. Waldmann, *J. Exp. Med. 163*, 1539 (1986).
64. A. Abuchowski, G. M. Kazo, C. R. Verhoest, T. Van Es, D. Kafke Witz, M. L. Nucci, A. T. Viau, and F. F. Davis, *Cancer Biochem. Biophys. 7*, 175 (1984).
65. M. S. Hershfield, R. H. Buckley, M. L. Greenberg, A. L. Melton, R. Schiff, C. Hatem, J. Kurtzberg, M. L. Markert, R. H. Kobayashi, and A. Abuchowski, *N. Engl. J. Med. 316*, 589 (1987).

11

Toxicity of Bilirubin and Detoxification by PEG–Bilirubin Oxidase Conjugate

A New Tactic for Treatment of Jaundice

HIROSHI MAEDA, MASAMI KIMURA, IKUHARU SASAKI, YOSHIHIKO HIROSE, and TOSHIMITSU KONNO

11.1. INTRODUCTION

Bilirubin, the end product of heme catabolism, is generally regarded as toxic and highly fatal in newborn infants and fulminant hepatitis. Bilirubin encephalopathy (kernicterus) is usually considered to be caused by the entry of circulating, free (albumin-unbound), unconjugated bilirubin into the cerebral tissue.[1,2] Bilirubin conjugation with glucuronic acid takes place in the liver and the process is impaired in liver diseases. Our tactic for the treatment of jaundice is to decompose toxic bilirubin by the enzyme bilirubin oxidase, and for that purpose the enzyme is made into a polymer conjugate to improve its pharmacological properties.[3]

For clinical application, it is essential to confirm that these final products are less toxic than bilirubin, or are nontoxic to the major vital organs, and that they will be excreted rapidly from the bloodstream. In this study, we clarified the biological characteristics of the degradation products *in vitro* and *in vivo*. To demonstrate the toxic effects of bilirubin in a live animal model is complicated, so we examined bilirubin cytotoxicity and detoxification by bilirubin oxidase (BOX) *in vitro*.[4]

HIROSHI MAEDA, MASAMI KIMURA and TOSHIMITSU KONNO • Kumamoto University Medical School, Kumamoto 860, Japan. IKUHARU SASAKI and YOSHIHIKO HIROSE • Amano Pharmaceutical Co., Ltd., Nagoya 481, Japan.

Poly(Ethylene Glycol) Chemistry: Biotechnical and Biomedical Applications, edited by J. Milton Harris. Plenum Press, New York, 1992.

Toxicity of bilirubin to cellular activities has been investigated by many workers: it uncouples oxidative phosphorylation,[5–7] it depresses both protein synthesis[8,9] and DNA synthesis,[10,11] it decreases ATP content,[9,12] and it increases potassium leakage in various tissue culture cells.[9] It is also recognized to be potentially immunosuppressive and immunotoxic by its direct cytotoxic effects on human lymphocytes,[13] granulocytes,[13] and macrophages.[14] Furthermore, bilirubin-mediated hemolysis resulting from bilirubin–erythrocyte membrane interaction has been reported.[15–17]

However, to our knowledge, no studies except ours[3,4] have been conducted on the cytotoxicity of the degradation products that result from treatment of bilirubin with BOX. If no toxicity of these degradation products of bilirubin by BOX is found, then the therapeutic rationales using BOX become obvious. In this study, we examined the cytotoxicity of these breakdown products using the C 1300 neuroblastoma cell line of mouse. Further, urinary excretion of the degradation products was also elucidated in a rat model.

Previously, various methods were used to remove bilirubin, such as plasma exchange, steroid therapy, and phototherapy, but none has proven to have therapeutic value as a first choice regimen.

BOX used for the present approach is a highly specific enzyme for bilirubin (with M_r 52,000), derived from the microorganism *Myrothecium verrucaria* MT-1.[18–21] Langer *et al.* reported a treatment for severe neonatal jaundice with an immobilized BOX column system.[22,23] However, the column system has many inconveniences, such as physical confinement, clotting problems, and high incidence of infections.

In the past several years the use of chemical modification of proteins,[24–29] particularly of enzymes, has grown rapidly. We and others have successfully synthesized a number of polymer-conjugated protein drugs, such as styrene-co-maleic acid conjugated neocarzinostatin (SMANCS),[24,25] poly(ethylene glycol) (PEG)–L-asparaginase,[26,27] PEG–superoxide dismutase,[28] PEG–adenosine deaminase (ADA),[29] and PEG–interleukin 2.[30] We have now modified BOX with PEG to increase its plasma half-life, to diminish its immunogenicity, and to make it injectable and more effective.

11.2. EXPERIMENTAL PROCEDURE

11.2.1. Bilirubin Oxidase (BOX) and Other Reagents

BOX, 3.5 U mg^{-1}, was obtained from Amano Pharmaceutical Co., Ltd., Nagoya, Japan. Crystalline bilirubin was obtained from Sigma Chemical Co., St. Louis, Mo. FDA (fluorescein diacetate) was obtained from Dojin Chemical Co., Ltd., Kumamoto, Japan. All other chemicals were from commercial sources. The bilirubin solution was prepared according to the methods described by Sugita *et al.*[30] with a slight modification: bilirubin, used as unconjugated form, was dissolved in 0.1 M

NaOH at various specific concentrations and then passed through a 0.2-μm pore filter for sterilization immediately before use. Culture medium containing bilirubin was prepared as follows: bilirubin was dissolved in 0.1 M NaOH at various concentration, and 10 μl of the solution was added to 2 ml of RPMI 1640 medium with 10% heat-inactivated fetal calf serum (FCS) with or without BOX (final concentrations, 50 μg ml^{-1}); the final bilirubin concentration was in the range of 0–2.5 mg ml^{-1} and the pH was 7.2–7.4. All reactions were conducted in the dark to avoid photooxidation.

11.2.2. Cell Culture and Growth Inhibition

C 1300 mouse neuroblastoma cells were maintained and grown in monolayer culture in RPMI 1640 medium supplemented with 10% FCS at 37 °C under a humidified atmosphere containing 5% CO_2. Single cell suspensions were obtained by treatment with 0.05% trypsin/0.02% EDTA at 37 °C. These suspensions were used for cell counting or inoculation. The cells were plated in 16-mm wells (Falcon 24-well plates) at a concentration of 1.6×10^5 cells per well in 1 ml of culture medium. After overnight incubation at 37 °C, samples of medium containing bilirubin solution, bilirubin plus BOX, or 0.1 M NaOH as control were added to the cells. Cell number and viability were determined by the trypan blue dye exclusion method using a hemocytometer.

11.2.3. Fluorescence Microscopy

This procedure is used to evaluate the membrane transport by using FDA as a probe that becomes fluorescent after uptake into the cell; the method is described in detail elsewhere.[32]

11.2.4. Assays for DNA and Protein Synthesis

The effect of bilirubin on DNA synthesis was quantified as incorporation of [^{3}H]thymidine into DNA as usual. C 1300 cells were seeded at a concentration of 1×10^4 cells per well in a 96-well dish and were allowed to grow overnight in RPMI 1640 medium containing 10% FCS. Then, to the medium was added a solution (experimental media) containing bilirubin or bilirubin treated with BOX (50 μg ml^{-1}); this mixture was incubated for specified intervals. The cells were then pulse labeled with [^{3}H]thymidine, 0.25 μCi per dish for 3 h in triplicates. At the end of the incubation period, the cells were collected onto glass fiber filters and thymidine incorporation was determined by a liquid scintillation counter.

The effect of bilirubin on protein synthesis was quantified by incorporation of [^{3}H]leucine into cultured C1300 cells, which were plated at 3.6×10^5 per 35-mm-well dish. After overnight incubation, the cells were treated with the experimental medium as described above for 3 h, and then 0.1 ml of [^{3}H]leucine (5 μCi ml^{-1}) was added to half of the cultures for 3 h. The other cultures were exposed to experimental medium for 6 h and used to measure total protein content. Incorporation of [^{3}H]leucine was determined by precipitating cells per protein with cold 10% trichloroacetic acid and collected onto the glass fiber filters, and quantified by liquid scintillation counting.

11.2.5. Urinary Excretion of the Degradation Products of BOX-Treated Bilirubin

Male Wistar rats each weighing about 300 g were obtained from SLC (Shizuoka, Japan) and fed a regular diet *ad libitum*. With the rats under general anesthesia (40 mg per kg of pentobarbital sodium given intraperitoneally), a polyethylene tube (0.8 mm in diameter) was inserted into the urinary tract for collecting urine samples at specific intervals. Rats were injected iv with 3 ml of 20% mannitol to facilitate urinary output. Ten min later 3 ml of bilirubin (1.0 mg ml^{-1} in 0.2 M borate buffer, pH 8.0) or its degradation product (1.0 mg bilirubin plus 4 mg BOX in the same buffer) was injected i.v. Urine samples were collected in minute centrifuge tubes and centrifuged. The supernatants were diluted with 0.2 M borate buffer (pH 8.5) 10 or 50 times and absorbance was measured at 440 nm for bilirubin or 320 nm for the degradation products recovered in the urine.

11.2.6. Preparation of PEG–BOX

An active ester of PEG was prepared by reacting PEG and *p*-nitrophenyl chloroformate (*p*-NPCF–PEG) according to Veronase *et al.*[33] This active ester derivative of PEG has numbers of advantages over the commonly used cyanuric chloride derivative as follows: milder reaction condition, nontoxic nature, better stability of the bond being formed, selective reactivity to amino groups, etc.[33] For preparation of *p*-NPCF derivative, PEG with a mean *Mr* of 5000 obtained from Polyscience Inc. (Warrington, PA) was used.

The general scheme of the reaction of *p*-NPCF–PEG and BOX is shown in Figure 1, where putrescine (diaminobutane) was first coupled with the protein carboxyl groups as a spacer side chain to provide reactive amino group to which active ester *p*-NPCF–PEG was reacted. Although substantial numbers of amino groups exist in BOX, we found they did not react readily with cyanuric chloride. When it reacted at higher pH the activity of BOX was extensively sacrificed (Sasaki *et al.*, unpublished). About 10 mg of BOX was dissolved in 1 ml of 20 mM HEPES buffer at pH 6.5, and then, relative to BOX, 1000 molar excess of putrescine sulfate, 100 molar excess of *N*-hydroxysuccinimide, and 200 molar excess of water-soluble carbodiimide [1-ethyl-3-(3-dimethylaminopropyl) carbodiimide–HCl] (EDC) were added, and reacted under stirring at 4 °C overnight. Then the reaction mixture with aminobutylated BOX was applied on a column of Sephadex G-25 which was equilibrated with 20 mM borate buffer (pH 9.0), and aminobutylated BOX was separated.

To one mole of this aminobutylated BOX a 200 molar excess of *p*-NPCF–PEG

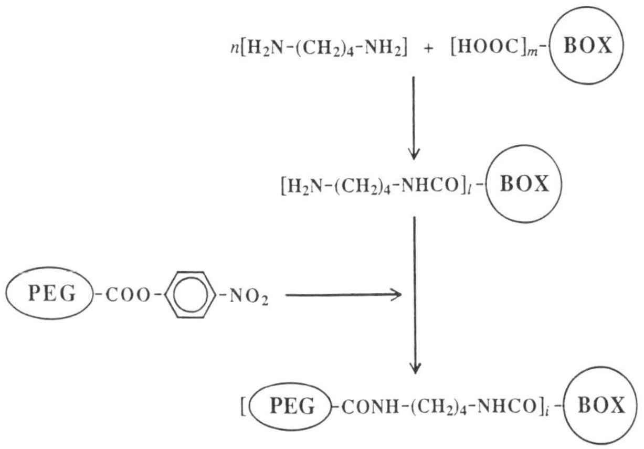

Figure 1. General scheme of conjugation reaction of bilirubin oxidase (BOX) with poly(ethylene glycol) (PEG). First step is the introduction of amino group spacer.

was added to react at 37 °C for 1 h and at 4 °C for overnight. Then, the reaction mixture was passed through a hollow fiber ultrafiltration to remove compounds below M_r 40,000, including unreacted PEG derivatives. Obtained conjugate was passed through a column of Sephacryl S-300 for further purification and concentrated peak was desalted and lyophilized. The conjugate was subjected to quantification of amino group with trinitrobenzene sulfonate, elemental analysis, and of the enzyme activity as described below. The decrease in amino group (11–12 moles per mole of BOX) was regarded as the number of amino groups modified by PEG. Change in N/C ratio in elemental analysis was also confirmed.

11.2.7. Measurement of Plasma BOX Activity and Plasma Clearance *in Vivo*

Rats were given 3.5 units of native BOX or PEG–BOX via the tail vein. At specific intervals, a 20-μl aliquot of blood was taken from the tail vein or the tail artery. The blood was added to 180 μl of phosphate buffered saline and centrifuged at 9500 × *g* for a few seconds. Then, 100 μl of supernatant was reacted at 37 °C with 1.0 ml of substrate solution composed of 2.0 mg of bilirubin dissolved in 100 ml of 0.2 M tris-HCl buffer at pH 8.4 in 9.0 mM sodium dodecylsulfate. Enzyme activity was determined as the decrease in absorbance of bilirubin at 440 nm for 1 min. A standard curve was obtained by using intact BOX at different concentrations.

11.2.8. Treatment of Jaundice with BOX in Icteric Rats and Decrease in Plasma Bilirubin

Native BOX or PEG–BOX (70 units in 3 ml) was injected into the tail vein of icteric rats which were surgically prepared by ligating and cutting the bile duct. Blood samples (0.5–0.8 ml) were collected at varying time intervals by cardiac or tail artery puncture with a heparin-coated syringe fitted with a 26-gauge needle; samples were immediately centrifuged at 9500 × *g* for 2–3 s. The supernatant was taken for bilirubin quantification.

11.2.9. Tissue Distribution of [^{51}Cr]-Labeled Native BOX and PEG–BOX in Rats

Radioactive BOXs were prepared by the method essentially described by Hanatowich *et al.*[34] which utilizes the bifunctional chelating agent DTPA, and the DTPA-tagged protein was chelated with radioactive chromium by using [^{51}Cr]Cl_3. The radiolabeled BOX (2 × 10^5 cpm in 1 ml) was injected into the tail vein of rats. The rats were killed under ether anesthesia at 48 h after intravenous injection of BOXs, and various tissues were removed to measure specific radioactivity.

11.2.10. Immunogenicity

11.2.10.1. Antigenicity

New Zealand White rabbits were immunized by intradermal administration as usual once every 2 weeks with 1.0 unit of native BOX or PEG–BOX together with Freund's complete adjuvant. We also tested the immunogenicity when BOX or PEG–BOX was injected i.v. without Freund's adjuvant. Ten days after the last immunization blood was withdrawn, and then Ouchterlony's immunodiffusion method was carried out to detect antibody formation.[35] We also examined the effect of antibody on the enzyme activity of native BOX and PEG–BOX by measuring the inhibitory effect of antibody.

11.2.10.2. Induction of Anaphylaxis Reaction in Guinea Pigs

Five male guinea pigs per group were used for tests at doses of 0.1, 1.0, and 10.0 mg per kg of test materials, and a tenfold diluted horse serum at 1 ml per kg (about 6 mg per kg), which were given i.v. on days 1, 3, and 5; and then 10 days after the last injection, blood was withdrawn by cardiac puncture and used to test for passive cutaneous anaphylaxis reaction. A systemic anaphylaxis reaction was induced by injecting the test materials into animals two weeks later at a dose of 1.0 mg per head i.v., and we observed the various symptoms for 30 min while the scores were recorded.

Passive cutaneous anaphylaxis was performed by injecting sc above blood samples (anti-serum) after serial dilution (see Section 11.3 below). Then, 4 h later, Evans-blue (5 mg) and BOX or PEG–BOX (1 mg each in 1.0 ml) was injected i.v., and the blue spots which appeared were quantified by extraction of dye from the skin specimens of blue spots.

11.3. RESULTS

11.3.1. Cytotoxicity of Bilirubin and Detoxification by BOX

The growth inhibitory effect of bilirubin and its reversal by BOX is shown in Figure 2a. Toxicity of bilirubin at 1.25 $mg\,dl^{-1}$ is significant but most of it was removed in the presence of BOX (Figure 2b). When the transport function of the cells were tested using FDA as a probe, we found that bilirubin did suppress uptake of FDA, but treatment with BOX restored this membrane transport function (Figure 3).

Effects of bilirubin on DNA and protein synthesis were similarly tested using C 1300 cells. Its toxic or suppressive effect on [^{3}H]thymidine and [^{3}H]leucine uptake was remarkable and dose-dependent (Figure 4). When BOX or PEG–BOX at 0.125 $U\,ml^{-1}$ was added, the toxic effects were removed. Albumin also reduced the toxic effect of bilirubin in a dose-dependent manner.

11.3.2. Excretion of the Degradation Products of Bilirubin Produced by BOX in the Urine

As shown in Figure 5, the degradation products were excreted efficiently in the urine, and a total recovery in two hours was about 80%. In contrast, free bilirubin was not excreted at all in the urine.

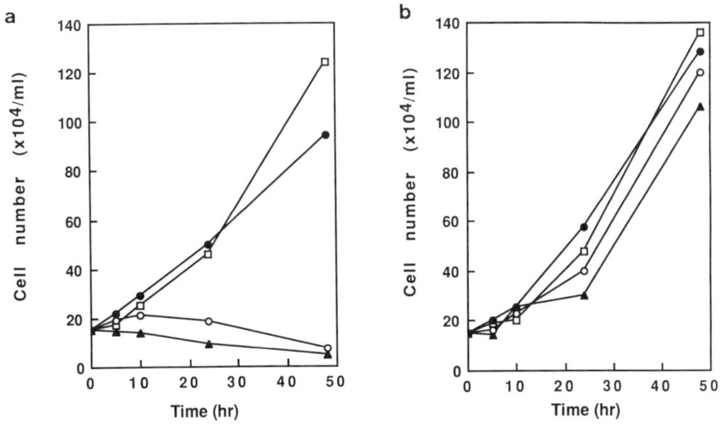

Figure 2. Growth inhibition of C 1300 neuroblastoma cells by bilirubin (a) and its degradation products formed by reaction with BOX (b). The concentrations of bilirubin were (□) 0, (●) 0.625, (○) 1.25, and (▲) 2.5 $mg\,dl^{-1}$ and BOX was 50 $\mu g\,ml^{-1}$ (0.175 $U\,ml^{-1}$) in b. From Kimura *et al.*[4]

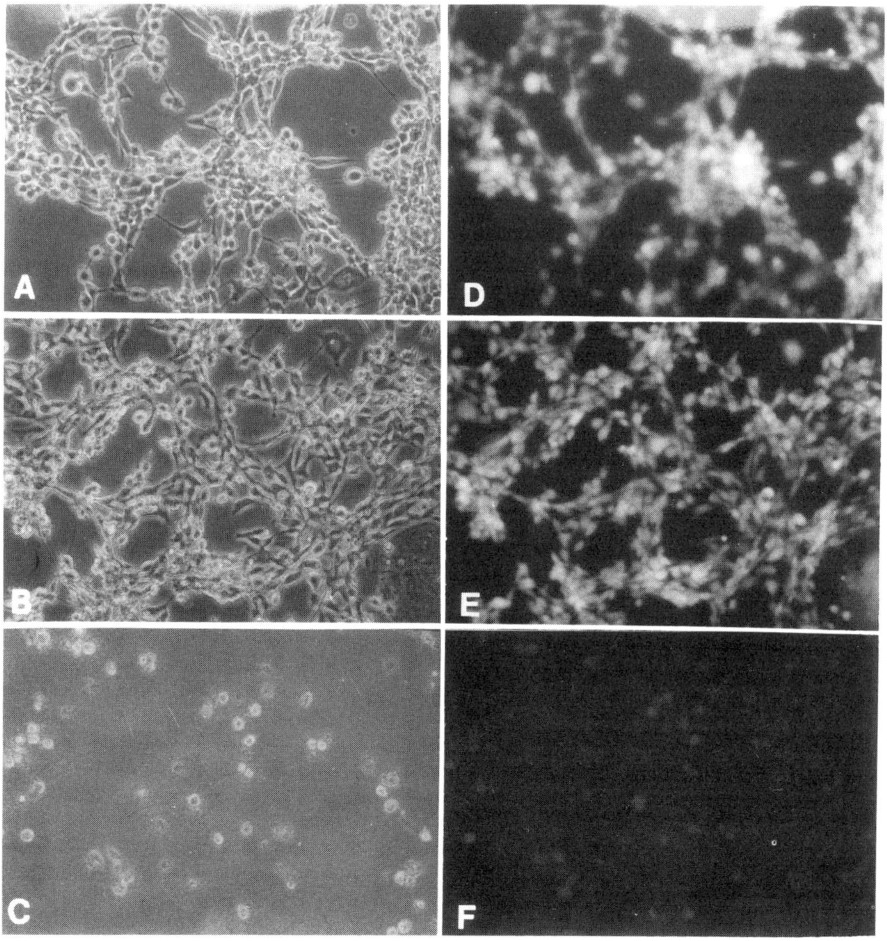

Figure 3. Photomicrographs of C 1300 cells treated for 24 h with bilirubin or with the degradation products produced by BOX under phase contrast (A–C) and fluorescence microscopy (D–F). A, D, control (no bilirubin or its degradation products added); B, E, 2.5 mg dl^{-1} plus BOX, 50 μg ml^{-1}; C, F, bilirubin, 2.5 mg dl^{-1} with no BOX. Magnification: × 194. (From Kimura *et al.*[4])

11.3.3. Properties of PEG–BOX Conjugate

Figure 6 shows the elution profile of PEG–BOX applied on Sephacryl S-300 and eluted with borate buffer. A clear difference is seen between BOX and PEG–BOX. Table 1 summarizes various chemical and enzymatic parameters of both native BOX and PEG–BOX. There is no remarkable difference in the stability between them at different pH and temperatures (not shown), namely, both of them are stable between

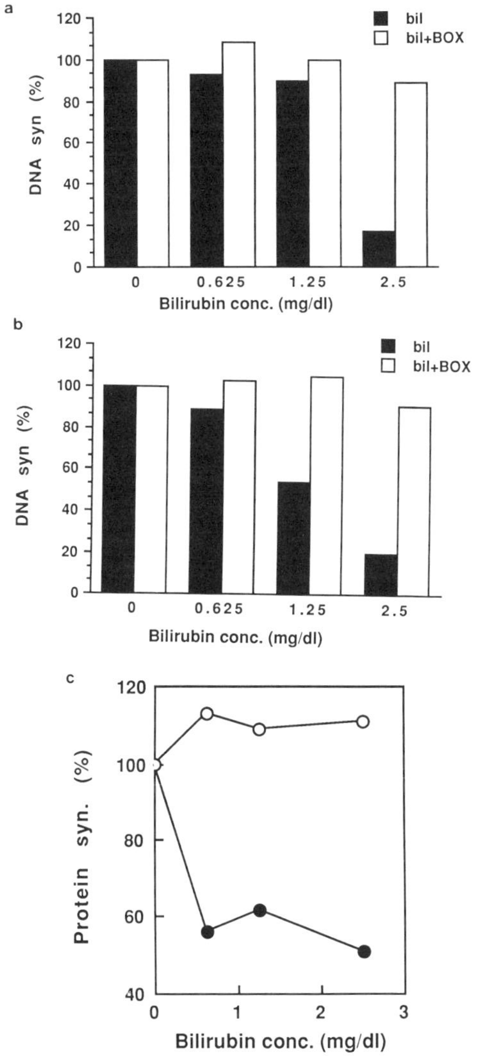

Figure 4. Effect of bilirubin or degradation products on DNA and protein syntheses in C 1300 cells. DNA synthesis with various concentrations of bilirubin (■) or degradation products (□) for 1 h (a) and 6 h (b). Cells were treated with [^{3}H]thymidine (0.25 μCi per well) for 3 h. Each value represents a mean of triplicate assays (percent to control). (c) Effect on protein synthesis. The cells were incubated for 3 h with bilirubin (●) or its degradation products (○) and then pulse-labeled with [^{3}H]leucine for another 3 h. The points are mean values from two wells. From Kimura *et al.*[4]

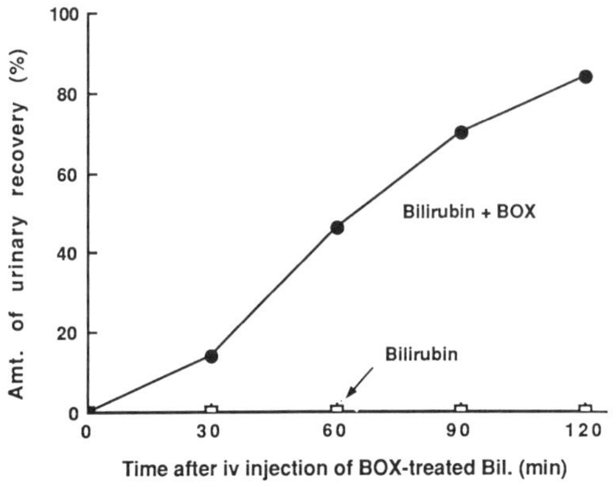

Figure 5. Urinary excretion of free bilirubin or BOX-treated bilirubin in the rat. Wistar rats were injected i.v. with free bilirubin (3 mg per rat) (□), or BOX-treated bilirubin (3 mg bilirubin plus 12 mg BOX per rat) (●) and urine samples were taken via catheter of the urinary bladder every 30 min for 2 h. The concentrations of bilirubin and degradation products in the urine were measured by absorbance at 440 and 320 nm, respectively, and the values represent a percent of the amount of urinary recovery against the injected dose. Mannitol (20% solution, 3 ml) was administered i.v. in both groups to facilitate urination (see text). This figure shows data of duplicate experiments. From Kimura *et al.*[4]

pH 7.0–9.5 or below 45 °C, but activity will be lost below pH 6 or above 10, or at a temperature above 45 °C. It is interesting to note that the K_m value of the conjugate is about 30% of the intact BOX, indicating higher affinity caused by the introduction of the amphipathic PEG chain. When the various proteinases were added to both BOXs, intact BOX showed gradual inactivation by subtilisin while PEG–BOX is very

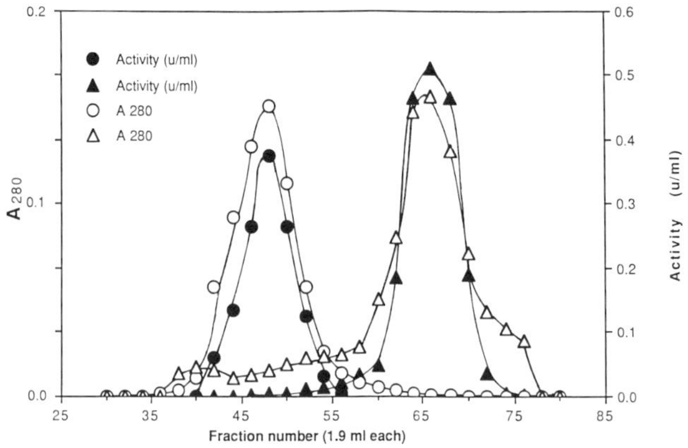

Figure 6. Elution profile of BOX and PEG–BOX on a Sephacryl S-300 column; column size, 1.5 × 95 cm; sample, 2 mg protein each; flow rate, 15 ml h^{-1}; elution buffer, 20 mM borate in 0.25 M NaCl pH 9.0. Open symbols, absorption at 280 nm; closed symbols, enzyme activity. Circle, PEG–BOX; triangle, native BOX. Two separate experiments were combined.

Table 1. Biochemical Properties of Bilirubin Oxidase (BOX) and Its PEG Conjugate

	BOX	PEG–BOX
Enzyme activity	100	40–65
Number of PEG introduced	None	11–12
Stable at		
pH	7.0–9.5	7.0–9.5
heat (°C)	below 45 °C	below 45 °C
K_m ($\times 10^{-4}$) M	1.9	0.59
Protease treatment and residual activity[a]		
Subtilisin	35	81
Thermolysin	98	100
Pronase	86	100
Trypsin	98	100
Chymotrypsin	98	100

[a]Enzyme treatment was conducted at 500 molar excess protease at 37°C, 1 h.

stable. Many other proteolytic enzymes show no remarkable effect due to less aggressive enzyme action (Table 1).

11.3.4. Plasma Clearance of PEG–BOX and Tissue Distribution

The enzyme activity of native BOX rapidly decreased after i.v. injection, while that of PEG–BOX decreased more slowly and was detectable even 48 h after injection (Figure 7). Plasma half-lives of native BOX and PEG–BOX were about 15 min and 5 hr, respectively. The area under the curve of the activity of PEG–BOX was 26 times greater than that of native BOX (Figure 7). The rapid drop in activity of native BOX from the circulation may indicate that it is trapped in some organs more readily than PEG–BOX. We then studied the accumulation of [^{51}Cr]-labeled native BOX and PEG–BOX in various organs and tissues. The results are shown in Table 2. Native BOX tended to be trapped more in the liver and kidney than PEG–BOX. PEG–BOX

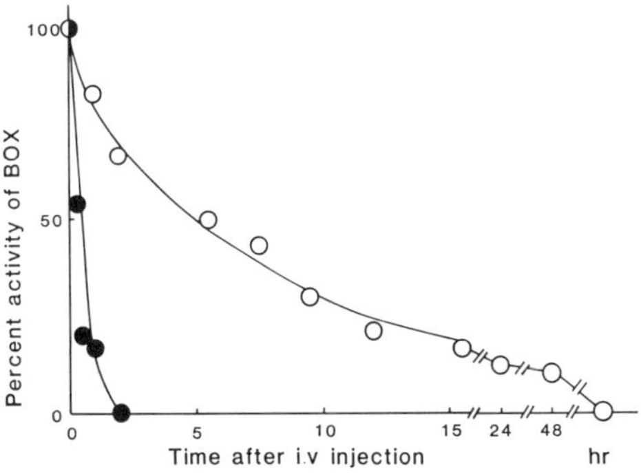

Figure 7. Residual activities of intact BOX (●) and PEG–BOX conjugate (○) in rat plasma after i.v. injection of 3.5 unit of each sample (see text). From Kimura *et al.*[3]

Table 2. Organ Accumulation of ^{51}Cr-Labeled Native BOX and PEG–BOX in Rats[a]

	Native BOX		PEG–BOX	
Organ	cpm per g	% dose	cpm per g	% dose
Blood	110 ± 25[b]	0.05 ± 0.01	466 ± 89	0.23 ± 0.04
Liver	4340 ± 1470	2.29 ± 0.94	1360 ± 208	0.67 ± 0.11
Kidney	3437 ± 320	1.72 ± 0.16	1429 ± 243	0.71 ± 0.15
Spleen	2350 ± 1240	1.17 ± 0.62	2730 ± 235	1.36 ± 0.12
Lung	100 ± 51	0.04 ± 0.02	401 ± 10	0.02 ± 0.01

[a]Values are measured 48 h after intravenous injection of ^{51}Cr-labeled native BOX or PEG–BOX into the tail vein of rats. Dose: 2×10^5 cpm per rat.
[b]Mean ± SD (from Reference 3).

accumulated more in the lung than native BOX, although the total amount is much smaller. A longer retention in plasma at a higher level was confirmed also in this experiment for PEG–BOX relative to BOX, similar to the result of enzyme activity (Figure 7).

11.3.5. Plasma Bilirubin Level after Injection of BOX and PEG–BOX Conjugates

11.3.5.1. Obstructive Jaundice by Dissection and Ligation of the Bile Duct in Rats

As shown in Figure 8a, after i.v. injection of native BOX (70 units per rat), plasma bilirubin levels decreased immediately to about 17% of the pretreatment level, but began to increase rapidly, and then 6–12 h after injection it became as high as the pretreatment level. In rats treated with PEG–BOX (70 units per rat) the bilirubin level dropped very rapidly and persisted at the low level for much longer (12–24 h) than that of the native BOX (Figure 8a).

11.3.5.2. Congenital Unconjugated Hyperbilirubinemia Model (Gunn Rat)

Plasma bilirubin levels were 6–8 mg dl^{-1} in 10-week-old untreated Gunn rats. When native BOX and PEG–BOX were administered i.v. at 70 units per rat, respectively, the plasma bilirubin level (as shown in Figure 8b) in the PEG–BOX group was more effectively reduced than with native BOX, similar to conjugated hyperbilirubinemia in Figure 8a.

11.3.6. Immunogenicity

The PEG conjugate showed no antigenicity when given i.v., while native BOX elicited antibody when given either i.v. or s.c. with Freund's complete adjuvant.

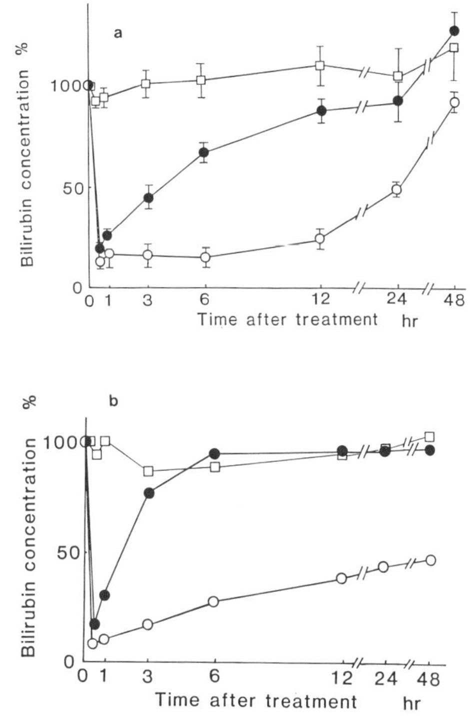

Figure 8. Effect of intact BOX (●) and PEG–BOX conjugate (○) on plasma bilirubin level: (a) obstructive jaundiced (Wistar) rat ($n = 3$); each value represents the mean ± SD of three rats; (b) constitutional jaundiced (Gunn) rat ($n = 1$). Each rat from both groups was injected i.v. with 70 units of enzyme activity, respectively, at time zero. (□), control group, which was injected with saline i.v. From Kimura *et al.*[3]

When the systemic induction of anaphylaxis by BOX and PEG–BOX was examined, BOX exhibited a strong reaction while PEG–BOX was very weak (Table 3). No passive cutaneous anaphylaxis was observed in PEG–BOX while BOX was much more remarkable (Table 4). Examination of cross-reactivity of PEG–BOX versus BOX, or BOX versus PEG–BOX, in the induction of anaphylaxis showed both groups became remarkable at the highest sensitization dose, namely, at 10 mg per kg (Tables 3 and 4). Neutralizing activity of BOX activity by either anti-PEG–BOX or anti-BOX antisera showed 82.2% inhibition of native BOX while enzyme activity of PEG–BOX was not inhibited or, if at all, very little. These results show that immunogenicity of PEG–BOX is eliminated by PEG-ylation.

11.4. DISCUSSION

The final objective of the present work is to develop a therapeutic agent for jaundice, namely, an injectable bilirubin oxidase. Our first step toward clinical application was to confirm the removal of toxicity of bilirubin by the enzyme treatment and the urinary excretion of the degradation products of bilirubin by the use

Table 3. Systemic Anaphylaxis Reaction

Group[a]	Dose of test sample used for sensitization	Severity of anaphylaxis reaction,[b] and numbers of animals					
		−	±	+	++	+++	++++
I. No treatment	—	5	0	0	0	0	0
II. PEG–BOX	0.1 mg per kg	0	2	2	1	0	0
III. PEG–BOX	1.0 mg per kg	0	2	3	0	0	0
IV. PEG–BOX	10.0 mg per kg	0	2	3	0	0	0
		[d]*0	0	0	4	1	1
V. Native BOX	0.1 mg per kg	0	0	0	0	0	5
VI. Native BOX	1.0 mg per kg	0	0	0	0	0	5
VII. Native BOX	10.0 mg per kg	0	0	0	0	0	5
		[e]*0	0	0	5	1	0
VIII. Horse serum	0.1 ml per kg[c]	0	0	0	0	2	3

[a]Each group consists of 5 male guinea pigs.
[b]Dose to induce anaphylaxis by each test material was 1.0 mg per kg, given i.v. Severity of anaphylaxis: − no symptom, ++++ death of animals.
[c]This same amount was used to induce anaphylaxis.
[d]*To this group 1.0 mg per kg of BOX was challenged to examine cross-reactivity.
[e]*To this group 1.0 mg per kg of PEG–BOX was challenged to examine cross-reactivity.

Table 4. Passive Cutaneous Anaphylaxis (PCA) Reaction

Group[a]	Dose of sensitization	PCA reaction,[b] in the skin and dilution					
		—	2	4	16	32	128[c]
I. No treatment	—	5	0	0	0	0	0
II. PEG–BOX	0.1 mg per kg	5	0	0	0	0	0
		[d]*5	0	—	—	—	—[f]
III. PEG–BOX	1.0 mg per kg	5	0	0	0	0	0
		[d]*5	0	—	—	—	—
IV. PEG–BOX	10.0 mg per kg	5	0	0	0	0	0
		[d]*1	0	4	—	—	—
V. Native BOX	0.1 mg per kg	5	0	0	0	0	0
		[e]*5	0	—	—	—	—
VI. Native BOX	1.0 mg per kg	0	0	0	5	5	0
		[e]*4	1	0	—	—	—
VII. Native BOX	10.0 mg per kg	0	0	0	5	5	0
		[e]*1	4	0	—	—	—
VIII. Horse serum	0.1 ml per kg[c]	0	5	4	1	0	0

[a]Each group consists of 5 male guinea pigs.
[b]Induction of anaphylaxis by each test material was 1.0 mg per kg or 0.1 ml per kg, given i.v.
[c]Dilution (fold) of test serum injected s.c.
[d]*Cross reactivity was tested PEG–BOX versus BOX for II, III, and IV. In these cases 1.0 mg per kg of test materials was injected i.v. to induce anaphylaxis.
[e]*Same as *d* except cross-reactivity tested was BOX versus PEG–BOX for V, VI, and VII.
[f]—, not done.

of this enzyme. Therefore, we investigated the cytotoxicity of bilirubin and its BOX-treated products to a mouse neural cell line by studying several parameters. Bilirubin had toxic effects on C 1300 neuroblastoma cells in a concentration-dependent fashion as expected.[8,10] On the other hand, we found the toxic effects were nullified by the addition of BOX or PEG–BOX (see Figures 2–4).[4] The breakdown products are quite hydrophilic, like biliverdine, which may be less toxic to cell membranes than lipophilic bilirubin. There may also be less internalization of bilirubin[36] or less interaction (perturbation) with the cell membranes by reduced lipophilicity.[37]

It is well known that serum proteins bind to unconjugated bilirubin and reduce the toxic effect of bilirubin on tissues *in vivo* and cells *in vitro*. In our experiments, when the concentration of fetal calf serum (FCS) used in culture was reduced from 10 to 7.5%, and then to 5.0%, the toxic effects of bilirubin became markedly intensified (data not shown). This observation is similar to the clinical manifestation in neonatal or premature kernicterus, in which the plasma albumin concentration is below 3%. These results confirmed that the toxicity was related to the concentration of free bilirubin rather than protein-bound bilirubin. When higher concentrations of bilirubin were used with a low percent of FCS in long-term incubation, some bilirubin precipitates were formed in the medium. When such precipitation occurred, the toxicity of bilirubin seemed to be reduced, as Lie and Brantlid[38] reported. Schiff *et al.*[10] reported that pH, oxidation, and uptake into the cells influenced the concentration of bilirubin in the media during the course of experiments with cells.

It was also known that BOX oxidizes free unconjugated bilirubin 100–200 times more than albumin-bound bilirubin *in vitro*. Thus, PEG–BOX may be able to reduce the toxicity of free unconjugated bilirubin and salvage neonatal jaundiced patients effectively.

Bilirubin is generally metabolized by the liver and excreted into the bile. In the case of obstructive jaundice, conjugated free bilirubin, but not unconjugated free bilirubin, is excreted by the kidney.[40] We investigated the renal excretion of unconjugated bilirubin and the oxidized products (Figure 5) and confirmed that unconjugated bilirubin is not excreted by the kidney.[4] On the other hand, soon after i.v. injection of bilirubin that had been treated with BOX, the color of urine changed to brown, the same color of the injected end products. This result indicates that the end products are more water-soluble and thus could be excreted by the kidney. These results of detoxification of bilirubin and excretion of the end products of BOX-treated bilirubin indicate the possible clinical application of administering PEG–BOX for the treatment of neonatal jaundice.

In this chapter we describe a new tactic for the treatment of jaundice by injecting a polymer-conjugated enzyme, PEG–BOX. Injection of PEG–BOX may be considerably superior to the extracorporeal systems, immobilized BOX column,[22,23] activated charcoal column, membrane dialysis, or plasma exchange. These methods involve a bleeding tendency because of extensive heparinization to prevent clotting, frequent possibilities of infection, physical confinement of patients, and thus longer time and higher expense may be required. Despite these risks, in most cases these methods are not curative. Injectable PEG–BOX is much simpler than the presently available methods by any criteria. The present data shown that this tactic appears to be more effective therapeutically. The major reason for this is the prolonged plasma half-life of PEG–BOX (Figure 7). We have demonstrated that a single i.v. injection of PEG–BOX can reduce the bilirubin value from 8 to 1.5 (normal range) in about 30 min, and this low level is maintained for more than 12 h. An additional injection resulted in normalization for about 48 h (not shown). Furthermore, lowering of the K_m value to about 30% of intact BOX (Table 2) indicates more effective enzyme action of PEG–BOX than the parental enzyme.

The reduced imunogenicity of PEG–BOX was confirmed (Tables 3 and 4). The immunogenicity became nullified when PEG–BOX was given s.c. without Freund's adjuvant or i.v. (not shown). Thus, possible adverse effects, namely, immunological problems, appear controllable and the clinical application of PEG–BOX seems feasible.

REFERENCES

1. G. B. Odell, *J. Clin. Invest. 38*, 823 (1959).
2. J. Bernstein and B. H. Landing, *Am. J. Pathol. 40*, 371 (1962).
3. M. Kimura, Y. Matsumura, Y. Miyauchi, and H. Maeda, *Proc. Soc. Exp. Biol. Med. 188*, 364 (1988).
4. M. Kimura, Y. Matsumura, T. Konno, Y. Miyauchi, and H. Maeda, *Proc. Soc. Exp. Biol. Med. 195*, 64 (1990).
5. R. Zeterstrom and L. Ernster, *Nature 15*, 1335 (1956).
6. M. G. Mustafa, M. L. Cowger and T. E. King, *Biochem. Biophys. Res. Commun. 29*, 661 (1967).
7. M. G. Mustafa, M. L. Cowger, and T. E. King, *J. Biol. Chem. 244*, 6463 (1969).
8. M. E. D. Notter and J. W. Kending, *Exp. Neurol. 94*, 670 (1986).
9. L. F. Rasmussen and R. P. Wennberg, *Res. Commun. Chem. Pathol. Pharmacol. 3*, 567 (1972).
10. O. Schiff, G. Chan, and M. J. Poznansky, *Pediatr Res. 19*, 908 (1972).
11. M. M. Thaler, *Nat. New Biol. 230*, 218 (1971).
12. M. L. Cowger, *Biochem. Med. 5*, 1 (1971).
13. I. Miler, M. Indrova, J. Bubenik, and J. Vondracek, *Folia Microbiol. (Praha) 30*, 272 (1985).
14. I. N. Dubin, B. Czernobilsky, and B. Herbst, *Arch. Pathol. 79*, 232 (1965).
15. R. Kaul, V. K. Bajpai, A. C. Shipstone, H. K. Kaul, and C. R. K. Murti, *Exp. Mol. Pathol. 34*, 290 (1981).
16. C. L. Kapoor, C. R. Krishna-Murti, and P. C. Bajpai, *Indian J. Med. Res. 60*, 918 (1972).
17. S. O. Lie and D. Brantlid, *Scand. J. Lab. Invest. 26*, 37 (1970).
18. S. Murao and N. Tanaka, *Agric. Biol. Chem. 45*, 2283 (1981).
19. S. Murao and N. Tanaka, *Agric. Biol. Chem. 46*, 2031 (1982).
20. S. Murao and N. Tanaka, *Agric. Biol. Chem. 46* 2499, 1982).
21. N. Tanaka and S. Murao, *Agric. Biol. Chem. 49*, 843 (1985).
22. C. Sung, A. Lavin, A. Klibanov, and R. Langer, *Trans. Am. Soc. Artif. Internal Organs 31*, 264 (1985).
23. A. Lavin, C. Sung, A. M. Klibanov, and R. Langer, *Science 230*, 543 (1985).
24. H. Maeda, T. Matsumoto, T. Konno, K. Iwai, and M. Ueda, *J. Protein Chem. 3*, 181 (1984).
25. H. Maeda, M. Ueda, T. Morinaga, and T. Matsumoto, *J. Med. Chem. 28*, 455 (1985).
26. Y. Kamisaki, H. Wada, T. Yagura, A. Matsushita, and Y. Inada, *J. Pharmacol. Exp. Ther. 216*, 410 (1981).
27. Y. Ashihara, T. Kono, S. Yamazaki, and Y. Inada, *Biochem. Biophys. Res. Commun. 83*, 385 (1987).

28. P. S. Pyatak, A. Abuchowski, and F. F. Davis, *Res. Comm. Chem. Pathol. Pharmacol. 29*, 113–127 (1980).
29. M. S. Hershfield, R. H. Buckley, M. L Greenberg, A. L. Melton, R. Schiff, C, Hatem, J. Kurzberg, M. L. Markert, R. H. Kobayashi, A. L. Kobayashi, and A. Abuchowski. *N. Engl. J. Med. 316*, 589 (1987).
30. N. V. Katre, M. J. Knauf, and W. J. Laird, *Proc. Natl. Acad. Sci. U.S.A. 84*, 1487 (1987).
31. K. Sugita, T. Sato, A. Fuse, and H. Nakajima, *Biol. Neonate 49*, 255 (1986).
32. H. Tsuda, H. Maeda, and S. Kishimoto, *Br. J. Cancer 43*, 793 (1981).
33. F. M. Veronase, R. Largajolli, E. Boccu, C. A. Benassi, and O. Schiavon, *Appl. Biochem. Biotech. 11*, 141 (1985).
34. D. J. Hanatowich, W. W. Layne, and R. L. Childs, *Int. J. Appl. Radiat. Isot. 33*, 327 (1982).
35. J. J. Munoz, *Methods Immunol. Immunochem. 3*, 146 (1970).
36. I. Diamond, and R. Schmid, *J. Clin. Invest. 45*, 678 (1966).
37. R. Broderson, *J. Pediatrics 96*, 349 (1980).
38. S. O. Lie and D. Brantlid, *J. Scand. Lab. Invest. 26*, 37 (1970).
39. C. S. Simionatto, K. E. Anderson, G. S. Drummond, and A. Kappas, *J. Clin. Invest. 75*, 513 (1985).
40. J. L. Golan, K. J. C. Dallinger, and B. Billing, *Clin. Sci. Mol. Med. 54*, 381 (1978).

12

PEG-Modified Hemoglobin as an Oxygen Carrier

KWANG NHO, SAMUEL ZALIPSKY, ABRAHAM ABUCHOWSKI, and FRANK F. DAVIS

12.1. INTRODUCTION

A need is strongly recognized for a safe red cell substitute that can carry oxygen to the hypoxic tissues of an anemic body. Banked blood is in short supply because of mounting fears of donors and limited storage life. A safe, stable, oxygen carrier would eliminate many problems such as cross-matching of blood types, danger of virus infections, short shelf life, and availability. Not only could this be used as an oxygen-carrying plasma expander for trauma victims and patients in surgery, such as cardiac bypass and angioplasty, this product could also be used as a perfusate for the preservation of isolated organs for transplantation. Cold storage is the popular choice for preservation at present, but there is an increasing demand for a safe organ perfusate that would promote the use of perfusion preservation to obtain a longer period of preservation.

In the light of these objectives a number of products have been developed and tested in animal models and, in some cases, the human clinical setting. Oxygen carriers so far developed can be categorized into two groups: chemical-based and hemoglobin-based. Chemicals such as perfluorocarbon have superb solubility for oxygen and carbon dioxide and have proved useful in limited applications such as radiography and coronary angioplasty. However, the overall safety in the case of massive transfusion is still in question.[1]

Hemoglobins as an oxygen carrier have been explored more thoroughly. As raw materials, human hemoglobin obtained from outdated banked blood and bovine

KWANG NHO, SAMUEL ZALIPSKY, ABRAHAM ABUCHOWSKI, and FRANK F. DAVIS • Enzon, Inc., South Plainfield, New Jersey 07080. *Present address for S. Z.*: Department of Chemistry, Rutgers–The State University of New Jersey, Piscataway, New Jersey 08855.

Poly(Ethylene Glycol) Chemistry: Biotechnical and Biomedical Applications, edited by J. Milton Harris. Plenum Press, New York, 1992.

hemoglobin have been routinely utilized. Pure hemoglobin, unmodified, has been tested *in vivo* and found to be ineffective because it is quickly cleared from the body due to its short circulation life (the molecular weight of hemoglobin is 64,500). The unmodified hemoglobin tends to dissociate into dimers quite readily under physiological conditions and its half-life turns out to be less than one hour in rats. This short retention time not only renders the material ineffective but also puts considerable strain on the organs that rapidly clear the hemoglobin breakdown product from the circulation. To overcome these problems, much effort has gone into modification of hemoglobin for the past two decades.

Modification of hemoglobin is done in a variety of ways: crosslinking of subunits to improve the tetrameric stability, attaching inert artificial polymers of high molecular weight onto hemoglobin to increase the total molecular weight, and polymerization of the hemoglobin molecules. Any combination of these three methods can also be implemented to take advantage of the benefits of each method. A spectrum of intramolecular crosslinking agents have been developed based on an earlier observation that tetrameric hemoglobin does not pass through renal glomerular tubules but is retained in the vasculature.[2] However, this type of modification seems to provide less than substantial extension of retention time. Moreover, recent work reveals that molecules of molecular weight of 64,500 can pass through renal tubules and appear in the urine.

Increasing the total molecular weight of hemoglobin through modification seems to be a better choice in prolonging vascular retention time. Polymers such as dextran,[3] hydroxyethyl starch,[4] dextran and inulin,[5] and polyethylene glycol[5–13] have all been bound to hemoglobin. These methods increased retention time by 5–10-fold in comparison to that of native hemoglobin.

Polyhemoglobin can be made by reaction with crosslinking agents, or either low or high molecular weight. Payne[14] first attempted to modify hemoglobin with glutaraldehyde. Pyridoxalated human hemoglobin polymerized with glutaraldehyde is commonly studied by many and *in vivo* performance and *in vitro* characteristics are well known.[15–20] This product is polydisperse and reproducibility would be difficult. The molecular weight range of this polyhemoglobin changes readily with slight alterations in the reaction conditions. Most of this heterogeneous product consists of approximately 80% polymerized hemoglobin and 20% unreacted tetramer. The molecular weight of the polymer ranges between 100,000 and several million with a peak at around 600,000. A much more controlled type of polyhemoglobin is produced by reaction with bifunctionally reactive poly(ethylene glycol) with molecular weight of resulting PEG–Hb ranging from 70,000 to 200,000.[6,12]

12.2. PEG–HEMOGLOBIN

Poly(ethylene glycol) (or PEG), also called polyoxyethylene (POE), is well known and has been used as a biocompatible modifier for a variety of enzymes and proteins to obtain an increased *in vivo* retention time and diminished immunogenicity and antigenicity.[21,22] PEG-modified biologics are found, in most cases, to maintain their intrinsic activities.[12] Similarly, PEG has been applied to hemoglobin for potential use in a physiological oxygen-carrying resuscitation fluid. To qualify as such, modified hemoglobin should possess properties such as extended plasma circulation time and adequate oxygen-carrying capacity. PEGs with molecular weights ranging from 750 to 6000 have been tried.[8,12] As a long, linear, hydrophilic molecule, PEG exerts significant influence on the properties of hemoglobin when attached. PEG–hemoglobins exhibit a wide range of differences in their properties depending on which molecular weight PEG is selected and how many PEGs are bound.

12.3. CHEMISTRY OF PEG CONJUGATION

Leonard *et al.*[9] reported a PEG–Hb conjugate containing an amide bond formed between amino groups of hemoglobin and a carboxylate group introduced onto

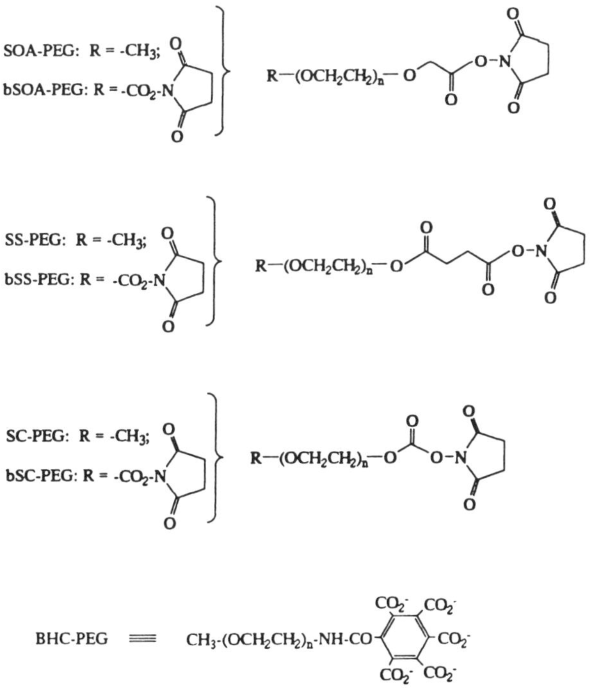

Figure 1. The activated PEGs used for modification of hemoglobin.

monomethoxy PEG, to make monomethoxypolyoxyethylene succinimidyl ester, which is also named in this text as PEG–succinimidyl oxyacetate (SOA–PEG) of molecular weight 5000 (Figure 1). The conjugates thus obtained, SOA–PEG–Hb, had a molecular size corresponding to that of a globular protein with a molecular weight of about 190,000. Later, Labrude *et al.*[8] achieved a different formulation by using same kind of activated PEG, SOA–PEG, with molecular weight of 1900. This newer type had 13–14 PEGs bound to one hemoglobin molecule making the molecular weight of PEG-hemoglobin 90,000.

PEG–bis(succinimidyl succinate), bSS–PEG, of molecular weight 3400 was synthesized by Iwasaki and Iwashita[6] and used to produce polymerized hemoglobin, bSS–PEG–Hb, which consisted of 19–21% unpolymerized hemoglobin, 29–31% dimer of hemoglobin, 32–34% tetramer, and 16–18% polymer whose molecular weight is higher than 600,000. From the number of PEGs attached to one hemoglobin molecule (which is four to five) and the weight-average molecular weight (which is 123,000), a large portion of the PEG-hemoglobin was monomer or dimer. The product contained none of the unmodified hemoglobin.

Succinyl PEG was later replaced by α-carboxymethyl Ω-carboxymethoxyl PEG as reported by Iwashita *et al.*[7] This activated PEG, which is named in this text as PEG–bis(succinimidyl oxyacetate), bSOA–PEG, of molecular weight 3600 was then conjugated to pyridoxalated hemoglobin to the point where 5–6 PEGs were bound to a hemoglobin molecule through an amide bond. The PEG–Hb thus produced consisted of 90% unpolymerized and 10% polymerized. The number-average molecular weight was about 90,000.

Another derivative, polyanionic PEG introduced by Zygmunt *et al.*[13] is a PEG possessing a polycarboxylate end prepared by coupling benzene hexacarboxylate (BHC) to amino–PEG by using 1-(3-dimethylaminopropyl)-3-ethyl-carbodiimide hydrochloride (EDCI). BHC–PEG thus produced is directly linked to hemoglobin by reaction of fixed BHC–carboxylate groups with amine functions of hemoglobin in the presence of EDCI. However, the detailed properties of BHC–PEG–Hb were not given.

Most recently developed is a PEG–succinimidyl carbonate (SC–PEG) which, upon reaction with lysines of hemoglobin, creates a very stable urethane linkage.[12] SC–PEG has been tested on a variety of enzymes and proteins, and the modified products maintained their intrinsic activities very well. Bovine hemoglobin (bHb) modified with SC–PEG of molecular weight ranging from 2000 to 5000 has been characterized.[23] With a similar method, both ends of PEG can be activated to yield PEG–bis(succinimidyl carbonate), bSC–PEG, having a molecular weight of 2000–6000. The bSC–PEG has also been used to prepare much more stable, crosslinked or polymerized hemoglobin.

The stability of the different linkages between PEG and amines of hemoglobin is not easily determined by their *in vivo* performance because it is known that the half-life is dependent on the dosage of injection.[25] However, the ester bond in the protein conjugated with SS–PEG is vulnerable to hydrolysis under physiological conditions while the urethane bond from SC–PEG is not.

12.4. PROPERTIES OF PEG–HEMOGLOBIN

SOA–PEG–Hb was prepared by Leonard and Dellacherie[9] with deoxyHb or oxyHb in the presence of inositolhexaphosphate (IHP) and 2,3-diphosphoglycerate (2,3-DPG). Apparently IHP was more effective than 2,3-DPG with regard to P_{50} (the oxygen partial pressure at which 50% of hemoglobin is oxygenated, for instance, the P_{50} of normal human red blood cell is 28 mm Hg), and modification of deoxyHb yielded higher P_{50} than modification of oxyHb. The SOA–PEG–deoxyHb, prepared in a solution containing 20:1 molar ratio of IHP to Hb and the pH of reaction varying from 5.8 to 6.6, yielded P_{50} values ranging from 23.4 to 25.8 mm Hg with a Hill coefficient (the degree of cooperativity in the binding of oxygen by hemoglobin; for instance, the Hill coefficient of normal human red cells is 2.8) of 2.3 to 2.5. The P_{50} was measured in a 0.1 M phosphate buffer with pH 7.15. The following conclusions were reached: (1) whatever the initial conformation of hemoglobin, the lower the pH of reaction, the better its oxygen-binding properties, (2) hemoglobin modified in the deoxygenated state exhibits better properties than the modified oxyHb, (3) in the absence of organic phosphates, the phosphate buffers are better reaction media than water or NaCl solutions for preserving the functional properties of hemoglobin, (4) the reaction of deoxyHb in the presence of IHP provides conjugates with P_{50} similar to that of unmodified pyridoxalated hemoglobin, (5) the P_{50} of pyridoxalated hemoglobin is completely preserved after its conjugation with SOA–PEG at low pH.

DeoxyHb, where the ε-amino groups of α-40 Lys, α-127 Lys, and the α-amino group of α-1 Val are engaged in salt bridges, has a higher pK_a, thus lowering reactivity toward the activated ester of PEG. Also, organic phosphates, such as IHP and 2,3-DPG, bind the deoxyHb more strongly by interaction with several amino groups located in the β-cleft, which are the ε-amino group of β-82 Lys, and the α-amino group of β-1 Val, and the imidazolyl groups of β-2 His and β-143 His. Moreover, the interaction is stronger with lower pH. Thus the use of deoxyHb complexed with organic phosphates prevents SOA–PEG from reacting with those amino groups that play an essential role in the oxygenation–deoxygenation cycle and these amines are left unmodified after the modification.

The bSS–PEG–Hb prepared by Iwasaki and Iwashita,[6] although very heterogeneous, exhibits biochemical parameters within narrow variations. P_{50} is 21.3 ± 1.4 mm Hg (measured in pH 7.4), viscosity 2.6 ± 0.4 cp, colloidal osmotic pressure 36.1 ± 2.4 mm Hg, hemoglobin concentration 6.0–6.5% (g dl^{-1}), Hill coefficient 2.2 ± 0.2. The P_{50} of bSS–PEG–hemoglobin reported is higher than that of polymerized hemoglobin which ranges between 16 to 21 mm Hg.[16,20] Moreover, the Hill coefficient of BSS–PEG–Hb is 2.2 ± 0.2 while that of polymerized hemoglobin is 1.5–2.0.[20] Based on these two indications the authors conclude that the oxygen delivery by bSS–PEG–hemoglobin is superior to that of polymerized pyridoxalated hemoglobin. bSOA–PEG–Hb developed later by this same group[7] exhibits similar oxygen-carrying properties to bSS–PEG–hemoglobin.

BHC–PEG–hemoglobin prepared by Zygmunt *et al.*[13] was reported to possess a maximum P_{50} of 25 mm Hg (as measured in 0.05 M Tris-buffer, pH 7.2, 25 °C) while under the same conditions the P_{50} of unmodified hemoglobin was 3 mm Hg and the maximum P_{50} obtained in the presence of 2,3-DPG was 10 mm Hg. It was noted that the high density of charges on the anionic group, which causes a more specific interaction with deoxyHb in the 2,3-DPG binding site, makes polyanionic BHC–PEG very active toward hemoglobin. Moreover, because of the extensive hydrophobic areas existing on each side of the 2,3-DPG-binding site, the presence of a hydrophobic ring surrounded by a highly hydrophilic PEG group can make the interaction inside this cavity even stronger. In other words, the authors concluded that the anionic nature of the modifier significantly influences the localization of their fixation to hemoglobin, and channels them toward the DPG-binding site in such a way that the amino groups of the hemoglobin β-cleft site can be involved in the covalent bond between the modifier and hemoglobin.

SC–PEG–bHb has merits in that, due to the strong urethane linkage, only moderate amounts of PEG need to be attached to achieve the desired goals in terms of P_{50} and *in vivo* half-life.[23,12] Attachment of minimal PEG on Hb is essential because P_{50} begins to decrease as more PEGs are bound. This seems reasonable considering that PEG modification is a kind of immobilization from the viewpoint of protein activity. Due to the presence of long strands of PEGs on the Hb surface, the freedom of subunit movement may suffer resulting in a loss of cooperativity and lowering of P_{50}. However, in case of SC–PEG–bHb, the intrinsic P_{50} of native bHb, which is about 28 mm Hg, has been maintained with up to 7–8 PEG molecules attached on one bHb molecule. Additionally, the decrease in P_{50} appeared to depend on the number of lysines modified, not on the molecular weight of the PEG used. SC–PEG–bHb with a molecular weight of 100,000 shows P_{50} of 26–28 mm Hg (as measured in 0.1 M NaCl buffer, pH 7.4, 37 °C), and viscosity of 6 g dl^{-1} 3.6 cp (at 37 °C), and Hill coefficient 2.0–2.2. Reduction in the alkaline Bohr effect (the influence exerted by carbon dioxide on the oxygen dissociation curve of blood) also corresponded to the degree of modification, but at this level of PEG attachment the Bohr effect almost remains unchanged.

Interestingly, crosslinked or polymerized bSC–PEG–bHb exhibits a higher P_{50} than that of native bHb. A lightly crosslinked sample with 1–3 bSC–PEGs bound intramolecularly to bHb shows P_{50} of 32–35 mm Hg (Table 1). Controlling the reaction between intramolecular crosslinking and intermolecular polymerization is possible by varying some of the reaction parameters such as bHb concentration, temperature, etc. However, the viscosity of the polymerized sample becomes unworkably high (up to 4.2–4.5 cp at 6 g dl^{-1} concentration).

Ajisaka *et al.*[24] also reported earlier the observation that the P_{50} of PEG–Hb changes depending on the number of amines modified with PEG. However, whether the molecular weight of PEG, with equal numbers of PEG bound, has an effect on the P_{50} was not clear.[24] Nonetheless, the P_{50} decreased as more PEGs were bound. Also, cooperativity of oxygen binding and alkaline Bohr effect suffered. Thus the authors suggested that the PEG attached preferentially to the amines of α-1 Val, β-82 Lys, and β-40 Lys, which are known to participate in cooperativity and the alkaline Bohr effect. Therefore, the PEG-hemoglobin modified with lower molecular weight PEG

Table 1. P_{50} of PEG–Hemoglobin

PEG–Hb	P_{50} (mm Hg)	Conditions of P_{50} measurement
SOA–PEG–Hb[20]	23.4–25.8	pH 7.15, 25 °C
bSOA–PEG–Hb[31]	20	pH 7.4, 37 °C
SS–PEG–Hb[22]	21–24	pH 7.4, 37 °C
bSS–PEG–Hb[14]	21.3 ± 1.4	pH 7.4, 37 °C
BHC–PEG–Hb[13]	25	pH 7.2, 25 °C
SC–PEG–bHb[32]	26 ± 2	pH 7.4, 37 °C
bSC–PEG–bHb[32]	32–35	pH 7.4, 37 °C
Native Hb	10–15	pH 7.4, 37 °C
Native bHb	28–32	pH 7.4, 37 °C

appears superior in carrying oxygen to that modified with higher molecular weight PEG. In the case of using lower molecular weight PEG, more PEG must be conjugated in order to achieve desired molecular size. This, the authors note, makes the conjugates unstable, thus prone to irreversible oxidation to MetHb. Hence the use of PEG of molecular range 4000–5000 is recommended.

12.5. EFFICACY OF PEG–HEMOGLOBIN

SS–PEG–Hb developed by Matsushita *et al.*,[25] with a molecular weight of 90,000, P_{50} 21–24 mm Hg, hemoglobin concentration 6%, and MetHb less than 5% has been tested in dogs. Hespan (6% Hetastarch solution, American Critical Care, IL) was used as control. Exchange transfusion (ET) was performed by first withdrawing 200 ml of blood, and following with infusion of the same amount of PEG–Hb solution. This procedure was repeated until the desired hematocrit was reached. Venous blood samples were collected to determine hematological parameters (total, RBC and plasma hemoglobin, RBC, WBC, and reticulocyte count), renal parameters (serum urea nitrogen and creatinine), hepatic parameters (serum glutamic oxaloacetic and pyruvic transaminase), coagulatory parameters (fibrinogen, prothrombin time, activated partial thromboplastin time, and platelet count), and serum electrolytes (sodium, potassium, and chloride). With respect to survival times after infusion of PEG–Hb, the dog which underwent 10% ET (final hematocrit 27.2%) was alive after 5 months, and the dog with a 50% ET (final hematocrit 19.6%) was alive after 5 months, and the dog with a 80% ET (final hematocrit 5%) was alive after 3 months. But dogs which received 80% ET with Hespan died after 7–10 days. The plasma half-life ($T_{1/2}$), which is the time for 50% of the initial plasma hemoglobin to disappear, was in the range of 32.1–45.4 hours depending on the amount of PEG–Hb infused. Hemoglobinuria was observed and urinary excretion of hemoglobin amounted to 8.6–13.9% of net hemoglobin infused for 48 hours after ET. Clearance of the rest of the infused hemoglobin was shown to be primarily accomplished through the reticuloendothelial system (RES).[26] Unfortunately, it was not determined whether the excreted hemoglobin was Hb dimer, Hb tetramer, PEG–Hb dimer, or PEG–Hb tetramer. In the PEG–Hb group, the hematocrit dropped after ET and then increased rapidly within the first 48 hours. By the fourth week, the hematocrit returned to the pre-ET range. For the Hespan group, which survived for 7 days, an increase in the hematocrit after ET was observed for only 24 hours and it decreased thereafter. RBC hemoglobin changed in parallel with hematocrit. WBC counts in the PEG–Hb group remained within the normal range. Serum electrolytes such as sodium, potassium, and chloride remained normal throughout the study period. Reticulocyte counts in the PEG–Hb group, which is an indicator of hematopoietic activity, showed an elevation after 5 days to 2 weeks. However, the Hespan group showed maximum elevation in reticulocytes at 2 days, which decreased thereafter. Thus it was suggested that bone marrow function was well preserved during the 2-week period post-ET with lower hemoglobin levels in the circulation. The long half-life of PEG–Hb, the authors also suggested, may be an effective stimulus for increased activity of erythropoietic function.

The influence of the two variables—rate of disappearance of plasma hemoglobin (in the form of PEG–Hb) and the rate of generation of new RBC—casts an interesting question regarding how long a half-life is desired for an appropriate red cell substitute such as PEG–Hb. The reported half-life of PEG–Hb is about 36–40 hours in dogs.[25,27] However, the efficacy of oxygen delivery was decreased due to the fact that PEG–Hb slowly turned into MetHb, which is not only incapable of carrying oxygen but also causes a substantial drop in P_{50}. This triple effect [i.e., the physical breakdown (as measured in half-life), the formation of MetHb, and the decrease in P_{50}] contributes to a loss in oxygen delivery. Matsushita *et al.*[27] observed that at the half-life (36 hours in dogs) the MetHb level rose to 25% (from 8%), and P_{50} declined to 12 mm Hg (from 22 mm Hg). Thus the effective oxygen-carrying capacity of PEG–Hb is not 50% of the initial at half-life, but is approximately 22% [= 50% × 12/22 × (100 − 25)/(100 − 8)]. In other words, the effective concentration of plasma PEG–Hb (assuming a 6% solution) at half-life is 1.32 $g\,dl^{-1}$ (= 6 $g\,dl^{-1}$ × 0.22). As the decline of P_{50} is dependent on MetHb formation, it becomes very important to keep MetHb formation low *in vivo*.

Nonetheless, according to Matsushita *et al.*,[25] RBC Hb concentration bounced back to 5–9 $g\,dl^{-1}$ (average 7 $g\,dl^{-1}$) at 36 hours after 80% ET. The sum of effective plasma Hb and regenerated RBC Hb is then 8.32 $g\,dl^{-1}$ (= 1.32 + 7). Although the contribution from the RBC Hb is substantial, total Hb concentration is still lower than 10 $g\,dl^{-1}$, which is considered to be essential by many to maintain health. Therefore, one can speculate that if the effective PEG–Hb is close to 50% of the original infusion at half-life (by maintaining the original P_{50} and MetHb), then the total effective Hb concentration will be 10 $g\,dl^{-1}$ [= 6 $g\,dl^{-1}$ × 0.5 (of PEG–Hb) + 7 $g\,dl^{-1}$ (of RBC Hb)]. Hence if the PEG–Hb does not autoxidize, the half-life of 36 hours in dogs is satisfactory.

On the other hand, this conclusion is deduced from the data at 80% ET which is not likely to occur in a real clinical situation. A more likely scenario in which a red cell substitute is wanted will be the case of 20–40% loss in blood volume, which may allow one to suppose that the current PEG–Hb products will work well.

A similar, but independent, study showed the comparable result[23,12] that 70% ET in rats with SC–PEG–bHb had a half-life ranging from 10 to 13 hours, and MetHb measured at half-life from 6–18% (Table 2). This work showed that with the uniquely stable linkage of the urethane bond, the integrity of PEG–Hb could be improved while attaching less PEG. SC–PEG–bHb also showed some improvement over other types of PEG–Hb with respect to the rate of MetHb formation. However, whether having less PEG on hemoglobin endows any benefit with respect to overall performance, including both efficacy and safety, has not yet been determined. Less PEG on the protein surface may raise the chance of exposure of the protein surface that may elicit an immune response. On the other hand, there is a limitation to the number of PEGS attachable because of the increase in viscosity and the drop in P_{50}.

12.6. TOXICITY OF PEG–HEMOGLOBIN

Since Rabiner *et al.*[2] reported that a solution of stroma-free hemoglobin does not exert adverse effects on renal functions, the possibility of using hemoglobin as an oxygen carrier for anemic patients has been actively investigated. Hemoglobin test samples were often contaminated with high levels of impurities and endotoxins mostly originating from red cell membrane particles, so the test results have been obscured and the conclusions controversial. However, after significant improvement in the purification process, most of the acute toxicity-related indications started to diminish. Exchange transfusion of one-third of the blood in rabbits with purified bovine hemoglobin, free of endotoxins, phosphatidyl ethanolamine, and any large polymers, showed no thrombocytopenia or leukopenia, no activation of coagulation or complement, and no hypoxemia, although hemoglobinuria was sustained for an hour.[28]

However, other investigations unveiled indications that hemoglobin itself may have a toxic influence on a variety of cellular mechanisms, including possible

Table 2. *In Vivo* Performance of PEG–Hb

PEG–Hb	Half-life (hours)	MetHb (%) at half-life	Animal studied
SOA–PEG–Hb[20]	N.A.[a]		
bSOA–PEG–Hb[31]	8.2–13.1		Rat
SS–PEG–Hb[22]	32.1–45.4	25	Mongrel dog
SS–PEG–Hb[14]	12.8		Rat
BHC–PEG–Hb[33]	N.A.		
SC–PEG–bHb[32]	9.6–13.0	6–18	Rat
bSC–PEG–bHb[32]	18.8	29	Rat

[a]Not available.

activation of the complement pathway, intravascular coagulation, cardiovascular vasoconstriction, and lipid peroxidation. These harmful effects were found to be caused by the primary oxygen metabolites such as the superoxide anion and hydrogen peroxide, which are intrinsically moderately toxic, while the true oxygen radical-mediated toxicity was observed when these species interacted in the presence of iron to form the highly toxic hydroxyl radical.[29] These findings vary in severity depending on the degree of hypoxicity and hypotension of the model under study. Nonetheless, it has been found that iron is the moiety responsible for the cellular damage since the iron binding protein, transferrin, and the iron chelator, deferoxamine, prevent hemoglobin from inducing such peroxidation.[30] It is these current findings of possible hemoglobin-related toxicity that prompted the U.S. Food and Drug Administration to recognize the need to point out details of studies to be observed by the developers of hemoglobin-based oxygen carrier.[31] However, in light of oxygen radical-related toxicity, the future administration of any hemoglobin-based oxygen carrier may require a co-infusion of superoxide dismutase (SOD) and/or catalase (CAT), or preferably, in a less immunogenic form as PEG-SOD and PEG–CAT.

Toxicity of PEG–Hb as an organ perfusate has been tested in a 24-hour hypothermic perfusion preservation of canine livers.[32] As a control, Oxypherol solution (perfluorotributylamine) was used. Following the removal of donor's liver, it was connected to a perfusion system, LPS-11 Organ Preservation System (Nikkisco Co., Japan). After 24 hours, allograft function was evaluated following indications such as bile excretion, consciousness level, activated coagulation time, and survival. The PEG–Hb used was prepared to contain 3.6–3.9 g dl^{-1} of hemoglobin, and 3.6% maltose and 1% glucose in physiological saline with 14.4 mM of potassium ion to ameliorate hypokalemia, which commonly occurs in this situation. Bile excretion started immediately after revascularization of the portal vein in all animals that received livers preserved in PEG–Hb, while 25% of the Oxypherol group demonstrated bile excretion. All animals in the PEG–Hb group maintained clear consciousness while 20% of the Oxypherol group were clear. Activated clotting time reverted to normal in the PEG–Hb group while the Oxypherol group showed little improvement. With respect to graft survival, all in the PEG–Hb group survived more than 24 hours and one survived for 7 days, while none of the Oxypherol group survived more than 24 hours. Histological findings showed that the Oxypherol group exhibited marked vacuolization and atrophy of hepatocytes, while in the PEG–Hb group cell structure was almost completely preserved, although there was mild hepatocellular atrophy and vacuolization.

Exchange transfusion-related toxicity has been studied in mongrel dogs where 80% of the blood was replaced with PEG–Hb (P_{50} = 22 mm Hg, $T_{1/2}$ = 36 hours, MW = 90,000).[10] Despite the high dosage of PEG–Hb, essentially no adverse effects were seen in hematological renal, hepatic, and coagulation system functions.[10,23] All the animals receiving PEG–Hb survived for more than 10 months. The reticulocyte count showed a high elevation in the early period, while the Hespan group (control) was slow in regaining normality in both reticulocyte count and hematological functions. However, at two weeks post-ET, vacuolized cells were observed in the renal tubular epithelium. But renal lesions were completely reversible and no abnormality was observed in renal functions.

12.7. CONCLUSION

As great as is the need for an oxygen-carrying red cell substitute, so also is the concern for the safety of the product paramount. Iron-related toxicity may be reduced by modifying the hemoglobin to reduce the tendency of iron to lose its place in the heme pocket. Possible vasoconstrictor activity of hemoglobin due to inhibiting vasorelaxation factors is still under investigation, and whether PEG–Hb will be able to attenuate this effect is not yet known. Other injuries caused by the breakdown products of PEG–Hb have been shown to be minimal and reversible, thus suggesting that the half-lives of currently-tested PEG–Hbs are probably long enough. However, there are no published results showing that a PEG–Hb with longer half-life than the current ones elicits less organ injury. The PEG–Hbs so far developed and tested appear to have more positive indications over other types of modified hemoglobin with respect to both efficacy and safety. PEG–Hbs may not have the highest P_{50} compared to other preparations, but when considered in conjunction with circulation time the overall oxygen delivery of PEG–Hb proves better. On the other hand, MetHb formed *in vivo* is a factor to be concerned about as oxygen delivery is correspondingly reduced. Further research on the formulation of PEG–Hb may be well rewarded if the efforts are so directed as to minimize the tendency for autoxidation.

REFERENCES

1. J G. Riess and M. LeBlanc, in *Blood Substitutes* (K. C. Lowe, ed.), pp. 94–129, VCH, New York (1988).
2. S. F. Rabiner, J. R. Helbert, H. Lopas, and L. H. Friedman, *J. Exp. Med. 126*, 1127 (1967).
3. S. C. Tam, J. Blumenstein, and J. T. F. Wong, *Proc. Natl. Acad. Sci. U.S.A. 73*, 2128 (1976).
4. J. E. Baldwin, B. Gill, and J. P. Whitten, *Tetrahedron 37*, 1723 (1981).
5. K. Ajisaka, *Ajinomoto KK., 82-04281E/03* (1982).
6. K. Iwasaki and Y. Iwashita, *Artif. Org. 10(5)*, 411 (1986).
7. Y. Iwashita, A. Yabuki, K. Yamaji, K. Iwasaki, T. Okami, C. Hirata, and K. Kosaka, *Biomater. Artif. Cells Artif. Org. 16(1–3)*, 271 (1988).
8. P. Labrude, P. Mouelle, P. Menu, C. Vigneron, E. Dellacherie, M. Leonard, and J. L. Tayot, *Int. J. Artif. Org. 11(5)*, 393 (1988).
9. M. Leonard and E. Dellacherie, *Biochem. Biophys. Acta 791*, 219 (1984).
10. M. Matsushita, A. Yabuki, P. S. Malchesky, H. Harasaki, and Y. Nose, *Biomater. Artif. Cells. Artif. Org. 16(1–3)*, 247 (1988).
11. A. Yabuki, K. Yamaji, H. Ohki, and Y. Iwashita, *Transfusion 30*, 516 (1990).
12. S. Zalipsky, R. Seltzer, and K. Nho, *Polymer Preprints 31(2)*, 173 (1990).
13. D. Zygmunt, M. Leonard, F. Bonneaux, D. Sacco, and E. Dellacherie, *Int. J. Biol. Macromol. 9*, 343 (1987).
14. J. W. Payne, *Biochem. J. 135*, 867 (1973).
15. K. Bonhard and V. Boyson, *Dtsch. Anslegeschrift 24*, 885 (1976).

16. F. DeVenuto and A. Zegna, *J. Surg. Res. 34*, 205 (1982).
17. H. I. Friedman, F. DeVenuto, B. D. Schwartz, and T. J. Nemeth, *Surg., Gynecol. Obstet. 159*, 429 (1984).
18. N. Kothe, B. Eichentoff, and K. Bonhard, *Surg., Gynecol. Obstet. 161*, 563 (1985).
19. D. H. Marks, J. E. Lynet, R. M. Letscher, R. P. Teneyck, A. D. Schaerle, and G. T. Makovec, *Military Med. 152*, 265 (1987).
20. L. R. Sehgal, S. A. Gould, A. L. Rosen, H. L. Sehgal, and G. S. Moss, *Surgery 95*, 433 (1984).
21. A. Abuchowski, T. van Es, C. Palczuk, and F. F. Davis, *J Biol. Chem. 252(11)*, 3578 (1977).
22. F. F. Davis, T. van Es, and N. C. Palczuk, *U.S. Patent #4,179,337* (1979).
23. S. Zalipsky, R. Seltzer, and K. Nho, *Polymeric Drugs and Drug Delivery Systems* (R. L. Dunn and R. M. Ottenbrite, eds.), pp. 91–100, ACS Symposium Series 469 (199).
24. K. Ajisaka and Y. Iwashita, *Biochem. Biophys. Res. Commun. 97(3)*, 1076 (1980).
25. M. Matsushita, Y. Iwashita, and K. Iwasaki, *Trans. Am. Soc. Artif. Internal Org. 32*, 490 (1986).
26. P. E. Keipert and T. M. S. Chang, *Trans. Am. Soc. Artif. Internal. Org. 29*, 329 (1983).
27. M. Matsushita and A. Yabuki, *Trans. Am. Soc. Artif. Internal Org. 33*, 352 (1987).
28. M. Feola, J. Simoni, R. Tran, and P. C. Canizaro, *Biomater. Artif. Cells Artif. Org. 16(1–3)*, 217 (1988).
29. B. Halliwell and J. M. C. Gutteridge, *Arch. Biochem. Biophys. 246*, 501 (1986).
30. A. E. Faassen, S. R. Sundby, S. S. Panter, R. M. Condie, and B. E. Hedlund, *Biomater. Artif. Cells Artif. Org. 16(1–3)*, 93 (1988).
31. *Draft Points to Consider in the Safety Evaluation of Hemoglobin-Based Oxygen Carriers*, U.S. FDA Center for Biologics Evaluation and Research (1990).
32. S. Fuchinoue, I. Nakajima, T. Agishi, S. Teraoka, T. Kawai, H. Honda, and K. Ota, *Trans. Am. Soc. Artif. Internal Org. 33*, 390 (1987).

13

Bovine Collagen Modified by PEG

W. RHEE, J. CARLINO, S. CHU, and H. HIGLEY

13.1. INTRODUCTION

Currently much interest has been shown in attaching poly(ethylene glycol) (PEG) to proteins as a means of increasing solubility and serum lifetime.[1–3] Modification of proteins by PEG has been used to reduce the immunogenicity of various enzymes.[4–6] PEG is well known to be nontoxic, nonantigenic, biocompatible, and soluble in water and organic solvents.[7–9]

Zyderm® Collagen Implant (fibrillar bovine atelocollagen) and Zyplast® Implant (glutaraldehyde crosslinked bovine atelocollagen) have become widely utilized biomaterials in dermatology and plastic surgery for soft tissue augmentation. Collagen Corporation's proprietary purification process produces injectable bovine corium collagen products with low immunogenicity capable of a clinically predictable, 6–9-month correction of facial rhytids and contour defects.[10] Nevertheless, approximately 3% of the general population exhibits a preexisting hypersensitivity response to bovine collagen on skin test,[11] and both patients and practitioners express interest in increasing the persistence of collagen implants. The covalent attachment of activated PEG to available lysines in peptides and proteins presumably masks antigenic epitopes and may mimic glycosylation in increasing clearance time in the circulation. To our knowledge pegylation of a major structural protein or extracellular matrix molecule has not been assessed previously.

In this study we have covalently attached PEG to fibrillar collagen in an attempt to generate biomaterials with greater persistence *in vivo* and lowered antigenicity compared to fibrillar collagen alone. Therefore, PEG with a molecular weight of 5000 was activated by two steps. In the first step, PEG–glutarate was prepared from PEG

W. RHEE • Research and Development, Collagen Corporation, Palo Alto, California 94303.
J. CARLINO, S. CHU, and H. HIGLEY • Celtrix Laboratories, Palo Alto, California 94303.

Poly(Ethylene Glycol) Chemistry: Biotechnical and Biomedical Applications, edited by J. Milton Harris. Plenum Press, New York, 1992.

and glutaric anhydride, then reacted with *N*-hydroxysuccinimide to form a succinimidylpoly(ethylene glycol) glutarate. The succinimidyl ester is then capable of reacting with a free amino group present on collagen to form a PEG–collagen conjugate. Physical, chemical, biocompatibility, and persistence analyses were performed on the composite to compare it to Zyderm® Collagen Implant (ZCI) and to Zyplast® Implant (ZI). Implantation experiments comparing these three materials in the rat subcutaneum and the porcine dermis were conducted to assess *in vivo* performance. The humoral and cellular immune responses to PEG–collagen were investigated in a guinea pig immunization and challenge model. The results of these experiments are presented here.

13.2. PREPARATION OF TEST MATERIALS

13.2.1. Preparation of Bovine Atelopeptide Collagen in Solution (CIS)

A highly purified, pepsin-solubilized, bovine corium collagen was used in these experiments. Bovine hide was softened and depilated by soaking in acetic acid. The hide was then comminuted and dispersed in HCl, pH 2. Pepsin was then added to the dispersion and the mixture was allowed to incubate for several days at approximately 20 °C. After the denatured enzyme was removed, the solution was then purified. The final concentration of solution was 3 mg ml^{-1} of atelopeptide bovine collagen in dilute aqueous HCl, pH 2. This solution is subsequently referred to as CIS.[13]

13.2.2. Preparation of Fibrillar Collagen

Fibrillar collagen was precipitated from CIS by rapid neutralization (pH 7.2) in dilute phosphate buffer as previously described.[12,13] The precipitated fibrillar collagen was separated from the supernatant by centrifugation and homogenized with phosphate buffered saline (PBS). The final protein concentration of the aqueous fibrillar suspension was 35 mg ml^{-1} in PBS at pH 7.2.

13.2.3. Preparation of Glutaraldehyde Crosslinked Collagen

This preparation has been described.[12,14] Briefly, glutaraldehyde was added to fibrillar collagen to provide a final glutaraldehyde concentration of less than 0.01%. This suspension was incubated for approximately 16 h at 15–17 °C. Following glutaraldehyde treatment, fibrils were harvested by centrifugation, washed in PBS at pH 7.2, recentrifuged, and finally suspended in PBS at a collagen concentration of approximately 35 mg ml^{-1}.

13.2.4. Preparation of PEG–Collagen

The method for the preparation of activated PEG followed the procedure of Abuchowski.[5] In brief, PEG was converted to the active ester in two steps. In the first step, PEG–glutarate was prepared from monomethyl–PEG (5000 dalton, Aldrich Chemical) and glutaric anhydride. The second step was the synthesis of *N*-hydroxy glutarate succinimide ester of PEG–glutarate. The final product, PEG–succinimidyl (PEG–sg), was obtained by recrystallization from cold benzene and petroleum ether.

Activated PEG reacts predominantly with lysine groups in the collagen molecule and its coupling process is illustrated in Figure 1. CIS was mixed with 0.2 M phosphate buffer to elevate the pH to 7.4. Next, a fourfold molar excess of PEG–sg to all possible lysine residues in collagen molecule was dissolved in water for injection and sterile-filtered. The PEG–sg solution was then added to the collagen solution, and the mixture allowed to stand at 20 °C for about 15 h. The solution was then centrifuged, and the resulting pellet of reconstituted fibrils collected and washed with PBS three times to remove residual PEG. The yield of PEG–collagen was 85% of the starting material (CIS). The final product was pure white and superficially resembled ZCI. However, it exhibited distinctive physicochemical characteristics.

13.3. PROPERTIES OF PEG–COLLAGEN

PEG-collagen prepared in line with Section 13.2.4. was characterized and compared with ZCI (Section 13.2.2) and ZI (Section 13.2.3.). The results of physical and chemical analysis of these materials are summarized in Table 1.

13.3.1. Extrusion Test

This assay measures the force required to extrude the test composition through a 30 gauge needle used in dermal treatment. The results are shown in Figure 2. As can

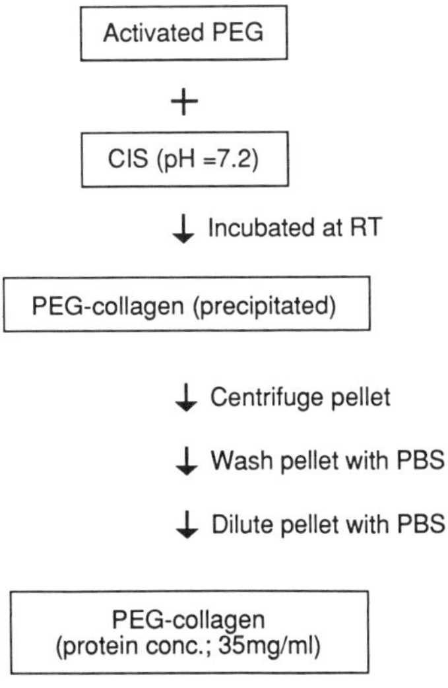

Figure 1. Flow chart of coupling process.

Table 1. Physical and Chemical Properties of PEG–Collagen versus ZCI and ZI

Tests	PEG–Collagen	ZCI	ZI
Extrusion plateau	10 N	20–30 N	10–15 N
Denaturation temperature (from differential scanning calorimetry) (°C)	58–60 singlet	45–55 multiplet	65–75 singlet
TNBS (free lysines per 30)	25	30	26–27
Trypsin sensitivity (% digested)	1–2	3–10	1–2
Collagenase sensitivity (% digested)	46	66	2

be seen from the graph, the extrusion force (in newtons) was monitored as a plunger travels at the speed of 5 cm per min (Figure 2). ZCI was extruded smoothly into air, requiring a force of about 20–30 newtons. ZI was extruded with a mild trace of spiking. At the plateau, ZI required about 10–15 newtons for extrusion. In contrast, PEG–collagen demonstrated a very low extrusion force (8–10 N), with little or no spiking, indicating even flow of the material with easier injectability.

13.3.2. Differential Scanning Calorimetry

Several investigators[15–17] have studied the denaturation of collagen using differential scanning calorimetry (DSC). In this study, the DSC was performed on a TA3000 calorimeter equipped with a DSC 20 cell (Mettler Instrument Corp., Hightstown, NJ) and calibrated using indium, lead, and zinc standards. Samples containing 1.0–1.2 mg collagen in about 30 μl were introduced into sample pans and sealed. PBS, which was used to suspend the fibrils, served as reference samples in these analyses. At heating rates of 10 °C per min, fibrillar collagen exhibited two

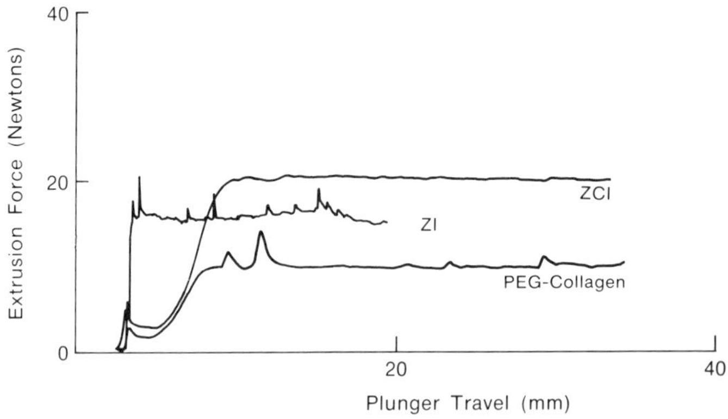

Figure 2. Extrusion of collagen formulations through fine-gauge needles. Extrusion through 1.25-cm-long needles, 150-μm ID into air. All materials contained collagen at 25 mg ml^{-1} in PBS.

major denaturational transitions at about 45 and 55 °C. In contrast, PEG–collagen and ZI exhibited single peaks at 58–60 °C and 65–75 °C, respectively (Figure 3). These results suggest that reconstituted fibrillar collagen contains a heterogeneous population of fibril sizes, possibly including molecules in a nonfibrillar state, while fibers of PEG–collagen and ZI are more uniform and stable.[17]

13.3.3. TNBS Assay

The reagent 2,4,6-trinitrobenzenesulfonic acid (TNBS) appears to offer a simple procedure for measuring available lysine in PEG-modified collagen due to its specificity of reaction with primary amino groups.[18,19] In this study, we have indirectly determined the number of reacted lysine groups in derivatized collagen samples by measuring free amino group concentration with TNBS. The reaction product absorbs light at 335 nm. The absorbances of the derivatized collagens (PEG–collagen and ZI) were compared with that of a nonderivatized collagen (ZCI). The results (Table 1) indicate that more than 15% of the lysine residues are pegylated when coupled with succinimidyl-activated PEG while about 10% of free lysine residues were derivatized in glutaraldehyde crosslinked collagen.

13.3.4. Protease Sensitivity

The purpose of these experiments was to compare the sensitivity of ZCI, ZI, and PEG–collagen to the proteases trypsin and collagenase as previously described.[14]

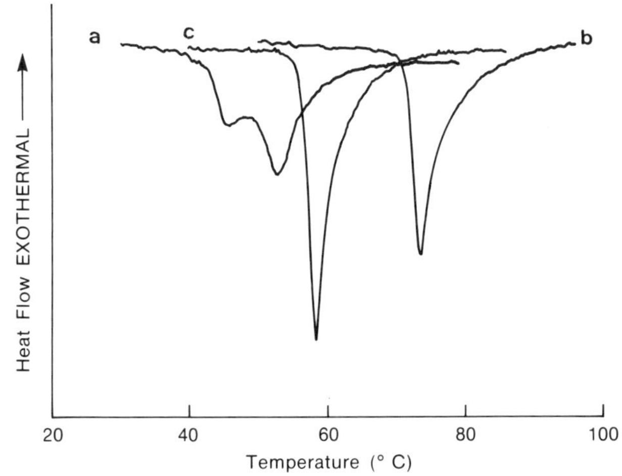

Figure 3. Heat denaturation of bovine collagen formulations: (a) ZCI, 35 mg ml^{-1} in PBS; (b) ZI, 35 mg ml^{-1} in PBS; (c) PEG–collagen, 35 mg ml^{-1} in PBS. Experiments were carried out on the Mettler TA 3000 calorimeter at a heating rate of 10 °C per min.

Such a comparison indicates the relative potential of the different collagens for degradation *in vivo* and thus, by inference, their relative clinical persistence.

Bacterial collagenase cleaves native (helical) and denatured collagens. The sensitivity of each composition to collagenase was measured.[14] The result shows that ZCI was 66% digested, compared to 2% for ZI and 46% for PEG–collagen. The drastically reduced sensitivity of ZI and PEG–collagen may be due to the highly crosslinked collagen matrix. Trypsin cleaves polypeptides at lysine and arginine residues. It is unable to attack the helical portion of collagen molecules in their native conformation, and thus sensitivity to this protease gives a measure of the degree of nonhelical collagen present in samples. ZCI was 3–10% sensitive, ZI was 1–2% sensitive, and PEG–collagen was about 1–2% sensitive. Sensitivity to trypsin may correlate with sensitivity to endogenous proteases following implantation.

13.3.5. Electron Microscopy

Transmission electron microscopy was performed to examine the PEG–collagen and to compare the ultrastructure of this material to ZCI. Samples of both materials were fixed in 2% glutaraldehyde in 0.2 M cacodylate buffer, post-fixed in 2% osmium textroxide, and stained en bloc with 2% uranyl acetate. They were then dehydrated in a graded series of ethanols and finally embedded in Maraglas. 50-nm sections were made, picked up on copper grids, and stained with 5% uranyl acetate prior to viewing in an electron microscope.

Native type I collagen has a very distinctive banding pattern, with repeating units of cross striations 64 to 67 nm along the length of the fibrils. This periodicity is a result of the overlapping of collagen molecules to form the fibrils. ZCI (Figure 4b), which has undergone processing to remove the telopeptides, still retains a banding pattern similar to native fibrils but can at times appear in an oblique pattern. In contrast, the PEG–collagen (Figure 4a) contains fibrils of a smaller diameter in comparison to ZCI. Also apparent is the lack of the characteristic bands and the association of small, electron-dense spheres in association with the PEG–collagen fibrils.

13.4. *In Vivo* BIOCOMPATIBILITY

Biomaterials are typically evaluated by implantation in experimental animals. The models discussed here have in our experience been useful to determine biocompatibility and have for the most part been good predictors of clinical performance and safety.

13.4.1. Rat Subcutaneous Model

To measure implant persistence and inflammatory potential *in vivo*, Sprague-Dawley rats were injected bilaterally in suprascapular subcutaneous sites with 0.5

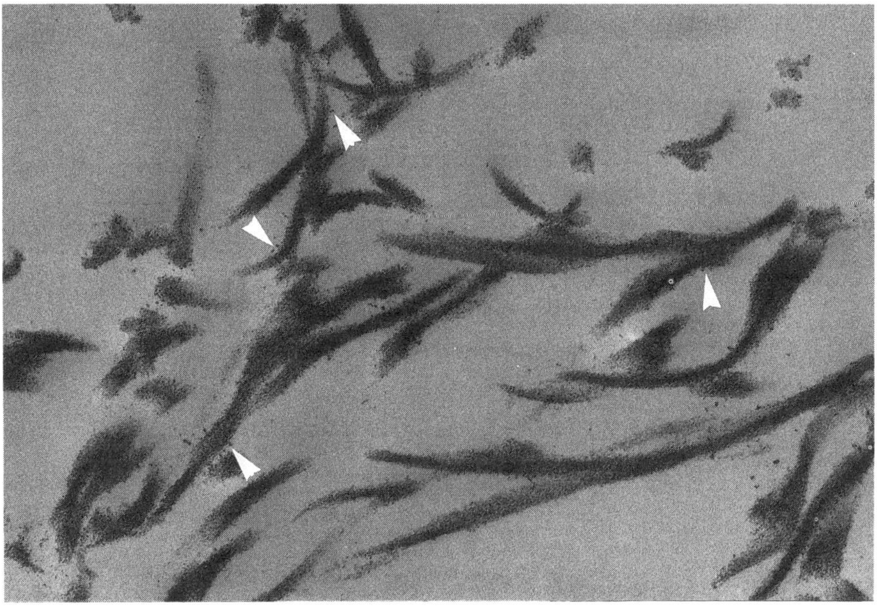

a

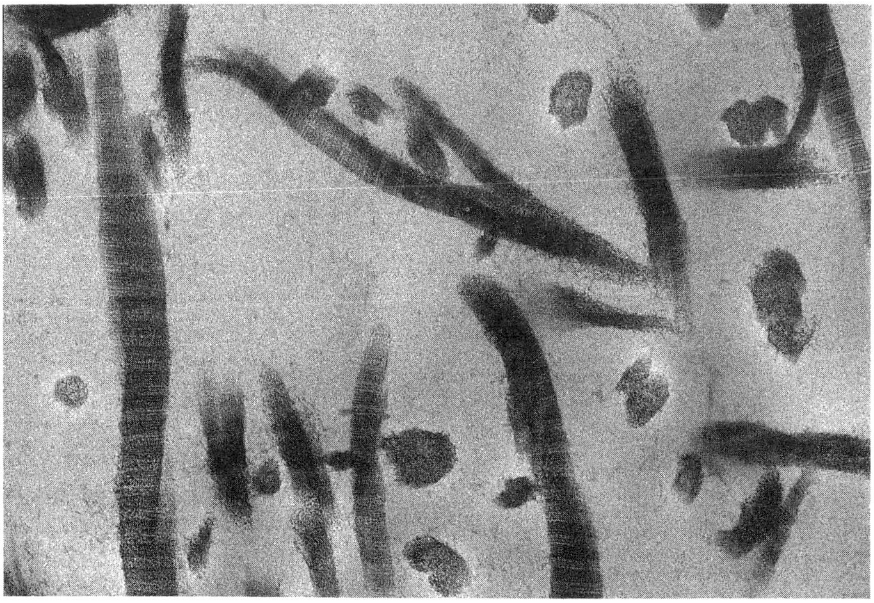

b

Figure 4. (a) Electron micrograph of PEG–collagen. Note the thin fibril diameter, indistinct banding pattern, and associated globular electron dense profiles (arrows). (b) Electron micrograph of ZCI. Compare fibril diameter and banding pattern. Uranyl acetate–lead citrate. 46,800×.

ml of PEG–collagen, ZCI or ZI delivered through a 30 gauge needle. Implants were removed at 7, 14, 30, 90, and 180 days post-implantation, weighed, fixed, and processed for conventional histologic evaluation.

In the rat subcutis, the PEG–collagen implants retained approximately 70% of their onset weight (Figure 5) over a 180-day period, an intermediate persistence value compared to ZI (>100%) or ZCI (<50%) implants. ZCI has been shown to exhibit a syneresis or compaction by buffer loss in this model, explaining the rapid loss of onset wet weight. Crosslinked collagens, like ZI, have been shown to be initially much more resistant to shrinkage in this model and, in addition, induce more new host connective tissue over time. The wet weight response of PEG–collagen may be partially explained by the affinity of the hydrophilic polymer component for the suspension buffer and tissue fluids. *In situ*, PEG–collagen implants were moderately colonized by host connective tissue and vasculature (Figure 6). All collagen-based materials induced a mild lymphohistiocytic inflammatory response that resolved within several days, eventually resulting in good integration into the surrounding host tissues. Of importance was the finding that there was no progressive or delayed inflammatory response to PEG–collagen even at 6 months, as might be speculated if a slow loss of "polymer-protection" exposed or altered collagen epitopes.

13.4.2. Porcine Intradermal Model

To assess the intradermal response of the PEG–collagen, young adult domestic pigs were anesthetized and injected intradermally in the flank with 0.1 ml of all three implant materials at days 0, 7, 14, and 30 before necropsy. At each time point all sites were examined to determine if the individual implant was visible or palpable. Finally, implant sites were harvested, fixed, and processed for histologic analysis.

In the pig, which has a skin thickness and appendageal structures comparable to the human, the intradermal position and appearance of PEG–collagen was similar

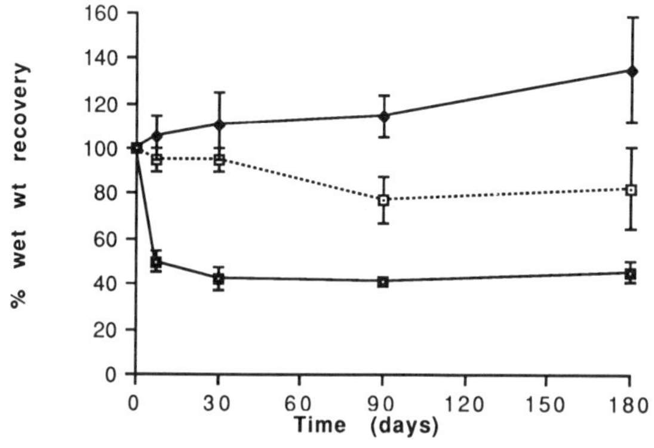

Figure 5. Wet weight persistence of: ---□, PEG–collagen; ■–■, ZCI; and ◆–◆ ZI in the rat subcutaneum.

to ZI. These implants formed superficial cohesive profiles on injection that were well retained within the dermis during the 30-day course of the experiment (Figure 7). In contrast, in this animal model, ZCI "fingers" down through the dermal collagen bundles into the subcutaneum within several weeks. Fibroblast colonization of all three materials was noted. There was some variable implant-directed histiocytic response to PGC in the porcine dermis, but this was moderate and seemingly not of a progressive nature.

13.4.3. Guinea Pig Immunologic Assessment

The preparation of ZCI includes the removal of the telopeptides from the collagen, thereby reducing the immunogenicity. This results in collagen which is only weakly antigenic when implanted in humans or laboratory animals. In order to preclinically evaluate whether pegylation of ZCI can further reduce the humoral and cellular immune responses to ZCI, these responses were amplified using adjuvant. When ZCI is mixed with adjuvant and implanted into guinea pigs, humoral and cellular responses can be readily detected. Therefore, the immunization of guinea pigs with adjuvant mixed with ZCI or PEG–collagen was used to assess whether pegylation reduces the antibody and delayed-type hypersensitivity (DTH) responses to bovine collagen.

Guinea pigs were given two intramuscular injections of ZCI or PEG–collagen in complete Freund's adjuvant on day 0. Serum samples were obtained by heart puncture prior to immunization and on days 14 and 30. Relative antibody levels were determined by a direct enzyme-linked immunosorbent assay (ELISA) using bovine collagen coated plates and a rabbit anti-guinea pig IgG-peroxidase conjugate. After development of the color reaction, optical densities (ODs) were read at 414 nm. Antibody titers were defined as the reciprocal serum dilution which yielded an OD above 0.1 OD units. DTH responses were evaluated at 24, 48, and 72 h after 0.1 ml intradermal injections of ZCI and PEG–collagen on both flanks of each animal. The diameter of the resulting skin wheal was measured with micrometer calipers. The sites were then excised, fixed, and processed for histologic analysis.

Pegylation of bovine collagen resulted in a decrease in both the humoral and cellular immune responses of guinea pigs to collagen. Moderate antibody titers to ZCI were apparent by day 14 in ZCI immunized animals (Table 2). In contrast, PEG–collagen immunized animals exhibited essentially no antibody response. Thirty days following immunization, anti-ZCI titers were significantly elevated in the positive controls, while the response in PEG–collagen treated animals remained modest to absent.

Normal, nonimmunized guinea pigs given an intradermal injection of ZCI exhibit no detectable DTH response. However, the cutaneous wheal and erythema characteristic of a DTH response is apparent in guinea pigs immunized with ZCI and adjuvant as described above. When the 24-hour DTH response to ZCI was compared to that of PEG–collagen, a lower response to PEG–collagen was observed in both ZCI and PEG–collagen immunized animals (Figure 8). In addition, the DTH

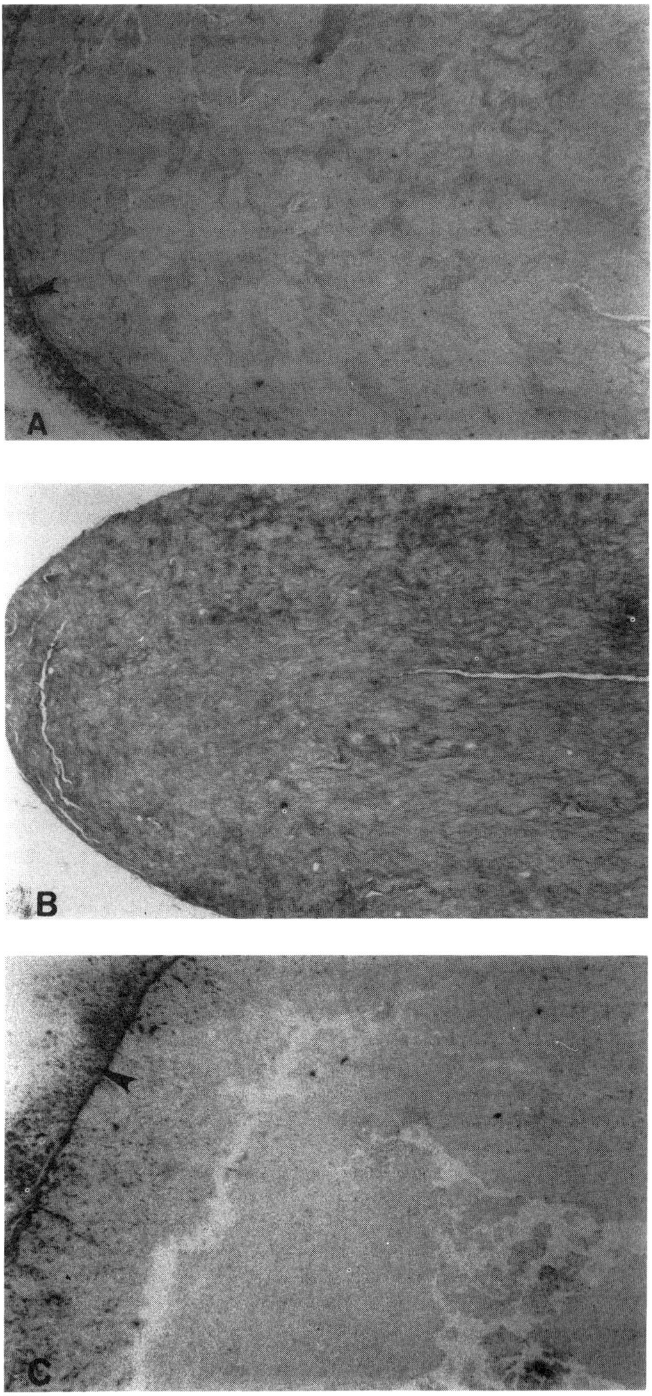

Figure 6. Histologic appearance of (A) PEG–collagen, (B) ZCI, (C) ZI after 14 days in the rat subcutaneum. Note the slight host connective tissue encapsulation (arrows), minimal inflammatory response. Trichrome. 40×.

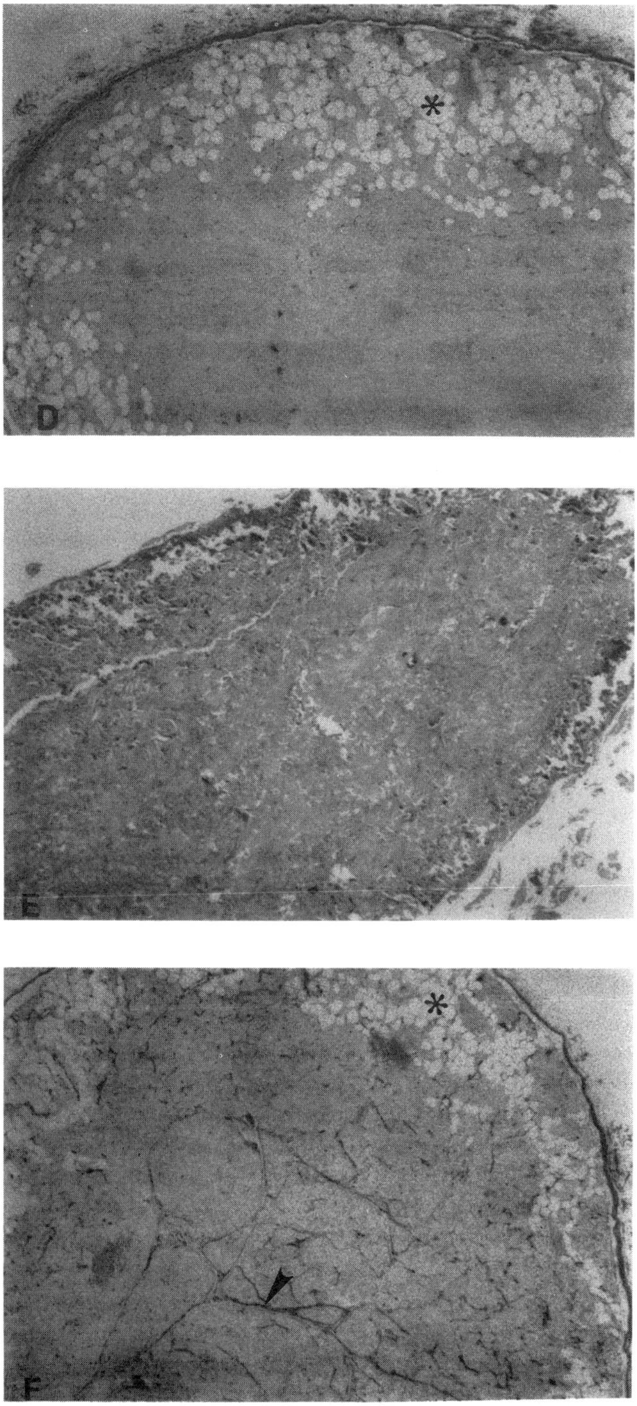

Figure 6. (*Continued*) Histologic appearance of (D) PEG–collagen, (E) ZCI, (F) ZI after 90 days in the rat subcutaneum. Note the adipose replacement of PEG-collagen and ZI(*), extensive connective tissue trabecular colonization of ZI (arrow), loss of fibrillar structure in ZCI at this time. Trichrome. 40×.

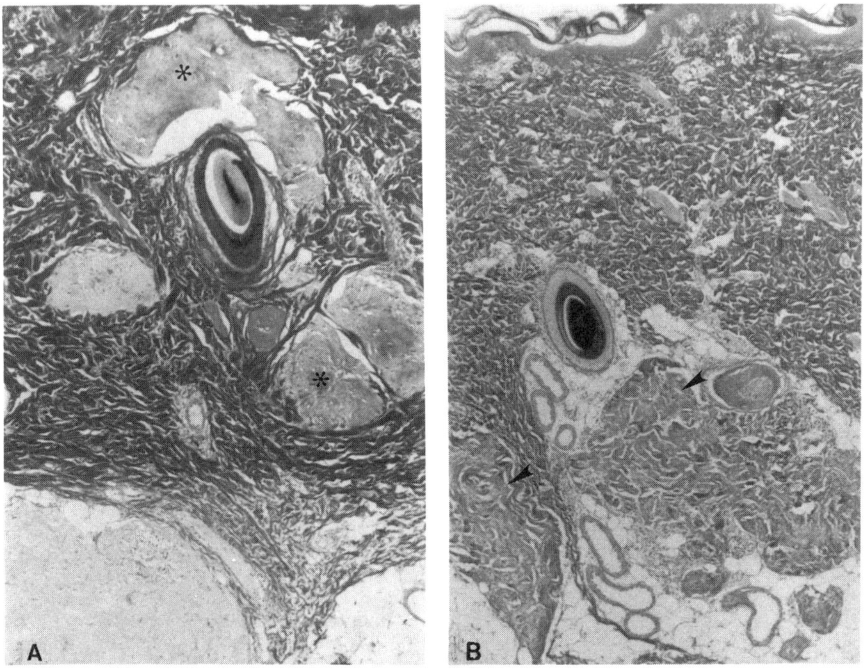

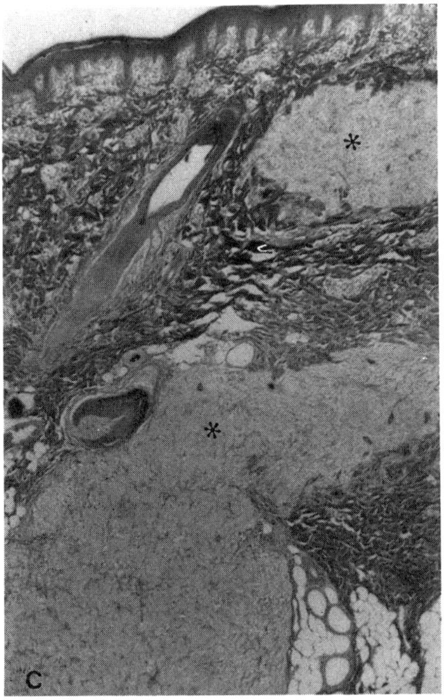

Figure 7. Histologic appearance of (A) PEG–collagen, (B) ZCI, (C) ZI after 30 days in the porcine dermis. Note retention of the PEG–collagen and ZI(*) in the superficial dermis and migration of the ZCI (arrows) into the subcutis at this time. Trichrome. 40×.

Table 2. Anti-Collagen Antibody Titers Are Reduced in Guinea Pigs Immunized with PEG–Collagen

Animal #	Treatment	Titer	
		Day 14	Day 30
1	ZCI	2560	>2560
2	ZCI	320	>2560
3	ZCI	320	>2560
4	ZCI	320	2560
5	ZCI	80	1280
6	PEG–collagen	40	640
7	PEG–collagen	0	640
8	PEG–collagen	0	160
9	PEG–collagen	0	20
10	PEG–collagen	0	0

response to ZCI in ZCI-immunized animals was vigorous, while ZCI sites in PEG–collagen immunized animals exhibited a reduced inflammatory response. Histologic evaluation of these DTH sites at 72 hours confirmed that animals immunized with PEG–collagen indeed had a reduced lymphohistiocytic infiltrate in skin test sites for both materials (Figure 9). Similarly, animals immunized with ZCI had a lower intradermal cellular immune response to PEG–collagen than to ZCI.

Thus, covalent attachment of PEG to bovine collagen significantly reduced both

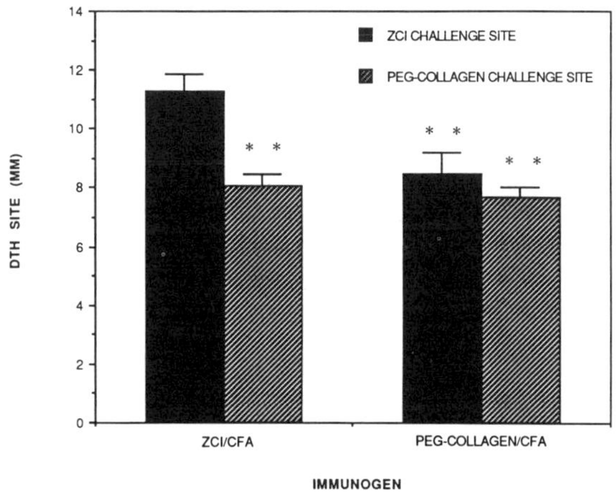

Figure 8. Decreased guinea pig DTH responses after immunization or challenge with PEG–collagen. Data represent measurements of two sites per challenge antigen on each animal (mean ± SEM) 24 h after challenge on day 31. Five animals per group. ** = significantly different from collagen-immunized, collagen-challenged group at $p \leqslant .01$.

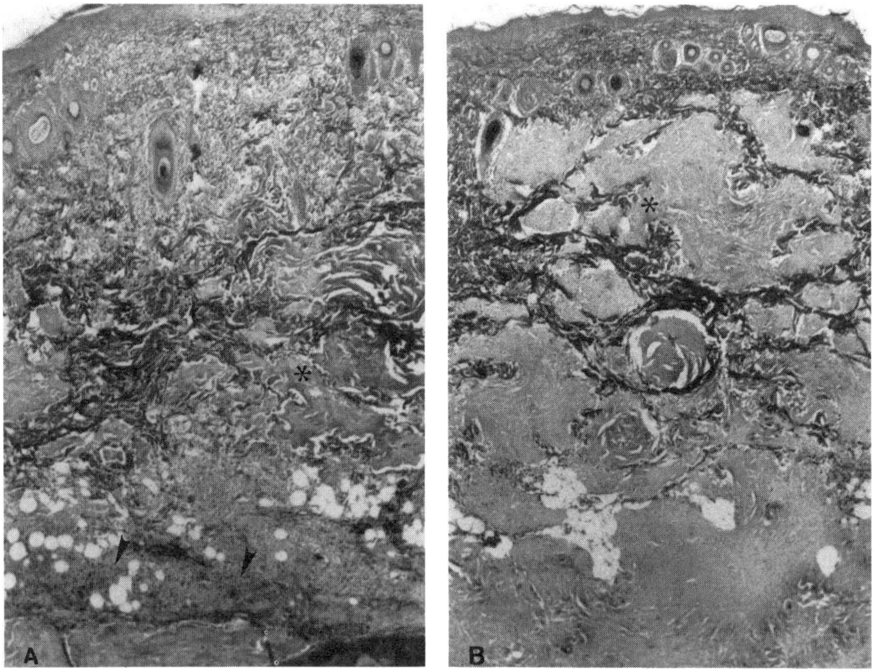

Figure 9. (A) 72-h ZCI DTH site from ZCI immunized guinea pig. Note considerable inflammatory infiltrate (arrow) near implant (*) and in subcutaneum and around dermal appendages. (B) 72-h PEG–collagen DTH site in ZCI immunized guinea pig. Note minimal inflammatory response (arrow) around implant (*) in dermis and subcutaneum. (C) 72-h ZCI DTH site in PEG–collagen immunized guinea pig. Note moderate inflammatory infiltrate (arrow) in subcutis associated with implant (*). (D) 72-h PEG–collagen DTH site in PEG–collagen immunized guinea pig. Note paucity of infiltrate associated with implant (*) in subcutaneum. Trichromes. 20×.

the humoral and cellular immune responses to bovine collagen elicited in guinea pigs by an aggressive immunization regimen.

13.4.4. Immunocytochemistry of Bovine Type I Collagen

Antibody probes specific to bovine type I collagen have been used to discriminate implant materials from other collagens and to trace the fate of ZCI and ZI in the dermis of other species.

Using immunocytochemistry, formalin-fixed, paraffin-embedded sections of onset PEG–collagen and ZCI were stained with a rabbit polyclonal anti-bovine type I collagen antibody. Prior to staining, the sample endogenous peroxidase activity was quenched with H_2O_2 and nonspecific staining blocked with normal goat serum. To unmask collagen antigenic sites denatured by fixation, the sections were digested

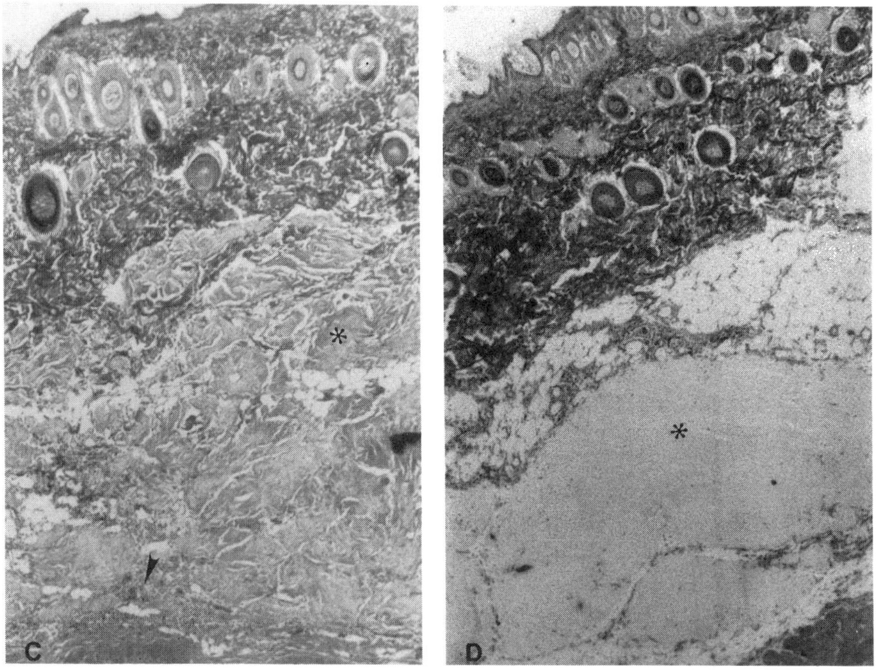

Figure 9. *(Continued)*

with 1% hyaluronidase for 1 h at 37 °C. Final detection of the antibody bound to the section was made with a biotinylated goat anti-rabbit/strepavidin peroxidase system.

Unsuccessful attempts to immunostain fixed embedded sections of onset PEG–collagen with anti-bovine type I collagen antisera demonstrated the initial unavailability of reactive epitopes in this material. ZCI, as expected, exhibited an intense reaction, indicative of bovine collagen antigen recognition. Shortly after implantation of PEG–collagen in experimental animals, however, the anti-bovine type I collagen immunoreactivity was restored, suggesting some loss of steric impedance to antibody *in vivo*.

13.5. SUMMARY

A bovine collagen–PEG polymer composite was produced with distinctive physical characteristics. Electron microscopic evaluation revealed material with a small fibril diameter, indistinct banding pattern, and electron-dense "decorations" thought to correspond to the PEG moieties. Extrusion testing, calorimetric, enzyme digestion, and lysine residue analysis suggested the unique nature of PEG–collagen by comparison to ZI, a glutaraldehyde crosslinked collagen and ZCI.

The results of the animal model testing suggest that PEG–collagen is biocompatible, persists well intradermally, and has a low inflammatory potential. PEG–collagen also was shown to have an immunogenicity even lower than the atelocollagen ZCI, probably due to some masking of the bovine collagen epitopes by the polymer. It appears that PEG–collagen and related materials are likely to be promising candidates for soft tissue augmentation with the potential for good clinical persistence and a reduced hypersensitivity response.

REFERENCES

1. J. M. Harris, *J. Macromol. Sci., Rev. Macromol. Chem. Phys. C25*, 325 (1985).
2. B. Z. Weiner and A. Zilkha, *J. Med. Chem. 16*, 573 (1973).
3. N. V. Katre, M. J. Knauf, and W. J. Laird, *Proc. Natl. Acad. Sci. U.S.A. 84*, 1487 (1987).
4. A. Abuchowski, T. van Es, N. C. Palczuk, and F. F. Davis, *J. Biol. Chem. 252*, 3578 (1977).
5. A. Abuchowski, J. R. McCoy, N. C. Palczuk, T. van Es, and F. F. Davis, *J. Biol. Chem. 252*, 3582 (1977).
6. Y. Ashihara, T. Kona, S. Yamazaki, and Y. Inada, *Biochem. Biophys. Res. Commun. 83*, 385 (1978).
7. K. Soehring, K. Scriba, M. Frahm, and G. Zollner, *Arch. Int. Pharmacodyn. Ther. 87*, 301 (1951).
8. C. G. Hunter, D. E. Stevenson, and P. L. Chambers, *Food Cosmet. Toxicol. 5*, 195 (1967).
9. H. F. Smyth, Jr., C. P. Carpenter, and C. S. Weil, *J. Am. Pharm. Assoc. 44*, 27 (1955).
10. A. Kligman and B. Pharriss (eds.), Symposium on Injectable Collagen Implants, *J. Dermatol. Surg. Onc. 14*, 1 (1988).
11. F. DeLustro, S. T. Smith, J. Sundsmo, G. Salem, S. Kincaid, and L. Ellingsworth, *Plast. Reconstr. Surg. 79(4)*, 581 (1987).
12. J. M. McPherson, D. G. Wallace, A. M. Conti, S. Sawamura, R. A. Condell, S. Wade, and K. A. Piez, *Collagen Rel. Res. 5*, 199 (1985).
13. D. G. Wallace and A. Thompson, *Biopolymers 22*, 1793 (1983).
14. J. M. McPherson, P. W. Ledger, S. Sawamura, A. M. Conti, S. Wade, H. Reihanian, and D. G. Wallace, *J. Biomed. Mater. Res. 20*, 79 (1986).
15. A. Finch and D. A. Ledward, *Biochem. Biophys. Acta 278*, 433 (1972).
16. F. Flandin, C. Buffevant, and D. Herbage, *Biochem. Biophys. Acta 791*, 205 (1984).
17. D. A. Wallace, R. A. Condell, J. W. Donovan, A. Paivinen, W. M. Rhee, and S. B. Wade, *Biopolymers 25*, 1875 (1986).
18. A. F. S. A. Habeeb, *Anal. Biochem. 14*, 328 (1966).
19. M. L. Kakade and I. E. Liener, *Anal. Biochem. 27*, 273 (1969).

14

Poly(Ethylene Oxide) and Blood Contact

A Chronicle of One Laboratory

EDWARD W. MERRILL

14.1. STUDIES OF POLYMERS OTHER THAN PEO IN BLOOD CONTACTING APPLICATIONS

This brief resumé is a personal view, offered as a historical narrative to complement the other contributions to this volume dealing with the role of PEO in blood–surface interactions.

Because of our interest in the polymer as a hemodialysis membrane, *cellophane* was studied by Lipps[1,2] and Britton[3] in 1965–1969 with heparin *ionically* bound to the surface. It was our naive hypothesis at that time that heparin would render a surface antithrombogenic. *Antithrombogenic* appeared as a description of these modified cellophanes, as assessed by prolongation of clotting via the intrinsic system. Cellophane may be considered a form of hydrogel held together by virtual (not covalent) junctions that are crystallites, in which hydrophilicity is imparted by the ether oxygens between carbons 1 and 4, 1 and 5, and the hydroxyls: secondary on carbons 2 and 3, primary on carbon 6. As became increasingly apparent through numerous studies in other laboratories, cellophane activates complement and attracts platelets, and cannot be described as a nonthrombogenic material, even with attached heparin.

Convinced at this time that high water content *hydrogel* might prove desirable if made from a synthetic polymer rather than nature's cellulose, we begin investigation of *polyvinyl alcohol* (PVA): its various forms crosslinked by aldehydes with or without heparin included in the reaction mixture. Wong[4–7] carried the major part in

---

EDWARD W. MERRILL • Department of Chemical Engineering, Massachusetts Institute of Technology, Cambridge, Massachusetts 02139.

Poly(Ethylene Glycol) Chemistry: Biotechnical and Biomedical Applications, edited by J. Milton Harris. Plenum Press, New York, 1992.

these studies, which led to materials of widely varying properties ranging from microporous tough spongelike forms to very fragile gels. Certain forms represented a set of properties, including bound heparin, that were sufficiently promising to be tested as vascular implants *in vivo* in canine models. Silliman[8] extended Wong's PVA materials to hemodialysis membranes. That occasional apparent success with the implants was less frequent than occlusion of the implant by thrombus in retrospect does not teach us much about *nonthrombogenicity*.

Suturing of the implant into the ends of the living vessel provokes the *extrinsic clotting system*, which cannot be affected by covalently bound heparin. Some antithrombogenic agent (some intervention in the extrinsic cascade) may be necessary, especially in small-diameter vascular implants. In other tests of these modified PVAs, the initial response to blood (platelet deposition and spreading) evolved usually into a stable protein layer that appeared to passify the material, as judged by transmission electron microscopy. We turned from aldehyde crosslinked PVA to radiation (electron beam) crosslinked PVA with the studies of Bray[9,10] and Peppas,[11–14] which led to very interesting, mechanically tough hydrogels possibly useful as articular cartilage. They turned out not to be nonthrombogenic, as judged by platelet deposition (*in vitro*). Parenthetically we note that in a sense PVA and poly(ethylene oxide) PEO are stoichiometric isomers, the repeat units represented by —(C_2H_4O)—, with the obvious difference that PVA owes its hydrophilicity entirely to its secondary hydroxyl (hydrogen bond *donor* and acceptor) groups, while PEO owes its hydrophilicity to its hydrogen bond *acceptor* ether oxygen atoms (neglecting the ends of the chain). We return to a comparison of PVA with PEO in Section 7.

The hypothesis that *hydrophilic* surfaces should be nonthrombogenic being refuted, if not generally, at least by these preceding examples, we examined *hydrophobic* surfaces. Weathersby[15] in 1969 showed that *poly(dimethyl siloxane)*, PDMS, crosslinked by electron beam radiation while under argon, gave longer whole blood clotting times (WBCT) than conventional silicone rubber. Weathersby analyzed surfaces by infrared and found that the polar groups, e.g., carbonyl, present in thermally cured PDMS or PDMS irradiated in air, were absent in the argon-blanketed materials, and attributed the observed prolongation of WBCT (up to 5 minutes) to this fact. Weathersby's studies were continued by Morgan[16] up to 1972. Dr. George Gifford of Children's Hospital Medical Center, Boston, and Morgan showed[17] that PDMS crosslinked by electron beam while under argon showed negligible *inflammatory* response after implantation in rats, while peroxide-cured PDMS produced significant inflammatory response. That inflammatory response does not necessarily correlate with thrombogenic potential is at present well-established, and from the perspective of 1990 we know that PDMS is not acutely thrombogenic nor is it nonthrombogenic, at least in respect to platelets which deposit and spread on even the most carefully prepared PDMS surfaces in *ex vivo* shunt studies.

Seeking a blood compatible surface on a mechanically strong elastomer, Sefton[18] transformed the surface of the hydrophobic styrene–butadiene–styrene triblock copolymer (Kraton® type rubber) by hydroxylation to make it hydrophilic. In effect Sefton created a layer of *hydrogel* at the surface of which the water-solvated chains, connected to nonsolvated polystyrene domains, could be described by the formula

(1) $—CH_2CH(OH)CH(OH)—CH_2—$

to be contrasted with a pair of PVA units

(2) $—CH_2CH(OH)CH_2CH(OH)—$

or with a pair of PEO units

(3) $—CH_2CH_2OCH_2CH_2O—$

Like PVA, the hydroxylated polybutadiene via its hydroxyl could have been used to fix heparin. The only blood contact evaluation made of the material was via a whole blood clotting test. The WBCT varied from 1.5 to 2.2 times that of a plain glass control, while the original unreacted triblock material showed values of 1.7 to 1.9. Judged by WBCT, surface hydroxylation produced no improvement, but the WBCT test is not a reliable indicator of thrombogenicity. At least one of its problems is the necessary existence of *two* surfaces contacting blood—the material surface and *air*. At the air interface, blood plasma proteins denature and form thin but rigid layers.[19] In the process of tilting the test tubes, these layers are disrupted and fresh blood comes to the free surface What happens to the denatured film depends on its adherence, or lack of it, to the material wall of the test tube during this manipulation. Hydrophobic walls (e.g., PDMS, wax) may lead to longer WBCT by immobilizing the flakes of denatured protein via hydrophobic interaction, delaying the activation of the intrinsic clotting cascade.

14.2. PEO AND BLOOD CONTACT—OVERVIEW

As we noted in a paper entitled *PEO as a Biomaterial*,[20] by 1980 various bits of evidence demonstrated conclusively that PEO adsorbed out of solution onto glass vessel walls prevented subsequent adsorption of protein or virus from their respective solutions, and it is impossible to name any other water-soluble polymer that is equally effective. Second, one cannot count upon simple adsorption of PEO to be indefinitely successful. Eventually it must desorb if the ambient solution has zero concentration of PEO. Therefore, some form of permanent immobilization is required in order to produce a useful blood contacting surface.

Convinced of the virtues of PEO, we first produced materials in which the PEO chains are *virtually* crosslinked by the phase separation of the hard segment of segmented poly(urethane ureas) (SPU). We then turned to *covalent* endlinking of PEO into a polysiloxane, and more or less in parallel, to radiation crosslinking of PEO in aqueous solution, which results in random covalent crosslinkages between the chains. Thus a hydrogel is formed which contains nothing but PEO.

Why we proceeded in this sequence and what we learned from it are described in this chapter.

14.3. PEO IN SEGMENTED POLY(URETHANE UREAS)

In the period of 1979–1982 Vera Sa da Costa[21–24] prepared segmented poly(urethane ureas), SPU, using various diisocyanates with three types of polyether chains: poly(tetramethylene oxide) PTMO, polypropylene oxide (PPO), and PEO, chain extruded by a diamine. Nargis Mahmud,[25,26] closely following Sa da Costa's studies, employed as the diisocyanate 1,4-*trans*-cyclohexane diisocyanate. The several SPU were studied by an *in vitro* platelet retention test using anticoagulated human blood (citrated), and the degree of thrombogenicity was measured as the fraction of platelets *retained* as the blood sample passed over a column packed with a constant volume of glass beads, their surfaces coated by solvent deposition of the SPU (or other polymer) under test. Sa da Costa and Mahmud relied heavily on X-ray photoelectron spectroscopy (XPS) to determine the *surface* composition (top 35 Å) of these two-phase materials. The fact that XPS requires hard vacuum while the sample is necessarily swollen by water, at least to some degree, when exposed to blood, caused only minor problems of interpretation with SPU because generally the concentration of the hard segment phase in the dry state *at the surface*, assayed as % nitrogen and determined by XPS, was always less, usually twofold or so less, than the bulk % N, determined by infrared spectroscopy (Table III in Ref. 24). We interpreted this to mean that the *polyether* represented the lower energy component (in contrast to the hard segment), therefore that even when dry the surface should be enriched in polyether, and that when soaked in water the slightly hydrophilic PPO and the very hydrophilic PEO would be at even a higher concentration at the surface, and thus as exposed to blood. Sa da Costa and Mahmud used the XPS signature of ethereal carbon —$\underline{\text{C}}$—O—$\underline{\text{C}}$— as a measure of polyether at the surface, the other possible carbon signatures being those of alkane carbon, and carbonyl carbon: thus they plotted platelet retention versus fraction of surface carbon ether bonded to oxygen. Their results are shown in Figure 1.

A striking fact apparent from their figure is the very high level of platelet retention on the material labelled HSA, "hard segment analogue," namely, a copolymer of ethylene diamine and MID 4,4[1] diphenylmethyl diisocyanate, at a value of $\Phi = 0$ (no ether carbon). This suggests that to the extent such material is present on the surface of a SPU, the higher will be the platelet deposition. As a contrary extreme, at a value of Φ about 0.82 we note a form of SPU constructed from PEO of 1500 mol wt and the recently introduced 1,4-*trans*-cyclohexane diisocyanate (CHDI) with a low platelet retention index (around 0.05).

Retrospectively, in the light of our recent studies on PEO–polysiloxanes and radiation crosslinked PEO, we note some factual errors or conceptual errors in our writings on the SPU. For example, in Figure 1, concerning two SPU utilizing PEO of 3500 or 4500 mol wt and reacted with isophorone diisocyanate, we wrote[24] "*that the unreacted diols, the hard segment analogue TDI–ethylene diamine copolymer and the SPU 24, SPU 3500 and SPU 4500 were crystalline at 37 °C in contact with water or blood was confirmed by their grossly observable turbidity.*" More recent experience suggests that the PEO of 1000, 3500, or 4500 mol wt upon contact with water swells and dissolves off a surface, or at least becomes an amorphous network. Thus

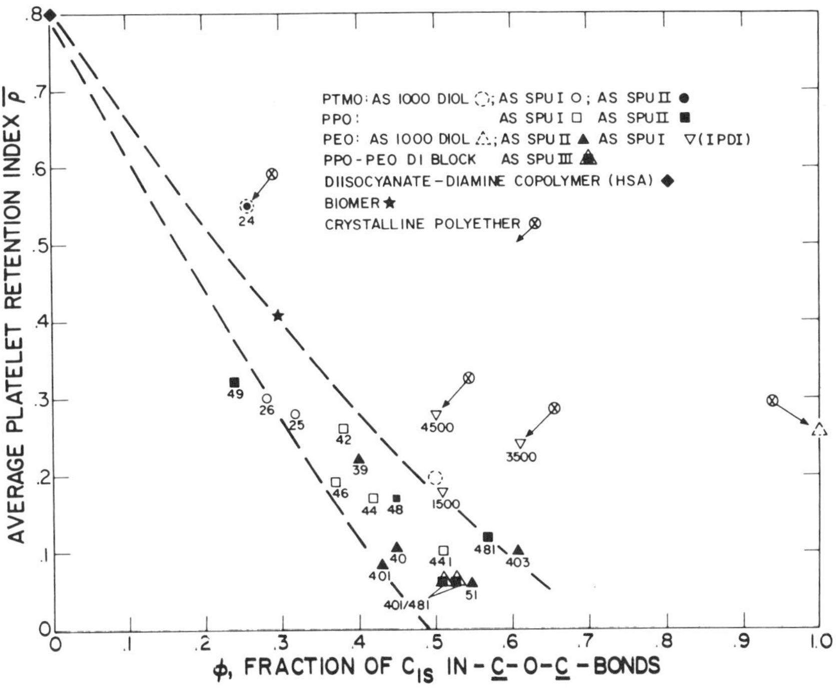

Figure 1. Average platelet retention index versus ϕ, where ϕ is the fraction of C_{15} in —C—O—C— bonds. Reprinted from Ref. 22 by permission of the American Chemical Society.

the data for allegedly pure PEO ($\Phi = 1$) probably was a test of partly uncovered glass beads, and the data for the isophorone diisocyanate (IPDI) SPU containing PEO of 3500 and 4500 represents some form of phase separation so gross as to exceed one-half the wavelength of incident light. The alleged crystallinity was never confirmed by DSC.

At this period MIT Professor of Biology David Waugh was associated with our efforts and was actively studying adsorption of thrombin to various surfaces, using a column packed with spheres overcoated with various segmented polyurethanes. Results are shown in Figure 2, for two SPU containing PEO as the polyether and two containing poly(tetramethylene oxide), PTMO.

The SPU containing PEO of degree of polymerization $n = 46$ (mol wt 2000), curve 1 is very close to the void volume curve, indicating negligible adsorption. These relations were compared in Table 1, wherein HSA refers to the hard segment analogue made from ethylene diamine and MDI (4,4[1] diphenylmethyl diisocyanate). We concluded that the HSA was the most adsorptive of thrombin, as it was of platelets. In this case, the measure of adsorption was Θ, fraction of surface covered, deduced from U, the moles of thrombin removed from solution per cm^2 of surface assuming 2150 $Å^2$ per thrombin molecule. We concluded that PTMO in crystalline form was more active than PTMO in amorphous form, $\Theta = 0.33$ versus 0.24; and that

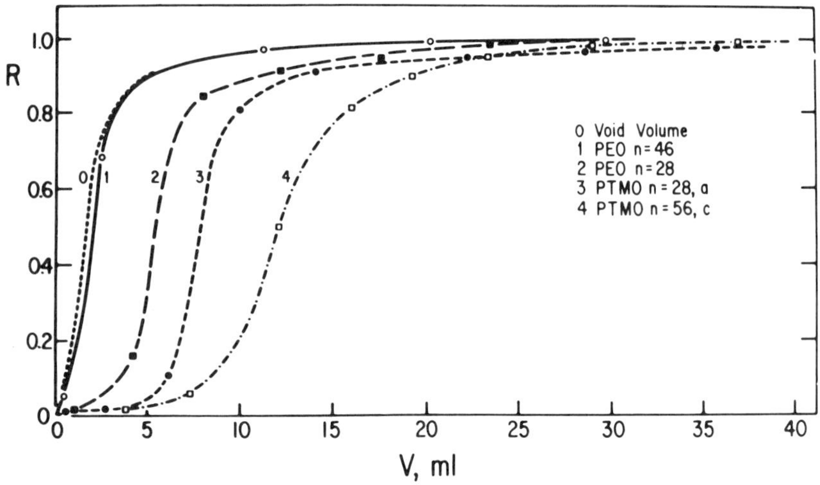

Figure 2. Ratio of thrombin concentrations (effluent/influent), *R*, as a function of volume (ml) supplied to column. [Reprinted from Ref. 21 by permission of Academic Press.]

PEO in a composition leading to 64% water was far less adsorptive ($\Theta = 0.04$). It was with this background of evidence on the role of the hard segment in SPU that we wrote[20]:

> We believe that the high water content PEO hydrogel networks offer promise as a blood contacting surface and that further improvement can be had by alternate methods to create networks *not using diisocyanates*.

For this reason we turned to the materials described in the next section.

Table 1. Thrombin Adsorption and Platelet Retention on Segmented Polyurethanes

				Thrombin		
Surface[b]	n[c]	a or c[d]	%H_2O[e]	$U \times 10^{12}$ (mol^{-2})	Θ[f]	1-PRI[g]
1 PEO	46	a	64	0.2	0.04	0.05
2 PEO	28	a	33	0.7	0.15	0.12
3 PTMO	28	a	28	1.1	0.24	0.29
4 PTMO	56	c	5	1.5	0.33	0.55
5 HSA	5	c	~0	2.29	0.48	~0.8

[a]Reprinted from Ref. 21 by permission of Academic Press.
[b]PTMO denotes poly(tetramethylene oxide), HSA denotes copolymer of diisocyanate and diamine.
[c]n is the degree of polymerization of PEO or PTMO soft segment; of the entire copolymer HSA.
[d]a or c: amorphous or crystalline.
[e]At equilibrium in phosphate buffered saline.
[f]Θ is fraction of surface covered versus projected area of thrombin = 2150 Å^2.
[g]PRI is the fraction of platelets *recovered*. Thus 1-PRI is the platelet retention index.

14.4. PEO AND POLY(ETHYLENE GLYCOL) MONOMETHYL ETHER (PEGME) IN NETWORKS CROSSLINKED BY POLYFUNCTIONAL SILOXANES: THE PEKALA STUDIES

As measured by platelet retention *in vitro*, the copolymer sequence produced from diamines and diisocyanates appeared disastrous. We were therefore skeptical about utilizing polyfunctional isocyanates to end link PEG by the formation of urethane bonds for the purpose of producing networks (although it is not established that a cluster of urethane linkages —NHCOO— would produce the same effect as urea sequences —NHCONH—).

The reputation in 1980 of the polysiloxanes, especially polydimethylsiloxane (PDMS), as bland materials in blood contacting applications led us to utilize polysiloxanes as the crosslinking agent for PEG. Pekala[27–29] was the first in our laboratory to study this method.

He synthesized as the polyfunctional crosslinker a molecule called poly(glycid-oxypropyl methyl siloxane), PGPMS, having the repeat unit

(4)

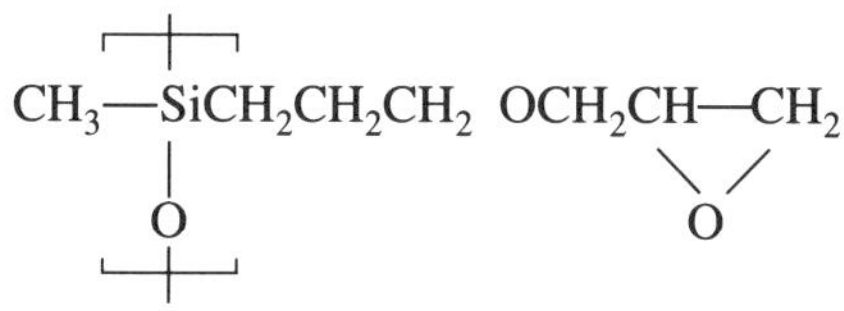

37 glycidoxy

It is interesting that the number of repeat units, thus the functionality, was so high, around 37. Pekala was also proceeding with the hypothesis that the extremely hydrophilic PEO, which as a network undergoes large degrees of swelling upon contact with water, might be replaced by polypropylene oxide (PPO) as the difunctional chain molecule (difunctional because of hydroxyl at each end). The PPO unit is

(5) $$—CH_2CH(CH_3)O—$$

The reaction by which the polyether became crosslinked involves boron trifluoride catalyzed ring opening of the glycidyl on PGPMS, and reaction with the hydrogen of the terminal hydroxyl on the polyether:

(6)

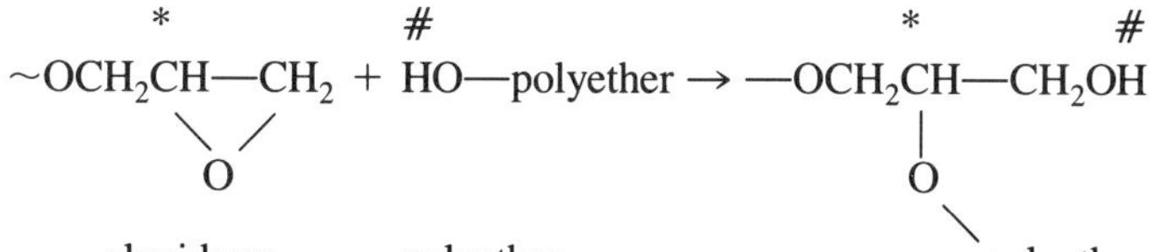

glycidoxy polyether polyether

The symbols * and # are used simply to clarify how the polyether becomes connected to the glycidoxy group. A primary hydroxyl group is regenerated as the polyether hydroxyl reacts, making this a peculiar reaction in that it cannot be considered simply a condensation of the type

(7) $$—A + B— \rightarrow —C—$$

In the presence of BF_3 a second competitive reaction occurs, cationic ring opening polymerization of the glycidyl groups:

(8) 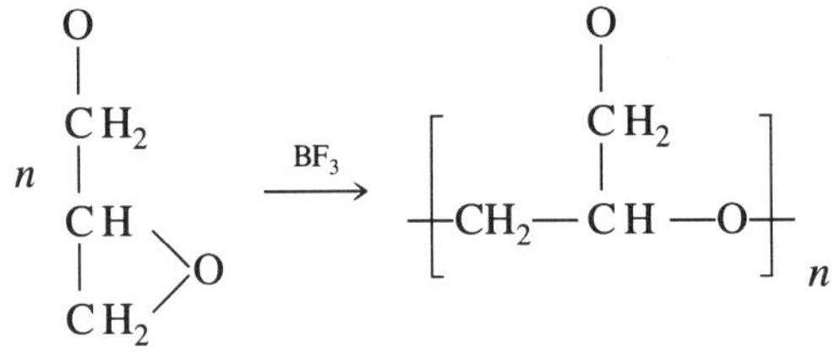

As had been the case with studies of segmented polyurethanes, platelet reaction with the surfaces created by Pekala was assessed by an *in vitro* platelet retention test using a column of beads coated with the test material. Fibrinogen adsorption at this time was also carried out *in vitro* using ^{125}I-labeled human fibrinogen.

As shown in Figure 3, Pekala[29] varied the mol fraction X of propylene oxide units (as PPO) in the networks, i.e.,

$$\frac{\text{number of (5)}}{\text{number of (5)} + \text{number of (4)}}$$

and found that as X was increased from 0.20 to 0.92, the concentration of adsorbed

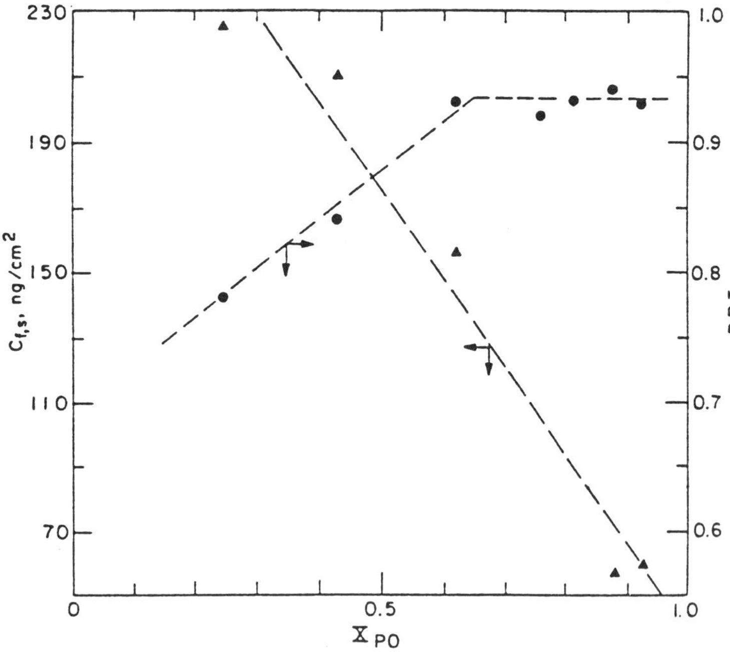

Figure 3. Surface fibrinogen concentration $C_{f,s}$ (ng cm^{-2}) and platelet recovery index PRI as functions of mol fraction propylene oxide units X_{PO}. $C_{f,s}$ was determined after 2 minute exposure to 500 μg ml^{-1} fibrinogen solution. [Reprinted from Ref. 29 by permission of Butterworth, Ltd.]

fibrinogen decreased linearly from about 225 to about 60 ng cm^{-2}, and the platelet retention index (fraction of platelets retained on the column) decreased from about 0.22 to about 0.07. At that time we believed that platelet retention indices of 0.2 represented rather bland surfaces.

In a continuing study Pekala incorporated PEGME

$$HOCH_2CH_2O(CH_2CH_2O)_nCH_2CH_2OCH_3 \tag{9}$$

into the networks created from PPO and PGPMS. He found significant lowering of fibrinogen adsorption as the amount of PEGME was increased. The control was the network made from PPG and PGPMS without any PEGME. What was particularly surprising at that time was the platelet response to the same networks. To quote Pekala[29]:

> The network synthesized from only PPG 2000 + PGPMS produced the least amount of platelet retention in the columns overall. Blood compatibility, defined as the ability to inhibit platelet retention in this experiment, has been affected adversely by the presence of PEGME in the other networks. There is a degree of donor-to-donor variability in the platelet retention index results which makes it essential to sample blood from multiple donors for comparison of polymers. Some donors were more reactive towards PEGME-containing networks than they were towards the glass controls. In contrast, all donors showed low reactivity with the model network (PPG 2000 + PGPMS).
>
> The polyether/PGPMS networks represent an interesting case in which the molecular response (i.e. protein adsorption) to blood has been affected in a consistent and predictable manner by the incorporation of PEGME. In general, PEGME shows a low affinity for binding proteins, and the above networks confirm this previous finding. The cellular response (i.e. platelet adhesion), however, does not correlate with the network's capacity to bind fibrinogen as has been reported by others.

Because of uncertainty as to the meaning of the data, platelet retention by networks made *only* from PEG (not PPG) was not published at that time. Table 2 (Table XX of Ref. 27) includes not only PEGME added to PPG–PGPMS, but includes a network made with the short-chain PEG, mol wt 1540, without any PPG. To call

Table 2. Platelet Retention Index Data for Various Polyether/PGPMS Networks[a]

Network Description	mol% $(CH_2CH(CH_3)O)^{(a)}$	mol% $(CH_2CH_2O)^{(b)}$	Platelet retention index donors #1	#2	#3	#4
PPG 2000 + PGPMS	80.8	0.0	0.31	0.06	0.17	0.08
PPG 2000 + PGPMS	62.2	0.0	0.31	0.19	0.21	0.03
PPG 2000 + PGPMS	43.3	0.0	0.34	0.10	0.25	0.04
PPG 2000 + PGPMS + PEGME 350	28.8	33.5	0.92	0.14	0.58	0.57
PEG 1540 + PGPMS	0.0	63.3	0.96	0.04	0.78	0.95
Glass control	—	—	0.44	0.30	0.39	0.34

[a]Thus the mol% $-O-Si(CH_3)(-)-CH_2CH_2CH_2O-CH_2CH(-O-)CH_2$ (epoxide) = 100 − sum of (a) + (b).

attention to the peculiarity of these data, very high retention values are underlined. These data not only illustrate donor-to-donor variability by the *in vitro* platelet retention test, but indicate very aggressive response to the PEG networks by the blood of two donors (#1, #4), i.e., virtually all platelets retained far in excess even of the glass control. On the other hand, the PEG–PGPMS network appeared remarkably bland toward the blood of donor #2. As will be noted below, subsequent evidence gained in our laboratory on similar materials leads us to conclude that this was the first positive demonstration of the possibility that PEO can enhance thrombogenicity, although at the time we failed to recognize it.

14.5. PEO IN NETWORKS CROSSLINKED BY POLYFUNCTIONAL SILOXANES: THE CYNTHIA SUNG STUDIES

Cynthia Sung[30] proceeded from the work of Pekala, but with a fundamentally different objective: study of the properties of such polyether–polysiloxane materials as drug adsorbing–drug releasing media. Elliot Chaikof[33] overlapped in time the research of Sung in parallel investigations, directed toward blood contact. Before describing the specific studies of Sung and their importance, we note the general differences between the studies of Pekala on the one hand, and Sung and Chaikof on the other.

(1) Sung and Chaikof utilized, and Chaikof synthesized *de novo* a variant of Pekala's PGPMS, this being PGPMS/DMS, i.e., poly(glycidoxypropylmethyldimethyl siloxane), of which the structure is

(10)

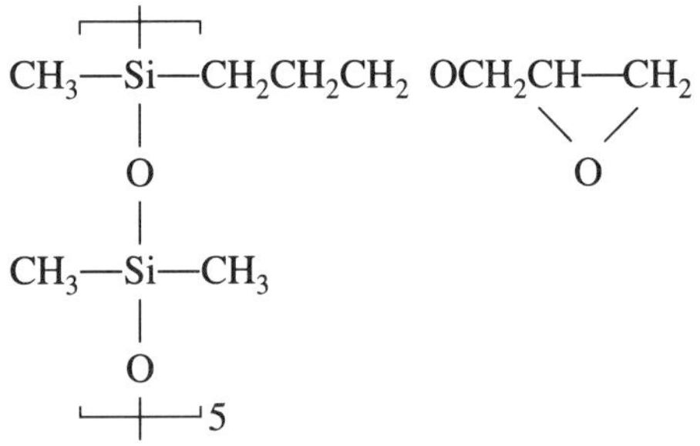

The polyfunctional siloxane of Sung and Chaikof thus had 5 glycidyl groups per molecule, while Pekala's had 37 (see equation 4 above). Second, PGPMS/DMS is a copolymer, approximately 50% of the units being dimethyl siloxane, while Pekala synthesized his material by reacting polymethylhydrosiloxane —HSi(CH_3)O— of molecular weight 2270, degree of polymerization 37, with allyl glycidyl ether, thereby producing a homopolymer, having a glycidoxy group on every silicon atom.

Thus Pekala's PGPMS was roughly four times larger than Sung's and Chaikof's PGPMS/DMS.

(2) Sung and Chaikof utilized only PEG, not PEGME, and PEG with a *minimum* mol wt of 2000 (cf Pekala's PEG of 1540 mol wt). Most of Sung's materials were produced with PEG of 8000 mol wt. Chaikof included PEG of 20,000 mol wt.

(3) During this time our medical collaborators were Dr. Allan Callow and colleagues at New England Medical Center/Tufts Medical School. Their evaluation of blood contact used methods different from any previously employed in our group:

(a) Platelet deposition was evaluated *in vivo* and *ex vivo* in baboon using ^{111}In-labeled platelets.

(b) Fibrinogen deposition was evaluated *in vivo* and *ex vivo* in baboon using ^{125}I-labeled fibrinogen.

(c) Complement activation (C3, C4, C5) was studied *in vitro*.

These studies of blood contact concerned Chaikof's studies primarily. The impetus for these studies came from an early *in vivo* test in baboon of one of Sung's materials.

(4) Both Sung and Chaikof went beyond Pekala in studying the network formed from the polysiloxane by BF_3 initiation without *any* PEO present. In notation used below, it is to be understood that PGPMS/DMS refers to this material. It may be described as a poly(ether-siloxane). The reaction involved is that shown as equation 8 in the preceding section.

Sung studied the partitioning of tricyclic antidepressants into PEO–PGPMS/DMS networks[30–32] as a function of their composition. A typical tricyclic antidepressant is protriptyline:

(11)

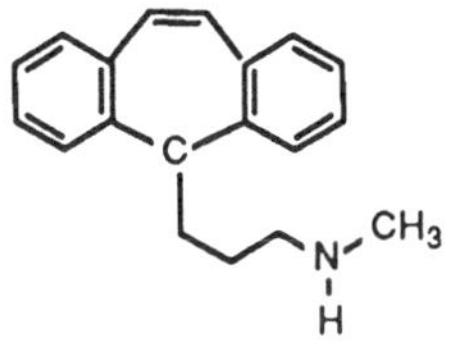

The hydrophobic fused ring "head" is thus connected to a "tail" which owes its hydrophilic properties to the secondary amine.

The partition coefficient is defined as

$$\frac{\text{concentration of protriptyline in the hydrated network}}{\text{concentration of protriptyline in bathing solution}}$$

Figure 4 shows[32] how this varies with wt% PEO incorporated into the network. To the far left, the bar refers to BF_3 crosslinked PGPMS/DMS with no added PEO. To the far right, "100% PEO" refers to radiation crosslinked PEO (4% to 7% PEO, 96% to 93% water), to be discussed in Section 14.7). It is apparent that the composite having 50% PEO, 50% PGPMS/DMS has the maximum partition coefficient for the drug, 10- or more fold *greater* than either extreme: the polysiloxane without PEO, or the PEO without polysiloxane. This appears to indicate that there is a special hydrophilic–hydrophobic balance in the region which connects PEO to the glycidoxypropylsiloxane, i.e.,

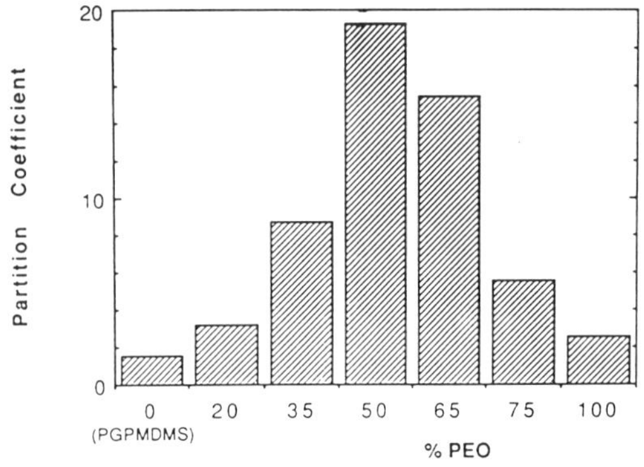

Figure 4. Partition coefficient of protriptyline in PEO hydrogels of varying wt% PEO. Zero percent refers to the reaction product of poly[glycidoxypropyl methyl dimethyl siloxane]. 100% refers to radiation crosslinked solutions of pure PEO. [Reprinted from Ref. 32 by permission of the American Pharmaceutical Association.]

(12)
(see equation 6)

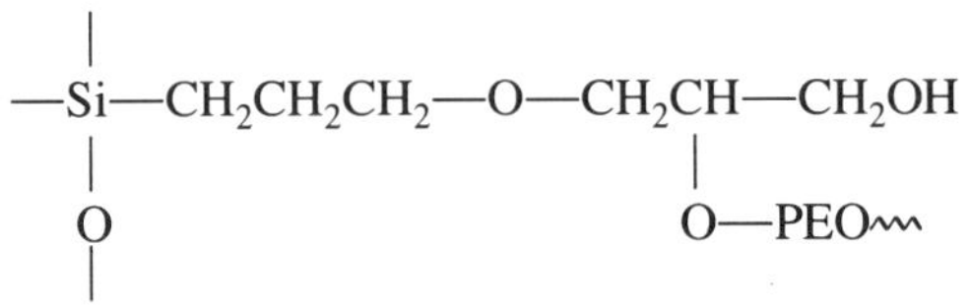

The content of this region is at a maximum around 50% PEO content. The focus of this paper being blood-contacting applications, these findings of Sung are of interest, as will be seen, in connection with unexpected behavior of PEO–PGPMS/DMS hydrogels toward blood and are incorporated into a hypothesis as to why, in certain circumstances, already foreshadowed by Pekala's work, PEO can enhance thrombogenicity.

We must also credit Sung for having the tenacity to evaluate three of her PEO–PGPMS/DMS materials in an *in vivo* baboon model, i.e., as coated on a filament and exposed to flowing blood for one hour, the blood containing ^{111}In-labeled platelets and ^{125}I-labeled fibrinogen. She found far less of either label on her materials than on expanded polytetrafluoroethylene, which served as a control. Proceeding from these encouraging results, Elliot Chaikof (next section) carried out a detailed study of PEO–PGPMS/DMS materials in contact with blood.

Sung writes:

> Recently, it was reported that the TCA's impramine and amitriptyline are potent inhibitors of platelet aggregation and adhesion on synthetic surfaces. [S. F. Mohammad, D. B. Olsen, and S. W. Kim, "Inhibition of Platelet Adhesion to Biomaterials by Imipramine and Amitriptyline," *Biomater. Trans. 11*, 324 (1988).] These TCAs, slowly releasing from PEO–PGPMDMS networks, may further enhance the blood compatibility of these materials. This concept of improving the blood compatibility of the PEO–PGPMDMS networks by using them also as drug releasing materials of anticoagulant drugs should be

> tested in future studies using not only the TCAs but also other well-known inhibitors of clotting such as heparin and prostaglandin PGI_2.

Sung's suggestions have yet to be tested experimentally.

14.6. PEO IN NETWORKS CROSSLINKED BY POLYFUNCTIONAL SILOXANES: THE CHAIKOF STUDIES

Already a surgical resident, Chaikof synthesized[33] PEO–polysiloxane networks with a wide range of PEO molecular weight (2000, 8000, 20,000) and PEO content (20%, 35%, 50%, 65% by weight on a dry basis—the remainder being the polysiloxane). For brevity the networks are coded to reflect PEO mol wt and wt%, e.g., 2K65 means PEO of 2000 mol wt constituting 65 wt% of the network. Bulk[34] and surface[35–38] properties were studied in detail. The response of blood to hydrated surfaces of these networks was examined by *ex vivo* platelet deposition,[38,39] *ex vivo* fibrinogen adsorption,[39] and *in vitro* complement activation.[33]

From the viewpoint of polymer structure and properties alone, the materials synthesized by Chaikof were very interesting. For example, despite the fact that the PEO molecules were fixed into a network by reaction of their hydroxyl ends onto the polysiloxane, *spherulitic crystallization* was observed by optical microscope in the 8K and 20K networks when *dry*, and the extent of spherulite formation increased as PEO content increased. The chain folding necessary for spherulite formation implies long-range displacement of the ends of network *chains*, which in turn suggests that the "junction" material, i.e., the PGPMS/DMS portion, is highly mobile.

Upon exposure to water or physiologic saline, crystallization is lost and the PEO part of the network swells to an extent depending on molecular weight and on PEO content, resulting in a swelling ratio of up to 3. The surface of the 8K and 20K series materials after hydration was observed by an Electroscan® microscope. The surface was found to be featureless, all evidence of spherulites having disappeared. By the use of a special technique of fixation,[38] the networks were stabilized in their noncrystalline, water-swollen state so that upon dehydration they neither collapsed nor crystallized and thereafter could be examined by conventional scanning electron microscopy. This allowed a detailed examination of blood elements deposited on their surfaces after *ex vivo* exposure.

XPS analysis of these networks was of special interest because of the contrast to be drawn with respect to the segmented poly(urethane ureas) of Sa da Costa and Mahmud: in the SPU the content of hard segment (the "junction") material was always lower in the surface than in bulk. Exactly the opposite was found by Chaikof: the percent silicon detected by XPS[37] (thus in the first 35 Å of surface) was always greater than in the bulk material, even for the 20K65 form. (Samples for XPS examination could not be fixed by the above-mentioned technique because that would have introduced silicon from silazane and aldehyde-derived carbon.) This means that the polysiloxane ("junction") part of the network has lower surface free energy than that of PEO, while in the SPU the PEO has a lower surface free energy than the *hard segment* ("junction") part of the network.

The least expected result from Chaikof's studies was the observation that, under certain conditions, PEO appeared to enhance thrombogenicity: when the PEO molecule was relatively short (molecular weight 2000) and the PEO content was high.

For example, Figure 5 shows the deposition of ^{111}In-labeled platelets on the lumen of the *ex vivo* baboon shunt as a function of time. While the *long* PEO chain 20K65 (PEO of 20,000 mol wt) material shows no platelet buildup, and thus appears nonthrombogenic by comparison with the 20K35 (35% PEO) network, the case is totally reversed with the short PEO chain (2000 mol wt) network, and now the higher the content of PEO, the more deposition occurs on the surface. We interpret this to mean that in the 20K65 network, the surface available to blood is only PEO, the polysiloxane being "buried," while at the surface of the hydrated 2K65 network there is a considerable content of the region in which PEO connects to the glycidoxypropyl siloxane, i.e.,

(12)

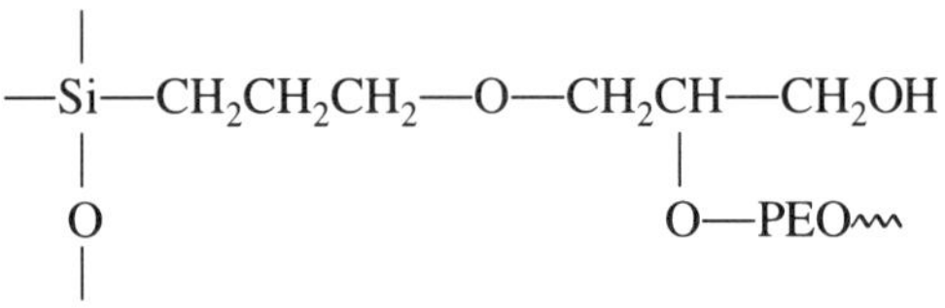

This is the same region to which we impute high affinity for the hydrophilic–hydrophobic tricyclic antidepressant drugs mentioned in the preceding section. It is

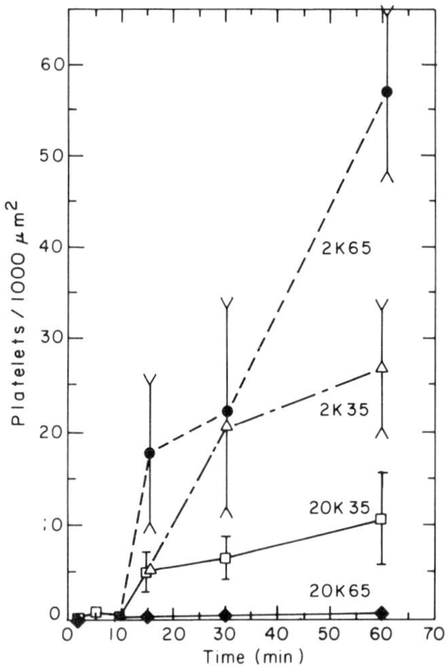

Figure 5. Deposition of ^{111}In-labeled platelets versus time of circulation in *ex vivo* baboon shunt for networks shown. 2K and 20K refer to PEO molecular weights, respectively, 2000 and 20,000. 35 and 65 refer to wt% PEO.

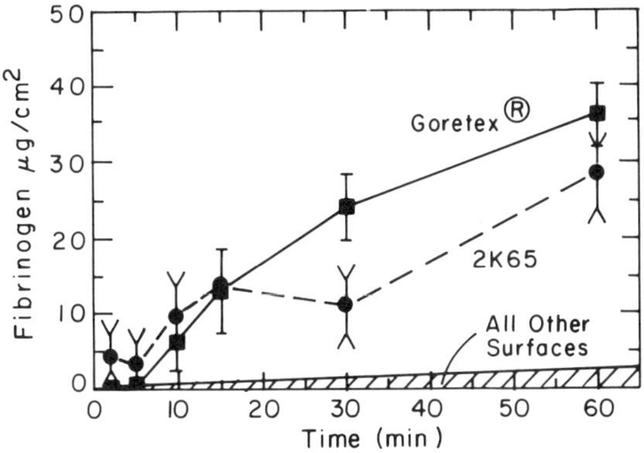

Figure 6. Adsorption of ^{125}I-labeled fibrinogen versus time of circulation in *ex vivo* baboon shunt. Goretex® refers to microporous PTFE. "All other surfaces" includes 2K35, 20K35, 20K65. See caption of Figure 5.

generally accepted that adsorbing proteins on a surface are precursors to platelet activation, and one important protein is fibrinogen—already discussed in connection with Pekala's work. Figure 6 shows ^{125}I-labeled fibrinogen uptake in the *ex vivo* baboon shunt as a function of time. Clearly the 2K65 surface is much more active than any other PEO–PGPMS/DMS network. It is our hypothesis that the molecular region shown above is responsible for reshaping adsorbed fibrinogen so that it becomes highly active toward platelets. We were therefore not surprised to find that the same pattern emerged in complement activation studies,[33] as shown in Table 3.

Herein the degree of *in vitro* activation of complement C3 is reported as a fraction of the maximum possible in the assay. On the one hand, a radiation crosslinked PEO hydrogel (to be described), the 20K65 PEO–PGPMS/DMS hydrogel, and polydimethyl siloxane PDMS all showed very low values, around 0.02. On the other hand, all 2000 mol wt PEO–PGPMS/DMS networks, i.e., 2K20, 2K35, 2K50, showed great activation of C3 (0.39, 0.30, 0.40).

Table 3. Complement C3 Activation

Material	$C3_a/C3_{a,max}$	Material	$C3_a/C3_{a,max}$
I. PEO–PGPMS/DMS types		II. Other	
2K20	0.39	PGPMS/DMS (no PEO)	0.02
2K35	0.30	PEO	0.02
2K50	0.40	PTFE	0.015
8K50	0.025	Mylar®	0.015
20K20	0.09		
20K35	0.06		
20K50	0.03		
20K65	0.02		

By a protocol of special fixation, previously mentioned, it was possible to carry out scanning electron microscopy[38] on the various PEO–PGPMS/DMS surfaces without the collapse of the material. The activity of the 2K65 surface is described thus:

> At 30 minutes scattered platelets, red cells, and fibrin strands were seen on the surface of 2K65, which progressed to a fibrin network with individual platelet aggregates following 60 minutes of blood exposure.

By contrast, the surface of 20K65 material after 30- or 60-minute exposure was clean, showing neither cells of any kind nor fibrin strands. We infer this surface to be 100% PEO, as a consequence of profound molecular rearrangement upon hydration, which "buried" the polysiloxane and made it inaccessible to blood elements.

We conclude that PEO appears to facilitate rearrangement of adsorbed protein *when the PEO chain is connected to some other molecular entity and their junction is accessible to the solution*, e.g., blood, carrying the protein. We believe this accounts in part for some of Pekala's observations, previously noted, regarding PEGME and short-chain PEO in his networks derived from PGPMS, as well as reports from other laboratories[40–42] indicating an enhancement of thrombogenicity by PEO under certain conditions.

Thus to avoid protein adsorption and activation it appears essential to present to the protein solution a surface which is exclusively PEO. Presumably, what is under the surface is irrelevant, provided the protein cannot get to it.

14.7. PEO NETWORKS BY RADIATION CROSSLINKING: THE DENNISON AND TAY STUDIES

Networks formed by radiation crosslinking of PEO in aqueous solution were studied by Kathleen Dennison,[43] following Pekala, but preceding Sung and Chaikof. We discuss her results somewhat out of order for the purpose of consolidating the findings of Pekala, Sung, and Chaikof, but Dennison's methods were used by both Sung and Chaikof to prepare "base line" PEO hydrogels representing one limit to their PEO–PGPMS/DMS materials (the other limit being PGPMS/DMS networks).

Dennison elucidated the relationships between PEO concentration in water, radiation dose, radiation dose rate, and the resulting material properties: (a) degree of crosslinking as measured by M_c (molecular weight of the chain between junctions), (b) degree of swelling in water after crosslinking, and (c) permeability to diverse solutes including isolated PEG molecules and globular proteins.

In general, radiation doses of 2 to 10 megarads suffice to create networks from aqueous solutions containing 3 to 20 wt% PEO. Table 4 shows these relationships. The higher the value of M_c, the molecular weight between crosslinks, the looser is the network (i.e., the more permeable). The higher the initial PEO concentration in solution, the higher is the dose necessary to achieve a given value of M_c. For example, only 5 megarads delivered to a solution containing 5% PEO produces about the same M_c (ca 5000) as 10 megarads delivered to a 15% solution (compare gels D and J).

Dennison[43] measured the diffusivity of five proteins, four of which are indicated in Table 5, in these hydrogels and established the ratio of diffusivity R in the hydrogel

Table 4. Equilibrium Volume Fractions of PEO in Hydrogels as a Function of Initial Concentration and Radiation Dose

	PEO hydrogel[a]	v_{2s}(Water)[b]	v_{2s}(PBS)[c]	M_c[d]
A	3%, 2.5	0.0172	0.0229	6926
B	3%, 5	0.0272	0.0349	3648
C	5%, 2.5	0.0137	0.0166	10425
D	5%, 5	0.0269	0.0386	5229
E	10%, 2.5	0.0074	0.0092	14655
F	10%, 5	0.0198	0.0232	10119
G	10%, 10	0.0461	0.0616	3752
H	10%, 15	0.0618	0.0849	2196
I	15%, 5	0.0187	0.0207	11671
J	15%, 10	0.0458	0.0616	4898
K	15%, 15	0.0639	0.0866	2898
L	20%, 5	0.0147	0.0175	13723
M	20%, 10	0.0483	0.0657	5436
N	20%, 15	0.0623	0.0840	3804

[a]First number is vol% PEO in solution, second number is radiation dose (megarad).
[b]Equilibrium volume fraction of PEO swollen in water.
[c]Equilibrium volume fraction of PEO swollen in phosphate buffered saline.
[d]Molecular weight between crosslinks M (g mol^{-1}) calculated from the Bray equation.[10]

[(swollen in phosphate buffered saline (PBS)] to the diffusivity of the same solute in PBS. The most "open" PEO hydrogel L, the least crosslinked (M_c = 13,723 from Table 5) offers significant permeability to molecules in the range of molecular weight 1000 to about 20,000 (or better, molecules having up to 22 Å effective radius). The least "open" PEO hydrogel, H, M_c 2196, is substantially impermeable to ovalbumin. A fifth protein, human serum albumin, mol wt 67,000 effective radius 36.1 Å, was excluded by all networks except L and I, in which R was around 10^{-3}.

These data of Dennison are important as a guide to the degree of crosslinking necessary to obtain protein exclusion when PEO is used as a coating on a support (e.g., catheter) intended to prevent access of proteins to the support itself. They also offer a guide to the diffusion coefficients that might be anticipated if the PEO hydrogels were to be used as drug-releasing media. For example, if heparin were to be released, one would expect that the usual range of molecular weight (5000 to 15,000 g mol^{-1}) would place the heparins in the *range of radii* (as coils) of 20 to 30 Å, corresponding to diffusivities found between lysozyme and chymotripsinogen in Table 5. Low-molecular-weight heparin fractions would evidently be more readily transported. The pioneering studies of Dennison were subsequently utilized by Sung and Chaikof, as already noted, but were almost simultaneously exploited by Sew-Wah Tay,[44] who worked contemporaneously with Dennison. She studied the immobilization of heparin onto two hydrogels, each crosslinked by electron irradiation of an

Table 5. Ratio of Diffusivity, *R*, of Protein in Hydrogel to Diffusivity of Same in Phosphate Buffered Saline, in Terms of Hydrogel Type, for Cyanocobalamin, Lysozyme, Chymotripsinogen, and Ovalbumin[a]

	PEO hydrogel	R_{Cyano}	R_{Lyso}	R_{Chymo}	R_{Oval}
L	20%, 5	0.867	0.296	0.121	4.82×10^{-3}
M	20%, 10	0.469	0.387	7.73×10^{-2}	1.95×10^{-3}
N	20%, 15	0.144	2.64×10^{-2}	2.83×10^{-3}	2.82×10^{-4}
I	15%, 5	0.249	0.485	0.140	1.14×10^{-2}
J	15%, 10	0.218	6.05×10^{-2}	6.24×10^{-3}	6.68×10^{-4}
K	15%, 15	0.128	2.89×10^{-2}	4.55×10^{-3}	4.86×10^{-5}
F	10%, 5	0.258	0.195	2.47×10^{-2}	1.35×10^{-2}
G	10%, 10	0.259	5.69×10^{-2}	2.68×10^{-3}	9.02×10^{-5}
H	10%, 15	0.117	3.92×10^{-2}	4.97×10^{-3}	0
B	3%, 5	6.67×10^{-2}	0.138	—	3.44×10^{-2}

[a]Key:

	10^{-3} mol wt	Effective radius (Å)	10^{6}·diffusivity in PBS ($cm^2 s^{-1}$)
Cyano = Cyanocobalamin	1.35	8.5	3.76
Lyso = Lysomzyme	17	20.6	1.04
Chymo = Chymotripsinogen	23	22.5	0.95
Oval = Ovalbumin	44	27.6	0.776

aqueous solution, one being poly(vinyl alcohol) PVA, and the other PEO. Tay utilized the earlier work of Bray[9,10] and Peppas[11-14] as a basis for making hydrogels by electron-beam crosslinking of water solutions of PVA.

Following formation of the gels, each was reacted with tresyl chloride (after solvent exchange) in order to bind heparin via an amino group. (The occurrence of amino groups in heparin is discussed extensively elsewhere.[45]) The activity of the bound heparin toward antithrombin and thrombin was then evaluated.[45] Comparing PVA and PEO hydrogels having about the same percent of grafted heparin (Table 6), Tay found that the heparin bound to tresylated PEO was about tenfold more active than heparin bound to tresylated PVA (necessarily, the equilibrium polymer content of the PEO hydrogel in water was about the same as the PVA hydrogel, i.e., around 10%). The improved response of PEO-bound heparin was ascribed to the "leash" effect: the heparin molecule, which can only be attached at the *end* of a PEO chain (at the other end of which PEO connected to a radiation-induced junction), had more degrees of freedom with which to interact with antithrombin III so as to render it active toward thrombin, while in PVA the heparin would be bound anywhere on the main chain. The average "leash" of the PEO hydrogel would approximate that of a chain having a molecular weight corresponding to M_c. In this case M_c for both PEO and PVA hydrogels was about 2000. From Dennison's work this value of M_c means that the networks were impermeable to molecules like antithrombin and thrombin, and would have had a very low diffusion coefficient for the type of heparin used. Thus the heparin bound was at or very near the surface, but it could have been bound rather

Table 6. Binding of Heparin to PVA versus PEO Hydrogel

	PVA	PEO
mg Heparin per g polymer	4.33	5.83
Units per g polymer	7.9	95.0
Specific activity (units per mg heparin)	1.82	16.3
% Retention of activity	1.15	10.3

tightly onto the tresyl activated second hydroxyls of PVA, and even by two or more sites per heparin molecule. Tay concludes by writing:

> These issues suggest the desirability of fractionating heparin to obtain fractions of finite amino content, preferably terminal, preferably one amino/heparin, and thereafter fractionating to obtain a subfraction active toward antithrombin.
>
> Then, bonding to hydrogel supports would permit much better quantitative interpretation of the effect of mode of surface binding. In any case, PEO is particularly interesting as a hydrogel support because of its negligible . . . binding of protein, a virtue which distinguishes it from practically every other synthetic polymer and from any other known hydrogel.

14.8. PEO STAR MOLECULES: METHODS OF IMMOBILIZATION AND POTENTIAL APPLICATIONS

The concepts evolving from the previously described work point toward the desirability of providing PEO in the form of star molecules, i.e., molecules having many PEO chains, thus potential "leashes," emanating from a small core. The hydroxyl ends of these PEO chains could then be activated for the purpose of binding biopolymers, achieving more easily the goal of long leashes with a higher concentration of ends in a given volume of polymer.

A particularly useful form of PEO star molecules has been prepared by Rempp[46] and colleagues,[47] shown schematically in Figure 7. The number of arms can vary from 10 to 100; the molecular weight of the arm can vary from around 3000 to more than 10,000. The divinyl benzene core of the star molecule amounts to less than 2% of its volume and, because of the attached arms, is inaccessible to proteins of the size of blood plasma proteins.

Radiation crosslinked solutions of PEO stars, using concentrations and radiation doses suggested by Dennison's work, were compared with radiation crosslinked linear PEO such as used by Dennison.[43] These PEO star gels were then evaluated[50] by the *ex vivo* baboon shunt described earlier by Chaikof[39] and co-workers. The response was like that of Chaikof's 20K65 hydrogel (Figure 5), i.e., no uptake of platelets with time. We conclude that the increased concentration of terminal hydroxyls of these star molecules, as compared to the low concentrations associated with linear PEG of equal molecular weight, does not lead to significant platelet retention. Of course, the disparity between hydroxyl contents of PVA and PEO hydrogels of the same polymer volume fraction differs by two or more orders of

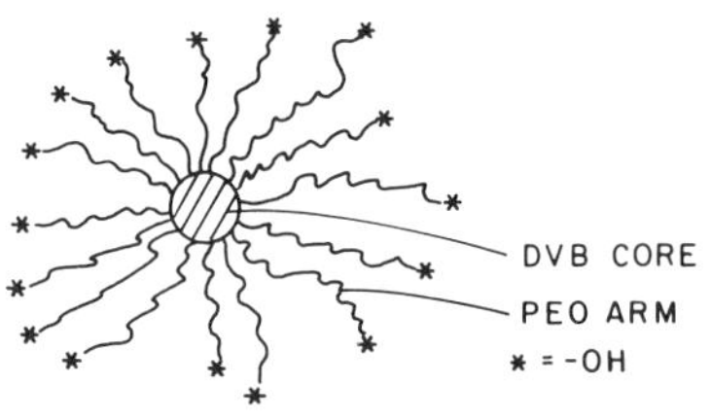

SET MOL.WT. OF PEO ARM M_{ARM}
BY ETHYLENE OXIDE TO K+ RATIO X
THUS: $M_{ARM} = 44X$

DETERMINE MOL. WT. OF STAR M_{STAR}
BY LIGHT SCATLERING

SINCE DVB << EO, $M_{STAR}/M_{ARM} \cong$ NUMBER OF ARMS

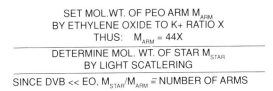

Figure 7. PEO star characteristics: molecular weight of, and number of, PEO arms.

magnitude, being one hydroxyl per unit (44 g) of PVA, but one hydroxyl per 100 units in PEO stars having 4400 molecular weight arms, a typical value.

Applications of these star molecules when fixed to a support are discussed in a recent reference.[48] The PEO stars can readily be reacted with tresyl chloride so as to activate the hydroxyls. When presented to a support containing a low concentration of amine or thiol groups, the stars become grafted by one or more arms, leaving the rest still accessible. Thus a dense monolayer of PEO stars is formed on the surface. Alternatively, a thin layer of a PEO star solution in water can be radiation-grafted to a support, followed by transfer to an organic solvent in which it can be tresylated. While the stars still maintain activity of the remaining tresylated ends (those not used to bind the star to the support), a biopolymer is introduced having amino or thiol groups, and becomes bound to the PEO star layer. Thus (a) the support, however thrombogenic, is excluded from contact by protein in external solutions; (b) the biopolymer, for example heparin, or a monoclonal antibody, is affixed to the end of a flexible leash; (c) the flexible leash is PEO, notable for absence of nonspecific binding of proteins.

Thus applications ranging from advanced protein purification systems via affinity ligand technology to antithrombogenic vascular devices can be envisioned.[48]

14.9. OVERVIEW

At this time it appears reasonable to suppose that PEO, *if alone at a surface*, renders the surface truly nonthrombogenic, although it has been argued that long-term exposure might lead to platelet consumption or to other as yet unobserved phenomena, and therefore that such experiments must be conducted. At the same time, it seems unmistakably clear that if PEO is attached to a different molecular entity which is also at the surface, accessible to blood elements (molecular and cellular), dramatic thrombogenicity can result. Thus the modification of surfaces by the attachment of PEG, PEGME for the purpose of producing nonthrombogenicity is no simple matter. The challenge is to insure that the surface is sufficiently covered by PEO in some form that under no conditions can the blood elements gain access to any other substance than PEO.

ACKNOWLEDGMENTS

This chapter is an account of studies undertaken by former graduate students while pursuing advanced degrees, in collaboration with medical colleagues in the Boston area, particularly Dr. Edwin W. Salzman of the Beth Israel Hospital and Harvard Medical School, and more recently, Dr. Allan D. Callow of New England Medical Center/Tufts University Medical School, and their respective associates.

The author pays tribute to and gives special thanks to each of his students cited for his or her participation in the intellectual expedition outlined above.

REFERENCES

1. E. W. Merrill, E. W. Salzman, B. J. Lipps, E. R. Gilliland, W. G. Austen, and J. Joison, *Trans. Am. Soc. Artif. Internal Organs 12*, 139 (1966).
2. E. W. Salzman, W. G. Austen, B. J. Lipps, E. W. Merrill, E. R. Gilliland, and J. Joison, *Surgery 61*, 1 (1967).
3. R. A. Britton, E. W. Merrill, E. R. Gilliland, E. W. Salzman, W. G. Austen, and D. S. Kemp, *J. Biomed. Mater. Res. 2*, 429 (1968).
4. P. S. L. Wong, E. W. Merrill, and E. W. Salzman, *Fed. Proc. 28*, 441 (1969) (abstract).
5. E. W. Merrill, E. W. Salzman, P. S. L. Wong, T. P. Ashford, A. H. Brown, and W. G. Austen, *J. Appl. Physiol. 29*, 723 (1970).
6. S. Berger, E. W. Salzman, E. W. Merrill, and P. S. L. Wong, in: *Platelets: Production, Function, Transfusion and Storage* (M. G. Baldini and S. Ebbe, eds.), pp. 299–312, Grune & Stratton, Philadelphia (1974).
7. E. W. Salzman, S. Berger, E. W. Merrill, and P. S. L. Wong, *Thromb. Diath. Haemorrh. 59* (supplement), 107 (1974).
8. J. E. Silliman, Ph.D. Thesis submitted to MIT (1972).
9. J. C. Bray and E. W. Merrill, *J. Biomed. Mater. Res. 7*, 431 (1973).
10. J. C. Bray and E. W. Merrill, *J. Appl. Polym. Sci. 17*, 3779 (1973).
11. N. A. Peppas and E. W. Merrill, *J. Polym. Sci., Polym. Chem. Ed. 14*, 441 (1976).
12. N. A. Peppas and E. W. Merrill, *J. Appl. Polym. Sci. 20*, 1457 (1976).
13. N. A. Peppas and E. W. Merrill, *J. Appl. Polym. Sci. 21*, 1763 (1977).
14. N. A. Peppas and E. W. Merrill, *J. Biomed. Mater. Res. 11*, 423 (1977).
15. P. K. Weathersby, M. S. Thesis in Chemical Engineering, MIT (1971).
16. M. S. Morgan, Ph.D. Thesis submitted to MIT (1972).
17. G. H. Gifford, Jr., E. W. Merrill, and M. S. Morgan, *J. Biomed. Mater. Res. 10*, 857 (1976).
18. M. V. Sefton, D.Sc. Thesis submitted to MIT (1974).
19. P. J. Gilinson, Jr., C. R. Dauwalter, and E. W. Merrill, *Trans. Soc. Rheol. 7*, 319 (1963).
20. E. W. Merrill and E. W. Salzman, *J. Am. Soc. Artif. Internal Organs 6*, 60 (1982).
21. V. Sa da Costa, D. Brier-Russell, G. Trudel, III, D. F. Waugh, E. W. Salzman, and E. W. Merrill, *J. Colloid Interface Sci. 76*, 594 (1980).

22. E. W. Merrill, V. Sa Da Costa, E. W. Salzman, D. Brier-Russell, L. Kuchner, D. R. Waugh, G. Trudel, III, S. Stopper, and V. Vitale, *Adv. Chem. Ser. 199* (Biomaterials: Interfacial Phenomena and Applications), pp. 95–107, A.C.S., Washington, D.C. (1982).
23. N. A. Mahmud, S. Wan, V. Sa da Costa, V. Vitale, D. Brier-Russell, L. Kuchner, E. W. Salzman, and E. W. Merrill, in: *Physico-chemical Aspects of Polymer Surfaces* (K. Mittal, ed.), Vol. 2, pp. 953–968, Plenum Press, New York (1982).
24. E. W. Merrill, V. Sa da Costa, E. W. Salzman, D. Brier-Russell, L. Kuchner, D. R. Waugh, G. Trudel, III, S. Stopper and V. Vitale, in: *Adv. Chem. Ser. 199* (Biomaterials: Interfacial Phenomena and Applications), pp. 95–107, A.C.S., Washington, D.C. (1982).
25. E. W. Merrill, E. W. Salzman, S. Wan, N. Mahmud, and L. Kuchner, *Trans. Soc. Artif. Internal Organs 28*, 482 (1982).
26. N. Mahmud, D.Sc. Thesis submitted to MIT (1984).
27. R. W. Pekala, Ph.D. Thesis submitted to MIT (1984).
28. R. W. Pekala, M. Rudoltz, E. R. Lang, E. W. Merrill, J. Lindon, L. Kushner, G. McManama, and E. W. Salzman, *Biomaterials 7*, 372 (1986).
29. R. W. Pekala, E. W. Merrill, J. Lindon, L. Kushner, and E. W. Salzman, *Biomaterials 7*, 379 (1986).
30. C. Sung, Ph.D. Thesis submitted to MIT (1988).
31. C. Sung, M. R. Sobarzo, and E. W. Merrill, *Polymer 31*, 556–563 (1990).
32. C. Sung, J. E. Raeder, and E. W. Merrill, *J. Pharm. Sci. 79*, 829 (1990).
33. E. L. Chaikof, Ph.D. Thesis submitted to MIT (1989).
34. E. L. Chaikof and E. W. Merrill, *New Polymeric Materials 2*(2), 125–147 (1990).
35. E. L. Chaikof, E. W. Merrill, S. L. Verdon, L. L. Hayes, R. J. Connolly, and A. D. Callow, *Polymer Commun. 31*, 182 (1990).
36. E. L. Chaikof, E. W. Merrill, S. L. Verdon, J. E. Coleman, L. L. Hayes, R. J. Connolly, K. Ramberg, and A. D. Callow, presentation at the 1989 International Chemical Congress of Pacific Basin Societies Symposium on Chain Dynamics at Polymer Interfaces, Session 2: Interface Characterization and Modification, Honolulu (December, 1989).
37. E. L. Chaikof and E. W. Merrill, *J. Colloid Interface Sci. 137*, 340 (1990).
38. S. Laliberté Verdon, E. L. Chaikof, J. E. Coleman, L. Hayes, K. Ramberg, R. J. Connolly, E. W. Merrill, and A. D. Callow, submitted for presentation at the Annual Meeting of the SEM Society of America (1989); *Scanning Microscopy 4*, 341 (1990).
39. E. L. Chaikof, E. W. Merrill, J. E. Coleman, K. Ramberg, R. J. Connolly, and A. D. Callow, *AIChE J. 36*, 994 (1990).
40. A. Z. Okkema, T. G. Grasel, R. J. Zdrahala, D. D. Solomon, and S. L. Cooper, *J. Biomater. Sci., Polymer Ed. 1*, 43 (1989).
41. D. W. Grainger, K. Knutson, S. W. Kim, and J. Feijen, *J. Biomed. Mater. Res. 24*, 403 (1990).
42. M. V. Sefton, oral communication in discussion following his presentation at ACS Boston Meeting, April 1990, Division of Polymer Chem., Session E, Symposium Honoring R. Langer, Paper #41.
43. K. A. Dennison, Ph.D. Thesis submitted to MIT (1986).
44. S.-W. Tay, Ph.D. Thesis submitted to MIT (1986).
45. S.-W. Tay, E. W. Merrill, E. W. Salzman, and J. Lindon, *Biomaterials 10*, 11 (1989).
46. P. Lutz and P. Rempp, *Makromol. Chem. 189*, 1051 (1988).
47. Y. Gnanou, P. Lutz, and P. Rempp, *Makromol. Chem. 189*, 2893 (1988).
48. E. W. Merrill, K. A. Wright, R. W. Pekala, K. A. Dennison, C. Sung, E. Chaikof, P. Rempp, P. Lutz, A. D. Callow, R. Connolly, K. Ramberg, and S. Verdon, in: *Polymers in Medicine: Biomedical and Pharmaceutical Applications* (R. Ottenbrite, ed.), Technomic Publishing Co., Lancaster, PA (in press) (1992).
49. P. Rempp, P. Lutz, and E. W. Merrill, *Polym. Prepr. Am. Chem. Soc. 31*, 215 (1990).
50. E. W. Merrill, P. Rempp, P. Lutz, A. Sagar, R. Connolly, A. D. Callow, K. Gould, and K. Ramberg, *Proceedings*, Society for Biomaterials Annual Meeting, Charleston, S.C. (May 1990).

15

Properties of Immobilized PEG Films and the Interaction with Proteins

Experiments and Modeling

C.-G. GÖLANDER, JAMES N. HERRON, KAP LIM, P. CLAESSON, P. STENIUS, and J. D. ANDRADE

15.1. INTRODUCTION

Poly(ethylene oxide), or as it is frequently denoted in the literature, poly(ethylene glycol) (PEG), is a nonionic, water-soluble polymer widely used for stabilizing colloids in food and paints and in formulating pharmaceuticals and cosmetics. The reason for the extensive use of this polymer is that it acts as a dispersant and yet is inert, e.g., it does not interfere adversely with other functional ingredients in the dispersion.

Recently, some other potential applications have been identified within the biotechnology area (see Figure 1). It has been shown that substances covered with a PEG coating do not show antigenic activity. This can be utilized for camouflaging drugs that may otherwise cause allergic reactions in the body. In particular, successive therapy has been reported with PEG-encapsulated enzymes such as adenosin-deaminase (immunodeficiency disease), superoxide-dismutase (kidney transplanta-

C.-G. GÖLANDER and P. STENIUS • Institute for Surface Chemistry, S-11486 Stockholm, Sweden. JAMES N. HERRON and KAP LIM • Center for Biopolymers at Interfaces, and Departments of Bioengineering and Pharmaceutics, University of Utah, Salt Lake City, Utah 84112. P. CLAESSON • The Surface Force Group, The Royal Institute of Technology, S-10044 Stockholm; and Institute for Surface Chemistry, S-11486 Stockholm, Sweden. J. D. ANDRADE • Center for Biopolymers at Interfaces, and Department of Bioengineering, University of Utah, Salt Lake City, Utah 84112.

Poly(Ethylene Glycol) Chemistry: Biotechnical and Biomedical Applications, edited by J. Milton Harris. Plenum Press, New York, 1992.

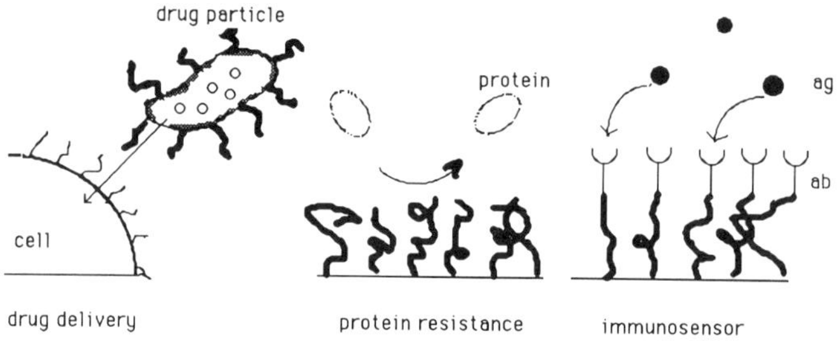

Figure 1. Three biotechnical applications for PEG coatings.

tion), and streptokinase (heart surgery). Other important areas are targeting of cytostatics in cancer therapy and the use of PEG/dextran mixtures in protein separation by affinity partitioning. In diagnostic assays and biosensors based on specific recognition of antigen–antibody (ag–ab) pairs, unspecific adsorption of proteins will cause a background noise. This can be avoided by covering the voids on the surface with PEG. Still another application of PEG is to incorporate this substance in a protein layer, which will keep the functional properties of the protein intact even when the protein film is stored in air. PEG here acts as a moisture preserver so that the native environment around the protein is retained. This property of PEG is also utilized in humectants.

It will be shown below that the molecular origin of these properties is quite intricate. We will start this PEG essay by describing some basic routes via which stable PEG films can be prepared and how the films can be studied physically and chemically by various surface analysis techniques. An important feature is that aqueous solutions of PEG show a lower consolute temperature (LCT), i.e., the solution splits into two phases (one with a very low concentration of PEG) above a critical temperature, called the cloud point. The correlation between this phase separation and the surface forces measured between two PEG-covered mica sheets will be discussed. The molecular behavior of PEG has also been studied by molecular dynamic simulations. The different theories for the molecular origin of PEG/water interaction will be briefly reviewed. Finally, we discuss how the properties of PEG are expressed in the interaction with proteins. A tentative explanation will be given as to the generally low protein adsorption found on PEG-coated surfaces.

15.2. PREPARATION OF PEG FILMS

The properties of PEG films, such as chemical stability, film thickness, and film composition, are influenced by the way they are prepared. It is difficult to cover a surface with a dense PEG film, because at temperatures below LCT, PEG molecules naturally repel each other in water. In most cases, other compounds, such as crosslinkers and film-forming agents, are also present in the film and may, particularly in thick films, induce chemical segregation both vertically and laterally in the film and also, in extreme cases, cause phase separation on the surface. Effects due to swelling in thick films may also obliterate the true PEG properties of the surfaces. In the following we discuss in some detail three different ways to realize a PEG film on a solid substrate: preparation of PEG hydrogels, chemical immobilization, and quasi-irreversible adsorption (see Figure 2).

15.2.1. PEG Hydrogels

A PEG hydrogel can be created by incorporating PEG in a polymerizable resin, which is first deposited on a substrate and then polymerized *in situ*. Acrylate or methacrylate resins, polymerized by a free radical mechanism, are suitable. The polymerization can be initiated thermally by azobisisobutyronitrile (AIBN) or peroxides (see, for example, Gregonis *et al.*[1]) or photochemically using photoinitiators like benzophenones, hydroxipropiophenones, or thioxanthones.[2] In many cases bifunctional (diacrylated) PEG has been incorporated into a crosslinked network. In such cases, as a consequence of the presence of the hydrophobic crosslinking agent, the final polymer backbone will have a mixed hydrophilic–hydrophobic character.

In order to obtain a coating with pendant PEG chains, we have utilized the monomethoxy PEG and prepared PEG-monoacrylate.[2] Coating resins containing

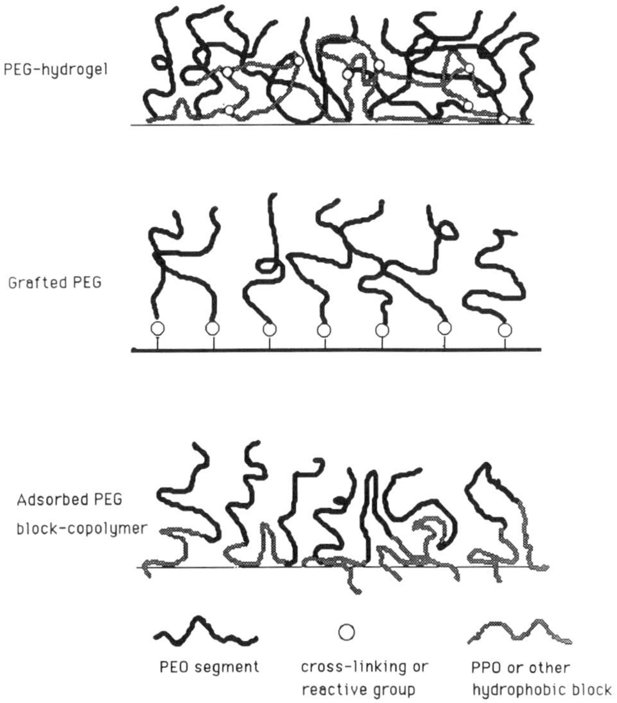

Figure 2. Schematic drawing showing the structural features of PEG layers obtained by different coating techniques.

either PEG-acrylate with molecular weight of 550, 1900, or 5000, or diacrylated ethoxylated trimethylolpropane [TMP $(EO)_{20}$], together with a crosslinker, hexanediol-diacrylate (HDDA), and 2-hydroxy-2-propiophenone (photoinitiator), are prepared in a toluene/ethanol/tetrahydrofurane mixture. A 1% solution was applied to a polymer surface with a 20-μm rod applicator. The film was allowed to dry and was then photocured. In order to maximize the ethylenoxide (EO) content in the film deposited on PVC, the molar ratio PEG-acrylate/HDDA was varied. The relative EO content was measured as the ratio between the electron intensity from carbon bound to a single oxygen (C—O) to that from CH_2 carbon in the C1s peak of the ESCA spectrum of the film. The EO content shows a maximum at a PEG/HDDA ratio between 1 and 2.[2] At higher ratios, the degree of crosslinking is lower. Consequently, film adhesion becomes poor. The EO content of the film increases with the molecular weight of the PEG monomer. However, beyond MW 1900, steric hindrance causes the EO content of the film to increase only slowly with molecular weight. Hence, molecular weights around 2000 seem to be sufficient for obtaining a high grafting yield and a high PEG content on the surface. At this MW the C—O/CH_2 ratio does not change much with the PEG/HDDA ratio. In all coatings prepared, the experimentally found C—O/CH_2 ratio was significantly lower than expected from stoichiometry. The most likely reason is that the polymer in the layer tends to minimize the free energy by migration of hydrophilic moieties to the bulk of the film, where a higher refractive index environment is offered than at the air interface. Consequently, hydrophobic segments will accumulate at the air interface. This process will be particularly pronounced in vacuum, where ESCA analysis occurs.

However, in water the situation is the reverse and polar groups migrate to the interface; this shows up in the low contact angles (around 15°) for these PEG hydrogels. The measured contact angle decreases as the C—O/CH_2 ratio increases.[2] The low contact angle measured for these hydrogels is also due to contributions from surface roughness, swelling, and capillary forces in porous structures. This is particularly clear when considering that higher values are measured for chemically immobilized monolayers of PEG (see below). For a constant C—O/CH_2 ratio, the contact angle decreased with increasing molecular weight of the PEG, which indicates that the PEG/water surface tension decreases in comparison to the PEG/air surface tension with increasing molecular weight (see below). Hydrogels are characterized by a high water uptake. The water content in swelled films of PEG 1900/HDDA 1:1 was around 45–50% w/w. A structural model of the PEG hydrogel is displayed in Figure 2.

To increase the EO content of the surface layer beyond the limits of the method described above, we have developed a two-step curing procedure shown in Figure 3. The coating is precured at low UV dosage to obtain a gel-like PEG coating, characterized by a polymer network with low crosslinking density, high mobility, and yet low water solubility. The substrate coated with the PEG gel is then exposed to water. This leads to migration of polar EO groups to the water interface. Finally, the layer is subjected to a high-dosage UV flash. The two-step procedure enhances the EO content at the interface, which increases the C—O/CH_2 ratio observed by ESCA approximately a factor of two for PEG gels with MW 1900.

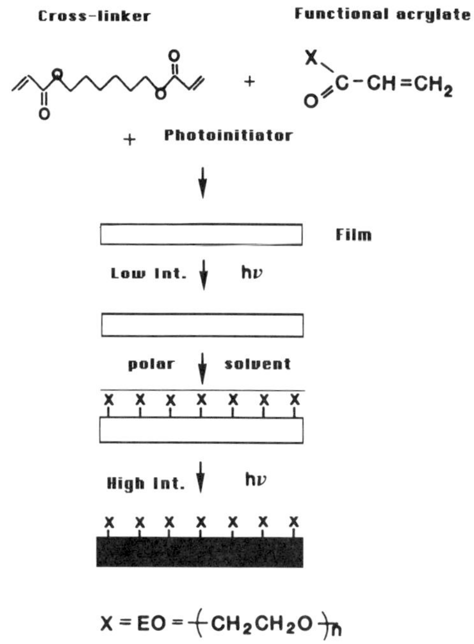

Figure 3. Two-step procedure for the photocuring of a PEG–acrylate layer for enhancement of the surface density of EO groups.

A layer containing equimolar amounts of PEG-acrylate (MW 1900) and HDDA coated onto PVC has excellent *barrier properties* against migration of hydrophobic substances like plasticizers. Hence, while the surface concentration of a dioctylphthalate plasticizer (measured by the fraction of CH_2 carbon in the C1s spectra) increased significantly with time on bare PVC (within hours at 20 °C and minutes at 60 °C) due to surface migration from the bulk, no change of the elementary composition with time occurred when PVC was covered with a PEG film.

15.2.2. Chemical Immobilization and Grafting

Various methods for covalent attachment of PEG to surfaces have been proposed. They usually require chemical derivatization of the terminal OH groups of PEG prior to reaction with a functionalized surface. Abuchowski *et al.*[6] introduced cyanuric chloride activation of PEG for reactions with enzymes in order to render them nonimmunogenic. Bückmann *et al.*[7] described the preparation of PEG ligands with bromide, amine, sulfonate and *N*-hydroxy succinimide for affinity partitioning of proteins, and Zalipsky *et al.*[8] prepared amino-, isocyanato-, and carboxylated PEG for attachment to drugs. Harris *et al.*[9] developed some improved and versatile methods to prepare PEG derivatives such as tosylate, amine, and aldehyde. They also studied how various PEG coupling reactions influence the activity of enzymes.[10] In one of our groups, Yee *et al.* tried to use the thiol–disulfide interchange reaction with a dithiolated PEG to immobilize a protein to quartz via a PEG spacer.[23] An excellent overview of various derivatization reactions has been given by Harris.[11]

We have used the aldehyde–amine reaction for immobilization of PEG–aldehyde to aminated solid surfaces. This reaction is convenient to use in aqueous media and can be driven to completion by addition of $NaCNBH_3$, a reducing agent that selectively reduces the imine product —CH=N—) in the presence of aldehyde. The only side reaction that occurs is polymerization of PEG–aldehyde by aldol condensation. PEG–aldehyde can be prepared by various reactions, some of which are shown in Table 1.

Of the first four reactions that we tried, the best yield (NMR) was found with the partial oxidation in acetic anhydride/DMSO (#4), proposed by Harris *et al.*[9,12] Harris has pointed out that the PEG–CHO is not chemically stable but may polymerize by aldol condensation. This does not occur with PEG–benzaldehyde or PEG–propionaldehyde (see reaction #5). However, the experimental results presented below are all based on PEG–CHO prepared by the #4 procedure.

A very important aspect of the properties of PEG surfaces is the surface density. The interpretation of results on PEG surfaces becomes ambiguous if only partial coverage of the surface is obtained. In order to increase the surface coverage we have performed the immobilization reaction under solution conditions close to the cloud point, where repulsion between PEG chains are small.

Figure 4 shows schematically the coupling procedure and Figure 2 the structural model of the PEG surface. The cloud point of PEG 1900 is around 180 °C. To induce clouding at realistic reaction temperatures, "salting out" with potassium sulfate was

Table 1. Some Derivatization Reactions for Preparing PEO–CHO

1. Oxidation Ce^{4+} or pyridinium chloroformate (PCC)

$$C_5H_5N + CrO_3 \xrightarrow{H} (PCC)$$

$$CH_3\text{—}(EO)_n\text{—}OH + PCC \xrightarrow{\text{toluene}} CH_3\ (EO)_n\text{—}O\text{—}CH_2\text{—}CHO + (CrO_2)$$

2. Choloracetaldehyde diacetal

$$CH_3\text{—}(EO)_n\text{—}OH + ClCH_2CH\ (OCH_3)_2 \xrightarrow{NaOH} CH_3\text{—}(EO)_n\text{—}O\text{—}CH_2CH\ (OCH_3)_2$$

$$\xrightarrow{H_2SO_4} CH_3\text{—}(EO)_n\text{—}O\text{—}CH_2CHO + 2CH_3OH$$

3. Aluminum tert-butoxide

$$CH_3\text{—}(EO)_n\text{—}OH + (CH_3)_2C{=}O \xrightarrow{Al(OC_4H_{10}\text{-}t)} CH_3\text{—}(EO)_n\text{—}CHO + (CH_3)_2CHOH$$

4. $DMSO/(CH_3CO)_2O$

$$(CH_3)_2S{=}O + (CH_3CO)_2O + CH_3\text{—}(EO)_n\text{—}OH$$

$$\rightarrow CH_3\text{—}(EO)_{n-1}\text{—}OCH_2CHO + 2CH_3COOH + (CH_3)_2S$$

5. Propanedithiol/chloropropionaldehyde diacetal

$$HS\text{—}CH_2CH_2CH_2\text{—}HS \xrightarrow{NaOCH_3} Na\text{—}S\text{—}CH_2CH_2CH_2\text{—}S\text{—}Na$$

$$\xrightarrow{Cl\text{—}CH_2CH_2CH\text{—}(OC_2H_5)_2} Na\text{—}S\text{—}CH_2CH_2CH_2\text{—}S\text{—}CH_2CH_2CH\text{—}(OC_2H_5)_2\ (+NaCl)$$

$$\xrightarrow{(EO)_n\text{—}OTs,H_2SO_4} (EO)_n\text{—}S\text{—}CH_2CH_2CH_2\text{—}S\text{—}CH_2CH_2\text{—}CHO\ (+NACl)$$

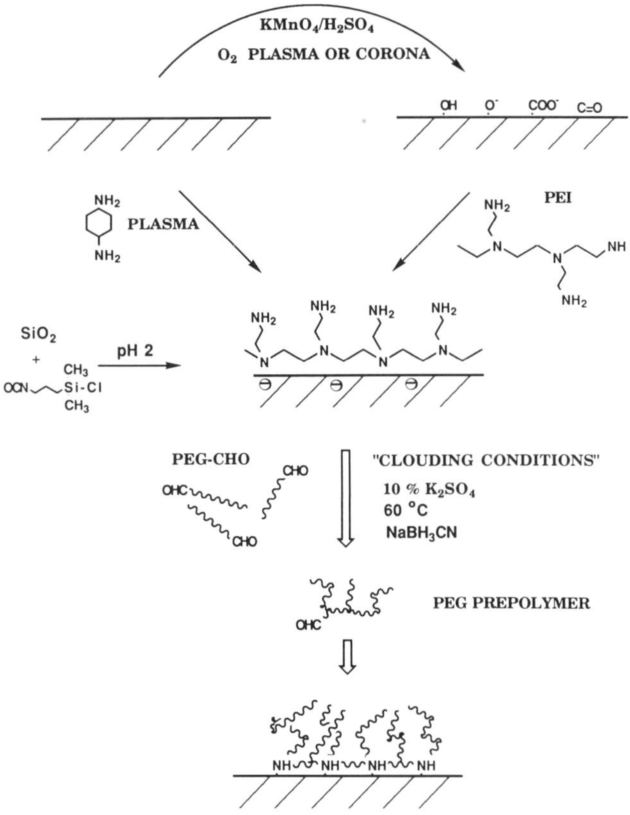

Figure 4. Grafting of PEG by the Schiff base reaction between PEG–CHO and surface-NH_2 on surfaces aminated in various ways.

used.[14] However, due to steric hindrance, the reaction is sluggish. In order to find the conditions that give maximum PEG coverage, we determined both the C—O/CH_2 ratio in the layer and the attenuation of the signals from substrate atoms in the ESCA analysis. The optimal conditions for coupling aldehyde–PEG to aminated surfaces were found to be pH6,[14] 60 °C, 10% K_2SO_4, and 40 hours reaction time.

Figure 5 compares the carbon 1s peak from PEG surfaces prepared under optimal conditions with the peak obtained when immobilization was accomplished in pure water at 20 °C and in a 10% w/w K_2SO_4 solution at 60 °C for the same reaction time. In both cases $NaBH_3CN$ (reducing agent) was added. The figure clearly shows the improvement in PEG grafting density (C—O/CH_2 ratio) when using close to phase separation conditions. These PEG grafted surfaces are chemically very stable in the sense that no significant change occurred in the ESCA spectra after prolonged rinsing with water, 0.1M NaOH, 0.1M HCl, ethanol, or 10% trifluoroacetic acid; see Figure 6.

To prepare a substrate surface with amino groups for PEG immobilization is not trivial. Yet it is critical, since a high density of amino groups is required if a dense

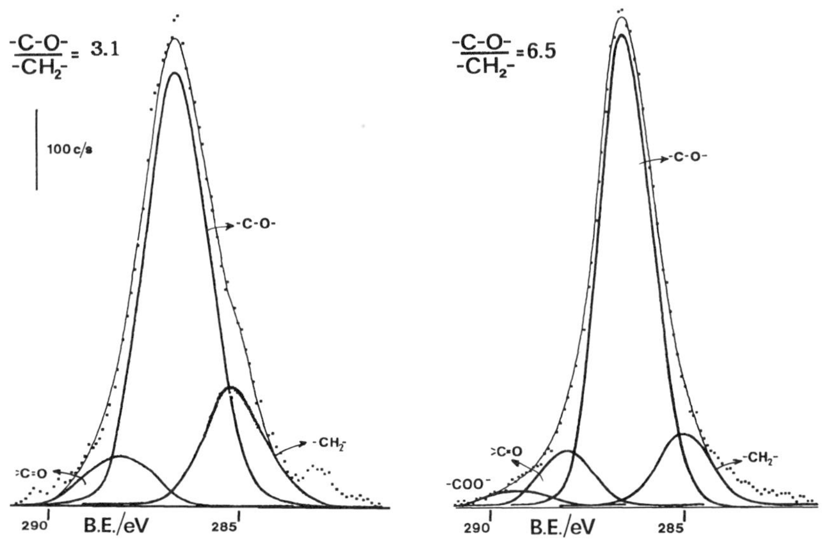

Figure 5. Carbon 1s spectrum from PEG films obtained by immobilization of PEG-aldehyde to an aminofunctional surface. Left, in pure water, 20 °C; right, in a 10% w/w K_2SO_4 solution, pH 6, 60 °C. The signal is dominated by a peak centered at 286.5 eV corresponding to C—O carbon in PEG. The C—C peak at 285 eV originates from the substrate (and contaminants) and the peak at 289.2 eV originates from carboxylate species.

packing of PEG is to be obtained (at least 1 NH_2 per nm^2). These groups must also be accessible for reaction. We have used three ways to achieve this (see Figure 4): *"irreversible" adsorption* of polyethyleneimine on an oxidized polymer surface,[16,17] *plasma polymerization* of diaminocyclohexane (DACH)[19] on any surface (allylamine was used by Gombotz *et al.*[18]) and *silanization* of silicon oxide with aminopropyl trimethoxy silane (APS)[21] or isocyanatopropyl dimethyl monochlorosilane (IPS), (hydrolyzed to $-NH_2$ after immobilization)[20] yielding densities well above 1–2 NH_2 per nm^2.

PEG–CHO was reacted in the way described above with the aminated surfaces.

The agreement between the amounts of immobilized PEG measured using various methods is good (see Table 2). However, measurements in wet and dry states give different thickness and refractive index values. The standard procedure using PEI as an amino functional substrate results in a layer with 2 mg per m^2 PEG or 1.5 nm^2 per molecule. In the dry state this means a thickness of approximately 2.5 nm and a refractive index which agrees with the bulk value for PEG (n = 1.44). The low refractive thickness (n = 1.36) and relatively thick film d = 10 nm measured in water indicates considerable swelling. The calculated water content in this PEG monolayer (from refractive index) is >70% w/w, which is considerably higher than for the hydrogels. The contact angles are much higher than for the PEG hydrogels: $\Theta_{adv} = 56 \pm 3°$ and $\Theta_{rec} = 35 \pm 2°$. The high values indicate a nonzero surface tension against water and/or a rather low surface tension against air. Hence, at 20 °C we obtained

Table 2. Quantification of Covalently Attached PEG Layers (MW 1900) on Various Substrates

Method	Substrate	NH_2 layer	Γ ($mg\,m^{-2}$)	*d* (nm)	*n*
Ellips[a] (dry state)	SiO_2	DACH[19], 5 nm	1.2	1.5	1.44
Ellips[a] (wet state)	SiO_2	PEI, 3 nm	2.0[b]	10	1.36
Ellips[a] (dry state)	SiO_2	PEI, 3 nm	2.0[b]	2.5	1.44
Ellips[a] (dry state)	SiO_2	IPS silane	1.2[b]	1.6	1.44
SFA[c] (wet state)	Mica	PEG–lysine	1.1[d]	3.5	
SFA[c] (wet state)	Mica–OH	IPS silane	1.1[d]	4.0	
ESCA	Mica–OH	IPS silane	1.2[d]		
QCM[e]	SiO_2	APS > 10 nm	7.0		

[a]Obtained from ellipsometry.[3]
[b]Calculated from Cuypers model.[50]
[c]Obtained from surface force measurement.[4]
[d]Obtained from quantitative ESCA analysis.[5] Agrees well with values obtained from ellipsometry on the same surface film on a silicon wafer.
[e]Obtained from measurements with a surface acoustic wave (SAW) device.

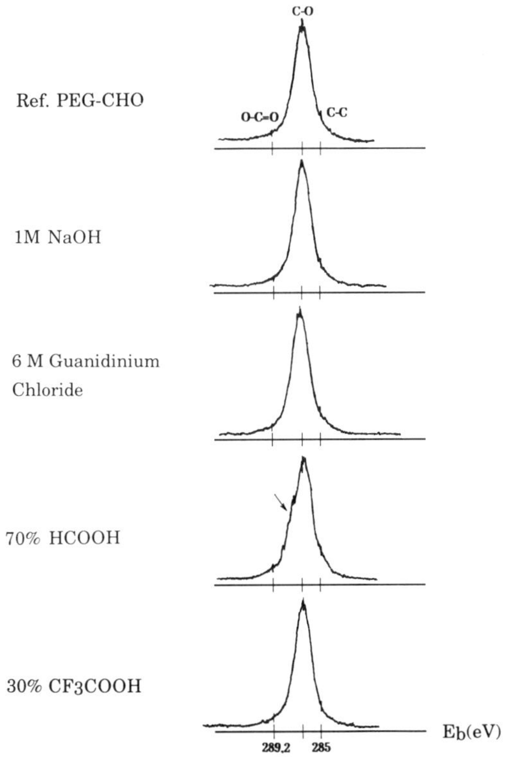

Figure 6. Carbon 1s spectra from a PEG film immobilized on polyethylene and rinsed in water 1 hour (Ref. PEG–CHO), 1 M NaOH, 6 M guanidinium chloride, 70% HCOOH, or 30% CH_3COOH. Only in the case of formic acid rinsing is there a small asymmetry in the peak indicating interaction between the formic acid and ether in PEG.

$\gamma = 44\ mN\,m^{-1}$, by straight line extrapolation from pendant drop measurements on melts of PEG, MW 1900.[51] The surface tension decreases with temperature to 36 $mN\,m^{-1}$ at 140 °C. We also measured a (small) surface tension increase with molecular weight [41 $mN\,m^{-1}$ for monomethoxy PEG (MW 550) and 47 $mN\,m^{-1}$ for PEG (MW 6000)]. This is most likely due to purely entropic factors.

Adsorption of a cationically modified PEG (PEG–lysine) to (negatively charged) mica at room temperature resulted in an adsorbed amount of 1.1 mg per m^2 PEG. An increase in temperature to 55 °C (close to the conditions used above during immobilization) increased the adsorbed amount to the vicinity of 2 mg per m^2, obtained above for grafted layers.

Table 2 shows that less PEG can be attached to IPS or DACH layers[19] (1.2 mg per m^2 or 2.5 nm^2 per molecule) although they carry >1 reactive amino group per nm^2 than PEI or APS layers. It is likely that, on the compact and rigid IPS or DACH films, the amino groups are less accessible to reaction than the mobile side chains in the PEI or APS films.[16]

The polymerizable APS results in an extremely thick (>20 monolayers) and rough layer. The large amounts of PEG in this layer indicate a three-dimensional PEG–APS network similar to a hydrogel.

To covalently attach PEG to mica surfaces is important, since this would allow direct measurement of the interactions between PEG layers under various conditions employing the surface force technique.[4] Several methods have been tried. One way is to first introduce silanol groups (SiOH) on mica by means of water vapor plasma treatment.[15] In the next step IPS vapor is allowed to react with the modified mica surface, the isocyanato group is converted to amine, and finally the PEG–aldehyde coupling reaction is carried out as described above. Alternatively, once the silane has been bonded to the mica surface, the isocyanate group may be reacted with a PEG melt (110 °C) or PEG solution, or the amine group can be reacted with a PEG–aldehyde melt or a PEG–aldehyde solution at elevated temperatures. The results obtained so far are summarized in Table 3.

Reaction with PEG–OH melt gave a slightly better yield than the solution reaction. However, PEG–CHO melt gave less yield than the solution reaction, probably due to oxidation of PEG–CHO at elevated temperature. PEG–benzaldehyde also gave less yield than the PEG–CHO reaction in solution. The comparatively low

Table 3. ESCA Quantification of PEG Layers Immobilized by the Aldehyde–Amine or Isocyanate–Hydroxyl Reactions on Mica

Sample	Reactant	Γ ($mg\,m^{-2}$)
Mica–IPS	PEO–OH melt	1.5
Mica–NH_2	PEG–CHO melt	0.6
Mica–NH_2	PEG–CHO solution	1.1
Mica–NH_2	PEGΦCHO solution	0.6

chemical "stability" of the monomeric PEG–CHO hence seems to be advantageous. A possible explanation is that the aldol condensation in solution leads to oligomers with residual reactive aldehyde groups, so that more PEG is bound by a single reaction event at the surface (Figure 4).

15.2.3. Quasi-Irreversible Adsorption

By this term we denote physical adsorption of high molecular weight copolymers of PEG which attach at multiple adsorption sites. Although the free energy of adsorption for each site may be relatively small, the attachment of a molecule to several sites leads to a multiplication effect, so that the total free energy of adsorption of a polymer becomes quite large. For this reason, polymers tend to be adsorbed either very strongly or not at all. Frequently, the adsorption isotherm will rise steeply at such low concentrations that, for all practical purposes, adsorption appears to be irreversible.

In one of our groups, we have studied how the structure of block-copolymers of PEO/PPO/PBO (polyethyleneoxide/polypropyleneoxide/polybutyleneoxide) influences the adsorption/desorption kinetics at the water/air and at the polyethylene/water interface.[24]

The block-copolymers that were used and the amounts adsorbed after 30 minutes at 20 °C, as well as the amount of polymer remaining on the surface after desorption for 30 minutes, are shown in Figure 7.

While the PEO/PBO/PEO triblock adsorbs in large amounts, the film also easily desorbs in water. This is most likely due to the self-aggregation properties induced by the hydrophobic PBO block. With preferentially hydrophilic PEO segments (which are very weakly attached to the surface) on the outside, these aggregates will readily desorb. For PEO/PPO star copolymers, close packing is severely restricted as a result of the high mobility of PEO tails. However, the (thin) films formed are quite stable against dissolution, evidently due to strong interaction between the hydrophobic core of the star polymer and the hydrophobic surface. A PEO/PPO alternate block obviously offers the largest hydrophobic contact area with the surface, so that a stable film with reasonable PEO density results. The PEO/PPO/PEO triblock hardly adsorbs at all.

15.3. INTERACTION BETWEEN PEG AND WATER

15.3.1. Theories and Models Describing the PEG–Water Interaction

The solubility of PEG in water is characterized by a closed immiscibility loop, i.e., there exists both a lower and a higher consolute temperature. This is the result of the unusual temperature dependence of the molecular interaction between PEG and water. We now proceed to discuss the origin of this behavior starting with a brief thermodynamical discussion.

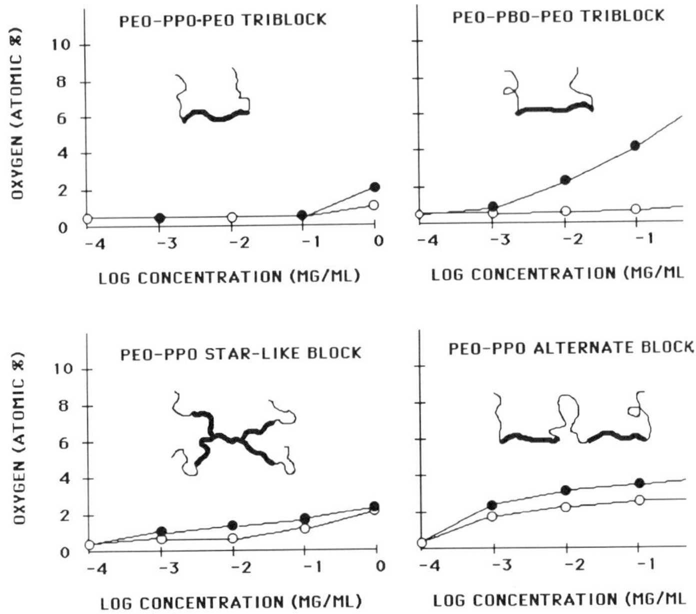

Figure 7. Relative adsorption of various PEG block-copolymers on polyethylene as determined from the relative oxygen content from ESCA analysis. The following block-copolymers were used: $PEO_{13}PPO_{30}PEO_{13}$ triblock; $PEO_{13}PBO_{25}PEO_{13}$ triblock; $(PEO_{26}PPO_{29})_2NCH_2CH_2N(PEO_{26}PPO_{29})_2$ star-like; $(PEO_{13}PPO_{30})_3$ alternate block. Redrawn from Ref. 24 and published with permission from John Wiley & Sons.

Hydrophobic molecules dissolved in aqueous solutions cannot form hydrogen bonds with water. As a consequence the surrounding water has to associate into "cage"-like structures in order to minimize the free energy of the system. The hydration of hydrophobic molecules results in an increase in water–water hydrogen bonds, a lowering of the entropy, and a lowering of the enthalpy. Consequently, when two hydrophobic units are brought together in water it results in a large increase in entropy ($\Delta S > 0$), a large compensating increase in enthalpy ($\Delta H < 0$), and a decrease in free energy ($\Delta G < 0$). This is the so-called "hydrophobic effect" which is also characterized by a decrease in heat capacity ($d\Delta H/dT<0$), i.e., as temperature increases the process becomes enthalpically more and entropically less favorable in a manner such that $\Delta G<0$ becomes almost independent of temperature.[25]

Kjellander and Florin-Robertsson[26] started from the idea that the PEG chain can be perfectly accommodated in a hexagonally water lattice. They argued that the hydration of a PEG chain is essentially hydrophobic, but a modification is imposed by the ether oxygen that can participate in hydrogen bonds with water. In Kjellander's model the enthalpy change (ΔH repulsive) and the entropy change (ΔS, attractive) caused by the association of two PEG chains are both large, but they nearly cancel in the expression for the free energy. To explain the phase behavior of the PEG–water system one must consider these two contributions and the ideal entropy of mixing

(ΔS_i). It is assumed that ΔS_i is small compared to ΔS at room temperature. Hence, the free energy change (ΔG) due to association can be written as: $\Delta G = \Delta H - T\Delta S - T\Delta S_i$.

At low temperatures (T) the free energy is negative and no phase separation takes place. As the temperature increases, the entropy term ($T\Delta S$) becomes more important and, above the cloud point, will cause the free energy of association to become negative. For PEG with a molecular weight of 1900 g mol^{-1} this occurs at a temperature of 180 °C.[27] It is important to realize that in Kjellander's model the phase separation of the PEG–water system is a consequence of the hydrated structure around the PEG chain. Only at temperatures considerably above the cloud point does the hydration of the PEG chain begin to vanish due to the increased thermal energy. When this occurs, both ΔH and $T\Delta S$ become less important. Instead the $T\Delta S_i$ term will eventually dominate, and the entropy of mixing will make the PEG–water system completely miscible again at high enough temperatures. Theoretical phase diagrams that agreed well with experiment could be calculated from Kjellander's model.

Two other models for the near-anomalous temperature behavior have been proposed (see Figure 8). Goldstein[28] suggested a two-state model with water hydrogen bonded or nonbonded to PEG, giving rise to repulsive and attractive domains in the PEG molecule. By treating the mixture by the Flory–Huggins statistical mechanical theory for solutions, with a temperature-dependent χ parameter, he was able to predict the presence of a solubility gap. At low temperatures repulsive hydrogen-bonded domains dominate, and at elevated temperature nonhydrogen-bonded domains dominate.

Karlström suggested a completely different explanation[29] for the phase separation of PEG in water and other polar solvents.[30] In his view, the change in interaction is related to temperature-induced conformational changes in the PEG chain. The oxygen atoms prefer a *gauche* conformation around the C—C bond and a *trans* conformation is preferred around the C—O bond. This leads to a high dipole moment for the segment, and consequently a strong interaction with water. As the temperature increases, more segments adopt conformations with smaller dipole moments, causing the interaction with water to become less favorable. Karlström's calculations predicted a phase diagram in reasonable agreement with experiment.

Karlström's model was further developed for application to PEG molecules *terminally attached* to surfaces. Using Scheutjen and Fleer's mean-field theory based on a lattice model and discrete description of layers, the same group[31] demonstrated that this two-state model gives a segment distribution where the nonpolar state dominates close to the surface while the polar state dominates further out. The segment density close to the surface and far out is large, and in the intermediate range it is lower. An increased grafting density results in a more extended conformation, which decreases the segment fraction close to the surface.

To model how adsorbed or *grafted PEG* layers interact, the scaling approach[32] based on self-similarity is very useful. It provides a simplified statistical mechanical description of adsorbed or grafted polymers without predicting precise numerical values for all adherend parameters. Considering van der Waals and steric forces and

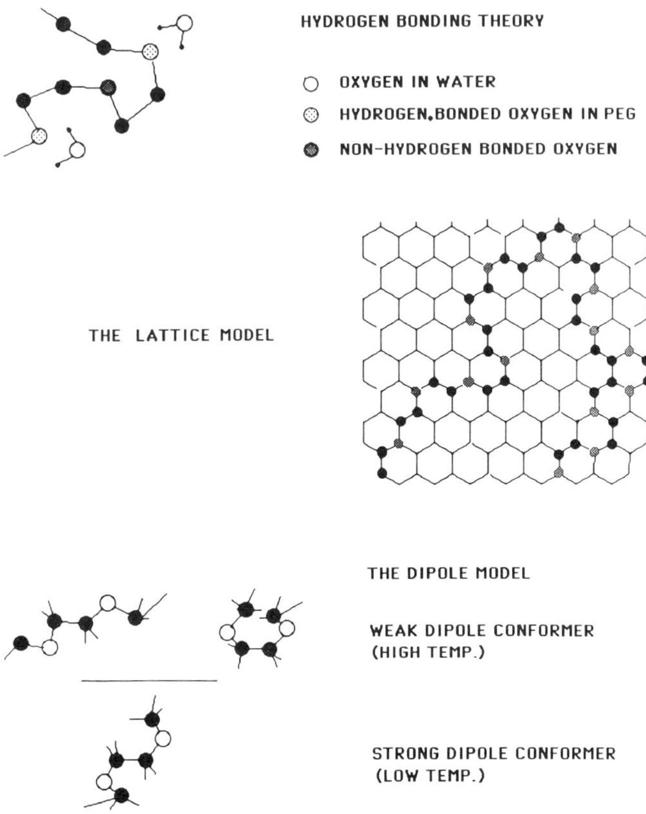

Figure 8. Simplified picture for three different models for the interaction between PEG and water as proposed in the literature; see text.

using this approach, Jeon *et al.*[33] could describe how the range of the steric repulsion between grafted polymer "brushes" of PEG varies as a function of packing density, thickness of the layer, and size of the PEG molecule (see Figure 9).

15.3.2. Computer Modeling of the Properties of PEG in Solution

Although statistical mechanical methods such as scaling concepts and self-consistent field theories of polymer chains are useful in predicting macroscopic properties of grafted PEG of high molecular weight[32] modeling of smaller PEGs at atomic scale can give further insight into the behavior of individual chains.

We have performed molecular dynamic (MD) simulations with 14 EO unit PEG chains and octadecane chains, respectively, at 300 K *in vacuo* and in water using the DISCOVER program (Biosym, USA).[34] The partial charge used was −0.30 eu for PEG oxygen and −0.82 eu for water oxygen. The nonbonding interaction is the most

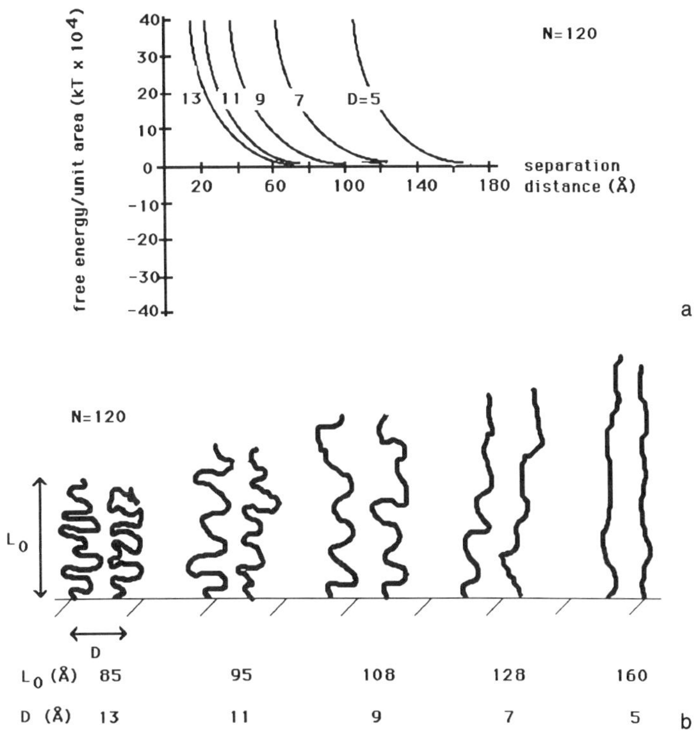

Figure 9. (a) Calculated force versus distance plots for the interaction between (b) grafted PEG-brushes obtained from de Gennes scaling theory, where N is the number of EO units in the molecule (redrawn from Ref. 33). D is the distance between attachment points and L_0 the extension of the PEG molecule from the surface.

difficult aspect of the potential function used in the molecular simulation. As with other molecular mechanics program, the DISCOVER program has the Lennard-Jones function and an electrostatic function to describe the nonbonding interaction.[38] The selection of parameters used in these calculations is important to describe the intermolecular interaction accurately. Although DISCOVER uses no explicit function for hydrogen bonds, it is at least partially considered in the electrostatic function.

The results for octadecane and PEG were compared in terms of chain flexibility and interaction with water. For an *octadecane* molecule *in vacuo*, with C—C bonds that are initially in the all-*trans* conformation, the torsional energy barrier is higher than the thermal energy level ($RT = 0.60$ kcal mol^{-1}). However, cooperative conversion between neighboring torsion bonds and deformations in bond length and bond angles may occur. This results in an overall oscillating motion of the hydrocarbon chain. In contrast, the motion of a *PEG* chain *in vacuo* is significantly different due to the fact that the C—O bond has a lower energy barrier between *gauche*$^-$, *trans*, and *gauche*$^+$ conformations (see Figure 4 in Chapter 3, this volume). However, the most favorable conformation for the C—O bond is *trans*. Since *in vacuo* a PEG chain can only interact with itself, each of the atoms in the molecule is attracted by others and soon the chain adopts a compact, coiled form. Simulations *in vacuo* thus demonstrate large differences in the internal structure of a PEG and an octadecane chain.

Differences of intermolecular interactions for octadecane and PEG can also be recognized with MD simulations in water. For the octadecane–water system, phase separation occurred in the simulation. The first layer of water molecules around the hydrocarbon chain was aligned parallel along the axes of the chain. The radial distribution function $g(r)$ of *octadecane in water* is illustrated in Figure 10a, as determined from the 10-picosecond molecular trajectories of one octadecane and 478 water molecules in a 20×20×35 Å box with periodic boundary condition. The first maximum of the pair correlation function $g(r)$ for C—O and C—H are located at almost the same distance, which indicates that the plane formed by the H—O—H bond angle in water is parallel to the axis of the hydrocarbon molecule. The low intensity of the two closest maxima as compared to that of the O—O pair correlation function for the water–water interaction shows that the density of water molecules in the first shell around octadecane is lower than that around a water molecule. These results are consistent with a previous Monte Carlo study of n-butane in water.[35]

The radial distribution functions for *PEG–water*, obtained by MD simulations, were (as shown in Figures 10b and c) different than those for octadecane–water. These data were obtained from a 10-picosecond simulation of one PEG and 579 water molecules in a 20×20×45 Å box with the periodic boundary condition. The structural order of the first shell of water around the carbon atoms of PEG is not as rigid as that for octadecane. Further, functions $g(r)$ for C—O and C—H are, as compared with the case of octadecane–water, more distinguishable from each other (Figure 10b) with a higher and narrower first maximum for C—O. The reason for the reduced ordering of water around the carbon atoms of PEG is the helical structure of the PEG chain and the presence of oxygen atoms in PEG. The PEG oxygen–water interaction disrupts the water structure around the carbon atoms in PEG. The small first intensity maximum in $g(r)$ for PEG oxygen and water shows that the hydrogen bond is not fully established between these two elements (Figure 10c). It has been shown that the greater partial charge of an alcohol group enables stronger binding of water.[36] The small distance between the adjoining pair of oxygen atoms of PEG (⩾2.84 Å) also contributes to the lack of full hydrogen bonding with water.[37] However, unlike polyoxymethylene (POM) that has only one methylene group between the adjoining pair of oxygen atoms, an additional CH_2 group in PEG prevents exclusive oxygen–oxygen interactions. *Gauche* conformation is preferred in POM due to this oxygen–oxygen interaction, while *trans* is preferred for C—O of PEG. These differences are reflected in molecular ball models where POM shows two "faces": one hydrophobic and one hydrophilic, while PEG is more symmetric with evenly distributed polar (C—O) and nonpolar (C—C) groups facing water.[46] Water molecules dampen the motion of PEG, as its conformation changed very little after a 20-picosecond MD run (Figure 11). Hence, simulation results with PEG and

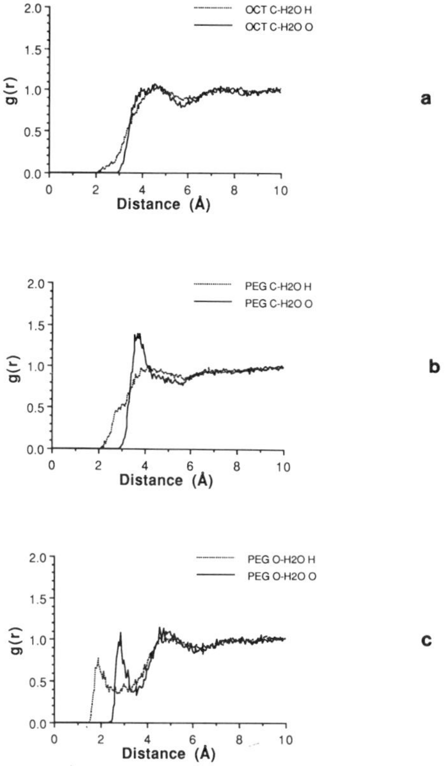

Figure 10. The radial distribution functions for PEG-water obtained from 10-ps molecular dynamics trajectories.

octadecane indicate that molecular differences of these two molecules can be demonstrated at least in a qualitative manner. Simulation of *grafted* PEG *in vacuo* shows that the PEG chains can adopt various equilibrium conformations with different densities depending on initial conformation, temperature, etc. This is described in more detail by Lim and Herron in Chapter 3 of this volume. Briefly, twenty 40-unit PEG chains fixed at one end collapsed to a thickness of 15–20 Å when the initial conformation was all-*gauche*$^-$. Such a thickness conforms to experimental data obtained from the air interface. The *gauche* conformation of the C—O bond is energetically unfavorable compared to *trans*, and when the PEG chains of this conformation interact with each other, the chains are forced to assume a more compact overall structure. However, when the initial PEG conformation is the crystal structure with *trans*–*gauche*$^+$–*trans* (CO as *trans* and C—C as *gauche*$^+$), the grafted PEG chain remains in its extended helical form throughout the duration of the simulation (the equilibrated thickness is 50 Å).

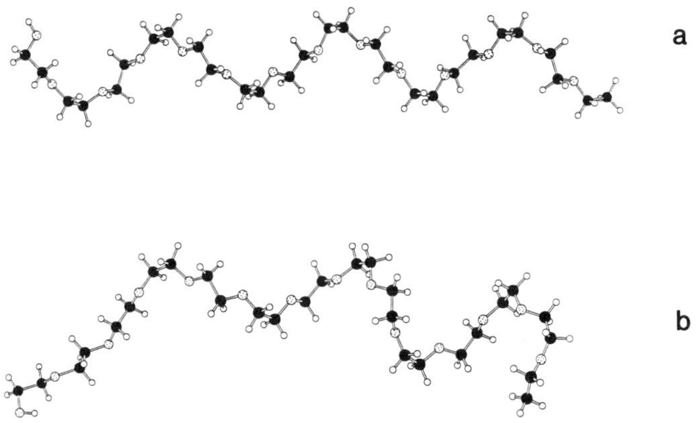

Figure 11. Conformation of a helical PEG molecule (a) before and (b) after a 20-ps MD run in water.

15.3.3. Direct Force Measurements between PEG Layers

Some investigations of the forces acting between PEG-coated mica surfaces have been reported. In a first study, PEG 1900 with a terminally attached positively charged lysine group, PEG–lysine, was adsorbed from solution.[22] The forces between mica surfaces without any PEG coating in dilute electrolyte solutions are dominated at large separations by a repulsive double-layer force. At a separation of less than 2–3 nm, this force is overcome by a van der Waals attraction in perfect agreement with predictions based on the DLVO theory.

In 10^{-4} M PEG–lysine, the long-range force is, at room temperature, still dominated by a double-layer force originating from an incomplete neutralization of the mica lattice charge. However, at separations less than 10 nm an additional repulsion appears. This repulsion, which completely overcomes the van der Waals attraction, is due to the interaction between adsorbed PEG chains, in particular the tails. The same forces were measured on approach and on separation, indicating a quasi-equilibrium situation.

The interaction becomes much more complicated when the PEG–lysine concentration is increased to 5×10^{-3} M (Figure 12). The force is purely repulsive and considerably stronger on compression than on decompression. The reason for this hysteresis can be either that PEG–lysine molecules loosely associated with the surface (e.g., as counterions) are forced to leave the gap between the surfaces as they are brought together, or that slow conformational changes take place in the adsorbed layer. For this low molecular weight PEG chain we favor the former interpretation. Klein and Luckham[39] have observed a similar hysteresis in the force curve between mica surfaces carrying adsorbed homopolymers of PEO (MW 40,000 g mol^{-1}).

Interestingly, when inorganic salt is added (0.1 M KBr) or when the temperature is increased to 55 °C, the hysteresis in the force–distance profile between PEG–lysine coated mica surfaces disappears. When salt is added, the range of the force decreases

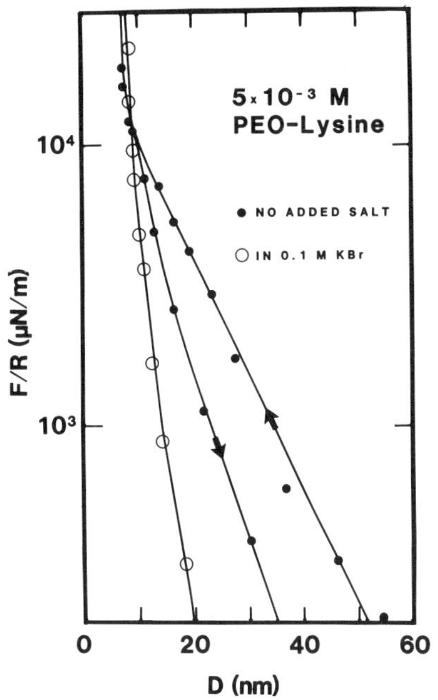

Figure 12. The influence of an addition of electrolyte on the force versus distance profile for electrostatically adsorbed PEG-lysin. (From Ref. 22, published with permission from Academic Press.)

dramatically from 50 nm to 20 nm (Figure 12). This is due to a reduction in the double layer and to a replacement of large PEG–lysine counterions for smaller potassium ions. This also explains the disappearance of the hysteresis in the force– distance profile. When the temperature is increased the PEG chain becomes more hydrophobic, which results in an increased adsorption and a more compact layer. The increased adsorption also results in a lowering of the surface charge density and less associated PEG–lysine counterions, which rationalizes the disappearance of the hysteresis.

For *grafted* PEG-OH 1900 (to isocyanatopropyl silanized water-vapor plasma treated mica surface), so far only forces in pure water at 20 °C have been investigated.[42] A repulsion similar to that observed between adsorbed PEG–lysine layers is present at separations below 30 nm. However, in comparison with adsorbed PEG, a less repulsive interaction was observed down to a separation of 9 nm (Figure 13). This supports the suggestion that hysteresis in the force–distance curve is largely due to displacement of polymers from between the surfaces. In fact, it is possible that the presence of a small hysteresis between surfaces with grafted PEG is due to the presence of some polymers not covalently attached to the surface (due to insufficient rinsing after the last reaction step). The compressed layer thickness (about 4 nm on each surface) is the same for adsorbed and grafted PEG.[41,42] The layer thickness is larger than twice the solution radius of gyration, R_g = 7–8 Å. Rather, it is in better agreement with the approximately $6R_g$ (here around 4 nm Å). This indicates a rather

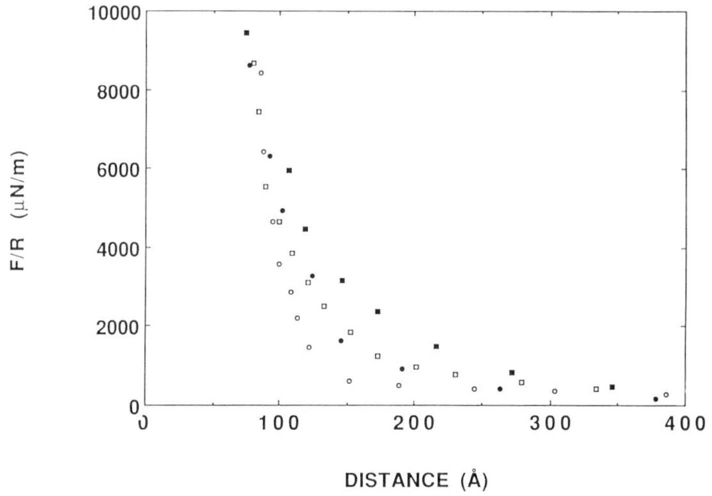

Figure 13. Normalized force (*F*/*R*) versus distance (*D*) measured between mica surfaces bearing anchored PEG layers with MW 1900. ●, ■, electrostatically attached PEG-lysine; ○, □, grafted PEG, MW 1900. Squares, measured on approach; circles: measured on separation. From Ref. 42. Published with permission from Steinkopf Verlag Darmstadt.

extended conformation, in qualitative agreement with the results obtained by Taunton *et al.*[47] and with de Gennes's scaling considerations for a polymer brush in a good solvent.[32,48]

The behavior of PEG 1900 close to the cloud point temperature (about 180 °C) could not be investigated for practical reasons. However, a similar study has been performed for a nonionic surfactant, $C_{12}EO_5$, which has a cloud point at 27 °C.[40] The force between $C_{12}EO_5$ layers adsorbed to mica which had been hydrophobed by Langmuir–Blodgett deposition of a layer of dioctadecyl dimethylammonium ions is, at large separations, dominated by a weak double-layer force which does not change with temperature. At separations larger than 4.5 nm, the force shows a dramatic temperature dependence as indicated in Figure 14. At 15 °C the measured repulsion becomes much larger than the extrapolated double-layer force at distances below 3.5 nm. At 20 °C a weak minimum occurs, but still in the repulsive regime.

As temperature increases, the short-range hydration repulsion decreases in range and the minimum shifts to smaller separations and becomes attractive at 30 °C. The layer thickness increases with temperature to approximately 2.6 nm at 37 °C. The most plausible explanation for the increased thickness is that decreased repulsion between the EO chains leads to a closer packing (increased adsorption) of the $C_{12}EO_5$ molecules at the surface. The decreased repulsion is also manifested in the temperature dependence of the relatively short-range interaction between the adsorbed layers. There is no general agreement in the molecular mechanism underlying this tempera-

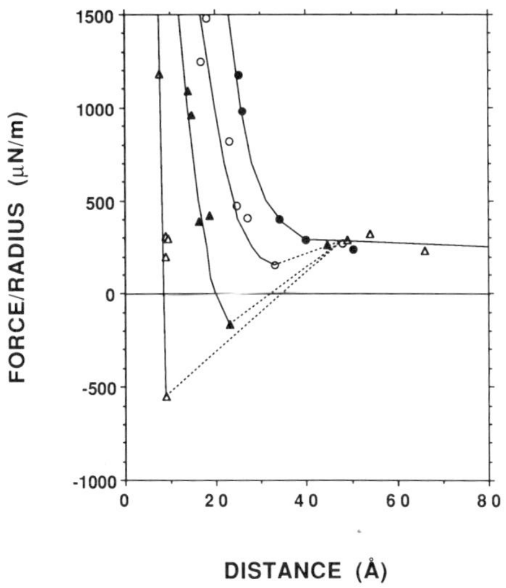

Figure 14. The force between $C_{12}EO_5$-coated surfaces as a function of separation at various temperatures (°C). ●, 15; ○, 20; ▲, 30, △, 37. (From Ref. 40.)

ture dependence. It is noteworthy, however, that it correlates directly with the cloud point of the surfactant/water system.

15.4. PROTEIN INTERACTION WITH PEG SURFACES

One of the first to study, in some detail, the adsorption of biomolecules to PEG surfaces was Nagaoka *et al.*[43] They found comparatively low platelet adhesion and protein adsorption to PEG hydrogels.

We have undertaken an extensive ellipsometry study on how proteins with isoelectric points in the range 4 to 10, i.e., with different charge at physiological conditions, adsorb on PEG grafted to silicon. The results were compared with the adsorption to hydrophobic PVC and to anionic polymethacrylate (PMA).[14] For these latter two surfaces, adsorption values in the order of a monolayer were generally found. For the (at pH 7) negatively charged (albumin) and neutral (IgG) proteins, adsorption is presumably driven by hydrophobic and van der Waals interactions while electrostatics is more important for the adsorption of poly-lysine to negatively charged PMA surfaces. Although PEG forms coaservate with polyacrylic acid, no such tendency was observed with the anionic protein on the PEG surface. Instead, generally, adsorption values below 0.1–0.2 mg m^{-2} (less than one tenth of a monolayer) were observed, indicating that steric repulsion dominates the interaction between PEG and proteins at physiological conditions. Also, in contrast to the other surfaces no pronounced plateau in the adsorption isotherms was found on PEG-coated surfaces. Rather, a slowly progressing increase in this "background level" adsorption was observed other the whole concentration range; see, for example, results for IgG in Figure 13.

The proteins that do adsorb to PEG-coated surfaces can not all be displaced by extensive rinsing. This may either be the result of interaction of protruding parts on the protein and bare substrate sites not covered with PEG, or due to mere physical entanglement of PEG and protein segments. Fibrinogen, the largest protein studied, gives slightly higher adsorbed amounts after rinsing compared to albumin, IgG, and poly-lysine. Similarly, low adsorption values on a PEG surface were found for complement proteins (C3 and C1q) when compared with adsorption values on a variety of other polymer films with different functionality prepared by plasma polymerization;[44] see Figures 15 and 16. Although one expects from studies of PEG–protein conjugates in solution that proteins on a PEG surface may be in a more native state than on hydrophobic or ionic surfaces, no such experimental evidence exists to date.

The adsorption of proteins to PEG surfaces decreases with increasing degree of polymerization. This is indicated by the N/C ratio from ESCA after adsorption of human albumin (0.1% w/w) to hydrogels with PEG 550, 1900, and 5000;[5] see Table 4. Similar results have been found for fibrinogen[45] and for the total amount of proteins adsorbed from blood plasma.[42]

The adsorption decrease with increasing PEG molecular weights is large up to 1500, above which only a marginal decrease is observed. This correlates with a decreasing grafting efficiency with increasing molecular weight, yielding a rather constant EO content at molecular weights above 2000.[5]

Protein adsorption on PEG surfaces increases with temperature, as shown in Table 5 for the total amount of protein adsorbed from plasma (ellipsometry) and for

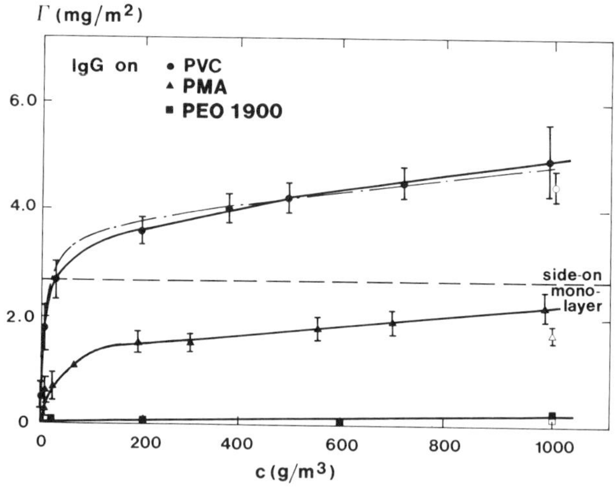

Figure 15. Adsorption isotherms at 20 °C for human IgG on various surfaces. (From Ref. 14.)

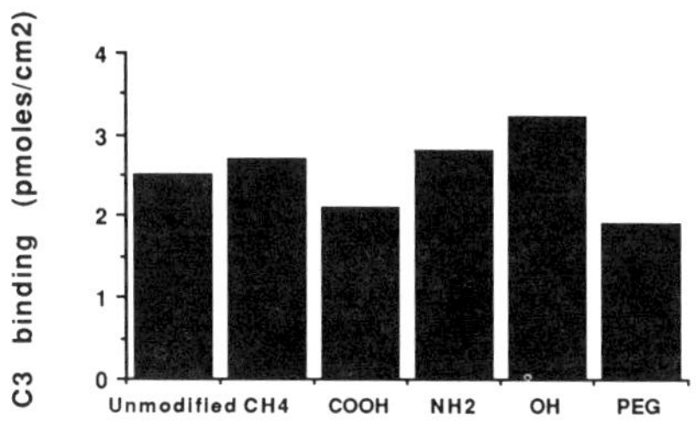

Figure 16. Comparison of the active adsorption (total minus unspecific adsorption) of radiolabeled complement factor C3 on a PEG surface and other functionalized surfaces as obtained from various gas-plasma treatments. Unspecific adsorption was measured after addition of 10 mM EDTA, i.e., measurements took place in the absence of Ca^{2+} and Mg^{2+} ions (essential for activation of complement) (see Ref. 44).

pure fibrinogen adsorption (0.1% in buffer, ESCA). The increase is almost linear over the whole temperature range studied (20–60 °C). This is most likely due to the temperature-dependent interactions of PEG, as discussed above. It is noteworthy that no dramatic changes of the fibrinogen structure are expected below 45 °C.

Finally, it should be mentioned that dextran also shows protein-repellant properties. However, in this case significantly higher molecular weight of the polymer must be used to obtain similar effects.

15.5. HYPOTHESIS ON THE PROTEIN INERTNESS OF PEG SURFACES

From this experimental and theoretical description, the following hypothesis evolves: The nonionic EO building unit in PEG is a dipole. The dipole moment depends on the conformation and an equilibrium between different states develops. At room temperature, the *gauche–trans–gauche* conformation dominates and gives rise to maximum dipole moment, a helical structure in vacuum,[49] and extensive hydration of the PEG chain in water. Since water is a better than theta solvent for PEG, this gives

Table 4. Relative Amounts of Human Albumin Remaining on PEG Surfaces after Rinsing as Determined by the Nitrogen/Carbon Ratio from ESCA Analysis

Sample	N/C (%)	Sample	N/C (%)
PVC	8.8	PEG 5000	0.9
PEG 550	5.5	TMP $(EO)_{20}$ *branched* 7.7	7.7
PEG 1900	2.2	Pure albumin	16

Table 5. The Influence of Temperature on the Adsorption of Proteins on PEG (MW 1900) Surfaces[a]

Temperature	Γ_{tot} (mg m^{-2})	$(N/C)_{fib}$ (%)
17	0.4	1.3
25	0.7	2.0
37	1.5	2.8
50	2.5	4.3
71	4.9	6.4

[a]The total amount of proteins adsorbed from serum (diluted 1/10) was determined by ellipsometry on PEG-treated silicon wafers. The relative amounts of fibrinogen adsorbed (from a single protein solution, c = 0.1% w/w) onto PEG-modified polyethylene was calculated from the nitrogen/carbon ratio obtained from ESCA analysis.

rise to a general and dominating steric repulsion force (osmotic and excluded volume contributions) between PEG and protein in aqueous solutions, since the extensive hydration causes the van der Waals force contribution to be very small and electrostatic interactions are not present. The effective range of the steric repulsion, reflected in lowering of the residual protein adsorption found on PEG surfaces, increases with molecular weight and surface coverage with PEG. A temperature rise as well as an addition of electrolyte changes the distribution of conformer states. As we have discussed above, this is accompanied by changes in the dipole strength of the EO units and a decreased repulsion between PEG chains which shows up as a decreased hydration and, eventually, phase separation. The consequence is that PEG is not such a good steric stabilizer at elevated temperature. As a result, protein adsorption increases with temperature and electrolyte addition.

REFERENCES

1. D. E. Gregonis, C. M. Chen, and J. D. Andrade, in: *Hydrogels for Medical and Related Applications* (J. D. Andrade, ed.), *ACS Symp. Ser. 31*, 88 (1973).
2. C.-G. Gölander, S. Jönsson, T. Vladkova, P. Stenius, and J.-C. Eriksson, *Colloids and Surfaces 21*, 149 (1986).
3. P. Drude, *Ann. Phys. 272*, 532, 865 (1889).
4. J. N. Israelachvili and G. E. Adams, *J. Chem. Soc., Faraday Trans. 1, 74*, 975 (1978).
5. P. C. Herder, P. M. Claesson, and C. E. Herder, *J. Colloid Interface Sci. 119*, 240 (1988).
6. A. Abuchowski, T. van Es, N. C. Palczuk, and F. Davis, *J. Biol. Chem. 252*, 3578 (1977).
7. A. F. Bückmann, M. Morr, and G. Johansson, *Makromol. Chem. 182*, 1379 (1981).
8. S. Zalipsky, C. Gilon, and A. Zilkha, *Eur. Polym. J. 19*, 1177 (1983).
9. J. M. Harris, E. C. Struck, M. G. Case, and M. S. Paley, *J. Polym. Sci. 22*, 341 (1984).
10. J. M. Harris, K. Yoshinaga, M. S. Paley, and M. R. Herati, in: *Advances in Separations Using Aqueous Phase Systems in Cell Biology and Biotechnology* (D. Fisher and I. A. Sutherland, eds.), Plenum Press, London (1988).
11. J. M. Harris, *Rev. Macromol. Chem. Phys. C25*, 325 (1985).
12. M. S. Paley and J. M. Harris, *J. Polym. Sci. 25*, 2447 (1987).
13. F. E. Bailey, Jr. and R. W. Callard, *J. Appl. Polym. Sci. 1*, 56 (1959).

14. E. Kiss, C-G. Gölander, and J. C. Eriksson, *Progr. Colloid & Polymer Sci. 74*, 113–119 (1987).
15. J. L. Parker, D. L. Cho, and P. M. Claesson, *J. Phys. Chem. 93*, 6121 (1989).
16. C.-G. Gölander and J.-C. Eriksson, *J. Colloid Interface Sci. 119*, 38 (1987).
17. J.-C. Eriksson, C.-G. Gölander, A. Baszkin, and L. Ter-Minassian-Saraga, *J. Colloid Interface Sci. 100*, 2 (1984).
18. W. R. Gombotz, W. Guanghui, and A. S. Hoffman, *J. Appl. Polym. Sci. 37*, 91 (1989).
19. C.-G. Gölander, M. W. Rutland, D. L. Cho, A. Johansson, H. Ringblom, S. Jönsson, and H. K. Yasuda, *J. Appl. Polym. Sci.*, submitted.
20. E. Kiss and E. Gölander, *Colloids and Surfaces 49*, 335–342 (1990).
21. C.-G. Gölander and E. Kiss, *Colloids and Surfaces*, submitted.
22. P. M. Claesson and C.-G. Gölander, *J. Colloid Interface Sci. 117*, 366 (1987).
23. John Yee, *Synthesis and Interfacial Coupling of Mercapto Activated Poly(ethyleneoxide) via Thiol–Disulphide Interchange*, M.Sc. Thesis, Dept. of Bioengineering, University of Utah, Salt Lake City.
24. H. L. Lee, J. Kopecek, and J. D. Andrade, *J. Biomed. Mater. Res. 23*, 351 (1989).
25. R. Silverstone and K. Kronberg, *J. Phys. Chem. 93*, 6241 (1989).
26. R. Kjellander and E. Florin-Robertsson, *J. Chem. Soc., Faraday Trans. 1, 77*, 2053 (1981).
27. S. Saeki, N. Kuwahara, M. Nakata, and M. Kaneko, *Polymer 17*, 685 (1976).
28. R. E. Goldstein, *J. Chem. Phys. 80*, 5340 (1984).
29. G. Karlström, *J. Phys. Chem. 89*, 4962 (1985).
30. A. A. Samii, B. Lindman, and G. Karlström, *Prog. Colloid Polym. Sci. 82*, 1 (1990).
31. M. Björling, P. Linse, and G. Karlström, *J. Phys. Chem. 94*, 471 (1990).
32. P. G. de Gennes, *Macromolecules 13*, 1069 (1980).
33. S. I. Jeon, J. H. Lee, J. D. Andrade, and P. G. de Gennes, in press.
34. *DISCOVER*. A molecular simulation program from Biosym Technologies, 10065 Barnes Canyon Road, San Diego, CA 92121.
35. W. L. Jörgensen, *J. Chem. Phys. 77*, 5757 (1982).
36. J. L. Valles and J. W. Halley, *J. Chem. Phys. 92*, 694 (1990).
37. P. J. Flory, *Statistical Mechanics of Chain Molecules*, Chapter 5, Hanser Publishers, New York (1989).
38. P. Dauber-Osguthorpe, V. A. Roberts, D. J. OsguthoIpe, J. Wolff, M. Genest, and A. T. Hagler, *Proteins, Structure, Function and Genetics 4*, 31 (1988).
39. J. Klein and P. F. Luckham, *Macromolecules 17*, 1041 (1984).
40. P. M. Claesson, R. Kjellander, S. Stenius, and H. K. Christensen, *J. Chem. Soc., Faraday Trans. 82*, 2735 (1986).
41. P. M. Claesson, D. L. Cho, C.-G. Gölander, E. Kiss, and J. L. Parker, *Prog. Colloid Polym. Sci. 82*, 330–336 (1990).
42. E. Kiss and C.-G. Gölander, *J. Colloid and Interface Sci. 117*, 366–374 (1987).
43. Y. Mori, S. Nagaoka, H. Takiuchi, T. Kikuchi, N. Noguchi, H. Tanzawa, and Y. Noishiki, *Trans. Am. Soc. Artif. Internal Organs 28*, 459 (1982).
44. K. Nilsson Ekdahl, B. Nilsson, C.-G. Gölander, B. Lassen, H. Elwing, and U. R. Nilsson, *J. Biomed. Mater. Res.*, submitted.
45. W. R. Gombotz, W. Guanghui, A. S. Hoffman, and T. A. Horbett, *Proc. The Third World Biomater. Congr.*, Kyoto, Japan (April 21–25, 1988).
46. Y-S. Yeh, Y. Iriyama, Y. Matsuzawa, S. R. Hanson, and H. Yasuda, *J. Biomed. Mater. Res. 22*, 795 (1988).
47. H. J. Taunton, C. Toprakciouglu, L. J. Fetters, and J. Klein, *Macromolecules 23*, 571 (1990).
48. S. Patel, M. Tirell, and G. Hadziioannou, *Colloids and Surfaces 31*, 157 (1988).
49. F. E. Bailey and J. V. Koleske, *Polyethyleneoxide*, Academic Press, New York (1976).
50. P. A. Cuypers, W. T. Hermens, and H. C. Hemker, *N.Y. Acad. Sci. 283*, 77 (1977).
51. E. Kiss and C-G. Gölander, *Colloids and Surfaces 58*, 263–270 (1991).

16

Protein Adsorption to and Elution from Polyether Surfaces

WAYNE R. GOMBOTZ, WANG GUANGHUI, THOMAS A. HORBETT, and ALLAN S. HOFFMAN

16.1. INTRODUCTION

Poly(ethylene oxide) (or PEO) surfaces represent an important class of biomaterials because of their low capacity for protein adsorption. A potentially wide range of applications exists for PEO surfaces including blood contacting devices, drug delivery systems, contact lenses, intraocular lenses, vascular grafts, catheters, immunoassays, biosensors, and media for protein and cell separations, to name a few.

Many groups have prepared PEO surfaces by a variety of techniques. The simplest, and probably least stable type of PEO surface for biomaterials applications, is one in which PEO or PEO copolymers are adsorbed directly to a substrate.[1–11] More stable systems have been made by covalently bonding PEO to a substrate.[12–24] PEO has also been grafted to surfaces via a backbone polymer[12,13,25–27] or incorporated as block segments in a variety of polymers, the most common type being the polyether urethanes.[28–33] Crosslinked PEO homopolymer networks have been prepared by radiation[34,35] or chemical crosslinking.[36]

All of the PEO surfaces mentioned above have been shown to exhibit low degrees of protein adsorption, yet the reasons for this phenomenon are not entirely understood. Several theories have been proposed to explain the low protein affinity of PEO both on surfaces and in solution. These include a rapid mobility of the hydrated

WAYNE R. GOMBOTZ • Bristol-Myers Squibb, Pharmaceutical Research Institute, Seattle, Washington 98121. WANG GUANGHUI, THOMAS A. HORBETT, and ALLAN S. HOFFMAN • Center for Bioengineering, University of Washington, Seattle, Washington 98195.

Poly(Ethylene Glycol) Chemistry: Biotechnical and Biomedical Applications, edited by J. Milton Harris. Plenum Press, New York, 1992.

PEO chains,[12,13] the large excluded volume of the PEO molecule,[37,38] a repulsive force that results from a loss of configurational entropy when a protein approaches a PEO molecule,[12] the low interfacial free energy at the PEO–water interface,[39] and the lack of protein binding sites (ionic and hydrophobic) on PEO.[15,34] Many of these theories may be related to the unique aqueous solution properties of PEO molecules.[40,41] PEO is soluble in water at room temperature for a wide range of molecular weights while related polymers such as poly(methylene oxide) and poly(propylene oxide) (PPO) are water insoluble.[41] Studies of PEO in aqueous systems suggest that a minimum of two and often three hydrogen-bonded water molecules are needed to satisfy the basic hydration of each ether unit of PEO[42,43] (see Chapter 2, this volume).

We recently completed a study in which different molecular weight PEO molecules were covalently immobilized onto poly(ethylene terephthalate) (PET) films.[44] Fibrinogen adsorption to these surfaces decreased with increasing PEO molecular weight up to 3500. A further increase in PEO molecular weight resulted in only slight decreases in fibrinogen adsorption. Gravimetric analysis of the films revealed there were many more moles of low molecular weight PEO molecules on the substrates than high molecular weight PEO. We concluded from this study that a low surface density of high molecular weight PEO is more effective in reducing protein adsorption than a high surface density of low molecular weight PEO. This protein repulsion phenomenon of high molecular weight PEO surfaces may be attributed to the way in which PEO interacts with and hydrogen-bonds water.

The work presented in this chapter further investigates the interactions of proteins with PEO surfaces. The PPO molecule has the same backbone structure as PEO with an additional methyl group in place of a hydrogen. This methyl group interferes with the hydrogen bonding of water to the ether groups and renders the PPO molecule much more hydrophobic than PEO. Varying the amount of PPO present on a PEO surface should influence the way in which proteins adsorb to that surface. PEO, PPO, and PEO/PPO random and block copolymers were covalently bound to allylamine plasma treated PET films. ^{125}I-labeled fibrinogen and albumin were adsorbed to these surfaces from buffer. Fibrinogen was also adsorbed from baboon plasma and eluted with a sodium dodecyl sulfate (SDS) buffer solution to investigate the strength of the interaction between the PEO surface and the protein.

16.2. EXPERIMENTAL: MATERIALS AND METHODS

16.2.1. Surface Synthesis and Characterization

The surfaces containing the PEO homopolymers were synthesized and characterized as described previously using an allylamine plasma derivatized PET film (Lux Thermanox coverslips purchased from Miles Scientific Laboratories) with subsequent cyanuric chloride activation (Figure 1).[20] The PPO homopolymers (Jeffamine D230, D400, and D2000) and PEO/PPO random copolymers Jeffamine (M2005, M2070, and ED2001) were provided as a gift from Texaco Chemical Co. The

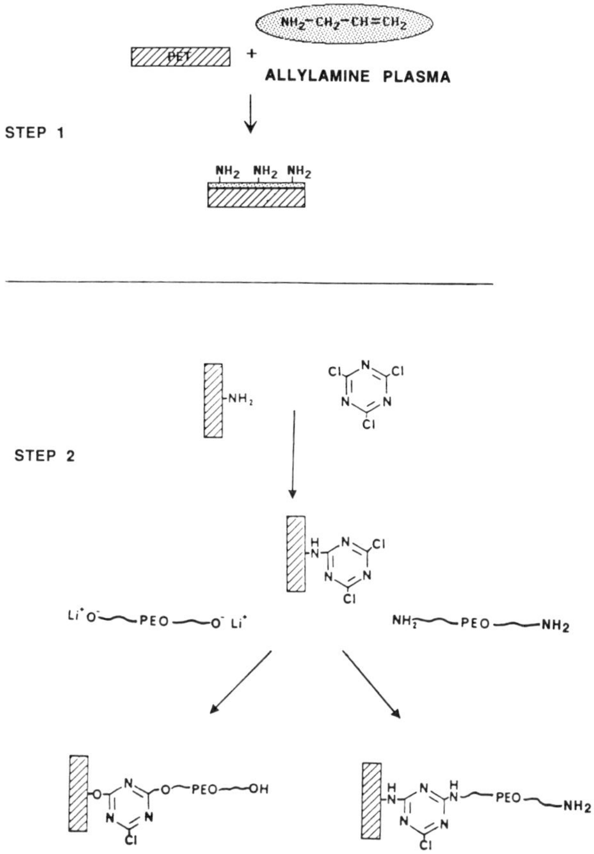

Figure 1. Immobilization of bis-amino and hydroxylated polyether molecules to PET films. In step 1 the PET is derivatized with amino groups by exposure to an allylamine plasma. In step 2 the films are activated with cyanuric chloride and then reacted with either bis-amino polyether molecules or the alkoxide ion derivative of hydroxylated polyethers.

Jeffamines were coupled to the films using the same reaction as described for the PEO homopolymers,[20] since the molecules contained at least one amino end group. Table 1 lists the characteristics of the Jeffamines and the concentrations used in the immobilization reaction. The PEO/PEO/PPO block copolymers (Pluronic F127 and F88) were supplied by BASF-Wyandotte (Table 2). Prior to immobilization onto the activated PET films, the alkoxide ion derivative of the hydroxyl terminated Pluronic molecule was formed by reacting with n-butyl lithium.[20] Control experiments were done by reacting the polyether molecules with PET/allylamine films that were not preactivated with cyanuric chloride. Upon completion of the reaction, the films were washed in benzene, acetone, and deionized and distilled water. Samples were characterized by ESCA and advancing water contact angle measurements.

Table 1. Properties of the Jeffamines as Supplied by the Manufactuer and the Concentrations Used in the Immobilization Reaction onto PET Films

Jeffamine[a] product	Approximate MW	Approximate mol% ratio EO/PO	Concentration ($mg\,ml^{-1}$)
D230	230	0/100	170
D400	400	0/100	170
D2000	2000	0/100	200
M2005	2000	6/94	200
M2070	2000	76/24	200
ED2001	2000	94/6	200

[a]Manufactured by Texaco Chemical Company, Bellaire, Texas.

16.2.2. Protein Adsorption

The protocol describing the adsorption of ^{125}I-labeled baboon fibrinogen and fraction V bovine serum albumin (BSA) (Sigma Chemical Co.) from buffer has been detailed in a previous publication.[44] Briefly, protein was adsorbed to prehydrated samples from 0.2 $mg\,ml^{-1}$ protein solutions containing ^{125}I-labeled protein (10^6–10^7 cpm per mg specific activity) for 2 h at 37 °C. The films were washed in buffer by dilution displacement rinsing and the remaining protein on the film was determined with a gamma counter. Fibrinogen was also adsorbed from citrated baboon plasma diluted in CPBSzI buffer (0.01 M sodium citrate, 0.01 M sodium phosphate, 0.12 M NaCl, 0.01 M NaI, 0.02% sodium azide, pH 7.4) to concentrations of 0.001, 0.01, 0.1, 1.0, 10.0, and 75.0%. The fibrinogen concentration in the 100% plasma was determined by a thrombin clottability assay and found to be between 2.0 and 5.0 $mg\,ml^{-1}$, depending on the animal from which the blood was taken. Adsorption was carried out for 2 h at 37 °C and terminated by the dilution displacement technique described earlier.

16.2.3. SDS Elutability Study

After washing with CPBSzI buffer, the samples used for the fibrinogen adsorption studies from plasma were immediately placed in 2.5 ml of sodium dodecyl

Table 2. Properties of the Pluronics as Supplied by the Manufacturer and the Concentrations Used in the Immobilization Reaction onto PET Films

Pluronic[a] product	Approximate MW	Approximate MW PEO/PPO/PEO	Concentration ($mg\,ml^{-1}$)
F127	11,500	4312/3886/4312	40
F88	10,800	4268/2223/4268	40

[a]Manufactured by BASF-Wyandotte Co, Parsippany, NJ.

sulfate (SDS) sample buffer (0.01 M phosphate, 1% SDS, pH 7.0) for 24 h at room temperature. Their radioactivity was measured on the gamma counter (eluted + retained radioactivity = $E + R$). After this initial count, the samples were rinsed in CPBSzI buffer and placed in new tubes containing 2.5 ml of CPBSzI. The amount of radioactivity on these samples was again determined (retained radioactivity = R). The fibrinogen elutability was calculated from $[(E + R) - R]/(E + R) \times 100$. These studies were run in triplicate.

16.3. RESULTS AND DISCUSSION

16.3.1. Surface Preparation and Characterization

Figure 2 shows the C1s ESCA spectra of the surfaces containing the Jeffamine ED2001, M2070, and M2070 molecules. There is a trend in increasing C—O peak size at 284.6 eV with increasing content of ethylene oxide in the copolymer. This is to be expected, since the copolymers containing more ethylene oxide have higher ratios of C—O to C—C bonds in the polymer backbone. Peaks are also present in the C1s ESCA spectra at higher binding energies. The chemical groups comprising the third largest peak at 287.5 eV cannot be identified without ambiguity, but it is likely that this peak is due in part to —C—N bonds in the plasma polymerized allylamine film and the cyanuric chloride molecules.[20] The peak located at 289.0 eV can be assigned to the ester group of PET. These two peaks are indicative of both the underlying allylamine and PET substrate. An angular-dependent ESCA study was conducted on these surfaces at three different sampling depths. There was no change in the relative sizes of the peaks in the C1s spectra with sampling depth. These data lead us to believe that the polyether surface coverage is incomplete. The nitrogen signal remained constant at all sampling depths, indicating that there are patches of plasma polymerized allylamine that are always exposed to the ESCA X-ray beam. If the allylamine layer had been completely covered with the polyether molecules, the nitrogen signal would have decreased with decreasing sampling depth.

Control allylamine surfaces (not activated with cyanuric chloride) that were exposed to the polyether molecules had no ether peaks present in their C1s ESCA spectrum. This suggests that there was no nonspecific adsorption of the polyethers to the surface and that these molecules were covalently bound by the cyanuric chloride chemistry to the plasma polymerized allylamine film.

Advancing water contact angles were also measured on the surfaces containing the PPO homopolymers (Figure 3). As the molecular weight of the PPO was increased the wettability of the films decreased. This may be a result of the increased hydrophobicity of higher molecular weight PPO molecules. Bailey and Callard studied the heat precipitability of PEO/PPO random copolymers from aqueous solution as a function of PPO content.[41] The observed decrease in the upper limit of solubility was attributed to the increased hydrophobic character of the molecule. All

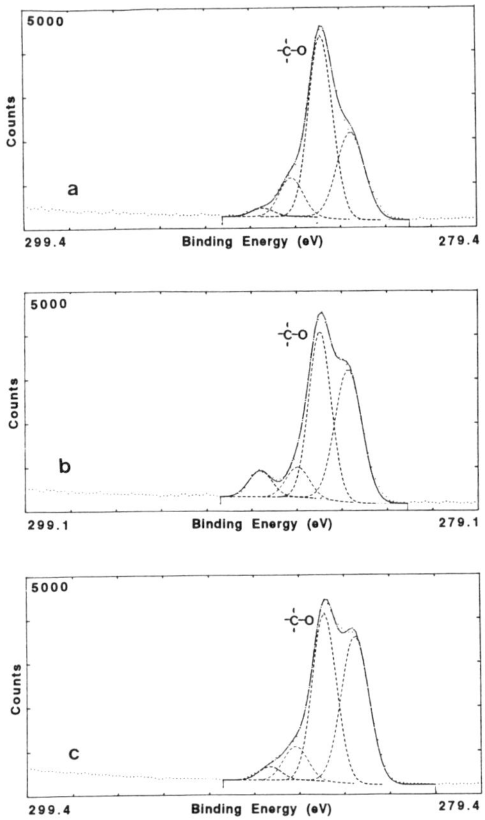

Figure 2. ESCA C1s spectra of surfaces containing immobilized PEO/PPO random copolymers. The PEO/PPO mole percent ratios of these copolymers are (a) 94/6 (Jeffamine ED2001), (b) 76/24 (Jeffamine M2070), and (c) 6/94 (Jeffamine M2005).

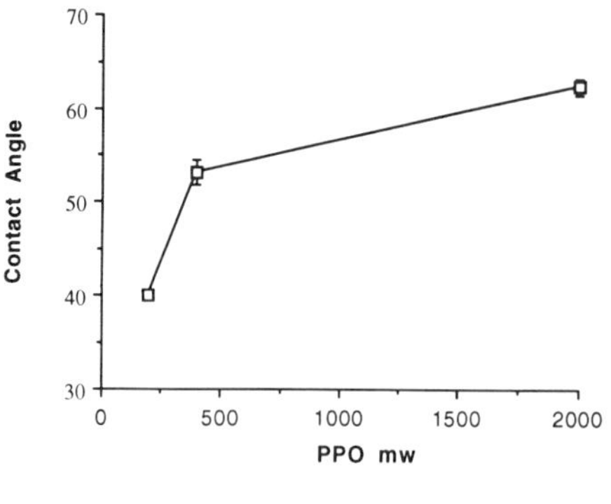

Figure 3. Advancing water contact angle as a function of the molecular weight of immobilized PPO on PET films exposed to an allylamine plasma.

surfaces containing PEO/PPO copolymers were more wettable than the PET/allylamine cyanuric chloride activated controls.

In summary, the results of the surface characterization studies indicate that the PEO and PPO homopolymers and the PEO/PPO copolymers were covalently bound to the surface of the allylamine treated PET by cyanuric chloride chemistry. There was, however, incomplete coverage of the surface by these molecules as evidenced by the higher than expected water contact angles on the PEO homopolymer surfaces[44] and the presence of underlying allylamine and PET peaks in the C1s ESCA spectra.

16.3.2. Protein Adsorption

16.3.2.1. Protein Adsorption from Buffer

The adsorption of fibrinogen from buffer to the PPO homopolymer surfaces is shown in Figure 4. There is a trend of increasing protein adsorption with an increase in molecular weight of the attached PPO. This trend appears to coincide with the water contact angle measurements (Figure 3). These results are in contrast to the fibrinogen adsorption studies on PEO surfaces, in which case the amount of protein adsorbed decreased significantly with PEO molecular weight.[44] The increased hydrophobicity of the higher molecular weight PPO surfaces may be contributing to the increased fibrinogen adsorption.

We next investigated the adsorption of fibrinogen and albumin to the 2000 molecular weight PEO/PPO random copolymers (Jeffamine ED2001, M2005, and M2070) as a function of the mole percent ethylene oxide in the copolymer. As the ethylene oxide content of these copolymers was increased, the amount of adsorbed fibrinogen and albumin decreased (Figures 5 and 6).

We recently suggested that PEO surfaces exhibit low degrees of protein adsorption due to the unique way in which the PEO molecule binds water.[44] The water molecules that are hydrogen bonded to the ether oxygens of PEO are believed to form a protective hydration shell around the PEO molecule.[42] This hydration shell, which has also been called structured water,[42,47,48] or bound water is believed in part to

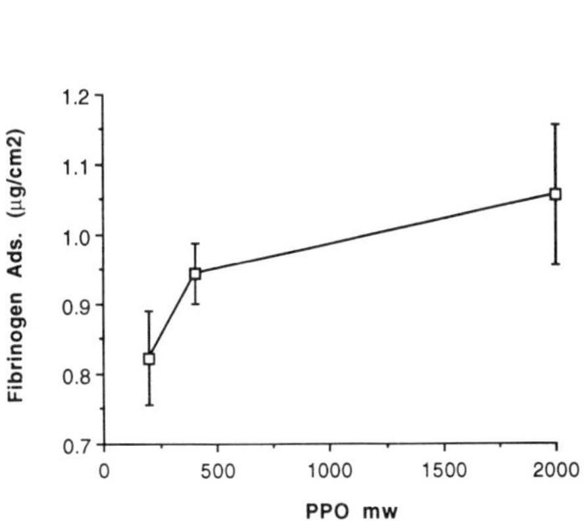

Figure 4. Fibrinogen adsorption from CPBSzI buffer as a function of molecular weight of PPO attached to a PET/allylamine film. The protein concentration was 0.2 mg ml^{-1} and adsorption was carried out for 2 h at 37 °C. Films were washed in buffer by dilution displacement. Each data point represents a mean (n = 3) ± s.d.

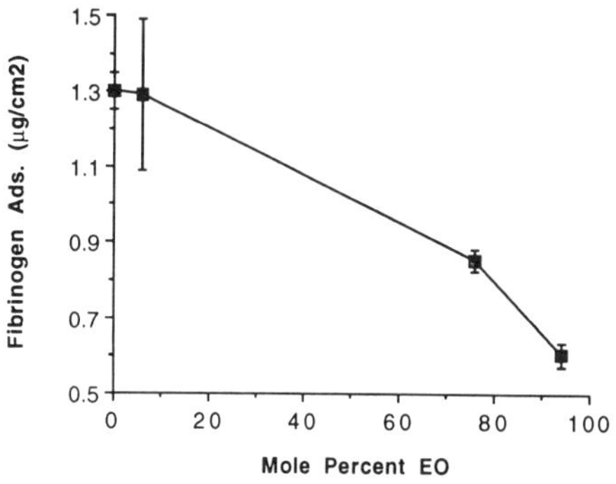

Figure 5. Fibrinogen adsorption from CPBSzI buffer to PEO/PPO random copolymers (molecular weight 2000) attached to PET/allylamine films as a function of the mole percent ethylene oxide in the copolymer. The protein concentration was 0.2 mg ml^{-1} and adsorption was carried out for 2 h at 37 °C. Films were washed in buffer by dilution displacement. Each data point represents a mean (n = 3) ± s.d.

be responsible for the inverse solubility–temperature relationship exhibited by PEO in aqueous solutions.[46] As the molecular weight of a PEO molecule is increased, the cloud point temperature of the PEO also increases. This suggests that the water structuring is a cooperative process.[45] In other words, the hydration shell becomes more stable as the number of hydrated ether linkages (PEO molecular weight) increases. Higher molecular weight PEO molecules may allow the water of hydration to better bridge the ether linkages.[49] Recent findings presented by Antonsen and Hoffman (Chapter 2, this volume) indicate that there is a gradual rise in the amount of bound water with PEO molecular weight beginning between 800 and 1000. They attribute this rise to an increased ability of the polymer chains to fold upon themselves and more easily share loosely-bound water between adjacent chains. This observation further substantiates the idea that the formation of a hydration shell around PEO molecules is a cooperative process that increases with PEO molecular weight. The hydration shell in turn may prevent the proteins from interacting with both the PEO molecule and the underlying substrate.

The decreases in protein adsorption with an increase in ethylene oxide content of the PEO/PPO copolymer surfaces described in this study may also be explained by

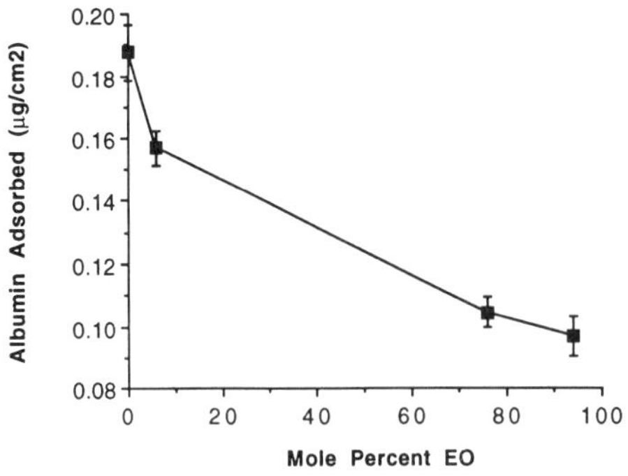

Figure 6. Albumin adsorption from CPBSzI buffer to PEO/PPO random copolymers (molecular weight 2000) attached to PET/allylamine films as a function of the mole percent ethylene oxide in the copolymer. The protein concentration was 0.2 mg ml^{-1} and adsorption was carried out for 2 h at 37 °C. Films were washed in buffer by dilution displacement. Each data point represents a mean (n = 3) ± s.d.

the formation of hydration shells around the PEO. Kjellander has noted that, despite the fact that PPO has the same backbone structure as PEO, high degrees of water structuring cannot be formed, since the methyl groups of the PPO constitute a steric hinderance.[42] The strain in water structure around the PPO leads to a smaller hydrogen bond energy than seen with water and a PEO molecule. As the amount of methyl groups in the PEO/PPO molecule increases, the amount of bound water or size of the hydration shell around the polymer would be expected to decrease. This decrease in bound water could in turn lead to a smaller excluded volume of the polyether and result in increased protein adsorption.

Figure 7 shows the amount of fibrinogen adsorbed to the two Pluronic surfaces and to the PET/allylamine. Both surfaces exhibit reduced fibrinogen adsorption when compared to PET/allylamine. The Pluronic molecules contain similar amounts of PEO but different amounts of PPO (Table 2). When the two Pluronic surfaces are compared, the surface with the higher molecular weight PPO adsorbed more protein. This difference in fibrinogen adsorption could be due to an increased hydrophobicity of the higher molecular weight PPO block of the copolymer and a decrease in the ability of this molecule to form a protective hydration shell.

16.3.2.2. Fibrinogen Adsorption from Plasma

Figure 8 shows the adsorption of fibrinogen from plasma to PET films and to films containing different molecular weights of immobilized PEO. Much less fibrinogen adsorbed to all of these surfaces when compared to fibrinogen adsorption from buffer. This can be explained by a competition for adsorption sites on the surface by other proteins in the plasma.

The fibrinogen adsorption to the PET and PEO 400 surfaces increased with increasing plasma concentration. The PEO 3500 and 20,000 surfaces had much lower amounts of fibrinogen adsorbed when compared to the PEO 400. This result is consistent with the trends in fibrinogen adsorption to PEO surfaces (MW 200, 400, 600, 1000, 3500, and 20,000) from buffer.[44] There appears to be an optimal PEO molecular weight of *ca* 3500 that is necessary for reduced protein adsorption. A further increase in the molecular weight does not result in a significant reduction in protein adsorption.

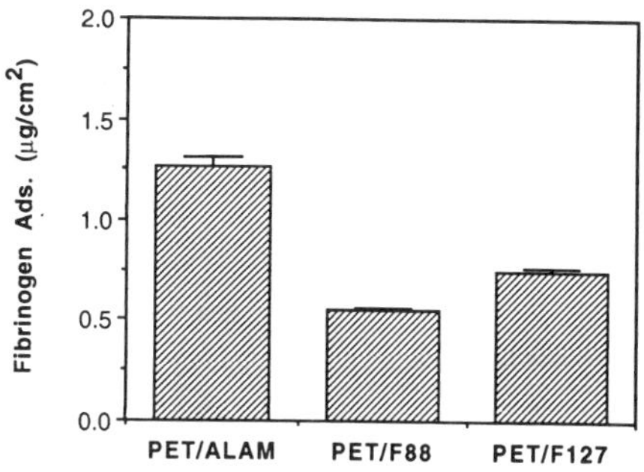

Figure 7. Fibrinogen adsorption to surfaces containing the Pluronics F88 and F127 and to a PET/allylamine surface. The protein concentration was 0.2 mg ml^{-1} and adsorption was carried out for 2 h at 37 °C. Films were washed in buffer by dilution displacement. Each data point represents a mean (n = 3) ± s.d.

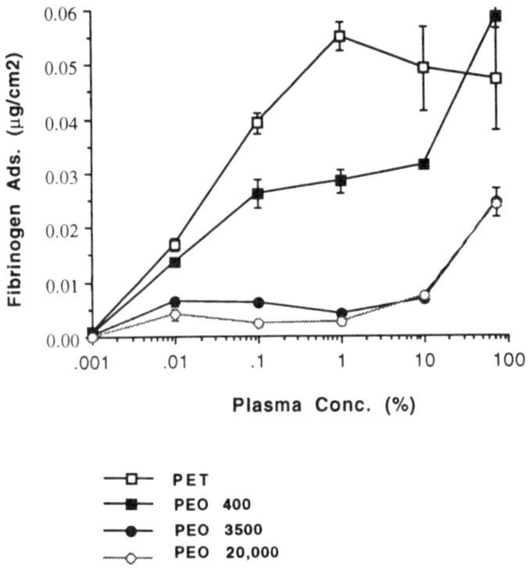

Figure 8. Fibrinogen adsorption to PET and films containing different molecular weight PEO molecules as a function of plasma concentration of the adsorption buffer. Adsorption was carried out for 2 h at 37 °C. Films were washed in buffer by dilution displacement. Each data point represents a mean ($n = 3$) ± s.d.

Figure 9 shows the fibrinogen adsorption from plasma to a PET/allylamine surface and films containing PPO 400 and 2000. All of these surfaces exhibit significantly higher fibrinogen adsorption than any of the PEO-containing surfaces (Figure 8). The PET/allylamine surface exhibited the highest fibrinogen adsorption of any of the substrates that we studied, Fibrinogen adsorption to the allylamine film showed a continual increase as the plasma concentration was increased. The high affinity of fibrinogen for this substrate may be due to a strong ionic interaction between the allylamine plasma polymerized film and fibrinogen. We have shown in a previous study that fibrinogen adsorption from buffer to this surface decreases with increasing buffer ionic strength, suggesting the presence of a positive charge on the substrates.[44]

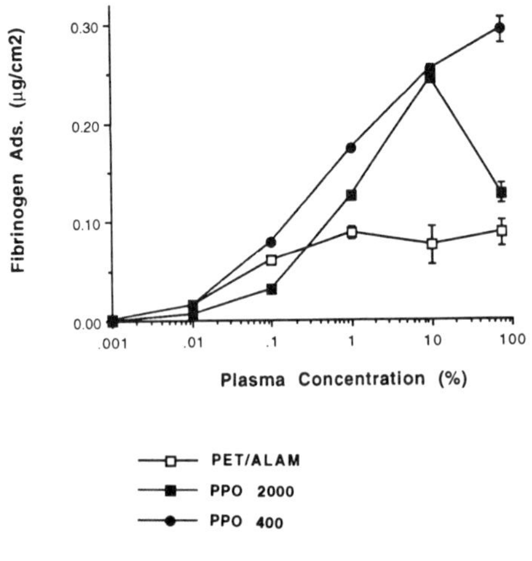

Figure 9. Fibrinogen adsoprtion to PET/allylamine films containing immobilized PPO 400 and PPO 2000 as a function of plasma concentration of the adsorption solution. Adsorption was carried out for 2 h at 37 °C. Films were washed in buffer by dilution displacement. Each data point represents a mean ($n = 3$) ± s.d.

16.3.2.3. Fibrinogen Elutability Studies

We have now established that the adsorption of BSA and fibrinogen from both buffer and plasma can vary markedly depending on the surface used. High molecular weight PEO surfaces, particularly those of 3500 or greater, exhibited the lowest levels of protein adsorption. The following results from the SDS elutability study give us a better understanding of the strength of the interaction between the surface and the protein.

Figures 10 and 11 are plots of the percent elutability as a function of plasma concentration for PET, allylamine, PEO, and PPO surfaces. The PPO 2000 surface had the lowest elutability of approximately 10%. This means that this sample retained about 90% of the fibrinogen which was initially adsorbed to its surface from the plasma. The fibrinogen could be interacting with the methyl groups of the PPO on the surface via hydrophobic forces. Since the PPO molecules would not be expected to form hydration shells around the polymer chains, the protein could also adsorb better to the underlying allylamine substrate.

The allylamine film had the next lowest elutability of about 20%. The high affinity and low elutability of this substrate may again be due to the strong ionic interactions between the fibrinogen molecule and the surface. All of the PEO surfaces showed higher elutabilities than the allylamine substrates on which they were immobilized. The PEO is therefore able to decrease the strong interactions between the fibrinogen molecule and the underlying allylamine substrate, while interacting very little with the protein molecules.

16.3.2.4. Summary

Figure 12 is a model which schematically depicts the protein repulsion mechanism of longer PEO chains on a surface. This model may be best understood in

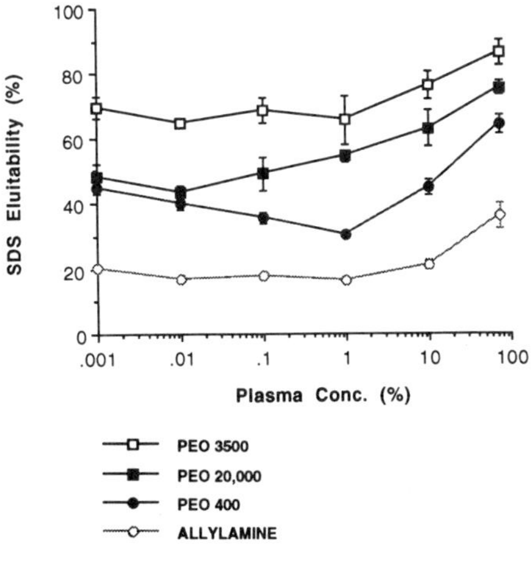

Figure 10. Percent elutabiity of fibrinogen adsorbed to PET/allylamine and PEO 200, 3500, and 20,000 surfaces as a function of plasma concentration of the adsorption solution. Adsorption was carried out for 2 h at 37 °C. Films were washed in buffer by dilution displacement and counted on a gamma counter while soaking in a 1% SDS buffer. After 24 h in SDS buffer, films were then dip-rinsed and recounted. Each data point represents a mean (n = 3) ± s.d.

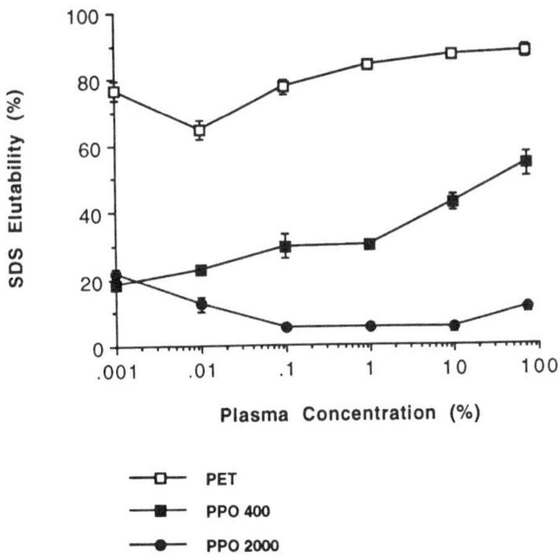

Figure 11. Percent elutability of fibrinogen adsorbed PET, PPO 200, and PPO 400 surfaces as a function of plasma concentration of the adsorption solution. Adsorption was carried out for 2 h at 37 °C. Films were washed in buffer by dilution displacement and counted on a gamma counter while soaking in a 1% SDS buffer. After 24 h in SDS buffer, films were then dip-rinsed and recounted. Each data point represents a mean (n = 3) ± s.d.

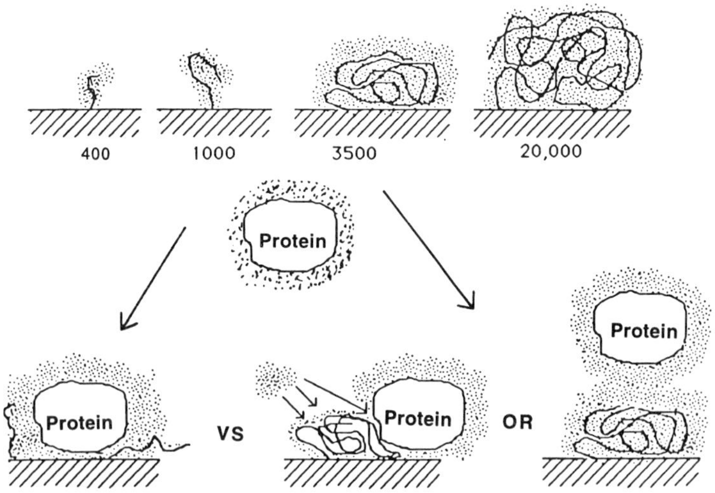

Figure 12. Schematic representation of hydrated low and high molecular weight PEO molecules, covalently immobilized on a surface in the presence and absence of protein molecules.

thermodynamic terms. The unperturbed PEO molecule on a surface will achieve its minimum free energy by binding water molecules to the ether groups (exothermic) and achieving as random (high entropy) a conformation as possible. When the PEO is above a certain molecular weight it will be able to fold on itself. This folding will permit a "loose" sharing of bound water molecules within the folds, averaging *ca* 2–3 water molecules per ether group (see Chapter 2, this volume). The sharing of bound water or hydration shell formation creates an excluded volume from which protein will be repelled. If a protein molecule attempts to compress this hydrated PEO coil, it will take energy to remove the ether bound water molecules, while at the same time the PEO coil will lose entropy. Thus, even though the released water molecules must be gaining entropy, the overall process apparently causes a rise in free energy and will not occur spontaneously. The "released" water molecules in such a process will be driven back into the PEO coil by an osmotic driving force. These concepts explaining the low protein adsorption behavior of PEO surfaces have been called the "excluded volume," "osmotic repulsion," or "entropic repulsion" theories. Our recent data on the PEO–water hydrate formation (see Chapter 2, this volume) support the molecular picture described above and fit well with our results on protein adsorption and elution.

16.4. CONCLUSIONS

A variety of amino or hydroxyl terminated polyether molecules can be covalently immobilized onto PET by radiofrequency glow discharge derivatization of the films. Our objective was to study the effects of polyether composition on protein adsorption to these surfaces. Fibrinogen and albumin adsorption from buffer decreased with increasing PEO molecular weight and increased with an increase in PPO molecular weight. In the copolymers, as the propylene oxide content was increased, fibrinogen adsorption also increased. Fibrinogen adsorption to PEO surfaces from plasma decreased with PEO molecular weight and much of the protein that was adsorbed to the higher molecular weight PEO surfaces was displaced by SDS buffer. This indicates that the interactions between the PEO molecules and protein are relatively weak and may be related to the hydration shell or bound water that forms around higher molecular weight PEO. Hopefully, these studies have shed more light on the *in vitro* interactions between proteins and PEO surfaces and can be applied to the design and development of useful biomaterials.

ACKNOWLEDGMENTS

The authors would like to thank the 3M Company and the Washington Technology Center at the University of Washington for their financial support and the NESAC/BIO Center (NIH Grant RR01296) for the use of the ESCA instrument. We would also like to thank Toray Industries Inc., Texaco Chemical Co., and BASF-Wyandotte, Inc. for kindly providing us with the polyether samples.

REFERENCES

1. J. N. George, *Blood 40*, 862 (1972).
2. Y. L. Cheng, S. A. Darst, and C. R. Robertson, *J. Colloid Interface Sci. 118*, 212 (1987).
3. W. Wasiewski, M. J. Fasco, B. M. Martin, T. C. Detwiler, and J. W. Fenton, *Thromb. Res. 8*, 881 (1976).
4. C. W. Hiatt, A. Shelokov, E. J. Rosenthal, and J. M. Galimore, *J. Chromatogr. 56*, 362 (1971).
5. D. H. Randerson and J. A. Taylor, in: *Plasmapheresis, New Trends in Therapeutic Applications* (Y. Nose, P. S. Malchesky, and J. W. Smith, eds.), pp. 69–80, ISAO Press, Cleveland (1983).
6. L. Illum, L. O. Jacobsen, R. H. Muller, E. Mak and S. S. Davis, *Biomaterials 8*, 113 (1987).
7. S. S. Davis, S. J. Douglas, L. Illum, P. D. Jones, E. Mak, and R. H. Muller, in: *Targeting of Drugs with Synthetic Systems* (G. Gregoriadis, J. Senior, and G. Poste, eds.), pp. 123–146, Plenum Press, New York (1985).
8. S. S. Davis and L. Illum, *Biomaterials 9*, 111 (1988).
9. J. B. Kayes and D. A. Rawlins, *Colloid Polym. Sci. 257*, 622 (1979).
10. J. H. Lee, J. Kopecek, and J. D. Andrade, *J. Biomed. Mater. Res. 23*, 35 (1989).
11. J. H. Lee, P. Kopeckova, J. Kopecek, and J. D. Andrade, *Biomaterials 11*, 455 (1990).
12. S. Nagaoka, Y. Mori, H. Takiuchi, K. Yokoata, H. Tanzawa, and S. Nishiumi, in: *Polymers as Biomaterials* (S. W. Shalaby, A. S. Hoffman, B. D. Ratner, and T. A. Horbett, eds.), pp. 361–374, Plenum Press, New York (1985).
13. Y. Mori, S. Nagaoka, H. Takiuchi, T. Kikuchi, N. Noguchi, H. Tanzawa, and Y. Noishiki, *Trans. Am. Soc. Artif. Internal Organs 28*, 459 (1982).
14. D. E. Gregonis, D. E. Buerger, R. A. Van Wagenen, S. K. Hunter, and J. D. Andrade, *Trans. 10th Annu. Meeting of the Society for Biomaterials*, p. 266, Washington, D.C. (1984).
15. C. G. Golander, S. Jonsson, T. Vladkova, P. Stenius, and J. C. Eriksson, *Colloids and Surfaces 21*, 149 (1986).
16. C. G. Golander and E. Kiss, *J. Colloid Interface Sci. 121*, 240 (1988).
17. E. Kiss, C. Golander, and J. C. Eriksson, *Prog. Colloid Polym. Sci. 74*, 113 (1989).
18. S. Winters, *Immobilized Heparin Via A Long Chain Poly(ethylene oxide) Spacer for Protein and Platelet Compatibility*, Doctoral Thesis, Department of Pharmaceutics, University of Utah (1987).
19. W. R. Gombotz, *Poly(ethylene oxide) Surfaces: Synthesis, Characterization and Biological Interaction Studies*, Doctoral Dissertation, Center for Bioengineering, University of Washington, Seattle, WA (1988).
20. W. R. Gombotz, W. Guanghui, and A. S. Hoffman, *J. Appl. Polym. Sci. 37*, 91 (1989).
21. S. W. Kim, H. Jacobs, J. Y. Lin, C. Nojori, and T. Okano, *Ann. N.Y. Acad. Sci. 516*, 116 (1987).
22. H. A. Jacobs, T. Okano, and S. W. Kim, *J. Biomed. Mater. Res. 23*, 611 (1989).
23. D. K. Han, K. D. Park, K. Ahn, S. Y. Jeong, and Y. H. Kim, *J. Biomed. Mater. Res. 23*, 87 (1989).
24. D. K. Han, S. Y. Jeong, and Y. H. Kim, *J. Biomed. Mater. Res. Appl. Biomat. 23*, 211 (1989).
25. Y. Sun, W. R. Gombotz, and A. S. Hoffman, *J. Bioactive and Compatible Polymers 1*, 316 (1986).
26. Y. Sun, A. S. Hoffman, and W. R. Gombotz, *Am. Chem. Soc., Polym. Prepr. 28*, 282 (1987).
27. Y. Sun, A. S. Hoffman, and W. R. Gombotz, *Proc. 13th Annu. Meeting of the Society for Biomaterials*, p. 266, Washington, D.C. (1987).
28. J. L. Brash, S. Uniyal, and Q. Samak, *Trans. Am. Soc. Artif. Internal Organs 20*, 69 (1974).
29. V. Sa Da Costa, D. Brier-Russell, E. W. Salzman, and E. W. Merrill, *J. Colloid Interface Sci. 80*, 445 (1981).
30. T. G. Grasel and S. L. Cooper, *Biomaterials 7*, 315 (1987).
31. D. W. Grainger, C. Nojiri, T. Okano, and S. W. Kim, *J. Biomed. Mater. Res. 23*, 979 (1989).

32. D. W. Grainger, K. Knutson, S. W. Kim, and J. Feijen, *J. Biomed. Mater. Res. 24*, 403 (1990).
33. J. G. Bots, L. van der Does, and A. Bantjes, *Biomaterials 7*, 393 (1986).
34. E. A. Merrill and E. W. Salzman, *ASAIO J. 6*, 60 (1983).
35. E. W. Merrill, E. W. Salzman, K. A. Dennison, S. W. Tay, and R. W. Pekala, *Progress in Artificial Organs*, 909 (1986).
36. N. B. Graham and M. E. McNeill, *Biomaterials 5*, 27 (1984).
37. J. Hermans, *J. Chem. Phys. 77*, 2193 (1982).
38. D. H. Atha and K. C. Ingham, *J. Biol. Chem. 258*, 5710 (1983).
39. D. L. Coleman, D. E. Gregonis, and J. D. Andrade, *J. Biomed. Mater. Res. 16*, 362 (1982).
40. F. E. Bailey and J. Y. Kolskc, *Poly(ethylene oxide)*, Academic Press, New York (1976).
41. F. E. Bailey and R. W. Callard, *J. Appl. Polym. Sci. 1*, 56 (1959).
42. R. Kjellander, *J. Chem. Soc., Faraday Trans. 2*, 2025 (1982).
43. K. J. Liu and J. C. Parsons, *Macromolecules 2*, 529 (1969).
44. W. R. Gombotz, W. Guanghui, T. A. Horbett, and A. S. Hoffman, *J. Biomed. Mater. Res. 25*, 1547 (1991).
45. G. G. Hammes and P. B. Schimmel, *J. Am. Chem. Soc. 89*, 442 (1967).
46. S. Saeki, N. Kawahara, N. Nakata, and M. Kaneko, *Polymer 17*, 685 (1976).
47. G. Karlstrom, *J. Phys. Chem. 89*, 4962 (1985).
48. R. Kjellander and E. Florin, *J. Chem. Soc., Faraday Trans. 77*, 2053 (1981).
49. R. A. Horne, J. P. Almeida, A. F. Day, and N. J. Yu, *J. Colloid Interface Sci. 35*, 77 (1971).

17

Poly(Ethylene Glycol) Gels and Drug Delivery

N. B. GRAHAM

17.1. INTRODUCTION

Poly(ethylene glycols) (or PEGs) have been utilized in various forms in pharmaceutical preparations for many years. They are used as additives to creams, as solubilizing agents, and as components of injectable formulations. In the form of surface active agents they are widely used in the formulation of a variety of products in both the foodstuffs and in the pharmaceutical industries. Their toxicity is considered to be adequate for many uses on or in the body. They thus provide a very attractive group of materials for the synthesis of novel "tailor-made" polymers for drug delivery or other therapeutic use. In the form of copolymers these materials have been licensed for use as both drug delivery and wound healing. This chapter describes work done to synthesize new and useful polymers based on this class of starting materials.

17.2. THE STRUCTURE AND PROPERTIES OF POLY(ETHYLENE GLYCOL)

Poly(ethylene glycols) are made from ethylene oxide by anionic polymerization as shown in Equation 1. The preparation leaves a hydroxyl on each end of the molecule

$$\overset{\displaystyle O}{\overset{/\quad\backslash}{CH_2CH_2}} \rightarrow HO\text{—}(\text{—}CH_2CH_2O\text{—})_n\text{—}H \qquad (1)$$

N. B. GRAHAM • Department of Pure and Applied Chemistry, University of Strathclyde, Glasgow G1 1XL, Scotland, U.K.

Poly(Ethylene Glycol) Chemistry: Biotechnical and Biomedical Applications, edited by J. Milton Harris. Plenum Press, New York, 1992.

which renders it reactive and able to be incorporated into a very large number of novel polymer structures. These polymers and the gels or networks which can be produced by a skilled synthetic polymer chemist can be well defined and reproducible. The derived polymers reflect many of the physical characteristics of the PEG from which they have been made and it is appropriate at this stage to consider some of these properties which make the derived polymers quite exceptional, if not unique.

17.2.1. The Hydroxyl End Groups

Poly(ethylene glycols) are not readily made at molecular weights above 20,000 daltons and those commercially available in large quantities are below 10,000 daltons. Novel synthetic polymers will thus utilize PEGs of molecular weights below this figure. It is noteworthy that in attempting to react the hydroxyls on the ends of PEGs it is quite difficult to obtain the exact stoichiometry which must be maintained to the standards of reproducibility required for a product to be used for pharmaceutical applications. For example, if one takes a PEG of molecular weight 8000 the hydroxyl content of the material is only 0.425% by weight. The reactants must thus be measured and mixed very precisely. In addition, this level of hydroxyl is equivalent to a water content of only 0.225% by weight and, as water can be a coreactant, great care must be taken to eliminate it from the reaction. It is not difficult to produce a solid lump of dry polymer, but it is not possible to obtain well-defined reproducible materials without attention to these and other details of the manufacture. This importance of the hydroxyl content means that the manufacturers' nominal molecular weight guide is of little practical use in the manufacture of the materials and the hydroxyl content must be measured in the laboratory by a well-tested and proven technique. As such techniques are often somewhat operator dependent, this is not a trivial undertaking.

17.2.2. The Crystallinity

The very regular and flexible backbone of the PEG, which is also free from substituents, provides a structure which can crystallize readily. The absence of substituents in such a flexible molecule provides melting points which are quite low. They are, however, conveniently close to readily accessible working temperatures. The polymers above number averages of approximately one thousand are normally highly crystalline with values up to a theoretical value estimated as 76 °C for a perfect infinite crystal.[1] In practice, real PEGs never exceed melting points of 68 °C. The crystallinity is rapidly destroyed in the presence of water, which allows the ready dissolution of the linear PEGs and the aqueous swelling of the xerogels (dry version of the hydrogel) made from them by their incorporation into more complex polymer networks. This presence of crystallinity in the final xerogel is highly beneficial, as it provides a significant enhancement of the physical strength without preventing their swelling in water, which is unusual.

17.2.3. The Glass Transition Temperature

The flexible backbone without substitution provides an amorphous material which is a low melting rubber in the absence of crystallinity. The glass transition temperature is below −40 °C and is modified in a complex manner when the linear polymer is mixed with various proportions of water.[2] PEG 5700 has a dry T_g of −49 °C, which decreases to −73 °C when one mole of water per ether oxygen is introduced but increases somewhat again with increasing water content. When the PEGs are incorporated into polymer networks by reaction of the terminal hydroxyls, the crystallinity and the glass transition temperature effects above are reflected in the properties of the polymers produced.

17.2.4. The Interaction of PEG and Its Copolymers with Water

The networks made from crystalline PEGs are themselves usually crystalline but of a lower degree than the equivalent linear polymer. Their level of crystallinity can be controlled by their composition and the degree of crosslinking. In general, their properties in the dry state are superficially similar to those of low-density polyethylene. They are opaque tough solids or membranes which, unlike poly(ethylene), swell in water to many times their dry weight, producing a transparent rubber. As these and related polymers derived from poly(ethylene glycols) appear to be of a most interesting degree of biocompatibility and haemocompatibility and this may well relate to the structure of water associated with the polymers, it is pertinent to consider what is the nature of this association with the poly(ethylene oxide) units in the polymers.

Among the ascending series of alkylene polyethers, poly(ethylene oxide) is the only one which has significant solubility in water at all molecular weights as in Table 1. The reason for this anomaly is not clear from a superficial scan of the structures in the series, but it is clear that PEG has a special position. Its interaction with water will be discussed further below.

The nature of the association of water with PEGs containing terminal hydroxyl groups and poly(ethylene oxide) units in which these hydroxyls have been removed during the incorporation into more complex networks, has been a subject of continuous investigation in our laboratories over a number of years. The literature is

Table 1. The Solubility in Water of Polyethers

Repeat unit	Solubility in water
$—OCH_2—$	No
$—OCH_2CH_2—$	Yes
$—OCH_2CH(CH_3)—$	No (above M_n *ca* 1000)
$—OCH_2CH_2CH_2CH_2—$	No

confusing as there is mention of mono-, di-, tri-, tetra- and hexahydrates.[3–8] Studies have been made on a wide variety of poly(ethylene oxide) structures as homopolymers, networks and PEG surfactants. On first consideration it appears that the most probable associations of these with water would be as either mono- or dihydrates as it is only possible to hydrogen-bond one or two water molecules to the ether group of the polymer.

The expected structure is represented in Figure 1 for the dihydrate. In our studies with PEGs of a range of molecular weights from a few hundred to thousands, no evidence supportive of the formation of a dihydrate was found. Rather, we always found strong indications of a trihydrate structure which seems to exist only at molecular weights in excess of approximately 600. The nature of the association of water with PEG thus appears to change markedly at around this molecular weight and it may well be that the apparent confusion in the literature is a result of the different results obtained with materials above and below this region of change. It would not be surprising if the biocompatibility results obtained with samples above and below this transition range differed. The trihydrate is evidenced by a break in the plot of water content against many physical properties. One such is shown in Figure 2.[9]

In this plot the relative viscosity of PEG/water mixtures is plotted against the composition. A clear discontinuity may be seen at close to three molecules of water per ether group and at 6 moles. Similar plots are observed with other physical properties such as refractive index and density. It is proposed that the trihydrate has a helical structure which can only be formed when a sufficient number of units in the chain are present to form at least two coils of the helix.[10] Below this critical molecular weight the association with water will be quite different. The trihydrate is very readily observed when a differential scanning calorimetric analysis of a mixture of an appropriate PEG and water is made. A typical result is shown in Figure 3. In this sample there was more than enough water to form a trihydrate and leave excess free water. The melting endotherms for both the trihydrate and for free water can be clearly seen.[2]

17.2.5. The Swelling in Water

The state of association of the PEG with water is clearly very temperature dependent as is revealed by the unusual and marked decrease in swelling of a typical

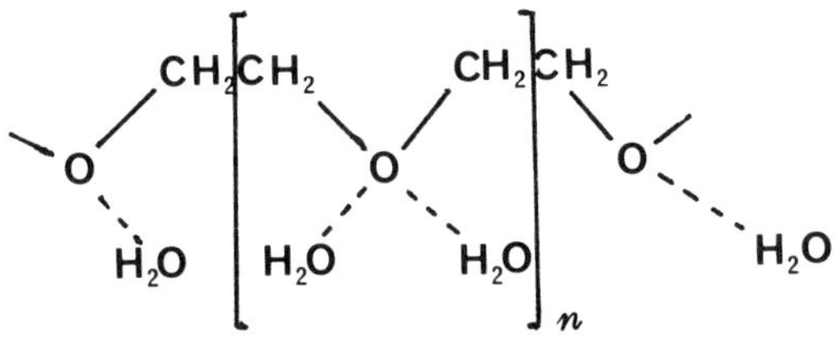

Figure 1. Structure of a hypothetical dihydrate of poly(ethylene oxide).

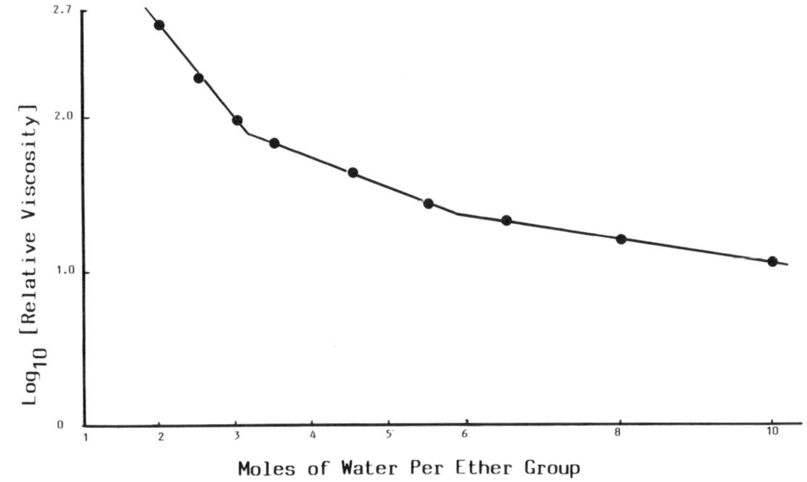

Figure 2. Plot of $\log_{10}$ [relative viscosity] at 52.6 against composition for mixtures of PEG 6000 and water showing a break at approximately 3 molecules of water per ether group.

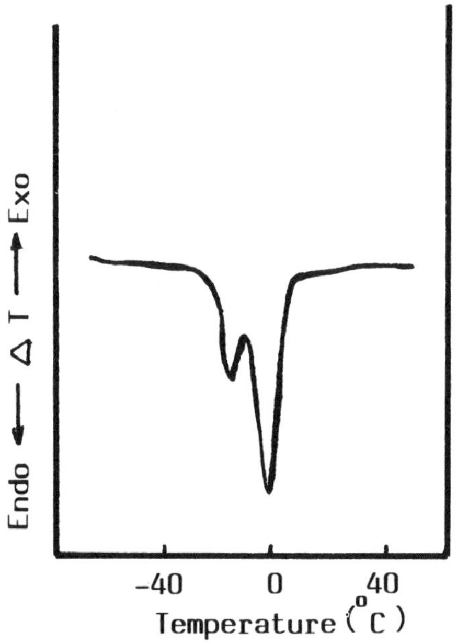

Figure 3. DSC scan of a fully swollen hydrogel based on PEG 6000 showing the free water and trihydrate peaks.

hydrogel as the temperature is increased, as shown in Figure 4. This means that the association of the water with the polymer is exothermic. This has been demonstrated in a number of calorimetric experiments, which are complicated by the fact that the polymer is initially partially crystalline. As the crystallites melt there is an initial endothermic process which, after approximately 30% by weight of water has been taken up by the polymer gel, becomes exothermic.[10] The decrease in swelling with increasing temperature is also a result of the fact that when dry poly(ethylene glycol) is dissolved in water, in contrast to the more expected situation which does prevail in most organic solvents, the net entropy change is negative rather than positive. This makes a larger positive contribution to the free energy of mixing as the temperature increases which, in turn, reduces the swelling.[11] The swelling would also be expected to be modified by the presence of salts or dissolved species which can compete for the water. This is in fact found to be the case. Salts, because of their osmotic competition with the swelling hydrogel, would be expected in all cases to cause a lower swelling by the salt solution than in pure water. It is found that there are the majority which, like sodium chloride, do behave in this manner but lithium chloride, for example, gives the opposite pattern of behavior and the gel swelling is higher in the lithium chloride solution that in pure water. Similarly, though probably for different reasons, the swelling is increased in solutions of urea. The problem of defining the structure of the water associated with the polymer *in vivo* is clearly complex.

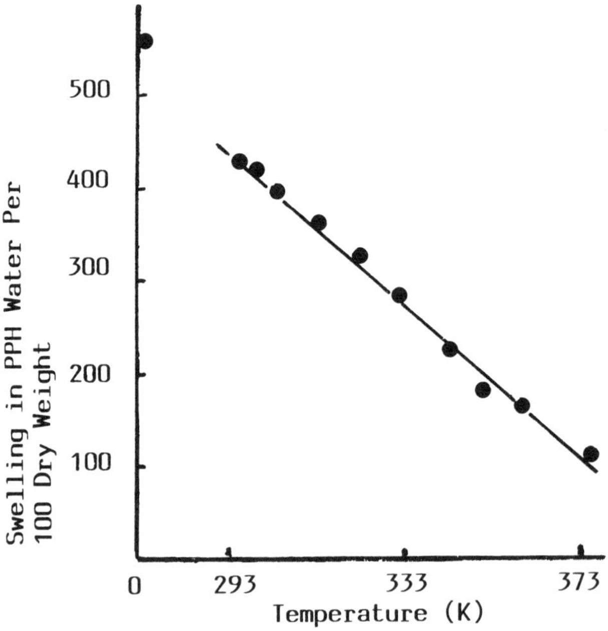

Figure 4. Decrease in swelling in water with increasing temperature

17.3. THE PREPARATION OF PEG HYDROGELS AND XEROGELS

17.3.1. Entanglement Crosslinking

Poly(ethylene oxide) (or PEG) xerogels have been made by a number of different techniques. An early approach was by "entanglement crosslinking" in which a poly(ethylene oxide) of high molecular weight in the hundreds of thousands to millions is mixed with a polyfunctional monomer with a free radical generating initiator and the monomer polymerized by heating to give an entangling network, which provides a swellable but not soluble xerogel.[12] Typically the PEO would be admixed with divinyl benzene, methylene bisacrylamide, or with acrylic acid; the latter, while only being difunctional, probably provides crosslinking by grafting onto the hydrogens of the PEO.[13]

17.3.2. Radiation Crosslinking

A more successful method of crosslinking was developed by The Union Carbide Company who filed a number of patents on the radiation crosslinking of high molecular weight PEO.[14] This technique produces material highly swollen with water in the presence of which it is desirable to perform the irradiation. It is thus acceptable for material which can be used as it is made in the fully swollen form as in wound healing dressings, but is difficult and expensive to dry. The material in the form of a crosslinked sheet has been licensed and commercialized as Vigilon™.

17.3.3. Chemical Crosslinking

Chemical crosslinking using the hydroxyl end groups of the PEGs is, in principle, a better way to obtain dry polymer directly from handlable starting materials. The techniques for obtaining reproducible polymers are somewhat demanding but the properties of the products are excellent and, as will be described below, they have some unexpected virtues when used for the delivery of contained drugs. Points requiring attention have been discussed above. The polymers are made from mixtures of the family of PEGs, a triol, and a diisocyanate. Optionally, other polyfunctional materials can be incorporated.[15] The molecular weight of the PEG, the nature of the isocyanate and polyol, along with the degree of crosslinking provide a large and versatile family of polymers which can be designed and optimized for the particular use. Of course in a licensable application it is necessary to restrict the formulations used to those which have received a Product License and this reduces the possibilities of varying the composition once the choice for licensing has been made. It may be desirable in some cases to utilize a catalyst in the crosslinking reaction of the formation of urethane groups from the isocyanate and hydroxyls. Conventional catalysts, such as tin of iron compounds, may be used so long as they are toxicologically acceptable for the application under consideration.

The simple process is represented as follows:

$$\text{Triol} + \text{PEG} + \text{Diisocyanate} \rightarrow \text{Crosslinked dry hydrogel}$$

A simple representation of polymer structure is shown in Figure 5. This polymer network represents that made using diphenylmethane-4,4′-diisocyanate. For biocompatibility the aliphatic dicyclohexyl-4,4′-diisocyanate is preferred. The products may be made in the shapes in which they are going to be used or else they can be cut or machined as desired.

17.3.4. Biodegradable Gels

In the design of devices for the delivery of drugs, especially when these are to be used as implants, it is highly desirable to be able to utilize biodegradable polymers which will be bioadsorbed and so not require removal from the body after they have served their purpose. We have studied such degradable hydrogels made by the incorporation of the compound C1 below which contains two 3,4-dihydro-2*H*-pyranyl rings.[16]

This monomer contains a carboxylic ester group which is an analog of a lactate and is prone to hydrolysis in water. Its scission over a period of time leads to the breakdown of polymers into which it is incorporated. It is possible to prepare polymers from this monomer by its direct polymerization under the influence of a Lewis or Brønstead type of acid as shown in Figure 6. The polymers produced are, however, glassy solids and not hydrogels. It is also possible to prepare a copolymer of this monomer by reaction with polyols. The reaction with a simple diol gives a linear polymer as shown in Figure 7. These polymers are inevitably of quite low molecular weight and, in general, stronger products are provided by the reaction with polyols, which provide crosslinking to yield a glassy amorphous matrix that has been shown to provide interesting biodegradable vehicles. Initially, attempts to produce hydrogels from these biodegradable backbones only gave very weak and unstable products. The possible compositions can be represented and understood by the use of a triangular composition plot in which the three components used are the dihydropyranyl monomer, a triol which in this example is 1,2,6-hexane triol, and poly(ethylene glycols), as given in Figure 8. It can be seen in this figure that the compositions which

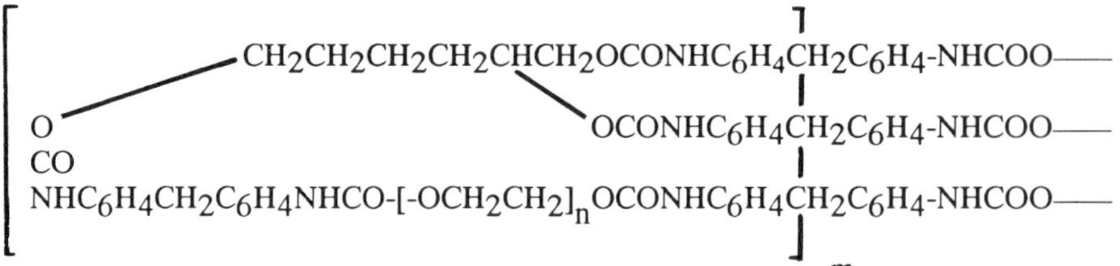

Figure 5. Representation of the repeat unit of a PEG urethane crosslinked hydrogel.

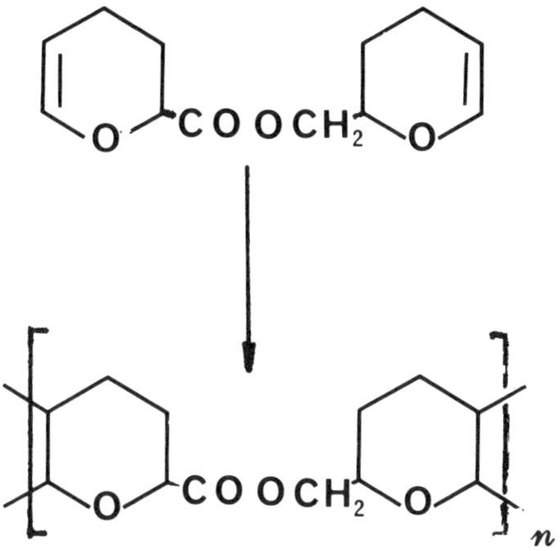

Figure 6. Polymerization scheme for the monomer C1.

will gel can contain either an excess of dihydropyranyl double bond or not. It was found that the use of a slight excess of the double bond over that required for equivalence to the hydroxyl produced a quite dramatic improvement in the resulting polymer physical properties as shown in Figure 9.[17]

This effect was believed to be due to a submicron phase separation which provided a reinforcement of the network and a significant improvement in the strength. These small domain reinforcements could be seen in the transmission electron micrographs of the compositions. In respect of their drug release, it was clearly demonstrated that the physical rather than the chemical properties of the

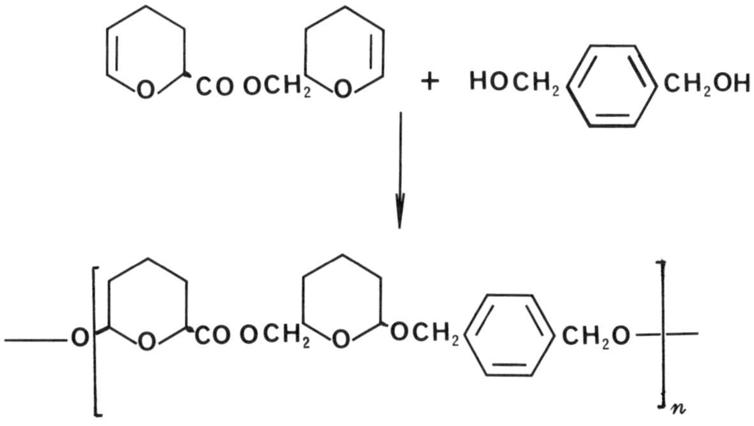

Figure 7. Polymerization of monomer C1 with 1,4-dimethylol benzene.

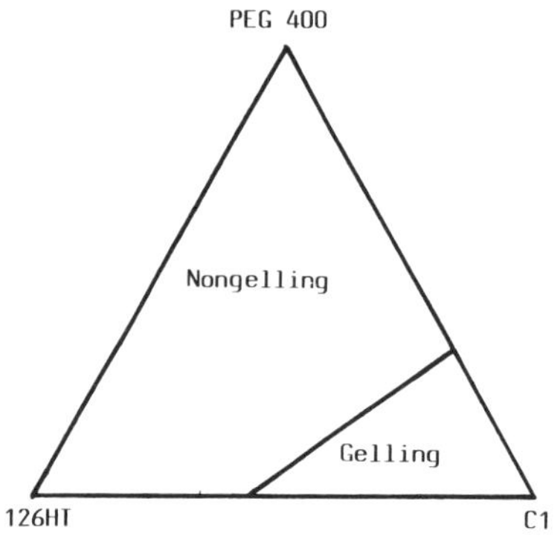

Figure 8. Triangular compositoin plot representing the compositions capable of forming crosslinked networks from 1,2,6-hexane triol (1,2,6HT), PEG400 and the monomer C1.

matrix were those which determined the profile of release of the contained drugs, and the release of prostaglandin E_2 from matrices of either polyurethane or polypyran were essentially identical so long as the swelling and crystallinity of the hydrogel matrices were the same. In related but unpublished studies done in collaboration with the World Health Organisation it was shown that polymers in this class containing some degree of water uptake due to the incorporation of a poly(ethylene glycol) could release the steroid levonorgestrel over a year in rabbits and at a faster rate than the composition in which no poly(ethylene glycol) was incorporated. The materials clearly show some promise for the sustained delivery as implants from relatively insoluble active agents. The release is shown in Figure 10.

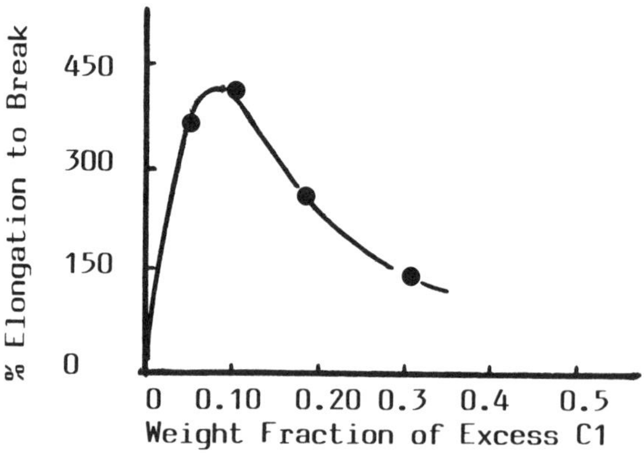

Figure 9. Plot of the elongation to break of hydrogels made from PEG7000, 1,2,6-hexane triol, and the monomer C1.

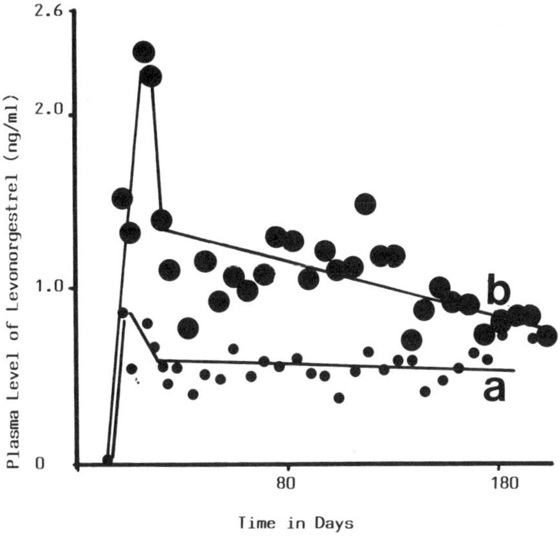

Figure 10. The blood levels of levonorgestrel in rabbits with implanted controlled release formulations based on: (a) formulation containing no PEG, (b) formulation containing PEG.

17.3.5. Block Copolymers

The use of crosslinked poly(ethylene oxides) places some limitations on the fabrication possibilities of the resulting materials, which are neither soluble in solvents nor thermoplastic. It is possible to prepare block copolymers of PEG which, in principle, can achieve both of these objectives. In an early patent to Fildes and Hutchinson of ICI the preparation of such materials and their proposed application to a veterinary drug delivery system was described.[17] The structure of a typical polymer from this patent is illustrated in Figure 11.

More recently a number of such materials have been revealed. In our own laboratories we have (as previously mentioned) used polymers based on blocks of PEG with poly(propylene glycol) linked by urethane–urea groups to produce membranes which are of promising hemocompatibility.[18] A group in Japan have also reported the use of block copolymers of PEG and their excellent drug release properties.[19] They consider the effect of the complex block domain structure on the release characteristics. This domain structure will also have an effect on the hemocompatibility and is a subject of considerable importance and current interest. Initially, the simple hypothesis was that, for hemocompatibility, one required merely a high water content in a material. This view has now been superseded by the proposal that some specific balance between size, concentration, and spacing of the hydrophilic and hydrophobic domains on the surface is likely to provide the best results. The family of block copolymers of poly(propylene glycol) and PEG provide materials which are soluble in solvents such as ethanol and methanol and can be conveniently

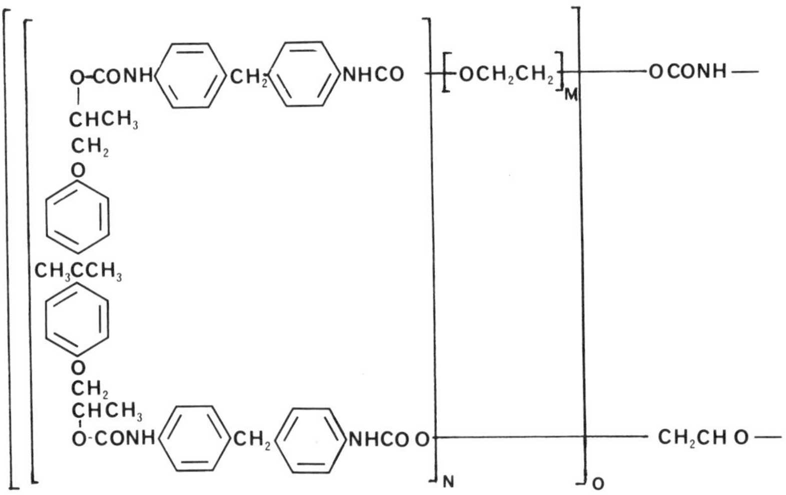

Figure 11. Typical repeat units of a PEG block copolymer as given in Reference 17.

fabricated into either supported or unsupported membranes by techniques such as rotational casting. The membranes so produced may have potential for improved renal dialysis membranes or for wound healing coverings.

17.4. THE USE OF PEG COPOLYMERS IN DRUG DELIVERY

17.4.1. The Demands of Controlled Delivery

During the 1950s the first major controlled release drug delivery systems were commercialized by Smith, Kline, and French as coated granules for the prolongation of the release of agents for alleviating the symptoms of colds, allergies, and various respiratory problems. These products were very successful and are still today major products at the basis of many hundreds of millions of dollars worth of business. At around the same period the Alza Corporation started to publicize the need for precise controlled drug delivery and must be given much of the credit for the current awareness of the field. The much advocated objective for controlled drug delivery systems during this period was that of constant or zero-order delivery. It was believed that better therapy would result from the attainment of more precisely constant plasma levels of the active agent from devices or delivery systems which delivered them to the body at an essentially constant rate. This has now been achieved and constant delivery can be achieved by a number of techniques among which is the use of swelling PEG hydrogels as described in this chapter. All of these systems which have achieved clinical application and the status of licensed products are in all probability going to be applied to the delivery of an increasing number of agents with the passage of time. Though the discussion in this chapter will deal with the attainment of constant delivery, it is necessary to say at this point that zero-order release is only one particular profile of the many which might be desirable. The future problems for the designers of delivery systems are those of achieving "programmed delivery" to any desired specifications. Five release profiles which are highly desirable are shown in Figure 12.

Profile A is the conventional delayed but not constant release. Profile B is constant or zero-order release. These two profiles are now common in commercial systems. Profile C is that of a substantial delayed release followed by a constant release of active agent. Such a system would be most useful for the delivery of active agents commencing at some period during the night. Profile D is that of a delay followed by a tight pulse of drug release. This again allows for nocturnal delivery or for the delivery of a hormone, which often requires pulsed rather than constant delivery. Finally, the last Profile E is that of multiple pulses at specified periods. For example, at one time the World Health Organisation were requesting contract submissions for such a system to deliver pulses of estradiol over three days each month for a succession of months. While such objectives present considerable difficulties, it has to be said that systems producing spaced sequential pulses of delivery are already marketed in the veterinary field for anthelmintics!

17.4.2. Sustained Delivery from PEG Hydrogels

The release of a dissolved diffusate from a homogeneous geometry is well defined in theory by Fick's laws of diffusion. From these laws the characteristics of the predicted release of a contained active agent from different monolithic shapes may be predicted. For an infinite flat slab charged with a dissolved diffusate in an initial amount M_0, which is numerically the same but of opposite sign to the amount of diffusate which is released at infinite time M_∞, the equation for the first 60% of the

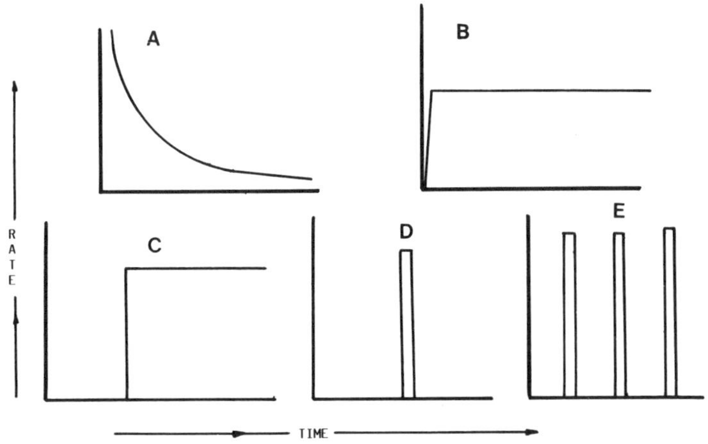

Figure 12. Five desirable controlled release rate/time profiles.

release is $M_t/M_\infty = 4(Dt/\pi l^2)^{0.5}$ where D is the diffusion coefficient, t is the elapsed time, l is the slab thickness, and π has its usual meaning. This equation may be rearranged and the values of t, M_t, and M_∞ substituted to provide the equation in terms of the half-life of drug release, $t_{0.5} = 0.049(l^2/D)$. From this it can be seen that the half-life of drug release is determined by the reciprocal of the diffusion coefficient and the square of the slab thickness. The diffusion coefficient is a function of the degree of water swelling (a particular characteristic of the hydrogel systems of this chapter) while the thickness is a purely geometric factor. Thus the design of suitable half-lives of release should be accomplished either by the control of the composition or by control of the geometry. This is very convenient as, though the synthetic polymer chemist can design a wide range of compositions which might have the possibility of meeting the release specification, once one has been successfully taken through the toxicology and the system is moving toward a product license, only the one composition on which the toxicology has been determined is permissible for use. The composition is usually frozen and the ability to utilize the geometry becomes invaluable. To test out the theory in practice, slabs of suitable dimensions that could approximate the "infinite" were made and the release of caffeine contained in them studied in the fully swollen state.[20] The effect of both the polymer composition and the geometry were determined. The half-life of release in all cases showed the expected proportionality to $t^{0.5}$ for the first 60% of the release. The half-life increased as expected, as the water content decreased with increasing level of crosslinking, and (as shown below) the half-life of release showed the expected straight-line relationship to the square of the slab thickness (Figure 13).

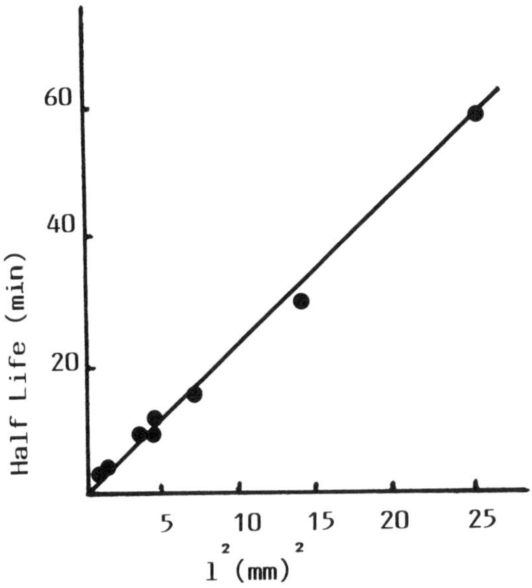

Figure 13. The half-life of caffeine release plotted against the square of the fully swollen slab thickness. From *J. Controlled Release* 5, 243, (1987) with permission of Elsevier Science Publishers.

These hydrogel systems thus provide a very precisely controllable example of a method of achieving release profiles of type A. However, these water-swollen systems suffer from three major faults. The first is that, though the release can be extended, it is necessary to use swollen devices that are at the limit of convenience in size for human use. The second is that the profile of release is that of A rather than the more desirable type B or zero order. Third, as water can evaporate the use of water-swollen devices is not conducive to obtaining good reproducibility on storage. One outstanding feature of these systems does, however, appear very attractive: the drug solution is trapped in a polymer matrix from which there is no possibility of it being "dose dumped." These systems are thus very appealing from the standpoint of safety. If a means could be found to overcome the deficiencies while retaining the safety feature, then the formulations would be attractive for potential development. This was achieved with the dried down slabs.

17.4.3. Release from Dry PEG Hydrogels

The PEG crosslinked hydrogels when made from the higher molecular weight PEG precursors are quite highly crystalline. While this contributes greatly to their physical strength, it also slows penetration of water and provides a rate of swelling which is found to have a most unusual dependence on the 2.6 power of the thickness. Thus for a small increase in the thickness one obtains a considerable prolongation of the swelling time and consequently of the release of any contained diffusate. The release of drugs from the dry slabs was therefore studied. The release of caffeine from dried down slabs showed, now, a much prolonged half-life of release, a constant or zero-order release for approximately half of the contained drug, and none of the storage stability problems which would accompany devices containing water.[21,22]

These promising delivery systems were first studied in the clinic in collaboration with Mostyn Embrey, Gynaecologist and Obstetrician at the John Radcliffe Hospital, Oxford, England. He suggested that the sustained vaginal delivery of prostaglandin E_2 (PGE_2) to women who at full term were suffering from an "unripened cervix" and who would be at risk of a very difficult and prolonged delivery with the significant possibility of the need for a caesarian section, might prove to be a very beneficial therapy. The idea was in principle very simple as these women were known as a group to have low levels of PGE_2 in their amniotic fluid and the PGE_2 was known to be an important link in the hormonal chain of events leading to the cervical ripening. The controlled delivery system merely provided replacement PGE_2 at a controlled rate under the control of the clinician and in an apparently very safe manner. The small lozenge-shaped devices containing 5 or 10 mg of PGE_2 were evaluated in the clinic and found to be very effective. The storage stability of the quite unstable PGE_2 was found to be surprisingly good when held in the dry hydrogel matrix and the product was commercialized, licensed, and introduced to the UK and Irish markets in 1989. A typical release from the lozenge-shaped device is given in Figure 14. The essentially constant release for some 50% of the contained drug is a characteristic feature of these systems.

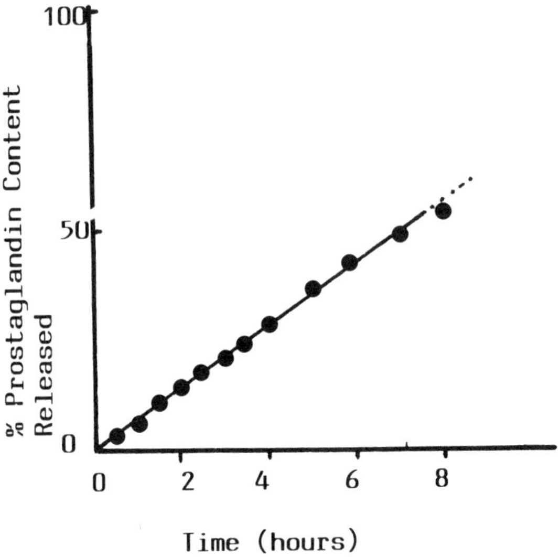

Figure 14. Percent release against time for prostaglandin E_2 (10 mg) from a lozenge-shaped crosslinked PEG pessary showing a constant rate of *in vitro* release for the first 50% of contained PGE_2.

17.4.4. Extended Constant Release from Nondumping Monoliths

The desirable release profiles above would be even better if the constant release was obtained for an even higher proportion of the contained drug. This should in principle be possible for systems in which an internal concentration profile is deliberately introduced thereby increasing the rate at the later stages to flatten the release profile. If a configuration could be found which only released from one surface of a slab, then incorporating an increased concentration at the surface away from the one in contact with the fluid should provide a flattened release profile. To achieve this a hollow cylindrical configuration was studied.[23] In collaboration with G. Smith and C. Hanning of the University of Leicester, Department of Anaesthesia, the possibility of a rectal delivery system to provide an essentially constant delivery of morphine over 12 hours was considered and developed.[24] A hollow cylinder of PEG hydrogel was made as in Figure 15. This could be first charged uniformly with a drug solution, then charged again with a higher concentration of drug in the interior wall by a second treatment with the drug solution. There are many variations on the method to obtain concentration profiles. The clinician requested a higher initial rate to achieve appropriate blood levels the more rapidly. The release, which is essentially constant over 80% of the contained drug, is shown in Figure 16.

These systems are under further development. If commercialized and granted a product license, they essentially become a throw-away infusion pump. The polymer, as expected, did not show any signs of tissue irritation of incompatibility in contact with the rectal tissues.

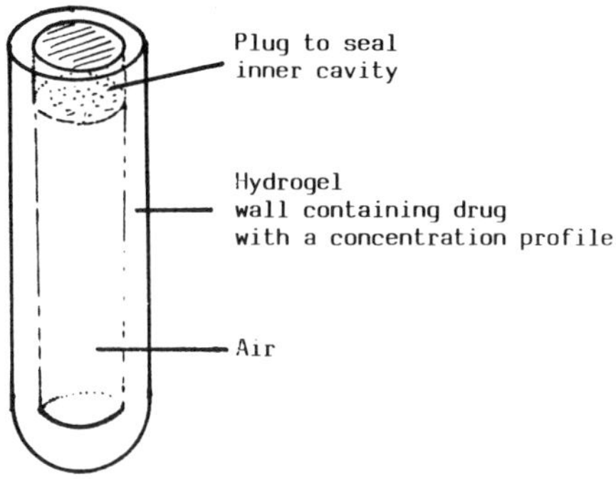

Figure 15. Typical arrangement of cylindrical hydrogel device for concentration profiled control of release.

17.4.5. The Use of PEG Hydrogels as Thick Membranes

The systems described above are essentially short term and the drugs can be delivered for periods up to 24 hours. While this is very suitable for normal pharmaceutical applications, it is necessary for many therapies to provide delivery from an insert or implant over periods varying from weeks to months. This can be achieved by using the PEG hydrogels as thick rate-controlling membranes.[25,26] The flux of diffusate across a membrane is defined by Fickian diffusion as $J = DKA\Delta C/l$, where J is the flux, D and l are as previously defined, K is the distribution coefficient, and ΔC is the difference in diffusate concentration between the aqueous solutions on each side of the membrane. Again it can be seen that the flux of diffusate depends on the diffusion coefficient and the membrane thickness.

Hydrogels are of considerable interest because they have useful fluxes for water soluble agents and the rate-controlling membranes can be thick enough to consider

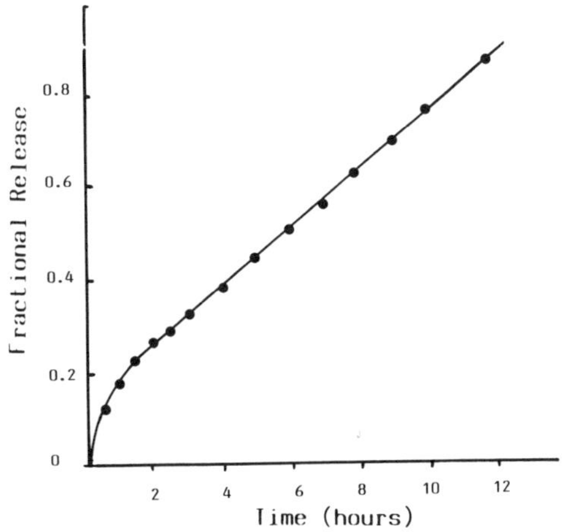

Figure 16. Typical profile of release for a water-soluble diffusate from a concentration profiled device as shown in Figure 15. The device is designed with an initially high rate to build up plasma levels rapidly.

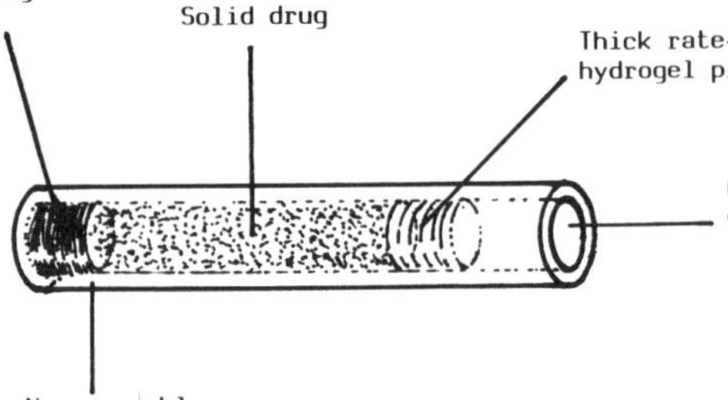

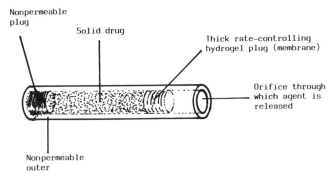

Figure 17. Implantable membrane cell utilizing a thick membrane of hydrogel for precise rate control.

that they might be made reproducibly. The approach taken has been to use the hydrogel in the equivalent of a tiny diffusion cell (Figure 17). In this a small cylindrical section of hydrogel is allowed to seal a tube made of a hydrophobic polymer. On swelling it provides a fluid tight seal. The release rate is determined by the composition, area, and thickness of the thick membrane while the length of release is determined by the amount of drug placed in the reservoir of the device. So long as there is at all times solid drug in the reservoir to provide a saturated solution, so maintaining a constant ΔC, the rate of release should in principle remain constant. This was in fact proven to be the case. These devices are very suitable for the prolonged delivery of small amounts of active materials. The first to be studied in animals in collaboration with R. Rodway of the Veterinary Department of the University of Leeds was the pineal gland hormone melatonin. To bring ewes into early estrus and, as a result, to produce early lambing and improved flock lambing control, implants were devised to deliver, at a constant rate, melatonin at various rates in the range from 0.2 mg per day to 5 mg per day. Using the thickness and the diffusing area as rate-controlling factors allowed such a variety of rate control to be achieved quite readily and rapidly.[27] The *in vivo* results were most encouraging and the lambing trials showed not only the desired increased efficiency and control of lambing, but also an increased lambing yield. More recently, closely related devices have been demonstrated to be effective for the delivery of peptides.

17.5. THE FUTURE PROSPECTS FOR PEG HYDROGELS

The proven safety and efficacy of the PEG hydrogels combined with their apparently excellent biocompatibility indicate that they are most likely to become one of the standard materials in the armoury of the biomaterials scientist. The other chapters of this book point to many of the advances being made in the laboratory and in commercial use. Taking into account the long history of the use of PEG linear polymers in conventional pharmacy it must be merely a matter of time before the use in many new fields becomes firmly established.

REFERENCES

1. D. R. Beech and C. Booth, *Polym. Lett. 8*, 731 (1970).
2. N. B. Graham, M. Zulfiqar, N. E. Nwachuku, and A. Rashid, *Polymer 30*, 528 (1989).
3. G. Boehmke and R. Heusch, *Fette Siefen Anstrichem. 62(2)*, 87 (1960).
4. M. Rosche, *Kolloid Z. 147*, 78 (1956).
5. M. Rosche, *Fette Siefen Anstrichem. 65*, 223 (1963).
6. J. L. Koenig and A. C. Angood, *J. Polym. Sci. A2, 8*, 1787 (1970).
7. K. J. Liu and J. L. Parsons, *Macromolecules 1*, 529 (1969).
8. J. Maxfield and I. W. Shepherd, *Polymer 16*, 505 (1975).
9. N. E. Nwachuku, Ph.D. Thesis, University of Strathclyde (1977).
10. N. B. Graham, M. Zulfiqar, N. E. Nwachuku, and A. Rashid, *Polymer 31*, 909 (1990).
11. N. B. Graham, N. E. Nwachuku, and D. J. Walsh, *Polymer 23*, 1345 (1982).
12. R. D. Lundberg, Canadian Patent, 756,190 (1967).
13. N. S. Chu, US Patent, 3,963,805 (1976).
14. P. A. King, US Patent, 3,149,006 (1968).
15. M. P. Embrey and N. B. Graham, US Patent 4,894,238 (1990).
16. N. B. Graham, M. E. McNeill, and A. Rashid, *J. Controlled Release 2*, 231 (1985).
17. J. T. Fildes and F. G. Hutchinson, US Patent, 4,202,880 (1980).
18. J. Yu, S. Sundaram, D. Weng, J. M. Courtney, C. R. Moran, and N. B. Graham, *Biomaterials*, March 1991, Vol. 12, 119–120.
19. N. Yui, K. Kataoka, A. Yamada, and Y. Sakurai, *J. Controlled Release 6*, 329 (1987).
20. N. B. Graham, M. E. McNeill, B. B. MacDonald, and M. Zulfiqar, *J. Controlled Release 5*, 243 (1987).
21. M. E. McNeill and N. B. Graham, *J. Controlled Release 1*, 99 (1984).
22. M. P. Embrey, N. B. Graham, M. E. McNeill, and K. Hillier, *J. Controlled Release 3*, 39 (1986).
23. M. E. McNeill and N. B. Graham, *ACS Symp. Ser. 348, Controlled-Release Technology Pharmaceutical Applications*, Chapter 12, p. 158 (1987).
24. G. Smith, C. Hanning, N. B. Graham, and M. E. McNeill, *Proc. Anaesthetic Research Soc.*, Belfast (1984).
25. N. B. Graham M. E. McNeill, and D. A. Wood, US Patent 4,814,182 (1989).
26. N. B. Graham, E. R. Huehns, and M. E. McNeill, UK Patent 2,153,675 (1987).
27. R. G. Rodway, R. R. Rajkumar, C. M. Argo, C. E. Webley, M. E. McNeill, and N. B. Graham, *Endocrinology 112*, Supplement 45 (1987).

18

PEO-Modified Surfaces—In Vitro, Ex Vivo, and In Vivo Blood Compatibility

KI DONG PARK and SUNG WAN KIM

18.1. INTRODUCTION

Thrombus formation is a serious problem in surgical therapy and clinical application of artificial organs. Therefore, the need for the creation of highly antithrombogenic biomaterials has been increasing. Although a substantial amount of work in the improvement of the blood compatibility of polymeric materials has been carried out, the results are still inconclusive. This is caused partly by the fact that the relationship between surface properties and surface-induced thrombosis has not been thoroughly evaluated.

Thrombus formation is caused by blood coagulation and platelet adhesion to and platelet activation on foreign surfaces.[1,2] Blood coagulation has been elucidated somewhat at the molecular level. Coagulation is triggered by both the intrinsic and extrinsic pathways and coagulation factors, i.e., clotting factors, are successively activated. Factor X, one of the coagulation factors, lies at the intersection of the two pathways. Activated (FX_a) converts prothrombin to thrombin. In the final stage of blood coagulation, thrombin catalyzes the proteolysis of fibrinogen to fibrin and a fibrin clot results. Protein adhesion onto foreign surfaces followed by platelet adhesion takes place as soon as the blood is in contact with the surfaces. The adhering platelets release constituents, such as ADP and serotonin, which apparently cause the subsequent platelet aggregation. The aggregated platelets release phospholipids,

KI DONG PARK and SUNG WAN KIM • Department of Pharmaceutics, Center for Controlled Chemical Delivery, University of Utah, Salt Lake City, Utah 84108.

Poly(Ethylene Glycol) Chemistry: Biotechnical and Biomedical Applications, edited by J. Milton Harris. Plenum Press, New York, 1992.

which potentiate the activation of the cascade reaction in the blood coagulation pathway.[3]

In the past, researchers have described the mechanisms of thrombus formation at foreign interfaces in terms of surface-related factors and hemodynamic effects. Although some bulk properties of the materials, such as mechanical compliance and water absorption, are considered to be important, the major material factors influencing blood interactions at polymer interfaces are surface properties which govern the initial events of thrombus formation and subsequent thrombogenicity.[4,5]

A variety of approaches have been taken to improve the blood compatibility of polymer surfaces. One approach involves surface modification by: (1) chemical modification by grafting of a hydrophilic component,[6,7] (2) surface modification incorporating bioactive agents such as fibrinolytic enzymes (urokinase and streptokinase),[8,9] various prostaglandins (PGE_1)[10,11] and potent anticoagulant (heparin),[12,13] through either physical or chemical coupling, and (3) biological modification using protein or cell seeding.[14,15]

In the authors' laboratory, a new approach to improve the blood compatibility of polymers has been developed. This approach was based on the following concepts: (1) Polyurethane is a useful and effective material with a proven track record in biomaterial research and clinical application. However, the inherent thrombogenicity of polyurethane remains a problem and limits its more widespread application. (2) The hydrophilic environment of the blood–material interface appears to reduce protein adsorption and platelet adhesion and can be achieved by the grafting of hydrophilic polymers, such as poly(ethylene oxide) (or PEO). The advantages of using PEO include a high water content of the surface, highly dynamic motion, and extended chain conformation at the blood–material interface. (3) Incorporated bioactive heparin on the surface can interact directly with the enzymatic and proteolytic cascades involved in thrombus formation.

Heparin has proven to be an effective agent for curtailing thrombosis and is effective when immobilized onto polymer surfaces.[16] The activity of immobilized heparin is dependent on the method of binding it to the polymer with particular regard to the role of the spacer type and length on subsequent heparin bioactivity.[17] In addition, it should be noted that sufficient mobility of bound heparin and its accessibility to physiological substrates are required for immobilized heparin to retain high bioactivity.[16,17]

Therefore, it is expected that polyurethane surfaces grafted with PEO and heparin will curtail surface-induced thrombus formation.

Hypotheses conceived with this research state that hydrophilic PEO spacers are expected to increase the bioactivity of heparin preventing fibrin formation, as well as reduce protein adsorption and platelet adhesion, due to low PEO interfacial free energy, lack of binding sites, and the highly dynamic motions of PEO spacers. The hypothesized action of heparin immobilized surfaces is shown in Figure 1.

In this chapter polymer surface modification, either by incorporation of PEO and bioactive heparin (in surface and in bulk) or by coating with PEO-containing triblock copolymer, is discussed.

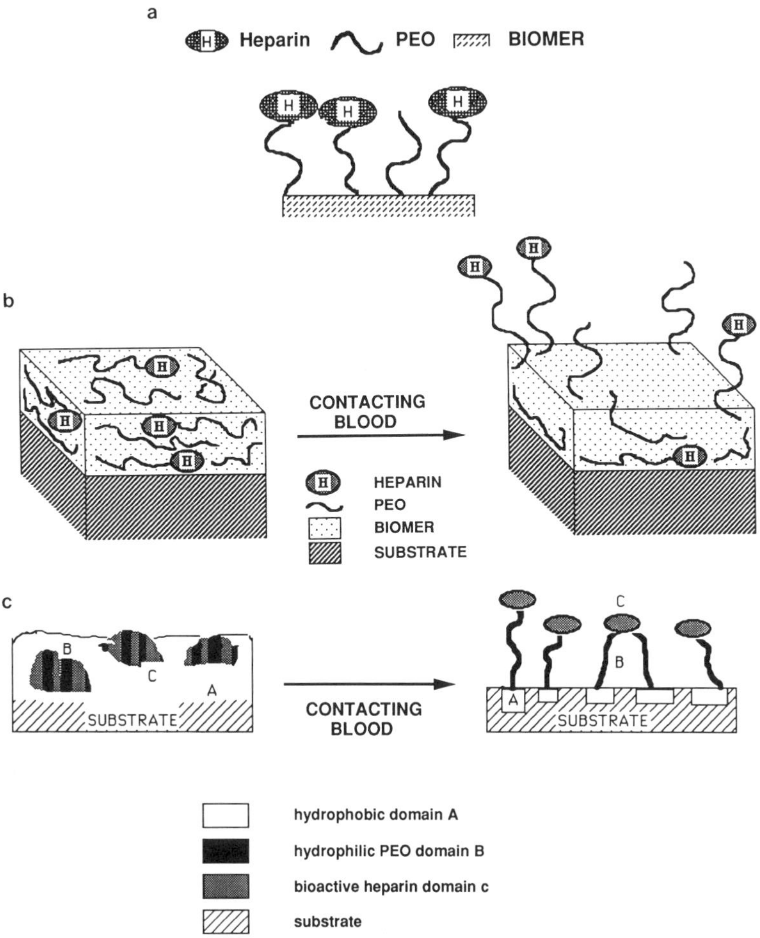

Figure 1. (a) Hypothesized action of heparin–PEO immobilized surface. PEO is positioned to minimize protein adsorption/cell adhesion; heparin prevents fibrin net formation. (b) Expected surface morphology of SPUU–PEO–heparin graft copolymer coating. (c) Possible surface morphology of PDMS–PEO–heparin triblock copolymer coating.

18.2. *IN SITU* SURFACE PEO GRAFTING AND BIOACTIVE HEPARIN IMMOBILIZATION

18.2.1. Chemistry

Heparin was immobilized onto segmented polyetherurethaneurea (Biomer®) surfaces using hydrophilic PEO spacer groups of different molecular weights (200, 1000, and 4000, designated as 0.2K, 1K, and 4K, respectively). The reaction process is shown in Figure 2.[18] The synthetic scheme involves the coupling of a telechelic diisocyanate-derivatized PEO to Biomer® through an allophanate/biuret reaction. The free isocyanate remaining on the spacer group is then coupled through a condensation reaction to functional groups (—OH, —NH_2) on heparin.

The first step in the procedure was to derivatize PEO polymers with diisocyanate functional groups. This was accomplished by reacting toluene diisocyanate (TDI) and PEO in a 2:1 molar ratio. TDI was first dissolved in benzene and a PEO solution in benzene was slowly added dropwise. The reaction proceeded under N_2 for 2–3 days at 60 °C. The TDI–PEO–TDI molecule was then purified through repeated precipitation in diethylether from benzene solution.

The TDI–PEO–TDI spacer groups were grafted onto the Biomer® surface through an allophanate/biuret reaction between the urethane/urea–nitrogen proton and the terminal isocyanate group of isocyanate-derivatized PEO. The TDI–PEO–TDI spacers were coupled to the surface of Biomer®-coated glass beads in the presence of a catalyst (0.1 v/v % dibutyltin dilaurate in benzene at 40–60 °C). The reaction was carried out over different times to vary the amount of PEO surface grafting achieved. The PEO grafted surface (B–PEO–NCO) was then thoroughly washed with benzene. Hexamethylene diisocyanate (HMDI) was also coupled to Biomer®-coated beads to serve as a control using the same procedure as with TDI–PEO–TDI spacers (B–C6–NCO). A portion of B–PEO–NCO and B–C6–NCO surfaces were then immersed in methanol to block the isocyanate groups remaining on the free ends of the grafts (B–PEO and B–C6). These surfaces, along with Biomer® surface, were used as controls in *in vitro* surface experiments. The amount of PEO and C6 spacers grafted onto Biomer®-coated bead surfaces were measured by acid–base back titration of the terminal free isocyanate groups. The inside surface of Biomer®-coated tubing was also modified under identical experimental conditions.

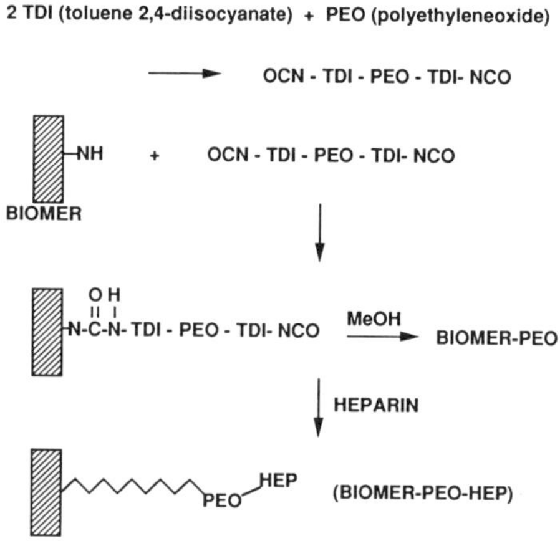

Figure 2. Synthetic scheme of heparin immobilized onto Biomer surface.

Heparin was covalently bound to the remaining B–PEO–NCO and B–C6–NCO surfaces through a coupling reaction between the free hydroxyl or amine groups on heparin and the free isocyanate group on the PEO or C6 spacer. Heparin was first dissolved in formamide and followed with the addition of 0.05 v/v % dibutyltin dilaurate and B–PEO–NCO or B–C6–NCO beads. The bead suspension was gently stirred at room temperature for 3 days. The B–PEO–HEP and B–C6–HEP beads were then thoroughly washed with water and acetone to remove unreacted materials and solvents. All samples were rinsed with distilled water until no further heparin could be detected in the wash solution by toluidine blue chromogenic assay.[19] Tubings grafted with B–PEO–NCO and B–C6–NCO were heparinized using a similar procedure and rinsed thoroughly to remove unreacted materials. Surface-immobilized heparin concentrations on both beads and tubings were determined by the toluidine blue chromogenic method.

18.2.2. *In Vitro* Bioactivity

Table 1 summarizes the total amount of heparin immobilized on the Biomer® surface as measured by the toluidine blue method. The bioactivity of immobilized heparin was determined by FX_a assay[20] and thrombin time (TT) measurements. Heparin catalytically potentiates the action of antithrombin III (ATIII) to inactivate thrombin, a critical process in suppressing the intrinsic blood coagulation cascade. Heparin contains specific binding sites for ATIII and thrombin, among other coagulation factors. Therefore, the orientation and exposure of heparin binding sites significantly affects the bioactivity of immobilized heparin. Generally, the bioactivity of heparin decreases after immobilization onto all of the surfaces studied, possibly because of chemical modifications or physical inaccessibility of heparin binding sites for coagulation factors. However, *ex vivo* results demonstrate that even a small fraction of bioactive immobilized heparin can have a significant effect on improving implanted shunt performance. By using a hexamethylene (C6) spacer between the substrate and bound heparin, immobilized heparin maintained $0.81 \pm 0.06 \times 10^{-2}$

Table 1. Heparin Activity of Immobilized Heparin on Biomer

Spacer	Spacer amount ($\times 10^{-8}$ mol cm^{-2})	Heparin amount[a] (μg cm^{-2})	Bioactivity ($\times 10^{-2}$ IU cm^{-2})	FX_a (%)[b]	Bioactivity ($\times 10^{-2}$ IU cm^{-2})	TT (%)[b]
C6	1.17	0.85 ± 0.12	0.81 ± 0.06[c]	5.32	0.75 ± 0.02[c]	4.93
PEO200	1.50	0.65 ± 0.04	0.86 ± 0.03	7.38	0.81 ± 0.03	6.96
PEO1000	1.30	0.50 ± 0.04	1.03 ± 0.02	11.50	0.95 ± 0.02	10.60
PEO4000	1.30	0.31 ± 0.05	1.06 ± 0.02	19.09	0.97 ± 0.02	17.47

[a]Toluidine blue method after washing thoroughly with PBS.
[b]Bioactivity ratio of immobilized heparin to the total amount.
[c]Mean ± S.D. (n = 3–5).

IU cm^{-2} bioactivity, which translates to 5.3% free heparin, as determined by ATIII complexation. The use of hydrophilic PEO spacers demonstrates that the bioactivity of immobilized heparin was consistently higher than that of the C6 alkyl spacer system. In addition, heparin bioactivity increased with increasing PEO spacer length. The heparin-immobilized surface using PEO4K maintained the highest bioactivity at approximately 1.06 ± 10^{-2} international units per cm^2 (19%), even though it possessed the least amount of heparin on the surface as detected by toluidine blue. These results indicate that the increasingly mobile nature of longer spacer chains permits a more bulk-like environment for heparin binding and enhances the observed bioactivity of immobilized heparin.

TT assays directly measure the inactivation of an excess amount of thrombin by heparin–ATIII complex through the prolongation or inhibition of clot formation. The results, as shown in Table 1, show behavior consistent with the FX_a assay and again show that the increasingly mobile nature of long hydrophilic spacer chains increases the observed bioactivity of immobilized heparin by providing a more bulk-like environment for heparin.

18.2.3. *In Vitro* Platelet Interaction

The relationship between platelet adhesion and PEO spacer length, as well as PEO amount, for B–PEO and B–PEO–Hep surfaces has been investigated.[18] As shown in Figures 3 and 4, PEO length and surface density are important factors. When either PEO0.2K ($0.7–2.5 \times 10^{-8}$ mol cm^{-2}) or PEO4K ($0.3–1.3 \times 10^{-8}$ mol cm^{-2}) chains were surface immobilized, PEO concentration had no significant effect on platelet adhesion. However, in the case of PEO1K ($0.4–1.9 \times 10^{-8}$ mol cm^{-2}), platelet adhesion decreased with increasing PEO concentration. Considering the chain length effect of PEO on platelet adhesion, PEO grafted surfaces indicated minimal platelet adhesion at PEO1K and minimal platelet activation at

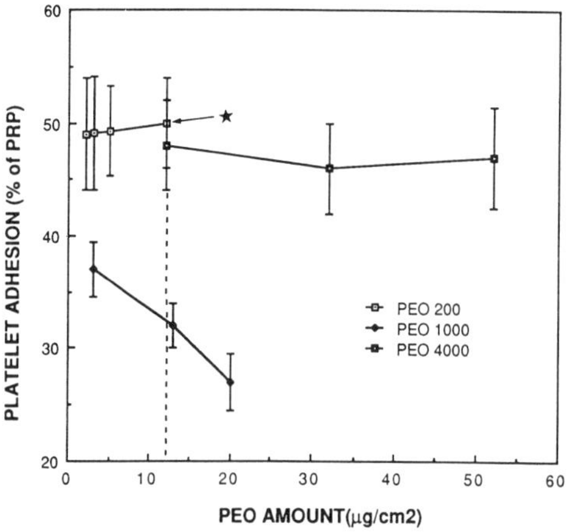

Figure 3. Effects of PEO amount as a spacer on platelet adhesion after 1 h incubation. Arrow indicates value determined by extrapolation (mean ± S.D., $n = 5$).

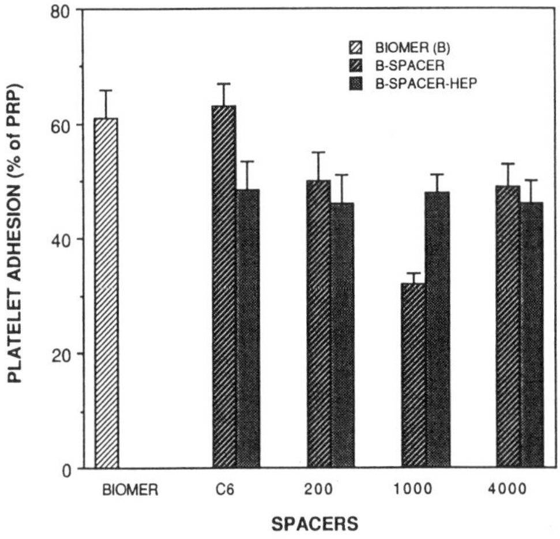

Figure 4. Effect of chain length of PEO on platelet adhesion at PEO amount 12 $\mu g\ cm^{-2}$ after 1 h incubation with PRP (mean ± S.D., $n = 5$).

PEO4K. Minimal platelet adhesion at PEO1K was unexpected. This aspect cannot be explained adequately and is at the present subject to further study. It was reported[21] that PEO surface–protein interactions are a function of surface density and chain length of PEO. In this system PEO1K might have optimum surface density however, PEO4K surface might not. PEO grafted surfaces have consistently lower platelet adhesion and activation (activation data not shown, see Ref. 22) than C6 grafted surfaces. This suggested that the PEO spacers, in general, suppress platelet adhesion and activation when compared to the hydrophobic alkyl spacer system. This can be explained by the interfacial free-energy concept[23] in which a decrease in interfacial free energy decreases protein adsorption, causing decreased platelet adhesion and activation. After immobilization of heparin, platelet adhesion was still decreased when compared with Biomer® surfaces. Platelet adhesion is nearly the same for different PEO spacer systems and for B–C6–Hep; a specific spacer length was not observed which is consistent with a previous report.[24]

18.2.4. *Ex Vivo* Shunt Studies

The nonthrombogenicity of modified polymer surfaces in whole blood was evaluated *ex vivo* by a newly developed low flow rate, low shear *ex vivo* rabbit A–A shunt method.[25] The procedure involved using surface-modified tubing (1.5 mm ID × 30 cm length) as a shunt in the carotid arteries of male rabbits. The experiment measured the time needed for the formation of a stable, nonembolized thrombus large enough to occlude the blood flow in the tube and this time was referred to as the occlusion time. The shunt flow was maintained at 2.5 $ml\ min^{-1}$ to minimize nonlaminar flow effects through the experiment. As shown in Figure 5, all of the heparinized surfaces prolonged occlusion time longer than controls. In addition, these

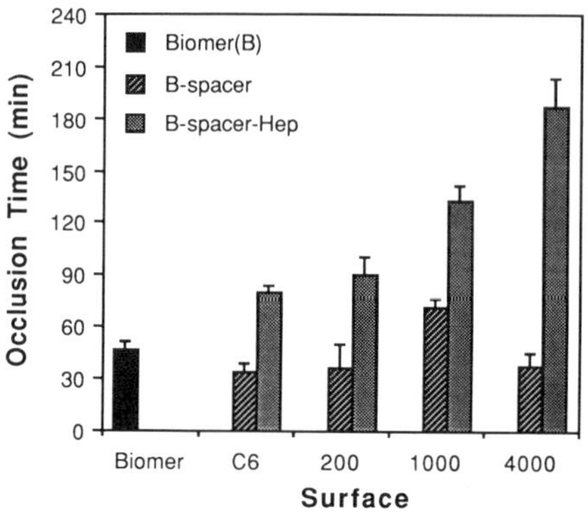

Figure 5. A–A shunt occlusion times for modified Biomer (mean ± S.D., n = 3–5).

surfaces showed increased occlusion times with increasing PEO spacer length. This suggested a specific spacer length effect on whole blood bioactivity of heparin and whole blood thromboresistance, in general. B–PEO surfaces, however, did not prolong occlusion times.

The improved *ex vivo* blood compatibility of B–PEO4K-*Hep*, compared with Biomer® and B–PEO4K, is due to the high heparin bioactivity achieved by using the long PEO spacer; consequently, fibrin net formation, followed by platelet adhesion, activation, and subsequent mural thrombus formation, are suppressed. Biomer® and B–PEO4K surfaces without heparin failed to prevent fibrin formation and fibrin-related platelet activation in whole blood, although B–PEO4K showed decreased platelet adhesion and activation in the *in vitro* study.

In extended A–A shunt studies[26] with varied flow rate (2.5, 5.0, 7.5, 15 $ml\,min^{-1}$), Biomer® and B–PEO4K showed increasingly prolonged occlusion times with increased flow rate while platelet count and aggregability decreased. The B–PEO4K surface demonstrated a prominent prolongation of occlusion time at a flow rate of 5.0 $ml\,min^{-1}$ with decreased platelet count and aggregability, indicating surface-induced platelet activation and systemic thromboembolization, possibly due to the PEO. In contrast, B–PEO4K–Hep surface showed the longest and most relatively flow-rate-independent occlusion time without detectable platelet activation.

18.2.5. *In Vivo* Canine Studies

Biomer®, B–PEO4K, and B–PEO4K–Hep grafts were studied *in vivo* using a dog model.[27,28] Biomer® vascular grafts (6 mm ID, 7 cm length) were fabricated by repeatedly dipping a glass rod in a Biomer® solution (15% w/v in DMAC). The grafts were removed from the glass rods by soaking in a 50% aqueous ethanol solution and gently slipping them off the rod. The luminal surfaces were modified by PEO grafting and heparin immobilization.

The grafts were implanted into the abdominal aortas of dogs. After times of 3 weeks to 3 months, the grafts were retrieved and processed for electron microscopy. The surface protein layer thickness was measured with a OsO_4 solution by transmission electron microscopy (TEM). Visualization of adsorbed proteins (albumin, IgG, and fibrinogen) was performed with scanning electron microscopy (SEM/TEM) using immunogold double antibody[29] and immuno-peroxidase[30] techniques.

B–PEO4K–Hep grafts demonstrated significant longer patency (3 months) than Biomer® or B–PEO4K controls (less than 1 month). The B–PEO4K–Hep grafts were patent at autopsy and there were neither detectable gross thrombi nor neointimal formation, except for the proximal anastomotic site shown in Figure 6. In contrast, retrieved grafts of Biomer® and B–PEO4K showed mural thrombi all along the graft length (Figures 7 and 8, respectively). These results correlated with the *in vitro* platelet study and *ex vivo* occlusion times using a rabbit A–A shunt system.

Biomer® and B–PEO grafts demonstrated a thick (1000 to 2000 Å) protein multilayer consisting mostly of fibrinogen and IgG and, to a minor extent, albumin. In contrast, B–PEO–Hep displayed a relatively thin (200–300 Å) protein layer with high concentrations of albumin and IgG, but low in fibrinogen. Biomer® and B–PEO grafts may lead to the denaturation of adsorbed proteins resulting in multilayer patterns. B–PEO–Hep may passivate the initial protein layer, preventing denaturation and resulting in a stable monolayer of protein. From a nonthrombogenicity point of

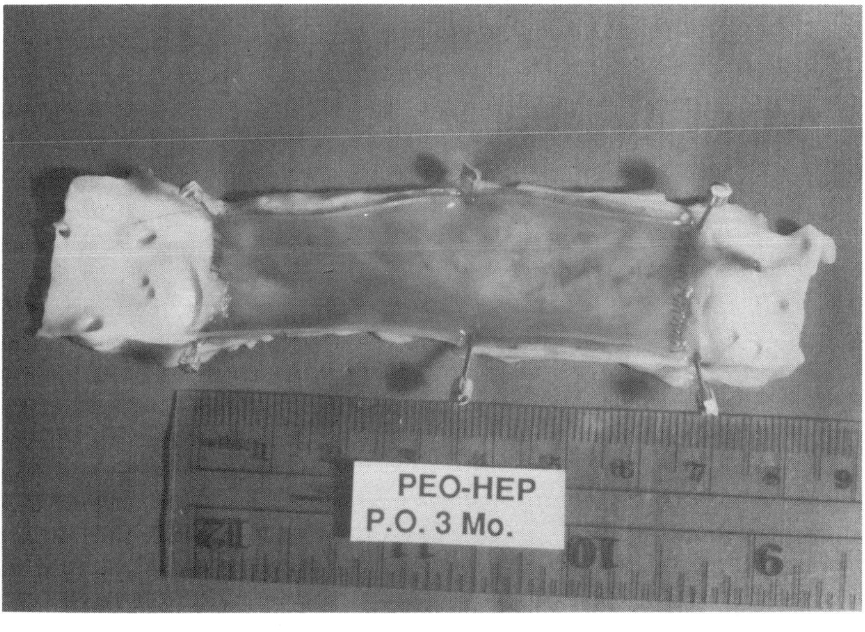

Figure 6. Luminal view of B–PEO4K–Hep graft after 3 months implantation.

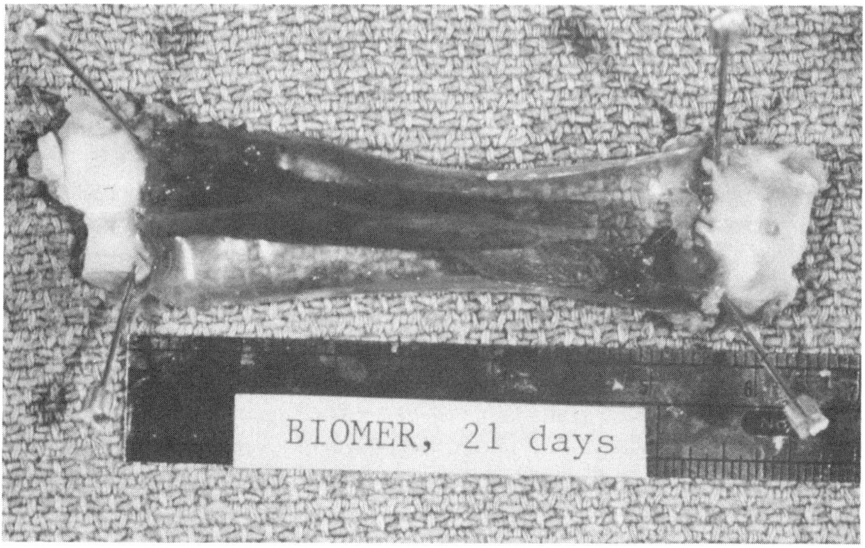

Figure 7. Luminal view of Biomer graft after 3 weeks implantation.

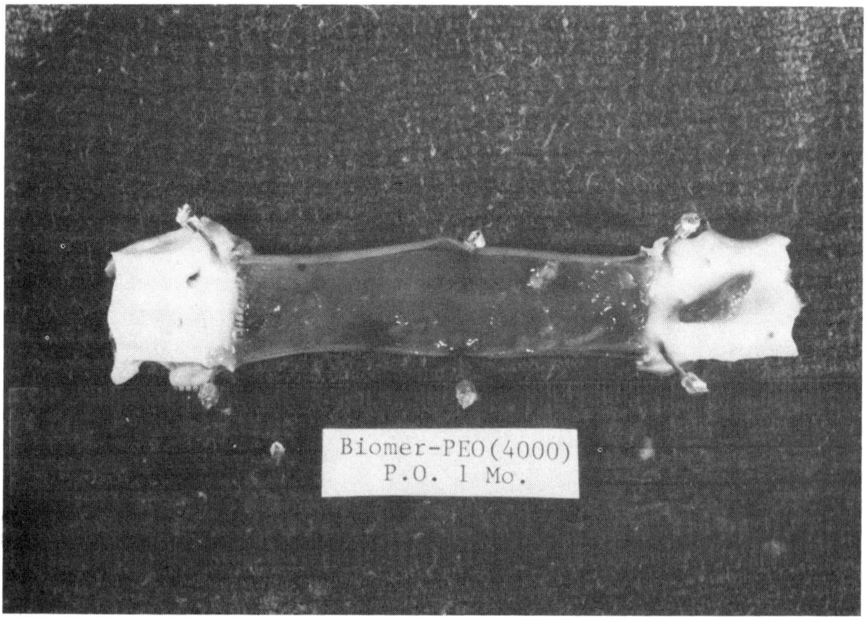

Figure 8. Luminal view of B–PEO4K graft after 1 month implantation.

view, a thinner protein layer *in vivo* appears to be a better surface. PEO surfaces have been shown to decrease *in vitro* protein adsorption.[6,7] The Biomer®–PEO graft copolymer also demonstrated decreased *in vitro* protein adsorption (as described in the next section). However, the multilayered-protein (adsorption observed in the long-term *in vivo* studies suggests that short-term *in vitro* protein adsorption may fail to predict long-term *in vivo* protein adsorption.

In general, the high activity of immobilized heparin found *in vitro*, in addition to a minimization of platelet adhesion and activation, correlated to the improved blood compatibility in *ex vivo* and *in vivo* whole blood studies. This was due to the prevention of fibrin net formation by immobilized heparin, resulting, as hypothesized, in a reduction of fibrin-related platelet aggregation and subsequent thrombus formation. PEO grafted surfaces without heparin failed to prevent thrombus formation in whole blood, although they did reduce platelet adhesion and activation *in vitro*. Table 2 summarizes the *in vitro*, *ex vivo*, and *in vivo* correlation of each surface.

In this study, the proposed hypothesis was proven both *in vitro* and *in vivo* and the obtained results can be utilized in the design of new blood contacting medical devices.

18.3. SPUU–PEO–HEP GRAFT COPOLYMERS

As described above, *in situ* surface immobilization of heparin (Hep) onto segmented polyurethaneurea (SPUU) surface using hydrophilic PEO spacers (surface reaction) was shown to be effective in improving the blood compatibility of the SPUU

Table 2. *In Vitro*, *Ex Vivo*, and *In Vivo* Correlations

Surface	*In vitro*	*Ex vivo*	*In vivo*
Biomer (B)	High platelet adh/act	Short occlusion time (O.T., 40 min). Flow-rate-dependent occlusion time. Hematological parameters change. Activated platelet and fibrin formation	Short patency (<1 month). Thick protein layer (>1000 Å)
B–PEO4K	Low platelet adh/act	Short O.T. (50 min). Flow-rate-dependent O.T. Hematological parameters change. Activitated platelet and fibrin formation. Thromboembolization	Short patency (<1 month). Thick protein layer (>1000 Å)
B–PEO4K–Hep	High hep. activity Low platelet adh/act	Prolonged O.T. (182 min). Relatively flow-independent O.T. No hematological parameters change. Minimal platelet activation and prevented fibrin formation	Longer patency (>3 months). Thin protein layer (200–300 Å)

surface. Based on these results, a new soluble SPUU–PEO–Hep graft copolymer (bulk reaction) was developed to show improved blood compatibility.[31] This new heparinized material can be applied as a coating over existing blood contacting devices without changing device bulk properties.

The expected biological response to this new polymer is the same as that to *in situ* surface immobilization. The PEO chain will enhance the bioactivity of heparin, as well as decrease protein adsorption, platelet adhesion and activation. Grafted heparin–PEO chains, as shown in Figure 1(b), are expected to reorient into the aqueous environment when contacting blood, providing the nonthrombogenic behavior observed in *in situ* heparin immobilization.[18,26–28]

Based on these perspectives, Biomer®–PEO–heparin (B–PEO–Hep) graft copolymers were synthesized and characterized. In addition, the chain length of the PEO spacer was varied to maximize heparin bioactivity. B–PEO graft copolymer and Biomer® were used as control materials. Biological responses to these polymers were evaluated *in vitro*, *ex vivo*, and *in vivo*.

B–PEO–Hep graft copolymers were synthesized via condensation reactions as shown in Figure 9, similar to those used in the *in situ* surface immobilization procedure.[31]

The procedure involved the coupling of HMDI to Biomer® backbone through an allophanate/biuret reaction. The free isocyanate groups attached to Biomer® were then coupled to a terminal hydroxyl group of PEO to form PEO-grated Biomer® (B–

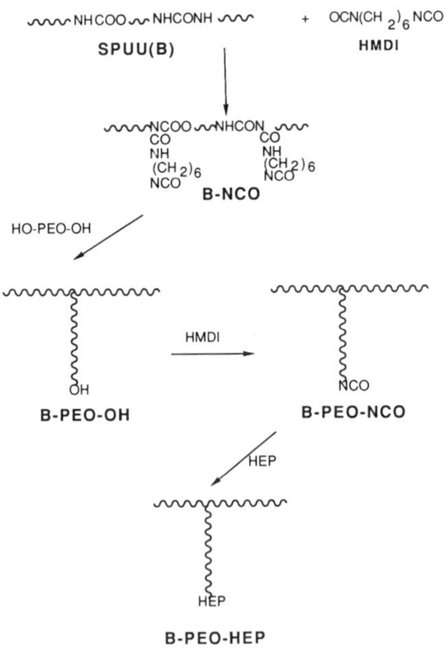

Figure 9. Synthetic procedure for SPUU–PEO–heparin graft copolymer.

PEO). The free hydroxyl groups of B–PEO were modified with HMDI to introduce a terminal isocyanate group. The NCO functionalized B–PEO was then coupled to functional groups on heparin (—OH, —HN_2) producing B–PEO–Hep graft copolymer. The spacer length of PEO was varied to maximize the bioactivity of immobilized heparin (PEO MW 1000, 3350, 7500 abbreviated 1K, 3.4K, 7.5K, respectively).

Synthetic intermediates were confirmed by FTIR, NCO group determination, and toluidine blue chromogenic assay. Physical characterization methods, such as contact angle measurements and DSC thermal analysis, were used to detail the properties of these graft copolymers containing covalently bound heparin.[31] These new heparinized copolymers can be applied as a coating on other existing blood contacting surfaces without changing bulk properties.

The expected biological responses to B–PEO–Hep graft copolymer coating were demonstrated by *in vitro*, *ex vivo* rabbit shunt, and *in vivo* dog implantation studies.[32,33] The bioactivity of immobilized heparin was shown to be a function of PEO spacer length, as summarized in Table 3. B–PEO3.4K–Hep copolymer-coated surface showed the highest bioactivity and these results correlated with the *in situ* immobilization system. All B–PEO surfaces showed less protein adsorption than Biomer® surfaces and the higher MW PEO system adsorbed less protein. B–PEO–Hep surfaces also demonstrated less protein adsorption than Biomer® and protein adsorption patterns similar to B–PEO surfaces. The relationship between platelet adhesion and PEO length for B–PEO surfaces as well as B–PEO–Hep surfaces is shown in Figure 10. Platelet adhesion to B–PEO surface was shown to be a function of PEO length with minimum adhesion at PE07.5K. Significant differences between chain lengths were not observed for heparin immobilized surfaces. The platelet release (activation) study demonstrated that both B–PEO and B–PEO–Hep surfaces inhibited platelet release, as compared to Biomer®, with no significant spacer length effect. In general, B–PEO and B–PEO–Hep surfaces prevented platelet adhesion and activation which correlated with the minimal protein adsorption observed.

In low flow rate *ex vivo* rabbit A–A shunt experiments (1.5 mm ID, 30 cm

Table 3. Heparin Activity of Surface Coated with Graft Copolymers

Surfaces	Surface heparin concentration[a] ($\mu g\ cm^{-2}$)	Bioactivity: TT ($\times 10^{-2}$ I.U. per cm^2) (%)[b]	Bioactivity: FX_a ($\times 10^{-2}$ I.U. per cm^2) (%)[b]
B–PEO (1K)–Hep	0.47 ± 0.08[c]	0.91 ± 0.03[c]	0.96 ± 0.02[c] (11.54)
B–PEO (3.4K)–Hep	0.28 ± 0.05	1.22 ± 0.02	1.24 ± 0.02 (22.17)
B–PEO (7.5K)–Hep	0.24 ± 0.04	0.80 ± 0.02	0.81 ± 0.04 (18.25)

[a]Toluidine blue method.
[b]Bioactivity ratio of immobilized heparin to the free heparin.
[c]Mean ± S.D. (n = 4).

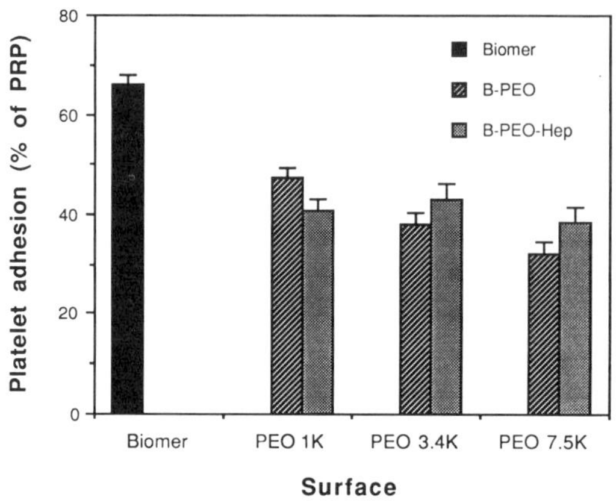

Figure 10. Platelet adhesion on polymer surfaces after 60 min incubation with PRP (mean ± S.D., n = 5).

length), all heparinized surfaces prolonged occlusion time more than controls (Figure 11). As observed in the *in situ* immobilized system, these surfaces prolonged occlusion times to a greater extent with increased PEO length. Maximum occlusion time was achieved with B–PEO3.4K–Hep. This is primarily attributed to the higher bioactivity of immobilized heparin with a longer PEO chain. Decreased protein adsorption, platelet adhesion and activation for B–PEO–Hep copolymer surfaces also correlated with prolonged occlusion time.

We investigated the influence of B–PEO3.4K–Hep and B–PEO3.4K copolymer-coated blood-contacting surface on patency and platelet deposition in small-diameter (4 mm ID) Biomer® grafts using dogs as an animal model.[33] Biomer® vascular grafts were fabricated by repeatedly dipping a glass rod in a Biomer® solution (20% w/v in

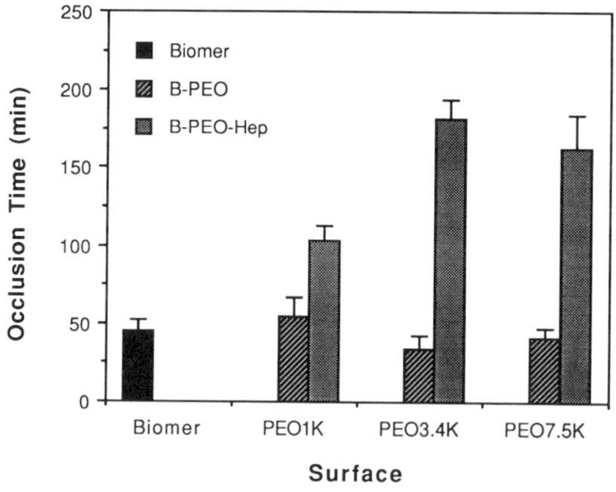

Figure 11. A–A shunt occlusion times for graft copolymer coatings (mean ± S.D., n = 5).

DMAC) and followed by soaking in a 50% ethanol–water solution to slip them off the rods. Solutions (1%) of B–PEO3.4K–Hep, B–PEO3.4K, and Biomer® were coated on the inner surface of the grafts. For patency and platelet adhesion studies, grafts were implanted in the bilateral carotid (2.5 cm length) and femoral arteries (3.5 cm length) in the dogs.

B–PEO3.4K–Hep copolymer-coated grafts still showed patency after three days. In contrast, all of the control grafts (B–PEO3.4K and Biomer®) occluded before 24 h, postoperatively. Scanning electron micrographs of the luminal surfaces of the control grafts demonstrated large amounts of adhered platelets with distorted morphologies while B–PEO3.4K–Hep grafts did not. This is consistent with the low patency seen with control grafts.

For the platelet study, ^{111}indium-labeled platelets were circulated through the experimental grafts for 15 min. Retrieved grafts were assessed for retained radioactivity using a gamma counter. B–PEO–Hep grafts showed significantly less platelet adhesion than either Biomer® or B–PEO grafts.

The improved nonthrombogenicity of B–PEO–Hep copolymer-coated surfaces was comparable to that achieved by *in situ* surface immobilization techniques and attests to the usefulness of this new procedure as a coating for improving the blood compatibility of blood-contacting surfaces.

18.4. PDMS–PEO–HEPARIN TRIBLOCK COPOLYMERS

Bioactive amphiphilic triblock copolymers consisting of heparin, PEO, and poly(dimethyl siloxane) (PDMS) have been synthesized and extensively characterized.[34–37]

Amino-semitelechelic PDMS was coupled to diamino-telechelic PEO using two different diisocyanate species to give block copolymers of PDMS–PEO–NH_2 (9500 MW PDMS, 4000 MW PEO). The terminal amino groups on this copolymer were then used to covalently bind heparin by two different procedures in tetrahydrofuran (THF):water solutions (Figure 12). In the first procedure, carboxylic acid groups on heparin were activated with 1-ethyl-3-(3-dimethylaminopropyl)carbodiimide hydrochloride (EDC) and subsequently coupled with amino-semitelechelic PDMS–PEO–NH_2 to give the heparinized product. In the second method, heparin was first derivatized by acid hydrolysis to give heparin with terminal aldehyde groups. This heparin form was then coupled to the copolymer via a Schiffs base complex using $NaCNBH_3$ as a selective reducing agent. Both reactions were performed in THF: water solution and yielded block copolymers having PEO–Hep moieties.

The PDMS–PEO–Hep triblock copolymers form a stable coating where the hydrophilic block may be anchored into a suitable existing hydrophobic substrate. For *in vitro* and *ex vivo* assessment, the triblock copolymer was coated over glass beads and polyurethane tubes (1.5 mm ID, 30 cm length), respectively.

The bioactivity of PDMS–PEO–Hep triblock copolymer coating assessed by toluidine blue chromogenic assay, TT, and Factor Xa assay is summarized in Table 4.

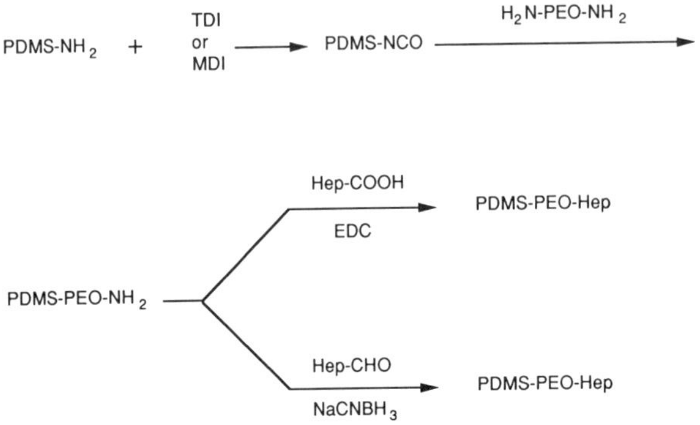

Figure 12. Synthetic scheme for triblock copolymers of PDMS–PEO–heparin.

In vitro comparisons of PDMS–PEO–Hep with its unheparinized analogs demonstrated a significant ability of the heparinized coating to neutralize clotting factors and prolong clotting times. Comparison bioactivity data with direct quantitation of surface-immobilized heparin in the polymer indicated that, although significant amounts of heparin were detected at the surface, only a fraction of these molecules were able to complex both ATIII and FX_a.

Platelet adhesion and release data suggested that PDMS–PEO–Hep reacted minimally with platelets in a platelet-rich plasma system and seemed to pacify surface activation when compared to Biomer®. This observation, combined with the effect of immobilized heparin on the surface to inactivate key contact activation factors, demonstrates the promise of this strategy for surface modification of existing hydrophobic substrates.

In low flow rate, low shear *ex vivo* A–A shunt experiments, control surfaces (Biomer®, crosslinked PEO and PDMS by triisocyanate) performed poorly with all of them demonstrating occlusion times of less than 50 min. However, occlusion times for the triblock copolymer coatings provided the longest occlusion time, presumably due to the presence of both PEO and heparin (Figure 13).

Table 4. *In Vitro* Quantitation of Surface-Immobilized Heparin on Coated Beads

	Heparin surface concentration		
Substrate	Toluidine blue ($\mu g\,cm^{-2}$)	Thrombin (units per cm^2)	Factor Xa (units per cm^2)
PDMS	0	0	0
PDMS–PEO	0	0	2.65×10^{-3}
PDMS–PEO–heparin	0.55	2.65×10^{-2}	2.66×10^{-2}

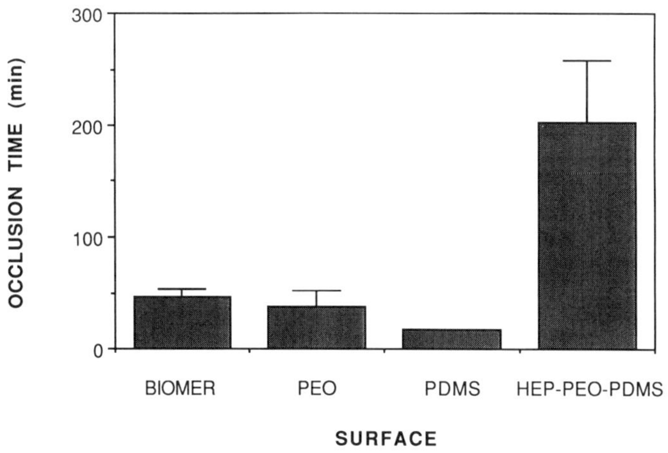

Figure 13. A–A shunt occlusion times for triblock copolymer coatings (mean ± S.D., $n = 3$). PEO and PDMS surfaces were prepared by using triisocyanate (Colonate®, Nippon Polyurethane Industrial Ltd., Tokyo, Japan).

These results were correlated with the *in vitro* evaluation through coagulation assays and platelet adhesion and activation assessments. It is likely that patency was a function of low platelet adhesivity and heparin activity.

Thus we conclude that the blood compatibility of polymeric surfaces can be improved through the use of an amphiphilic triblock copolymer, consisting of a hydrophobic block of PDMS, a hydrophilic spacer block of PEO, and a pharmacologically active block of heparin as a coating.

18.5. CONCLUSION

Polyurethane surfaces can be modified with hydrophilic PEO and bioactive heparin to improve its blood-compatible properties. Heparin-immobilized surfaces, either by PEO spacer (*in situ*) or by coating PEO–heparin graft copolymers or PEO containing triblock copolymers (in bulk) demonstrated significant improvement of blood compatibility.

In this study, the proposed hypothesis has been proven both *in vitro* and *in vivo*. The high activity of immobilized heparin by the use of PEO spacer, in addition to the minimization of protein and platelet interactions *in vitro*, correlated with the improved blood compatibility seen in *ex vivo* and *in vivo* whole blood. This is due mainly to the prevention of fibrin net formation by immobilized heparin, resulting in a reduction in fibrin-related platelet aggregation and subsequent thrombus formation, and also due to a reduction in protein adsorption and platelet interaction by a controlled hydrophilic environment. The results obtained can be utilized in the design of blood contacting medical devices.

ACKNOWLEDGMENT

This work was supported by NIH Grants HL 20251 and HL 17623. The authors thank Professor J. Feijen, Dr. E. Mack, Dr. T. Okano, Dr. D. Grainger, and Dr. C. Nojiri for their contributions.

REFERENCES

1. M. Szycher, in: *Biocompatible Polymers, Metals and Composites* (M. Szycher, ed.), p. 1, Technomic Pub. Co., Inc., Lancaster, PA (1983).
2. D. Ogston, in: *The Physiology of Hemostatis*, p. 1, Harvard Press, Cambridge, MA (1983).
3. M. Wintrobe, *Clinical Hematology*, p. 405, Lea and Febiger, Philadelphia, PA (1981).
4. S. W. Kim and J. Feijen, in: *CRC Critical Reviews in Biocompatibility* (D. Williams, ed.), Vol. 1, p. 229, CRC Press, Boca Raton, FL (1985).
5. J. D. Andrade, S. Nagaoka, S. L. Cooper, T. Okano, and S. W. Kim, *ASAIO J. 10*, 75 (1987).
6. B. D. Ratner, A. S. Hoffman, and J. D. Whiffen, *J. Bioeng. 2*, 313 (1978).
7. Y. Mori, S. Nagaoka, H. Takiuchi, T. Kikuchi, N. Noguchi, H. Tanzawa, and Y. Noshiki, *Trans. ASAIO 28*, 496 (1982).
8. H. P. Kim, S. M. Byun, Y. I. Yeom, and S. W. Kim, *J. Pharm. Sci. 72*, 225 (1983).
9. T. Ohshiro and G. Kosaki, *Int. J. Artif. Organs 4*, 58 (1980).
10. C. D. Ebert, E. S. Lee, and S. W. Kim, *J. Biomed. Mater. Res. 16*, 629 (1982).
11. J. Y. Lin, T. Okano, L. Dost, J. Feijen, and S. W. Kim, *Trans. ASAIO 31*, 468 (1982).
12. P. W. Heyman and S. W. Kim, *Makromol. Chem. 9*, 119 (1985).
13. S. W. Kim, *Med. Dev. Diag. Ind. 8*, 99 (1984).
14. W. E. Burkel, L. M. Graham, and J. C. Stanley, *Ann. N.Y. Acad. Sci. 516*, 131 (1986).
15. M. A. Packman, G. Evans, M. F. Glkynn, and J. F. Mustard, *J. Lab. Clin. Med. 73*, 686 (1979).
16. N. A. Plate and L. I. Valuev, *Biomaterials 4*, 14 (1983).
17. C. D. Ebert and S. W. Kim, *Thromb. Res. 26*, 43 (1982).
18. K. D. Park, T. Okano, C. Nojiri, and S. W. Kim, *J. Biomed. Mat. Res. 22*, 977 (1988).
19. P. K. Smith, A. K. Mallia, and G. T. Harmanson, *Anal. Biochem. 109*, 466 (1980).
20. A. N. Teien, M. Lie, and N. Abildaard, *Thromb. Res. 8*, 413 (1976).
21. S. I. Joen and J. D. Andrade, *J. Colloid Interface Sci. 142*, 159 (1991).
22. K. D. Park, Ph.D. Dissertation, University of Utah (1990).
23. J. D. Andrade, *Med. Instrum. 7*, 110 (1976).
24. S. W. Kim, C. D. Ebertm, J. C. McRea, C. Briggs, S. M. Byun, and H. P. Kim, *Ann. N.Y. Acad. Sci. 416*, 513 (1983).
25. C. Nojiri, T. Okano, D. Grainger, K. D. Park, S. Nakahama, K. Suzuki, and S. W. Kim, *Trans. ASAIO 33*, 596 (1987).
26. C. Nojiri, T. Okano, K. D. Park, and S. W. Kim, *Trans. ASAIO 34*, 386 (1988).
27. C. Nojiri, K. D. Park, T. Okano, and S. W. Kim, *Trans. ASAIO 35*, 357 (1989).
28. C. Nojiri, T. Okano, H. A. Jacobs, K. D. Park, S. F. Mohammad, D. B. Olsen, and S. W. Kim, *J. Biomed. Mater. Res. 24*, 1151 (1990).
29. K. Park, S. R. Simmons, and R. M. Albrecht, *Scanning Microsc. 1*, 339 (1987).
30. Y. Noishiki, *J. Biomed. Mater. Res. 16*, 359 (1982).

31. K. D. Park, A. Z. Piao, H. Jacobs, T. Okano, and S. W. Kim, *J. Polym. Sci., Part A 29*, 1725 (1991).
32. K. D. Park, W. G. Kim, H. Jacobs, T. Okano, and S. W. Kim, *J. Biomed. Mater. Res.* (in press).
33. W. G. Kim, S. F. Mohammad, K. D. Park, and S. W. Kim, *Trans. ASAIO 37*, M148 (1991).
34. D. Grainger, J. Feijen, and S. W. Kim, *J. Biomed. Mater. Res. 22*, 231 (1988).
35. D. Grainger, K. Knutson, S. W. Kim, and J. Feijen, *J. Biomed. Mater. Res. 24*, 403 (1990).
36. D. Grainger, T. Okano, S. W. Kim, D. B. Castner, B. D. Ratner, D. Briggs, and Y. K. Sung, *J. Biomed. Mater. Res. 24*, 547 (1990).
37. A. Z. Piao, C. Nojiri, K. D. Park, H. Jacobs, J. Feijen, and S. W. Kim, *J. Biomater. Sci., Polym. Ed. 1*, 299 (1990).

19

Immobilization of Proteins via PEG Chains

KRISTER HOLMBERG, KARIN BERGSTRÖM, and MAJ-BRITT STARK

19.1. INTRODUCTION

Grafting of poly(ethylene glycol) (or PEG) to solid surfaces has been recognized as a technique for obtaining low protein adsorption and low cell adhesion characteristics.[1,2] For instance, PEG coating is reported to give a marked suppression of plasma protein adsorption and platelet adhesion leading to reduced risk of thrombus formation, as demonstrated both *in vitro* and *in vivo*.[3–5]

The inert character of PEG surfaces is believed to be due to the solution properties of the polymer, its molecular conformation in aqueous solution, and the fact that it is completely noncharged. The oxyethylene unit, $—OCH_2CH_2—$, has a mean-square dipole moment of 1.13 mD^2 and a C—O—C bond angle of 110°, not far from the H—O—H angle.[6] The structural similarity with water and the strong hydrogen bonding to the ether oxygen atoms explain its complete solubility in water. (Interestingly, its homologs polyoxymethylene and polyoxypropylene, as well as its isomer polyacetaldehyde, are nonsoluble in water.)

PEG's ability to prevent proteins and other biomolecules from approaching the surface can be considered a steric stabilization effect. Steric stabilization usually has two contributions, an elastic term and an osmotic term. The elastic, or volume restriction component, results from the loss of conformational entropy when two surfaces approach each other, caused by a reduction in the available volume for each polymer segment.[7] Thus, when a protein approaches the PEG-modified surface a

---

KRISTER HOLMBERG and KARIN BERGSTRÖM • Berol Nobel, S-44485 Stenungsund, Sweden. MAJ-BRITT STARK • Department of Histology, University of Göteborg, S-40033 Göteborg, Sweden. *Present address for K.H.*: Institute for Surface Chemistry, S-11486 Stockholm, Sweden. *Present address for M.-B.S.*: Astra Hässle AB, Kärragatan 5, S-43183 Mölndal, Sweden.

Poly(Ethylene Glycol) Chemistry: Biotechnical and Biomedical Applications, edited by J. Milton Harris. Plenum Press, New York, 1992.

repulsive force develops due to loss of conformational freedom of the poly(ethylene glycol) chains.

The osmotic interactions, or mixing interactions, arise from the increase in polymer concentration on compressing two surfaces.[8] When a protein or another large molecule in water solution approaches the surface, the number of available conformations of the PEG segments is reduced due to compression or interpenetration of polymer chains and an osmotic repulsive force develops.[9] Whether interpenetration or compression or both occur depends on the density of PEG chains. If the PEG grafting is dense it is probable that compression is preferred to interpenetration, while if the grafting is less dense interpenetration is likely to dominate.

For a number of applications attachment of proteins and other biomolecules to PEG-grafted surfaces is of interest. In solid-phase immunoassay and in extracorporeal therapy, antibodies or other bioactive molecules immobilized to a support interact with cells or molecules in a specimen, e.g., a body fluid such as blood, plasma, or urine. In bioorganic synthesis an enzyme is linked to a solid matrix which may subsequently be used as a slurry, packed into columns or used in a membrane reactor. Implants and artificial organs need to be biocompatible and tissue- and osseointegration may be improved by attachment of growth stimulation agents, such as collagen, to the surfaces.

The applications mentioned above, immunoassay, extracorporeal therapy, enzymatic synthesis, and implants, are all based on a specific interaction between a biomolecule, usually a protein, and a cell or another biomolecule in an aqueous solution. When a solid-phase technique is employed it is important that the biomolecule is attached to the support in such a way that its biological properties are not destroyed. For instance, enzymatic activity can be completely lost if the coupling affects the active site,[10] and the antigen-recognizing ability of an antibody can be destroyed by binding to the F_{ab} segment.[11] Furthermore, it is an advantage if the underlying surface, the background, is inert in the sense that strong interaction with the immobilized molecule is avoided and that nonspecific adsorption of molecules from solution is minimized. Strong attraction forces, e.g., due to hydrophobic interactions, between a protein and a surface could eventually lead to such a strong change in protein conformation that the biological activity is lost. Nonspecific adsorption is a well-known problem in solid-phase diagnostics and is the prime reason for the lack of accuracy of some of these tests. (See Section 19.4.1 for further discussion).

Immobilization of the biologically active molecule to free ends of grafted PEG chains offers an attractive way of avoiding the problem of attraction by the underlying surface. Both gradual deformation of the attached molecule and nonspecific adsorption of other molecules or particles will be minimized. Provided that the immobilization procedure used gives minimal distortion of the biological properties and that the layer of immobilized molecules is not so dense that crowding phenomena appear,[12] the procedure would be expected to lead to products with high activity and good stability (see Figures 1 and 2).

The characteristic features of the PEG layer (i.e., hydrophilicity, freedom of

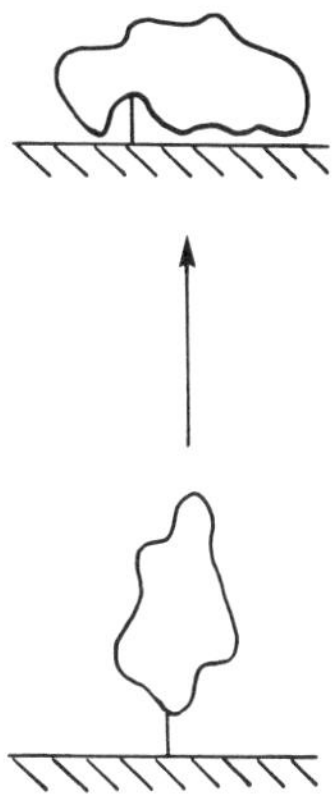

Figure 1. A protein directly attached to a hydrophobic or a highly charged surface will eventually undergo severe conformational changes due to attraction by the surface.

electrostatic charges, and rapid motion of the hydrated chains), being valuable in terms of avoiding interaction between proteins and the surface, cause considerable difficulties when it comes to immobilization efficiency. Even with very reactive end groups on the PEG chains, coupling of the biomolecule in solution is rendered difficult by the energy barrier involved in close approach of a protein to a dense PEG layer.

Thus, coupling of proteins and other biomolecules to PEG-grafted surfaces is not trivial. This paper discusses two approaches to the problem and gives some examples of procedures used.

19.2. IMMOBILIZATION ABOVE THE CLOUD POINT OF THE GRAFTED LAYER

Some aqueous nonionic polymer systems exhibit a lower critical solution temperature, i.e., a polymer-rich phase separates on heating. The phase separation temperature is normally referred to as the cloud point since the process is manifested by a strong clouding of the solution. Well-known examples of such polymers are poly(ethylene glycol), nonionic cellulose ethers, and hydrolyzed poly(vinyl acetate).[13] Similar anomalous temperature effects are also observed for nonionic surfactants of the oligooxyethylene monoalkylether type.[14]

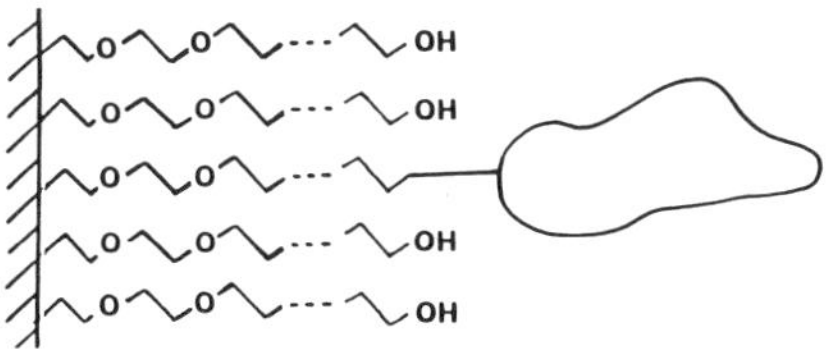

Figure 2. The PEG surface does not attract proteins in aqueous media. The attached protein moves away from the surface.

The decreased solubility of PEG and PEG derivatives in water upon heating—ultimately leading to phase separation—is ascribed to changes in either solute–solute, solute–water, or water–water interactions, resulting in increased attraction between PEG chains at elevated temperature.[15,16] The cloud point of pure PEG in water is above 100 °C, i.e., the polymer is completely water soluble at temperatures up to the boiling point. Introduction of less hydrophilic blocks, such as poly(propylene glycol) (PPG), into the polymer molecule is a convenient way to lower the cloud point. As can be seen from Table 1, ABA block copolymers of ethylene oxide and propylene oxide can be made with cloud points covering a broad temperature interval. Lowering of the cloud point by an increased proportion of PPG in the polymer can be interpreted as an increased attractive interaction between the chains.

The cloud point of a polymer in solution is also sensitive to the presence of electrolytes. The effect on the cloud point is related to the salting in/salting out effect. Lyophilic salts, such as Na_2SO_4, are consequently effective cloud point depressants while lyophobic electrolytes, such as NaSCN, may cause an increase of the cloud point.[13,17] The effect of a few common salts on the cloud point of an ABA block copolymer of ethylene oxide and propylene oxide is given in Table 2.

The cloud point phenomenon applies not only to polymers in bulk solution but to polymers attached to solid surfaces as well. By using a surface force technique, allowing measurements of the force acting between two surfaces as a function of their separation, a gradual decrease in hydrophilicity on temperature rise has been documented for PEG derivatives[18] and a cellulose ether[19] attached to hydrophobic surfaces. For hydrophilic polymers attached at one end to the surface, the temperature at which two such surfaces exhibit a net attraction corresponds well with the cloud point of the polymer in solution. The situation is more complex for surfaces covered with a surface-active polymer by multipoint attachment. The coated surface then remains hydrophilic at temperatures well above the cloud point. This is probably due to a segregation of hydrophobic segments close to the surface and hydrophilic segments toward the bulk solution. The orientation of hydrophilic groups toward the solution increases the local cloud point, thus maintaining a hydrophilic character of the surface also above the cloud point in solution.[19]

For poly(alkylene glycol) derivatives attached by one end to a hydrophobic surface, the transition temperature from hydrophilicity to hydrophobicity corre-

Table 1. Cloud Points (Cp) in Water for a Series of EO/PO Block Copolymers with Varying Composition

Cp (°C)	EO (%)
23	10
33	20
59	40
>100	80

Table 2. Cloud Points (Cp) for an EO/PO Block Copolymer in Different Salt Solutions

	Cp (°C)
Water, electrolyte-free	39
1% NaCl	35
5% NaCl	27
1% Na_2SO_4	26
5% Na_2SO_4	22

sponds well with the cloud point in solution. Thus, by choosing the appropriate PEG-to-PPG ratio and electrolyte concentration, the transition temperature of the surface can be predetermined with considerable accuracy.

The cloud point can be used for immobilization of proteins to a surface grafted with poly(alkylene glycol) chains, provided one end of the chain is attached to the surface and the other contains a free reactive group. The immobilization is carried out just above the cloud point where there is no energy barrier for the protein to approach the surface. After reaction, the temperature is reduced below the cloud point where again the grafted layer becomes hydrophilic. Adsorbed proteins can now be washed away since there is a net repulsion between the protein and the background surface. Presumably, covalently bound protein molecules will move away from the surface. That the grafted layer contracts above the cloud point and expands below it, can be shown by ellipsometry.[20] The principle of the immobilization procedure is shown schematically in Figure 3.

The procedure works well in the sense that a relatively high loading of protein on the surface can be obtained. The drawback of the method, however, is that also below the cloud point the poly(alkylene glycol) layer is not entirely hydrophilic. As shown in Figure 4, the protein rejection properties of an ethylene oxide/propylene oxide block copolymer are not as good as those of an ethylene oxide homopolymer.

The example below illustrates the immobilization of immunoglobulin G to a block-copolymer grafted surface.[21]

Procedure: A polyethylene surface is made aminofunctional by plasma polymerization of allylamine.[22] An EO/PO block copolymer with a molecular weight of 3000

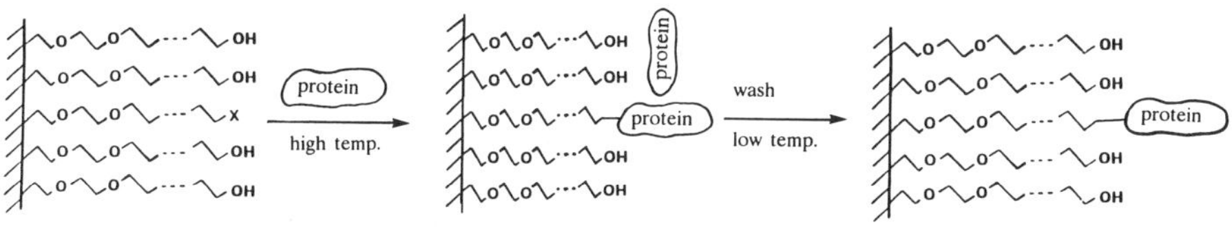

Figure 3. Immobilization of protein above the cloud point of a poly(alkylene glycol) grafted surface.

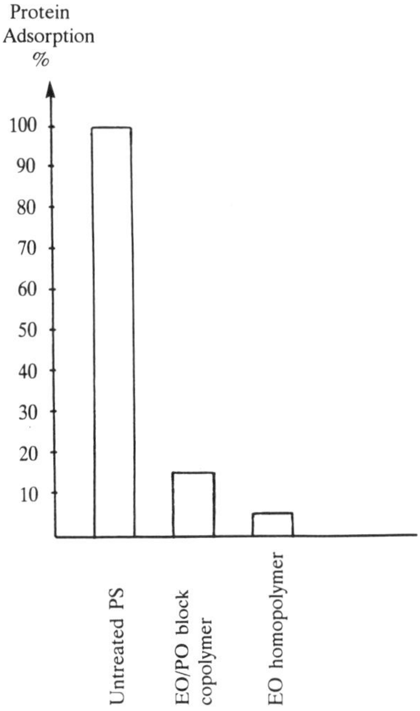

Figure 4. Protein adsorption to polystyrene (PS) surfaces grafted with EO/PO block copolymer and EO homopolymer, respectively. Untreated PS surface is used as a reference.

and a cloud point of 38 °C, carrying oxirane end groups, is reacted with amino groups at the surface. The reaction is carried out below the cloud point, at 35 °C, pH 9.5, for 24 h. After rinsing thoroughly with water a solution of 0.020 mg ml^{-1} immunoglobulin G in 0.05 M carbonate buffer of pH 9.5 is added and kept at 40 °C for 18 h. The temperature is reduced to 20 °C and the surface is rinsed repeatedly with water.

19.3. IMMOBILIZATION FROM NONPOLAR MEDIA

If a PEG-treated surface is exposed to a worse than theta (θ) solvent, the grafted chains will not reach out into the liquid phase but will form a compressed surface layer. (A θ solvent is a solvent that yields solutions of a polymer in its θ state, i.e., the polymer molecule exhibits its unperturbed dimensions. In a worse than θ solvent, interaction between polymer and solvent is unfavorable leading to restricted conformational freedom of the polymer chains.) In such an environment the poly(ethylene glycol) chains will not induce repulsion of particles or macromolecules in solution. Also, close approach by a protein to such a surface may be possible since the interaction between the PEG chains and the approaching biomolecule may be attractive. This phenomenon is analogous to steric stabilization of polymer-coated particles where it has been shown that when the solvency of the dispersion medium is changed (e.g., by addition of a nonsolvent, until the θ point is reached) the particles come in close contact and flocculation occurs.[23]

Aliphatic hydrocarbons are examples of worse than θ solvents for PEG. By using the microemulsion concept, varying amounts of water can be solubilized into the nonpolar solvent. A microemulsion is defined as a system of water, oil, and amphiphile which is a single, optically isotropic and thermodynamically stable liquid solution.[24] A microemulsion will remain nonpolar until considerable amounts of water have been added. For instance, a microemulsion based on a hydrocarbon containing 30 weight % aqueous buffer has a dielectric constant, ϵ_R, of 3–5 (relative permittivity at 50 MHz), which should be compared with ϵ_R around 70 for water solutions.[25,26] Such microemulsions are also nonsolvents for PEG.

Microemulsions have recently been found to be excellent reaction media for many enzymatic processes. The enzymes studied (such as lipases, phospholipases, and alcohol dehydrogenases) have been found to be active and stable in the microemulsion medium.[27–29]

We have now found that proteins can be immobilized to a PEG-grafted surface from a microemulsion having a low dielectric constant.[30] After coupling, the surface is washed with water whereby the PEG layer becomes hydrated and protein-rejecting. The principle is outlined in Figure 5.

A PEG-coated surface which has been exposed to a microemulsion or another nonpolar liquid can be hydrated by simple soaking in water. As is shown in Figure 6, transformation into an inert surface is not instantaneous, however, probably because of a slow reorientation of the poly(ethylene glycol) chains from a compact entangled structure to an extended conformation. After soaking in water overnight complete hydration seems to have occurred, as judged from protein adsorption properties.

The use of a microemulsion as solvent seems to be an efficient way to overcome the problem of contact between the PEG chains and proteins in solution. As is shown below, proteins, such as enzymes and antibodies, can be covalently bound to reactive, terminal groups on the PEG chains.

There are two prerequisites of a successful result, however: the solid support must be resistant to the microemulsion and the protein must not be deactivated. The first issue is trivial and easily tested. With a few exceptions, such as poly(methylmethacrylate), common plastics are unaffected by a few hours exposure to a microemulsion based on an aliphatic hydrocarbon as main component.

The question of protein deactivation is more critical. Interaction of the protein

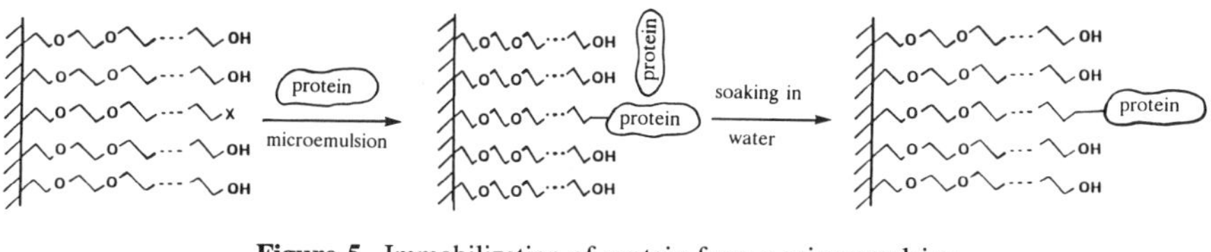

Figure 5. Immobilization of protein from a microemulsion.

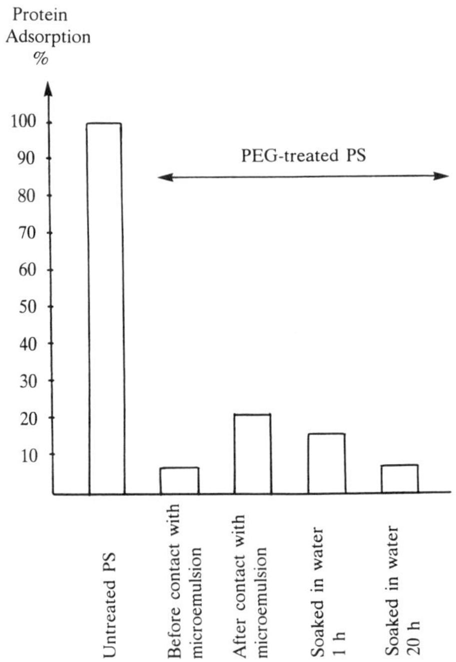

Figure 6. Protein adsorption to a polystyrene (PS) surface grafted with PEG. Measurements are made immediately after PEG grafting, after microemulsion treatment, and after recontact with water. The microemulsion consisted of 10% water, 85% isooctane, and 5% of the anionic surfactant sodium bis(2-ethylhexyl)sulfosuccinate.

with the surfactant could have deleterious influences since many detergents are also strong protein denaturants, producing conformational changes at very low concentrations upon binding.[31] In a microemulsion medium, surfactant will largely be found at the interface between water and hydrocarbon microdomains while proteins are likely to reside mainly in the water pools. Some surface-active proteins, such as lipases, may prefer the interface.[32] In our hands, the effect of the surfactant on protein activity and stability is highly individual and varies with both protein and surfactant. In most cases nonionic surfactants of the alcohol ethoxylate type and branched anionic surfactants, such as sodium bis(2-ethylhexyl sulfosuccinate), commonly known as AOT, have a negligible effect on activity of most proteins.

Another interesting feature of microemulsions as media for protein immobilization is the possibility of affecting the coupling pattern. Three factors are believed to govern which amino acid residues of proteins will participate in covalent attachment to a solid support.[10,33] The relative concentrations of amino acids are obviously of importance. Ser, Lys, and Thr, normally being the most abundant residues, will be favored on this account.

The degree of hydrophilicity of a residue and its immediate environment is the second factor, the availability of a functional group being dependent on whether it is situated on an amino acid residue which is exposed on the outside of the protein or buried inside the molecule. In aqueous media a protein will fold so as to expose its more hydrophilic regions. Hydrophobic residues, such as Tyr, Met, and Thr, are therefore more likely to be internalized than, for example, Lys and Asp.[34]

The relative reactivity of an amino acid residue is the third factor. It is, of course, not possible to compare accurately residues in terms of reactivity without specifying a particular reaction. However, for nucleophilic substitution reactions involving amino, thiol, or phenolic hydroxyl groups of the protein, reactions with Lys, Cys, and Tyr are likely to dominate.

If one considers the factors mentioned, relative concentration, hydrophilicity, and reactivity, Lys is predicted to be the most likely coupling residue under aqueous conditions, followed by Cys, Tyr, and His.[33]

In nonpolar media, in contrast, coupling to Lys is likely to be suppressed on availability grounds since the protein will adjust its conformation so that the more hydrophobic residues, like Tyr, will be exposed on the surface. Lys, like other more hydrophilic residues, will be found predominantly in the interior of the protein.

Furthermore, the relative reactivity of the residues is likely to depend on solvent polarity, since Lys contains an uncharged nucleophile, the γ-amino group, while Cys and Tyr react in the deprotonated, ionic form. The reaction between an electrophilic group at the PEG chain end (R—X) and a nucleophile (Y) will proceed via transition states of different charge types, depending on whether or not the nucleophile is charged:

$$(1)\qquad R{-}X + Y \rightarrow \left[Y^{+\delta}{-}R{-}X^{-\delta}\right]^{\ddagger} \rightarrow R{-}Y^{+} + X^{-}$$

$$(2)\qquad R{-}X + Y^{-} \rightarrow \left[Y^{-\delta}{-}R{-}X^{-\delta}\right]^{\ddagger} \rightarrow R{-}Y + X^{-}$$

In reactions of type (1) the reactants are uncharged, but the transition state has built up a charge. Such reactions are aided by polar solvents which reduce the energy of an ionic transition state. A change from a polar solvent to a nonpolar one will thus lead to a reduced rate of reaction. In type (2) reactions, on the other hand, the initial charge is dispersed in the transition state, so that the reactions are favored by more nonpolar solvents.[35]

Immobilization via lysine residues is a type (1) reaction, while coupling to cystein or tyrosine residues represents a reaction of type (2). A change of reaction medium from water to a microemulsion of low polarity is thus likely to reduce attachment via lysine amino groups. We have recently demonstrated this effect in model experiments using free amino acids. The order of reactivity Lys > Cys > Tyr in aqueous buffer changed to Cys > Lys > Tyr in a microemulsion of low water content.[36]

Thus, immobilization of proteins from water-in-oil microemulsions instead of aqueous buffers is likely to favor coupling via Cys for reactivity reasons and Tyr for protein conformational reasons. The net result will be that attachment via Lys is minimized. Such a change in immobilization pattern could be of practical interest. First, enzymes which have lysine residues at or close to the active site may exhibit a higher activity in the immobilized state if coupling is preferentially performed so that Lys—NH_2 is unaffected. Second, antibodies are often rich in —S—S— groups in the F_c region and the disulfide bonds may be transformed into thiol groups by careful reduction. Immobilizing antibodies via SH groups is thus a way of preferentially binding to the F_c portion while leaving the antigen binding F_{ab} segment intact. An example of comparison between immobilization of an enzyme from water and from a microemulsion is given in Section 19.4.4. Work with site specific immobilization of antibodies using the microemulsion concept is underway in our laboratory.

The example below shows immobilization of fibrinogen from an AOT-based microemulsion.

Procedure: A polyethylene surface is made aminofunctional by plasma polymerization of allylamine.[22] A poly(ethylene glycol) with a molecular weight of 4000, carrying terminal oxirane groups, is reacted with the amino groups, at the surface. The reaction is carried out at 40 °C, pH 9.5, for 24 h. After rinsing thoroughly with water, 0.05 mg ml^{-1} fibrinogen is coupled to free epoxide groups by reaction in a microemulsion containing 5% AOT, 85% isooctane, and 10% carbonate buffer of pH 9.5. After 4 h at 40 °C the surface is rinsed repeatedly with water. To regain full protein rejection of the surface the sample is soaked in water for 24 h.

19.4. APPLICATIONS

19.4.1. Solid-Phase Immunoassay

The enzyme-linked immunosorbent assay (ELISA) technique, described in 1971,[37] is a popular and powerful tool in medical diagnostics. The method uses two components, an enzyme conjugate and a solid phase to which an immunoreagent is attached. The enzyme has traditionally been horseradish peroxidase, which is covalently attached to a specific antibody. The enzyme chosen should be stable and have a high turnover rate and substrates should be readily available which produce a dense, easily visible color.

The ELISA solid phase is usually a microtiter plate made from polystyrene. The plates are coated with an immunoreagent which is either an antibody or an antigen. Two principles of the technique are outlined in Figure 7.

The improvements of the ELISA technique that have taken place since 1971 have concentrated on development of more specific antibodies with higher binding affinities for their antigens, enzymes demonstrating a faster turnover rate of their sub-

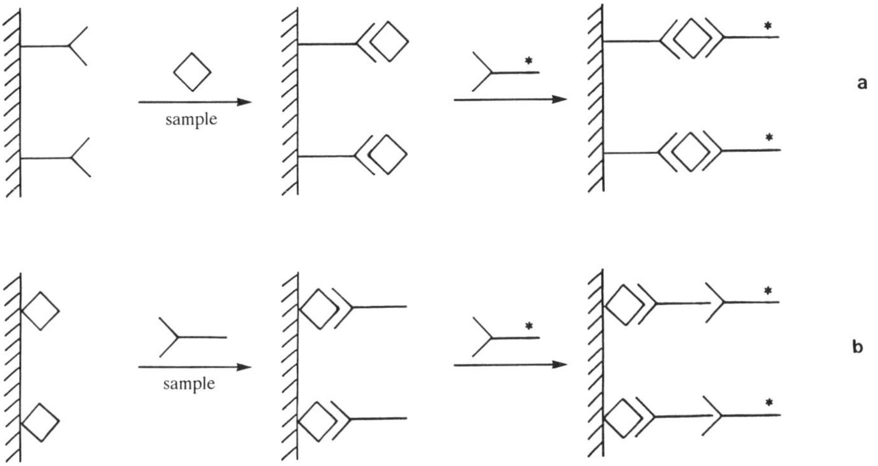

Figure 7. Two principles of ELISA. In (a) the surface is coated with antibodies against a target antigen. The presence of antigens on the surface is subsequently detected by an enzyme-linked antibody. In (b) the surface is coated with antigens and detection is made by an enzyme-linked anti-antibody.

strates, and better means to link antibodies and enzymes with retention of activities of both partners.

The sensitivity of an immunoassay can be defined by the ratio of the specific signal generated to the background noise of the system. Nonspecific binding of an antibody to the solid phase is the most important factor responsible for assay noise. Typically, additives such as detergents, nonreactive proteins, in particular BSA, or salts are added to the assay matrix to reduce nonspecific binding of the enzyme–antibody conjugate.[38]

Although such additions improve sensitivity and accuracy of the test, assay noise is still unsatisfactorily high in many tests. Both surfactants and nonreactive proteins bind to the surface reversibly and may be displaced by IgG antibodies.[39,40]

Coating of ELISA plates with an immunoreagent (antibody or antigen) is routinely done by simple adsorption from aqueous solution. The reagent usually adheres strongly to polystyrene, but some molecules, particularly those carrying long carbohydrate chains, may eventually desorb. The tendency to desorb is particularly strong when the antibody–antigen complex has been formed on the surface.

Another disadvantage with the simple adsorption is that the immunoreagent may gradually undergo a change in conformation on the surface, eventually leading to loss of immunoreactivity. This results in a short shelf-life of the coated plates.

The problems associated with the hydrophobic solid phase (i.e., nonspecific adsorption, desorption of the antibody–antigen complex, and gradual denaturation of the immunoreagent on the surface) can all be minimized by the technique of covalent binding of the immunoreagent to a PEG surface. This is illustrated schematically in Figure 8.

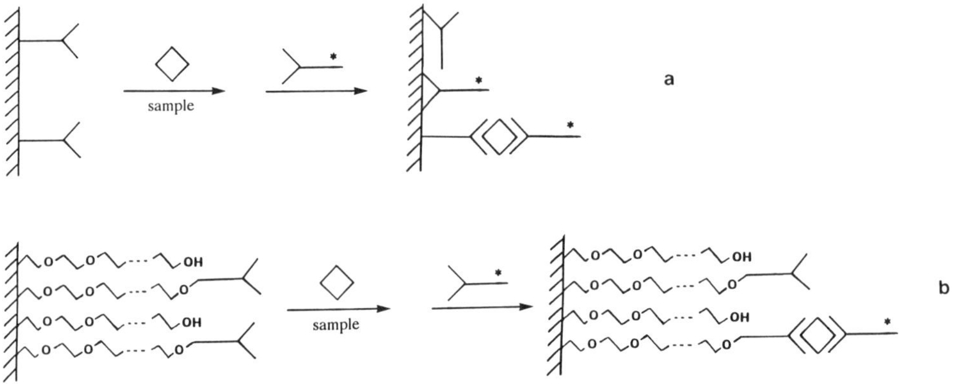

Figure 8. Path a shows a normal ELISA illustrating both nonspecific adsorption of the enzyme-linked antibody and change in conformation of the antibody used for coating. In path b where a dense layer of PEG chains is introduced, the tendency for antibodies to interact with the surface is reduced.

A comparison between the standard ELISA procedure (i.e., coating by adsorption to nontreated polystyrene) and the technique of covalent coupling to a PEG layer is given below.[41]

The microemulsion method was used for immobilization of *Borrelia* antigen to PEG. As can be seen in Figure 9, the amount of covalently coupled antigen is 80% of the amount that could be physically adsorbed. The background (represented by the negative serum) on the PEG-greated surface is reduced 8 times compared to the untreated polystyrene surface. This means that the sensivity of the assay (the ratio between positive/negative serum) is considerably improved.

19.4.2. Extracorporeal Therapy

Interleukin-2 (IL-2) is a lymphocytotrophic hormone, pivotal for the generation and regulation of the immune response. IL-2 has been found to have antitumor activity and it is one of the prime candidates for immunotherapy of cancer.[42] Clinical trials have shown that therapy with high-dose IL-2 can mediate the regression of established metastatic disease of patients with advanced malignancy. Toxicity from IL-2 treatment is high, however, resulting in organ dysfunction and even therapy-induced deaths.[43] Since it is believed that IL-2 functions by interaction with a specific receptor on activated T cells, without being internalized into the cell,[44] hemoperfusion over immobilized IL-2 in an extracorporeal device would appear to be an attractive way of administrating the agent. Apart from the possibilities of avoiding or reducing toxicity problems, the use of covalently bonded IL-2 would be expected to prolong the half-life of the peptide and thus, hopefully, improve its efficiency. The circulatory half-life of IL-2 is only a few minutes,[45] and since plasma clearance is mainly due to filtration in the kidney,[46] the use of the agent in immobilized form would be expected to alter the pharmacokinetics completely.

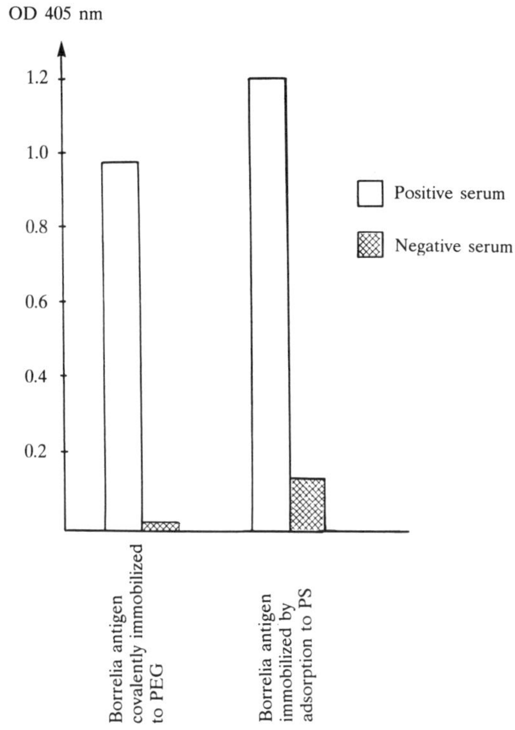

Figure 9. Borrelia antigen covalently immobilized to a PEG-grafted polystyrene (PS) surface and adsorbed to non-treated PS. Anionic groups are introduced on the surface by $KMnO_4/H_2SO_4$ treatment. Poly(ethylene imine) (PEI) is adsorbed and PEG diepoxide is reacted with amino groups of the attached PEI. Borrelia antigen is coupled to residual epoxide groups by reaction in a microemulsion medium.

The IL-2 receptor consists of one larger and one smaller protein and the binding sites of the two chains of the receptor are independent but co-operative.[44] The IL-2 molecule binds simultaneously to the two receptor proteins and, in doing so, it needs to attain a folded tertiary structure (see Figure 10).

The complicated receptor–hormone interaction has consequences for the immobilization strategy:

1. The coupling must not destroy the conformational freedom of the molecule.
2. The binding must not involve sites needed for receptor interaction.
3. Interaction with the solid surface must not cause a conformational change of the peptide.

Binding of IL-2 via PEG chains is attractive on several accounts. The long, flexible spacer molecules will allow maximum freedom of motion and the underlying PEG layer will minimize interaction with the surface. Furthermore, it has been demonstrated that PEG modification of IL-2 does not severely affect its bioactivity provided the degree of substitution is moderate.[47] To test this concept we have immobilized IL-2 to an extracorporeal chamber containing 23 serially coupled plates, 0.9 mm thick and spaced 0.7 mm from each other. Two outer plates, each 12 mm

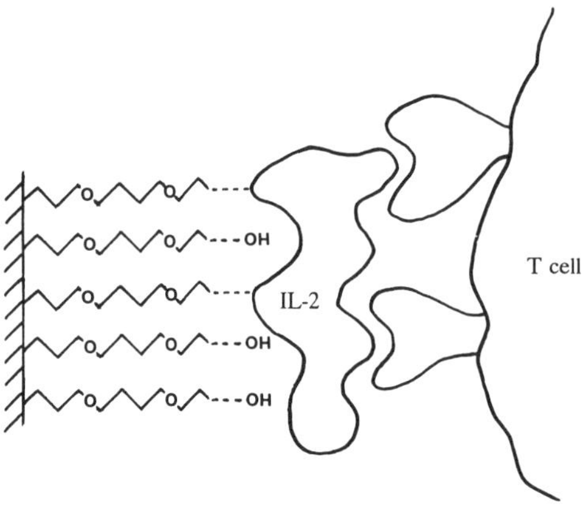

Figure 10. Immobilized IL-2 binds to its receptor on a T-cell surface. The IL-2 receptor consists of two proteins not linked via disulfide bonds. The two receptor proteins must interact with IL-2 simultaneously for high affinity binding to occur.

thick, stabilize the construction. The device holds 150 ml of fluid and has a surface area of 0.5 m^2. Use of the chamber for immunotherapy of cancer using Poly I:C and interferons has been described.[48]

The walls of this chamber, as well as the plates, are made from poly(methylmethacrylate) (PMMA), which is a hydrophobic polymer with excellent mechanical properties. However, the PMMA surface is chemically inert and does not contain functional groups suitable for a direct coupling to proteins and other biologically active substances. An immobilization procedure must therefore be preceded by some kind of activation of the surface of the carrier polymer. A further limitation with PMMA with regard to immobilization is its low tolerance to organic reagents and solvents. In practice, the reaction medium is restricted to aqueous solutions for all coupling steps involving the carrier. This fact limits available immobilization procedures.

The procedure for grafting PEG chains and covalent coupling of IL-2 to epoxy-terminated spacers is outlined in Figure 11. Immobilization of the peptide presented particular problems, since neither of the two above-mentioned methods of coupling to a PEG-modified surface could be applied. The PMMA surface does not stand exposure to a microemulsion or any other type of water-poor medium, thus preventing use of nonpolar reaction media. Also, immobilization above the cloud point of the hydrophilic layer is excluded since IL-2 is heat-sensitive and cannot be reacted at a temperature much above its working temperature at 37 °C.

To avoid these problems we adopted the alternative procedure of immobilizing IL-2 to an epoxy functional hydrophilic surface in a simple, straightforward way in aqueous solution. As mentioned above, this results in a low loading of biomolecules on the PEG surface. In this case a low immobilization density was believed to be acceptable, since immune system stimulation with IL-2 is performed with only minute amounts of active agent.

Although the immobilization density of IL-2 was low, the coupling efficiency

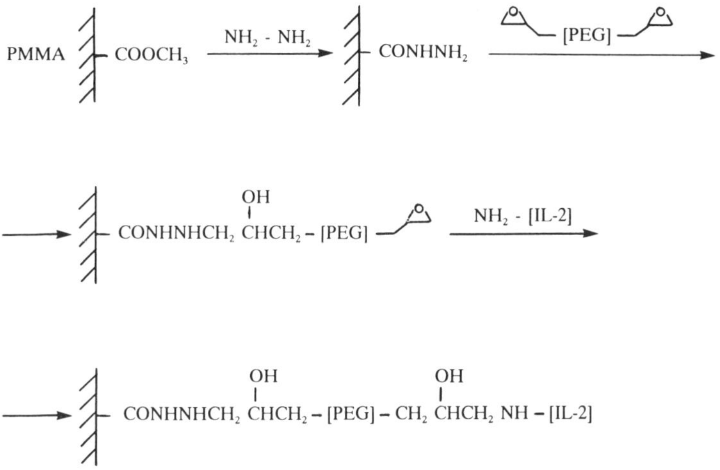

Figure 11. Immobilization of IL-2 on PMMA. The hydrazide is formed by 4 h treatment with a 25% solution of hydrazine hydrate in a 1:1 water–methanol mixture. After rinsing with water a 5% solution of PEG 4000 diepoxide is added and pH is adjusted to 9.6. After 16 h reaction at 20 °C the plates are rinsed repeatedly with water. A solution of IL-2 in a 0.1% solution of PEG 4000 in 0.05 M carbonate buffer of pH 9.6 (100,000 U r-IL-2 from Amgen per 100 ml) was added and kept at 20 °C for 24 h. IL-2 is assumed to react via its lysine NH_2 groups.

was good with more than 80% of the IL-2 added being bound to the plates after reaction. Extensive leakage tests were performed with the loaded devices, but no detachment of IL-2 could be demonstrated with immunoassay using antibodies against IL-2.

The activity of immobilized IL-2 was tested in *in vitro* studies. Human peripheral blood mononuclear cells from healthy volunteers were obtained by gradient centrifugation. Cells were counted and diluted to $5 \cdot 10^6$ cells per ml. The cell suspension was added to the plates containing immobilized IL-2. After 5 days incubation at 37 °C in a humidified CO_2 atmosphere the activity of IL-2 was determined by analysis of interferon-γ generated. As a reference free IL-2 was used in an amount equivalent to that immobilized. IL-2 immobilized directly onto the hydrazide functional surface (i.e., without the PEG spacer) was included for comparison. As can be seen from Table 3 IL-2 immobilized via the PEG spacer gave 40% of the activity of free IL-2, as determined by interferon-γ production. Immobilization without the grafted hydrophilic layer gave much less activity.

An *in vivo* study on 13 patients with advanced tumors was carried out with the extracorporeal device carrying IL-2 immobilized via PEG spacers. No decisive results in terms of tumor regression or immune system stimulation were obtained, however.[49] No toxic side effects from the treatments were obtained, indicating that toxicity can be minimized by the use of the immunotherapeutic drug in immobilized form.

Table 3. Production of Interferon-γ from Mononuclear Cells Incubated with Il-2 for 5 Days at 37 °C

IL-2 preparation	Interferon-γ ($ng\ ml^{-1}$)
Free IL-2	1.7
IL-2 immobilized via PEG spacer	0.7
IL-2 immobilized without PEG	0.2

19.4.3. Implants

Biomaterials made of collagen have been used extensively to augment or replace damaged soft tissue structures and to support the growth of cells. Collagen-based materials in the form of sponges and coatings have been used as implants and as a supporting matrix for growth of cultured skin components such as epidermal and fibroblast cells.[50,51]

It has been shown that, while a nontreated plastic implant may lead to abnormal regeneration in the connective tissues and epithelium, collagen coating of the graft may allow fibroblasts to produce normal connective tissue substances.[52] Studies of adhesion and growth of cells on collagen-coated surfaces are obviously of relevance to implants of collagen-based materials. Such *in vitro* experiments are considered necessary in designing biomaterials, although considerable limitations exist in extrapolating from the *in vitro* to the *in vivo* situation.[53]

In order to study the effect of a PEG tether on the biological activity of immobilized collagen, the following experiments were carried out[54]: Collagen of type 1 was coupled to a PEG-coated polyethylene (PE) surface containing terminal epoxide groups in the hydrophilic chains. The microemulsion method (see above) was used for immobilization. The immobilization route is outlined in Figure 12.

Coupling efficiency was tested by an ELISA technique using goat antiserum against collagen type 1. As can be seen from Figure 13, coupling in microemulsion by reaction with PEG–epoxide groups gives a loading of collagen on the surface equivalent to 80% of the amount of collagen that adsorbed directly to a nontreated PE surface. (One should keep in mind that the ELISA technique gives a measure of immunoactive protein—not total protein—on the surface. In most instances ELISA can be used with good accuracy also to determine the amount of surface-bound protein. Should collagen adsorption give a multilayer of proteins on the surface, the immunoassay technique is no longer suitable for determination of total loading on the surface, however.)

It can also be seen from Figure 13 that coupling of collagen from buffer solution to the hydrophilized oxirane-functional surface gives an extremely poor yield. This experiment is a nice illustration of the efficiency of the microemulsion technique for immobilization.

The hydrophilized plates carrying covalently bound collagen were tested for culture of HeLa cells. As comparison, nonhydrophilized plates with adsorbed

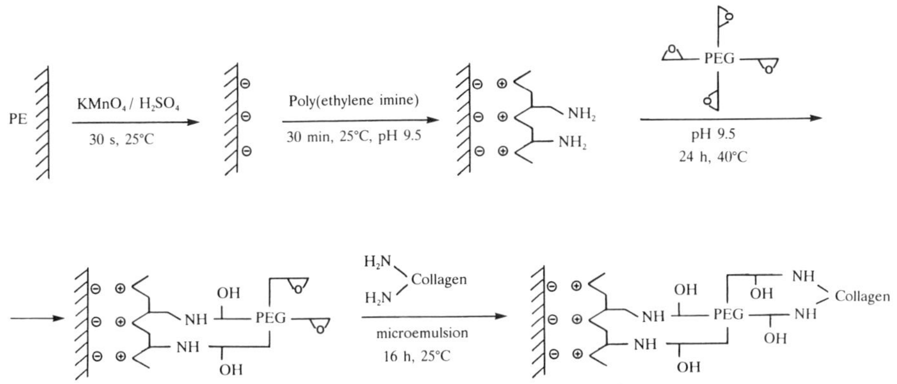

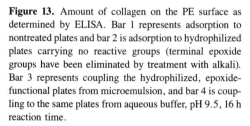

Figure 12. Immobilization of collagen on a polyethylene (PE) surface. The surface is treated with $KMnO_4$/H_2SO_4 to introduce anionic groups. Poly(ethylene imine) is adsorbed and branched PEG carrying terminal oxirane rings is reacted with amino groups of the surface-bound polyimine. Collagen is coupled to free epoxide groups by reaction in a microemulsion containing 10% AOT, 10% aqueous buffer of pH 9.5, and 80% nonane.

collagen, as well as PEG-treated plates without collagen, were used. The culture medium was MEM F 15 with 10% foetal calf serum added. Table 4 gives the increase in the number of cells after 3 days of culture. It is clearly seen that collagen attached to a hydrophilic layer is advantageous for cell growth. The observation should be of interest for collagen-coated artificial implants.

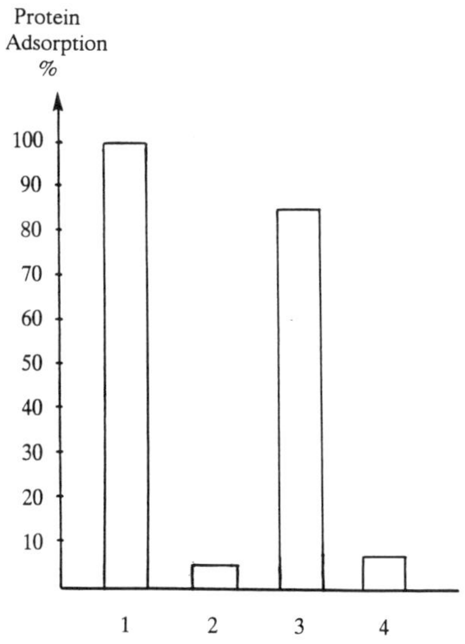

Figure 13. Amount of collagen on the PE surface as determined by ELISA. Bar 1 represents adsorption to nontreated plates and bar 2 is adsorption to hydrophilized plates carrying no reactive groups (terminal epoxide groups have been eliminated by treatment with alkali). Bar 3 represents coupling the hydrophilized, epoxide-functional plates from microemulsion, and bar 4 is coupling to the same plates from aqueous buffer, pH 9.5, 16 h reaction time.

Table 4. Increase in Number of HeLa Cells after 3 Days of Culture

Surface treatment	Increase of number of cells
PEG–collagen	20 time
Collagen (adsorbed)	8 times
PEG	3 times

19.4.4. Bioorganic Synthesis

The emerging industrial interest in ester synthesis has led to a need for immobilization methods for lipases which give high activity, as well as good stability, of the bound enzyme. Direct adsorption of lipases on carriers such as celite or silica often leads to a substantial drop in activity.

In order to minimize enzyme–matrix interaction, PEG chains grafted to silica via an aminopropylsilane coupling agent were used as linkers for a lipase. Tresylate was used as the reactive electrophilic group at the free ends of the poly(ethylene glycol) chains. The reaction sequence is outlined in Figure 14.[55]

Immobilization was performed in aqueous buffer, as well as in a microemulsion consisting of 94 weight % n-hexane, 1 weight % carbonate buffer, and 5 weight % of the nonionic surfactant tri(ethylene glycol) monododecyl ether. The activity of the immobilized 1(3)-specific lipase was tested with regard to both hydrolysis and transesterification of a triglyceride. As can be seen from Figure 15, the activity of the two lipase preparations was the same in the hydrolysis reaction. The transesterification (i.e., substitution of one fatty acid residue in a triglyceride by another) gave a completely different picture. While the preparation from microemulsion gave acceptable activity, that from aqueous buffer gave no reaction at all, as is shown in Figure 16.

It should be noted that the same amounts of immobilized enzyme were used in all experiments. The results from the transesterification reaction indicate that coupling from the different media gives differences in the immobilization pattern. A tentative explanation for the lack of activity of the aqueous preparation in the transesterification but not in the hydrolysis reaction is that, in the former reaction, the steric requirement on the active site is larger since a bulky nucleophile, a diglyceride, must be able to approach the enzyme-*O*-acyl intermediate (see Figure 17).

As mentioned above, immobilization from microemulsion is likely to favor coupling to thiol groups, while reaction performed in aqueous media favors binding to amino groups. It may be that, with the lipase, coupling in microemulsion affects the active site less than does aqueous coupling and that this difference shows up only in the sterically more demanding transesterification reaction.

As a comparison, the same lipase was also covalently bound to the silane-

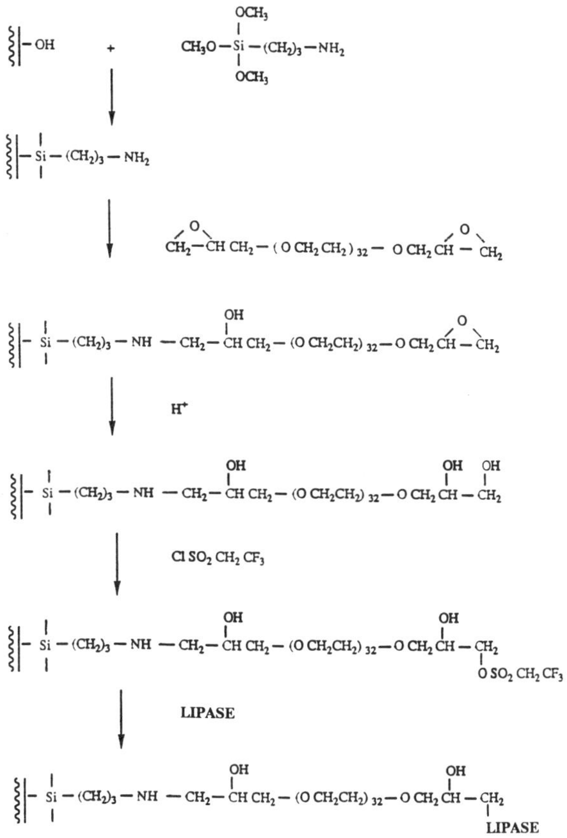

Figure 14. Immobilization of lipase to silica grafted with PEG chains.

derivatized silica without the long PEG spacer. The activity and stability of the two preparations of immobilized lipase were evaluated and compared with free lipase and lipase physically adsorbed on silica. Only preparations made from aqueous buffer solutions were compared. The results are shown in Figure 18. While the hydrolytic activity of both the free and the adsorbed lipase preparations deteriorate rapidly, with 90% of activity lost in less than three weeks, the covalently bound enzyme preparations retained their activity completely. It can be seen from Figure 18 that the lipase immobilized via the PEG spacer surpassed the free lipase in activity after only 10–11 days. This finding could be of considerable practical importance, since enzyme stability is vital to all industrial processes.

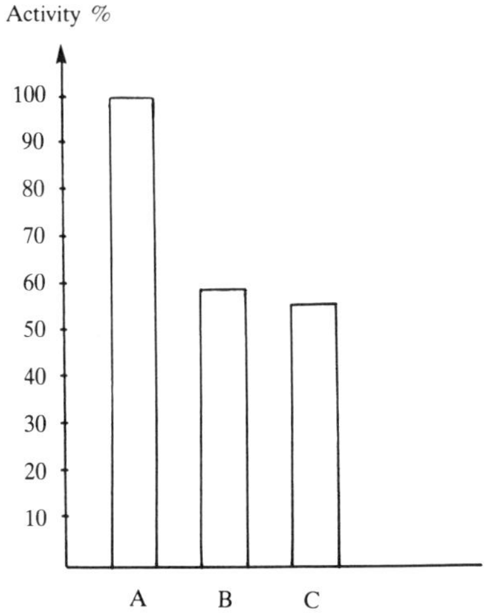

Figure 15. Hydrolytic activity of lipases immobilized from aqueous buffer (B) and from microemulsion (C) compared with free lipase (A).

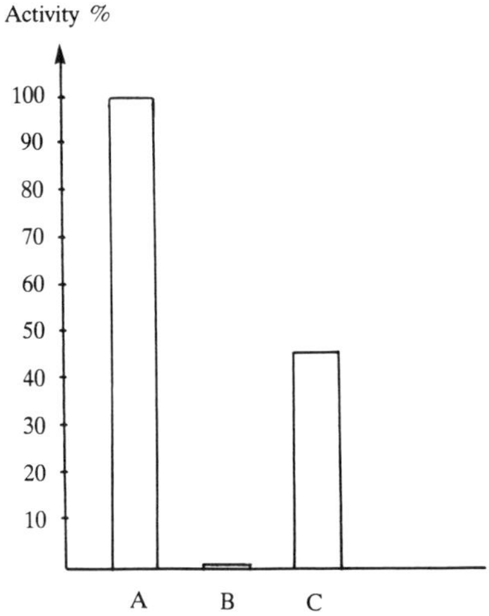

Figure 16. Activity of lipases in a transesterification reaction using enzyme immobilized from aqueous buffer (B) and from microemulsion (C) compared with free lipase (A).

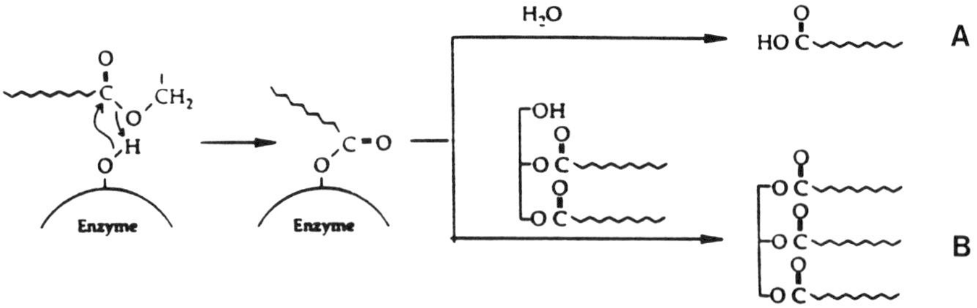

Figure 17. Schematic illustration of the pathways leading to enzymatic hydrolysis (A) and transesterification (B) of a triglyceride.

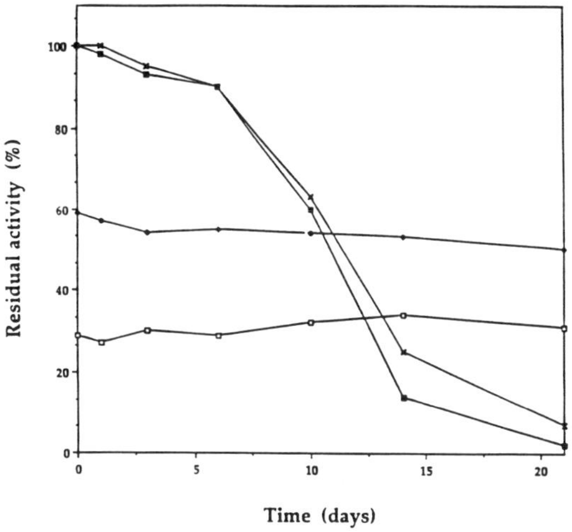

Figure 18. Stability of lipases: (×) lipase adsorbed onto celite, (■) free lipase, (□) lipase bound directly onto silica, and (♦) lipase immobilized onto PEG–silica.

REFERENCES

1. E. W. Merrill and E. W. Salzman, *ASAIO J. 6*, 60 (1983).
2. J. H. Lee and J. D. Andrade, in: *Polymer Surface Dynamics* (J. D. Andrade, ed.), p. 119, Plenum Press, New York (1988).
3. S. Nagaoka, Y. Mori, H. Takiuchi, K. Yokota, H. Tanzawa, and S. Nishiumi, *Polym. Prepr. 24*, 67 (1983).
4. S. Nagaoka, *Trans. Am. Soc. Internal Organs 10*, 76 (1988).
5. S. Nagaoka and A. Nakao, *Biomaterials 11*, 119 (1990).
6. F. E. Bailey, Jr. and J. V. Koleske, *Poly(ethyleneoxide)*, Academic Press, New York (1976).
7. D. W. J. Osmond, B. Vincent, and F. A. Waite, *Colloid Polym. Sci. 253*, 676 (1975).
8. T. F. Tadros, in: *The Effect of Polymers on Dispersion Properties* (T. F. Tadros, ed.), p. 1, Academic Press, New York (1982).
9. D. Knoll and J. Hermans, *J. Biol. Chem. 258*, 5710 (1983).
10. P. A. Srere and K. Uyeda, in: *Methods in Enzymology, Vol. 44*: Immobilized Enzymes (K. Mosbach, ed.), p. 11, Academic Press, New York (1976).
11. G. A. Quash, V. Thomas, G. Ogier, S. E. Alaoui, J.-G. Delcros, H. Ripoll, A.-M. Roch, S. Legastelois, R. Gibert, and J.-P. Ripoll, in: *Covalently Modified Antigens and Antibodies in Diagnosis and Therapy* (G. A. Quash and J. D. Rodwell, eds.), p. 155, Marcel Dekker, New York (1989).
12. J. Turková, in: *Methods in Enzymology, Vol. 44*: Immobilized Enzymes (K. Mosbach, ed.), p. 66, Academic Press, New York (1976).
13. G. Karlström, A. Carlsson, and B. Lindman, *J. Phys. Chem. 94*, 5005 (1990).
14. T. Nakagawa, in: *Nonionic Surfactants* (M. J. Schick, ed.), Chapter 17, Marcel Dekker, New York (1967).
15. R. Kjellander and E. Florin, *J. Chem. Soc., Faraday Trans. 1 77*, 2053 (1981).
16. G. Karlström, *J. Phys. Chem. 89*, 4962 (1985).
17. E. Florin, R. Kjellander, and J. C. Eriksson, *J. Chem. Soc., Faraday Trans. 1 80*, 2889 (1984).
18. P. M. Claesson, R. Kjellander, P. Stenius, and H. K. Christenson, *J. Chem. Soc., Faraday Trans. 1 82*, 2735 (1986).
19. M. Malmsten, P. M. Claesson, E. Pezron, and I. Pezron, *Langmuir 6*, 1572 (1990).
20. F. Tiberg, C. Brink, M. Hellsten, and K. Holmberg, *Colloid Polym. Sci.*, in press.

21. C. Andrén, K. Holmberg, B. Lindman, and M. Malmsten, Swedish patent application 8904397-0 (1989).
22. W. R. Gombotz and A. S. Hoffman, *J. Appl. Polym. Sci. 42*, 285 (1988).
23. D. H. Napper, *J. Colloid Interface Sci. 58*, 390 (1977).
24. I. Danielsson and B. Lindman, *Colloid Surf. 3*, 391 (1981).
25. M. Kahlweit *et al.* (17 coauthors), *J. Colloid Interface Sci. 118*, 436 (1987).
26. B. Gestblom and J. Sjöblom, *Langmuir 4*, 360 (1988).
27. K. Holmberg, *J. Surface Sci. Technol. 5*, 209 (1989).
28. K. M. Larsson, P. Adlercreutz, B. Mattiasson, and U. Olsson, *Biotech. Bioeng. 36*, 135 (1990).
29. P. D. I. Fletcher, B. H. Robinson, R. B. Freedman, and C. Oldfield, *J. Chem. Soc., Faraday Trans. 1 81*, 2667 (1985).
30. K. Bergström and K. Holmberg, Swedish patent application 8904396-2 (1989).
31. S. Lapanje, *Physicochemical Aspects of Protein Denaturation*, Wiley, New York (1978).
32. E. Österberg, C. Ristoff, and K. Holmberg, *Tenside 25*, 293 (1988).
33. W. H. Scouten, in: *Methods in Enzymology, Vol. 135*: Immobilized Enzymes and Cells, Part B (K. Mosbach, ed.), p. 30, Academic Press, New York (1987).
34. C. Tanford, *J. Am. Chem. Soc. 84*, 4240 (1962).
35. J. March, *Advanced Organic Chemistry*, 3rd ed., p. 316, Wiley, New York (1985).
36. K. Holmberg and M.-B. Stark, *Colloids Surfaces 47*, 211 (1990).
37. E. Engvall and P. Perlmann, *Immunochemistry 8*, 871 (1971).
38. A. Voller and D. E. Bidwell, in: *Alternative Immunoassays* (W. P. Collins, ed.), Chapter 6, Wiley, New York (1985).
39. A. Gardas and A. Lewartowska, *J. Immunol. Methods 106*, 251 (1988).
40. G. E. Kenny and C. L. Dunsmoor, *Isr. J. Med. Sci. 23*, 732 (1987).
41. K. Bergström and K. Holmberg, *Colloids Surfaces*, in press.
42. R. K. Oldham, J. R. Maleckar, J. R. Yannelli, and W. H. West, *Cancer Treat. Rev. 16*, 5 (1989).
43. S. A. Rosenberg, M. T. Lotse, and J. J. Mulé, *Ann. Intern. Med. 108*, 853 (1988).
44. K. A. Smith, *Annu. Rev. Cell Biol. 5*, 397 (1989).
45. S. C. Saris, S. A. Rosenberg, R. B. Friedman, J. T. Rubin, D. Barba, and E. H. Oldfield, *J. Neurosurg, 69*, 29 (1988).
46. T. Maack, V. Johnson, S. T. Kau, J. Figueiredo, and D. Sigulem, *Kidney International 16*, 251 (1979).
47. N. V. Katre, M. J. Knauf, and W. J. Laird, *Proc. Natl. Acad. Sci. U.S.A. 84*, 1487 (1987).
48. H. Hydén and K. Holmberg, *Arzneim.-Forsch./Drug Res. 36*, 120 (1986).
49. K. Bergström, K. Holmberg, H. Hydén, A. Hyltander, and K. Lundholm, unpublished work.
50. S. Srivastava, S. D. Gorham, and J. M. Courtney, *Biomaterials 11*, 162 (1990).
51. F. Grinell, *Int. Rev. Cytol. 53*, 65 (1978).
52. K. Hirai, Y. Shimizu, and T. Hino, *J. Exp. Pathol. 71*, 51 (1990).
53. S. Srivastava, S. D. Gorham, D. A. French, A. A. Shivas, and J. M. Courtney, *Biomaterials 11*, 155 (1990).
54. K. Bergström and K. Holmberg, to appear.
55. M.-B. Stark and K. Holmberg, *Biotech. Bioeng. 34*, 942 (1989).

20

Polystyrene-Immobilized PEG Chains

Dynamics and Application in Peptide Synthesis, Immunology, and Chromatography

ERNST BAYER and WOLFGANG RAPP

20.1. INTRODUCTION

Poly(ethylene glycol) (or PEG) is very compatible with peptides and proteins. It is soluble in water and almost all organic solvents, with the exception of aliphatic hydrocarbons and ether. This polymer was shown to be a valuable support for peptide and nucleotide synthesis in homogeneous solution (liquid-phase method[1–5]) as an alternative to the solid-phase method of Merrifield.[13] In general, PEG of molecular masses 3000–20,000 daltons are used in liquid-phase peptide synthesis. Even insoluble free peptides often are solubilized, if covalently linked to PEG. On the other hand, the conformation of the peptide bound to PEG is the same as the conformation of the free peptide in the same solvents. A synthetic cycle using the liquid-phase procedure is shown in Scheme 1. The couplings are carried out in homogeneous solution.

Kinetic investigations have shown[6] that amino acid–PEG esters have the same coupling rates as the low molecular mass amino acid esters (Table 1). Diffusion phenomena play no role in this technique, which is a major improvement over solid-phase methods. Instead of simple filtration used in solid-phase synthesis, membrane

ERNST BAYER and WOLFGANG RAPP • Institute for Organic Chemistry, University of Tübingen, D-7400 Tübingen, Germany.

Poly(Ethylene Glycol) Chemistry: Biotechnical and Biomedical Applications, edited by J. Milton Harris. Plenum Press, New York, 1992.

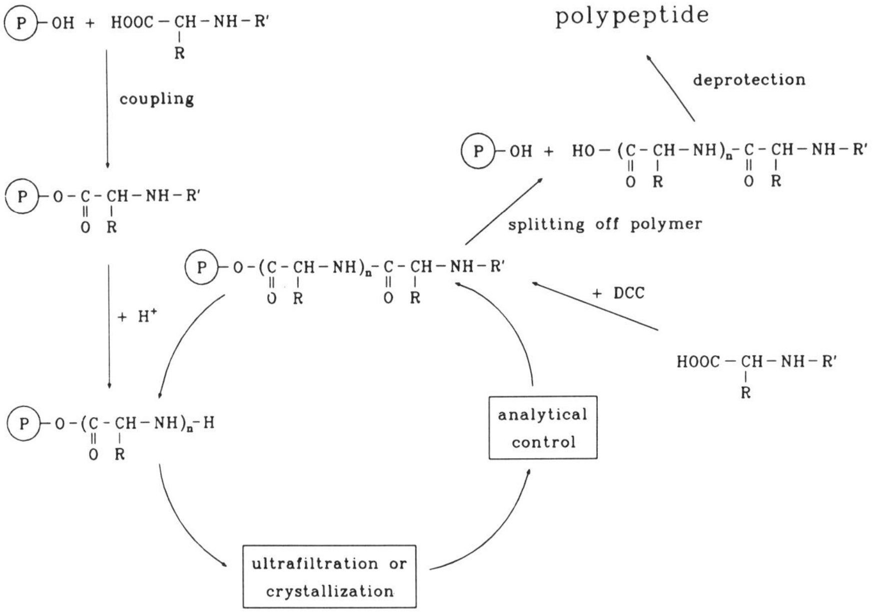

Scheme 1. Standard cycle for liquid-phase peptide synthesis.

filtration is applied in this procedure. This is possible because polymer-bound peptides of molecular masses of 3–20 kilodaltons and low molecular mass reagents show considerable differences in the molecular size. Recrystallization of peptide–PEGs can also be used as a valuable alternative procedure for purification purposes.[7]

A substantial advantage of the liquid-phase method is due to the solubility and the possibility to carry out more suitable analytical controls. In this synthesis, spectroscopic methods such as NMR and CD can be applied routinely. This method has also been widely employed to study changes in peptide folding during stepwise synthesis.[8–11]

Table 1. Comparison of the Rate of Coupling Glycine Active Esters to Glycyl–Polyethylenglycols (Gly–OPEG) in Solution and to the Glycyl-Graft Copolymers (Gly–OPEG–PS)

Polymeric support	k ($mol^{-1} \cdot s^{-1}$)	meq Gly per g polymer
Gly–$OPEG_{2000}$–PS[a]	0.037	0.20
Gly–$OPEG_{3300}$–PS	0.032	0.12
Gly–$OPEG_{5700}$–PS	0.029	0.08
Gly–$OPEG_{2000}$	0.016	0.49
Gly–$OPEG_{4000}$	0.017	0.38
Gly–$OPEG_{6000}$	0.013	0.27
Gly–OMe	0.012	

[a]The numbers represent molecular weight of the PEG chains.

Disadvantages of liquid-phase synthesis are, however, the problem of automation and the time-consuming operations.[12] Membrane filtration often takes very long. For the crystallization process, the product must be precipitated, washed, and redissolved. These operations practically render an untroublesome automation impossible.

Moreover, the solubilization effect of PEG in the case of longer peptides is often insufficient. It was therefore plausible to believe that this problem could be circumvented by immobilizing long-chain PEG on an insoluble support, thus combining the advantages of solid- and liquid-phase methods. Such immobilized PEGs also would be interesting in many other areas, like phase transfer catalysts, stationary phases for chromatography and immunology.

20.2. SYNTHESIS AND PHYSICAL PROPERTIES OF POLYSTYRENE–POLY(ETHYLENE GLYCOL) GRAFT COPOLYMERS

Since crosslinking of linear PEG would restrict mobility of the polymer chains, it was aimed to graft it on an insoluble matrix. Inorganic solids such as silica gels could not be considered, because modified silica gels do not possess enough stability under the conditions which are usually applied during peptide synthesis (such as alkaline and acidic treatments). The simplest immobilization procedure would have been the coupling of one of the two terminal hydroxyl groups of PEG to chloromethylated polystyrene according to the classical ether synthesis.[14,15]

The main problem of this procedure is, however, that when long-chain PEGs (> 800 daltons) are used, the yields are unsatisfactory. Moreover, the dihydroxyl PEGs could react to a large extent at both ends to give cyclic polyethylenglycols and additional crosslinking. Consequently, the number of the free hydroxyl groups will be reduced with concomitant lowering of the capacity of the graft copolymer for binding peptides. We have found that by means of anionic polymerization of ethylene oxide, PEG chains of molecular masses up to 20 kilodaltons can be immobilized on crosslinked polystyrene containing hydroxyl functional groups.[16–21] According to Scheme 2, the potassium salt of poly(styryl-methyl-tetraoxyethylene ether), which is

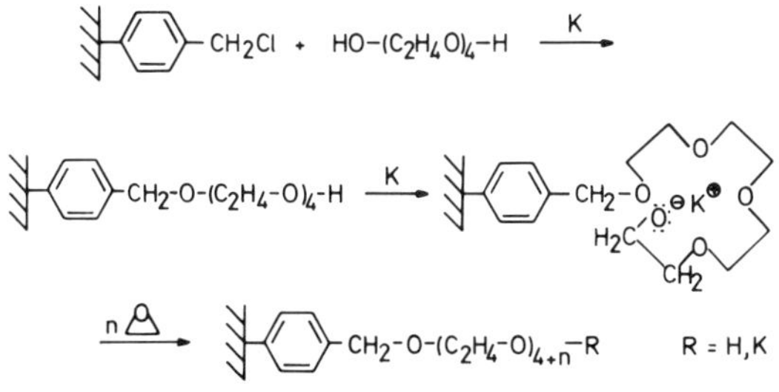

Scheme 2. Synthesis of PEG–PS graft copolymer.

present in the form of an open crown ether, is allowed to react. For carrying out this grafting reaction, functionalized polystyrene beads of any degree of porosity can be employed. By using nonporous beads, the ethylene oxide can only be grafted on the surface. In this case, the polyoxyethylene chains extend outward from the matrix like tentacles. Such polymers are interesting as separation materials for HPLC; however, they are not suitable for peptide synthesis because loading of peptides on them remains low. For peptide and nucleotide synthesis, the graft copolymers based on gelatinuous and porous polystyrene are more appropriate.

Graft copolymers with PEG chains of about 3000 daltons proved to be optimum. These copolymers contain about 70% linear PEG and about 30% crosslinked polystyrene matrix. Therefore, the properties of these polymers are to a large extent determined by the PEG portion of the molecules. This is shown especially by the swelling properties. Table 2 shows a comparison of swelling factors of the graft copolymers and polystyrene. The graft copolymers swell in all solvents which dissolve PEG. On the other hand, swelling is negligible in solvents which do not dissolve PEG, like aliphatic hydrocarbons and diethylether. Good solvation leads to a high degree of mobility of the PEG chains and the peptides bound to them. This mobility can be monitored by ^{13}C NMR relaxation time measurements.[22] Figure 1 demonstrates an increasing average mobility of the immobilized PEG chain up to approximately 3000 dalton, while free PEG in solution shows a decrease in mobility, as ^{13}C-relaxation times, T_1, reveal.

Therefore the PEG of 3000 daltons is optimal in spacer length using the PS defined in Figure 1. Of course crosslinking and degree of functionalization influence the mobility and hence the relaxation time, as is shown in Table 3. Polymers 5–10 in Table 3 show an intermediate loading of PEG, corresponding to one PEG chain for every eighth aromatic ring of PS. With increasing spacer length the mobility of the PEG chain increases up to a maximum of 2000–3000 daltons. Polymers 1–4 have the same matrix as polymers 5–10 but the degree of substitution is smaller. In these cases

Table 2. Comparison of Swelling Factors ($V_{swollen}/V_{dry}$) of the Crosslinked Polystyrene–1% DVB (PS) and the Graft Copolymer Poly(ethylene glycol)–Polystyrene–1% DVB (PS–PEG_{3000})

Solvent	PS	PS–PEG
Water	—	2.5
MeOH	0.95	2.5
EtOH	1.05	1.2
CH_2Cl_2	5.2	3.0
Toluene	5.3	3.1
DMF	3.5	3.2
MeCN	2.0	3.0
THF	5.5	3.4
Dioxane	4.8	3.7
Ether	2.5	1.1

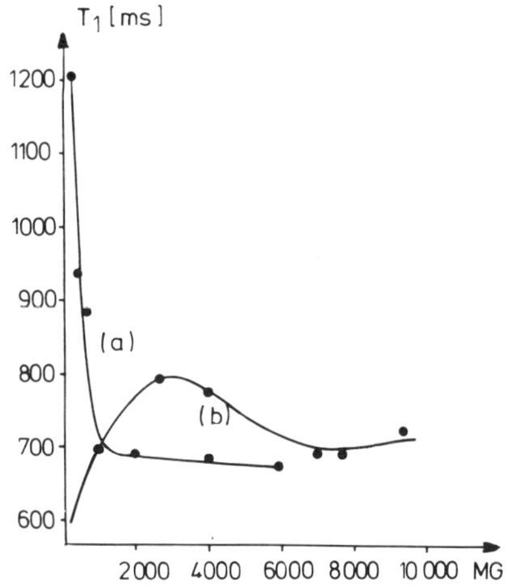

Figure 1. ^{13}C-NMR-relaxation times T_1 of free soluble PEG (a) and PS-immobilized PEG (b) as dependent on molecular weight.

only every 21th aromatic ring is substituted by PEG. This low degree of substitution results in an increased mobility because of reduced steric hindrance and reduced microviscosity within the beads. The maximum mobility is then shifted to higher molecular weights.

Higher crosslinking generally decreases the mobility of the PEG (polymers 11–14). However, this effect is much more pronounced in the case of shorter PEG chains (i.e., polymers 11 and 13 versus polymers 1 and 5 in Table 3). If the PEG spacer is longer the reduction of mobility is relatively smaller. This indicates that generally the

Table 3. ^{13}C-Relaxation Times (T_1) of PEG–PS Polymers with Different Spacer Length and Degree of Functionalization

Polymer	Crosslinking (% DVB)	Substitution of PS matrix (%)	MW PEG (dalton)	T_1 (ms)
1	1	5.7	194	500
2	1	5.7	2000	937
3	1	5.7	4000	1053
4	1	5.7	7500	1000
5	1	13.5	194	500
6	1	13.5	950	770
7	1	13.5	2000	790
8	1	13.5	3300	690
9	1	13.5	5100	650
10	1	13.5	7300	690
11	2	4.7	194	360
12	2	4.7	5600	940
13	20	22.9	194	194
14	20	22.9	3600	500

spacer length becomes more important if more rigid support materials are used. Therefore the decrease of T_1 and hence PEG mobility from polymer 8 with only 1% crosslinking and intermediate substitution of 13.5% to T_1 of polymer with 14 with 20% crosslinking and a substitution degree of 22.9% is much smaller, as in the case of the shorter PEG chain in polymer 13.

Figure 2 shows that ^{13}C-NMR-relaxation times of PS-bound peptide are lower than those bound to PEG and PEG–PS. The values for the peptides bound to the last two supports are about the same.

By NMR spectroscopic measurements in suspensions, PEG chains as well as peptides bound to them give sharp signals (Figure 3). These sharp signals are normally not expected from measurements of solid materials in suspensions. On the other hand, in solvents such as ether, which do not solvate PEG, broad NMR signals,

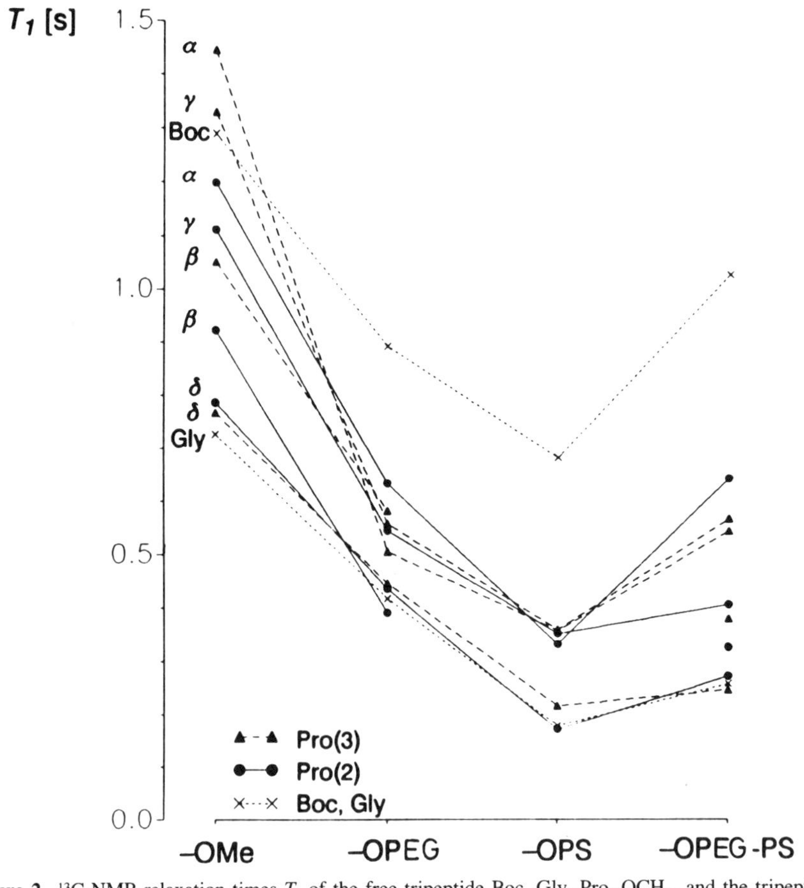

Figure 2. ^{13}C-NMR-relaxation times T_1 of the free tripeptide Boc–Gly–Pro–OCH_3, and the tripeptide covalently linked to PEG_{3000}, polystyrene (PS) and PEG–PS.

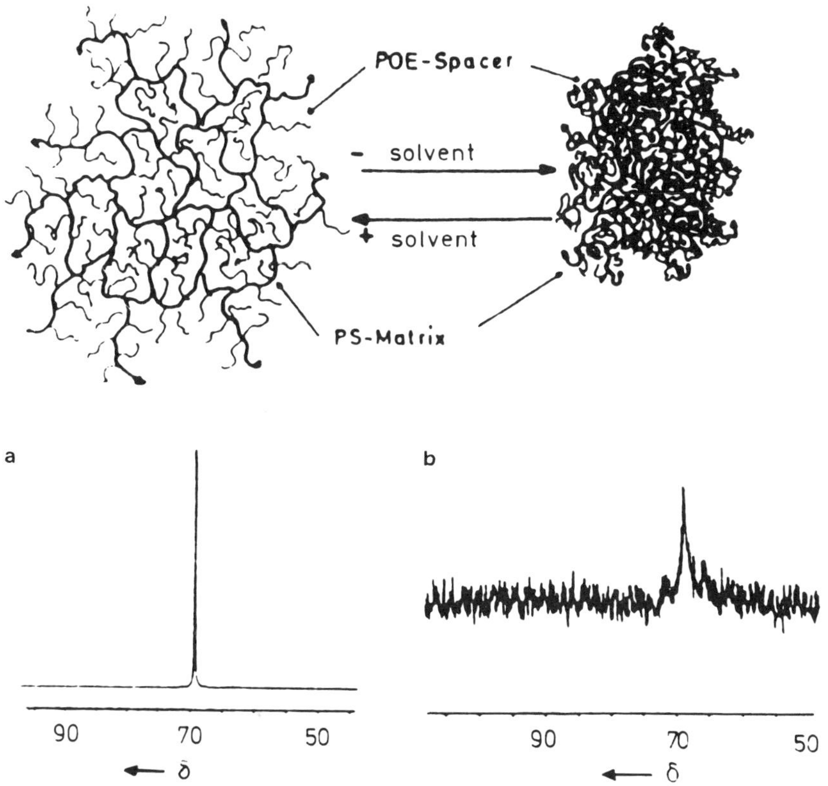

Figure 3. Conformations of the graft copolymer PEG–PS: (a) extented and solvated conformation: sharp ^{13}C-NMR signals in spite of measurement in suspension; (b) shrinked conformation in nonsolvating solvent diethylether, usual band broadening of solid in suspension.

typical of the measurement of solids in suspensions, are obtained. We can therefore distinguish between a solvated extended tentacle and a shrinked dense conformation of the graft copolymer as seen schematically in Figure 3. On this account, these polymers have received the trivial name tentacle polymers.[23]

Figure 4 shows the homogeneous distribution of the PEG chains on the surface and inside. Under the conditions of the electron microscopic procedure, the graft copolymers exist in the shrinked coil conformation. Remarkable are the observed comparable swelling factors of the tentacle polymers in different solvents (Table 2). This property allows the use of the graft copolymer for operations like the continuous flow synthesis in a column, packed beds for catalysis, or phases for chromatography. As seen, applying different solvents during the synthesis has little impact on the swelling of the copolymer and packing of the column. The swelling factors remain also largely unchanged if the terminal hydroxylic group is functionalized by even longer peptide chains.

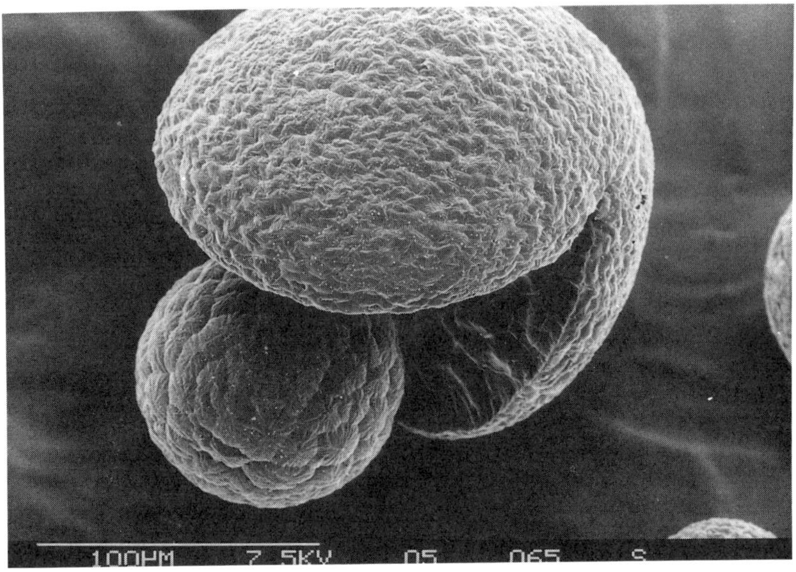

Figure 4. Electron micrograph of the graft copolymer PEG–PS.

An additional property of the copolymer which is seldom possessed by other polymers further promotes its use for continuous flow operations. Because of the excellent solvation, the PEG–PS beads are stable under pressures up to 200 bar, the condition under which relatively high flow rates can be acquired. This in turn shortens the required time for a coupling cycle in peptide synthesis. The advantages of the continuous flow synthesis will be discussed again in a subsequent section.

Since, according to the NMR investigations, the mobility of the PEG chains of the graft copolymers are about the same as the mobility of free PEGs, it can be assumed that these copolymers, after being solvated, behave like a homogeneous phase. The same conclusion can be reached by noting that the rate constants of peptide coupling on the graft copolymer have the same order of magnitude as those of the liquid-phase methode (Table 1). According to these considerations, the tentacle polymers as hydrophilic supports are suitable for rapid peptide synthesis and column operations such as chromatography and catalysis.

20.3. MONODISPERSED TENTACLE POLYMERS

For the optimization of continuous flow peptide synthesis, and mass transfer in chromatography, diffusion phenomena have to be taken into account. These are of prime importance in shortening the duration of processes such as the cycles of synthesis or retention time in chromatography. Since the reactions have a heterogeneous nature, particle diameter plays a decisive role in the diffusion processes. As particle diameter becomes larger, the contribution of diffusion becomes more extensive. The commercially purchasable resins used for the Merrifield synthesis and affinity chromatography are polydispersed, i.e., they show a wide range of diameter (1–200 μm). It is obvious that reaction time of synthesis is influenced by the largest particles. These also contain a relatively higher number of functional groups in comparison with the small particles, as shown in Figure 5. A distinct shortening of reaction time for every cycle would therefore be possible if monodispersed and uniform particles, possibly with small diameter, were available.

Most of the polymerization procedures, however, produce polydispersed materials. After detailed examinations of diffusion processes, Ugelstad *et al.*[24,25] have investigated the parameters for preparation of monodispersed polystyrene beads. Based on this work, preparation of monodispersed and crosslinked polystyrene beads with diameters of 5–20 μm was achieved. In accordance with Scheme 3, first of all, a "seed" is prepared in a homogeneous solution using an initiator.[26] In order to obtain particles with equal diameters, it is important to produce the seed in a short time and to work in the presence of bulky stabilizers which suppress formation of aggregates

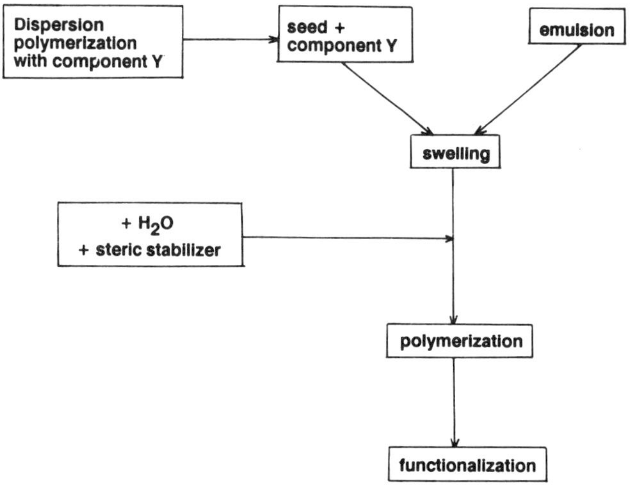

seed:
. monomer
. initiator
. component Y
. org. solvent
. steric stabilizer

emulsion:
. monomer
. cross linker
. monomer soluble initiator
. H_2O
. inert additive
(only for macroporous beads)

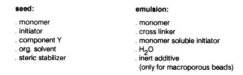

Scheme 3. Synthesis of monosized PS beads.

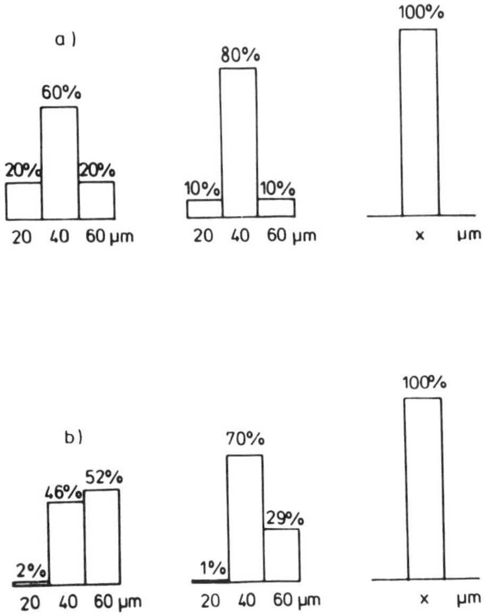

Figure 5. Size distribution of polydispersed polymer beads (a) and percent of the functional groups in the beads with different diameters (b).

and micelles. The particles of the "seed" are still not crosslinked and are then brought into an aqueous emulsion of monomeric styrene, the crosslinker divinylbenzene, and the initiator. In a diffusion-controlled process in the presence of water-soluble and sterically hindered stabilizers, the "seed" is then swollen and cross-linked according to Figure 6. Figure 7a shows the monodispersed and crosslinked polystyrene beads obtained. The deviation of diameters is less than 2% (e.g., 10 ± 0.02 µm). These monodispersed and crosslinked polystyrene beads can now be functionalized for different applications. It is also possible to graft PEG chains to these beads as already described. Figure 7b shows these monodispersed graft copolymers. The PEG chains of the above polymers exist in ordered spherical forms under the conditions of electron microscopy.

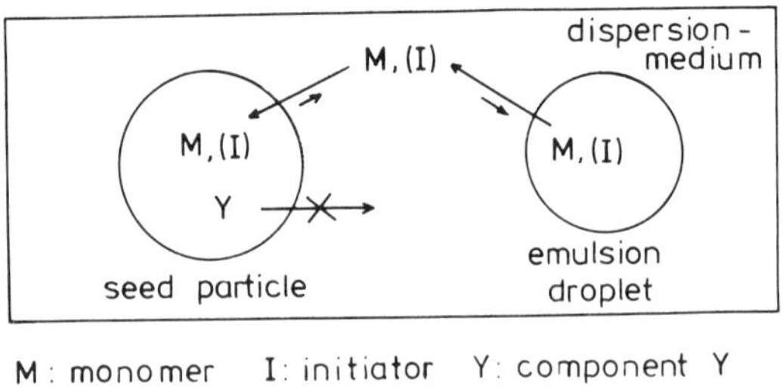

Figure 6. Swelling of monodispersed seed.

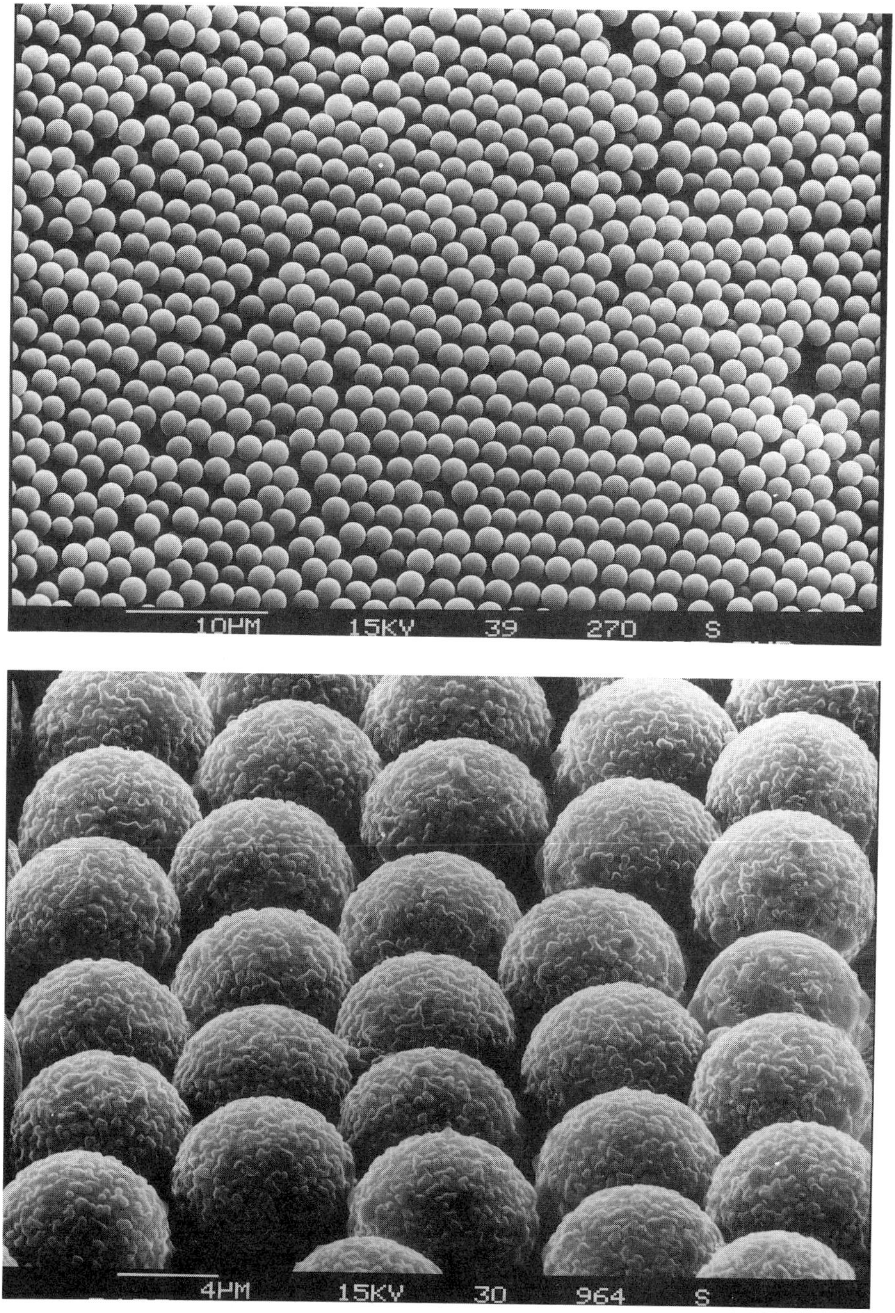

Figure 7. Electron micrographs of monodispersed polystyrene (top) and graft copolymer PEG–PS (bottom).

Sorption and diffusion properties of the PEG–PS graft copolymers influence the overall reaction rates and are important also for their use as supports for chromatography and catalysis. Sorption is dependent on matrix parameters such as degree of crosslinking, particle size, and polarity.

The degree of sorption of reagents in the beads is of prime importance. Normally, a large excess in the fluid medium is used to complete reactions in the beads. However, for the effective reaction rates[27] the concentration of reactants in the beads, which is expressed by the partition coefficient k in Table 4, must be considered. A sorption resp. partition coefficient > 1 indicates higher concentrations inside the bead compartment in comparison to the surrounding liquid medium. With increasing PEG content of the beads also the partition coefficient increases and simultaneously contributes to the higher coupling yields observed in using the tentacle polymers for peptide synthesis.

Table 4 shows the partition coefficient of BOC–Gly–ONP ester between differently crosslinked polystyrenes, PEG–PS graft copolymers, and the surrounding liquid. The sorption and partition coefficients were measured by using a standard procedure, developed by Hori *et al.*[28,29]

In Figure 8 the sorption of oracett blue 2 R on 1% crosslinked polystyrene of 44 μm particle size (curve 1) and the same polystyrene matrix, but grafted by PEG with molecular weight 3000 dalton (curve 2), is shown. Seventy-five percent of the graft copolymer consists of PEG, and the particle size is increased to 80 μm. Although the particle size of PEG–PS is approximately twice that compared to polystyrene, the sorption rate is four times higher. Besides the increase of sorption time, there is also a threefold increase in sorption capacity. Using 1% crosslinked 7.5 μm monosized PEG–PS graft copolymers the sorption increases dramatically (curve 3). Even higher crosslinked monodispersed graft copolymers (curve 4: 5% and curve 5: 10% crosslinking) show higher sorption rates compared to polystyrene.

To summarize these results, PEG–PS generally show higher sorption capacities and higher sorption rates compared to polystyrene. As a consequence more reactive

Table 4. Partition Coefficients of BOC–Gly–ONP between Various Polymer Beads and Acetonitrile Solution Phase at 25 °C

Polymer	Crosslinking of polystyrene (%)	Molecular mass of PEG (dalton)	PS matrix substitution (%)	Partition coefficient k ($kg\,L^{-1}$)
BOC–Gly–PS	1	—	13.5	1.11
BOC–Gly–PEG–PS (5)	1	194	13.5	1.30
BOC–Gly–PEG–PS (8)	1	3300	13.5	1.57
BOC–Gly–PEG–PS (13)	20	194	22.9	1.39
BOC–Gly–PEG–PS (14)	20	3600	22.9	1.68
BOC–Gly–MOPS[a]	1	—	13	1.14
BOC–Gly–PEG–MOPS[a]	1	194	13	1.37
BOC–Gly–PEG–MOPS[a]	1	2000	13	1.58

[a]Monosized matrix.

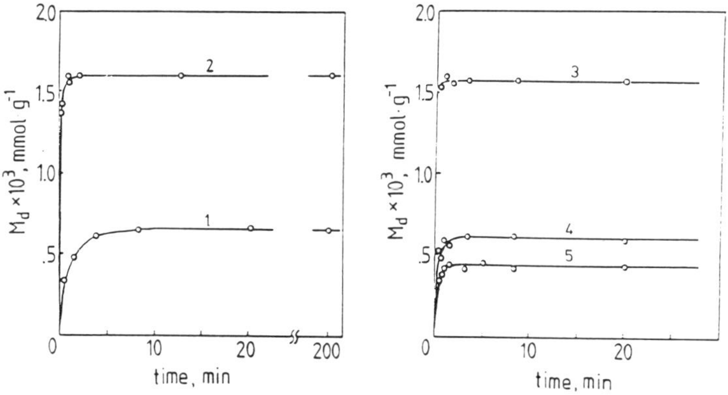

Figure 8. Sorption curves for 1.3×10^{-4} mol L^{-1} oracet blue 2B in acetonitrile (25 °C) on various polymer beads: (1) polydispersed 1% crosslinked PS, 44 μ average; (2) polydispersed 1% PS–PEG_{3000}, 80 μm average; (3) monodispersed 1% crosslinked PS–PEG_{2000}, 10 μm; (4) monodispersed 5% crosslinked PS–POE_{4000}, 10 μm; (5) monodispersed 10% crosslinked PS–PEG_{3000}, 10 μm. Weight ratio polymer:solvent = 1:120.

compounds come inside the beads within shorter times, resulting in an increase of overall reaction rate. This increase of concentration in the beads is important, because each individual bead represents the reaction compartment and not the liquid phase.

With these pressure-stable and hydrophilic polymeric supports, diffusion problems can be minimized. The new supports cannot only be used for the synthesis of peptides and nucleotides, but also as stationary phase for HPLC, and as supports for affinity chromatography and catalysts.

20.4. APPLICATION OF PEG–PS GRAFT COPOLYMER FOR PEPTIDE SYNTHESIS

The tentacle PEG–PS polymers are pressure stable and show almost equal swelling factors in different solvents. These two properties make them suitable for packing columns and carrying out continuous peptide synthesis with high flow rates. Because of that, the tentacle polymers distinguish themselves from other polymers. The continuous flow peptide synthesis has advantages over the batchwise solid-phase peptide synthesis.

The graft copolymer of PEG and crosslinked polystyrene also shows high chemical stability, and can be prepared in spherical form and equal bead size. All of these properties are required for rapid peptide synthesis by the continuous flow method. Since its introduction, it has been shown in many cases that this procedure is superior to the normal batchwise synthesis.[30,31]

A substantial advantage of the continuous flow method is that the large excess of coupling components and solvents normally used in solid-phase synthesis can be drastically reduced. Moreover, the coupling and deprotection reaction can be monitored by UV measurements at the inlet and outlet of the column. In this respect the UV active protecting groups, such as fluorenylmethyloxycarbonyl (Fmoc), are specially suitable. The lowering of the amount of excess reagents in comparison with the batchwise method results from carrying out the reaction by continuous flow. Coupling components are dissolved in a small volume with a relatively high concentration and are allowed to pass through a column. In this case concentration profiles are developed; they are shown in Figure 9. While moving the band of reagents through the column, a relatively small amount of support comes in contact with the whole excess of coupling components. The concentration of excess reagents in every column segment is larger than in the case when the same amount of reagents would have been in contact with the whole amount of support, as is the case in batch procedures. The narrower the coupling components band, the larger the excess of reagents in a given column segment. Thereby the reaction can proceed more completely and the amount of statistical failure sequences will be lowered. By using polymeric beads with diameter of about 80 μm, a cycle for coupling one amino acid requires about 6–20 min. Even if double couplings are used, a peptide with 30 amino acid residues can be synthesized in a few hours.

Figure 10 shows the monitoring of a coupling cycle with the time scale of the different operations for coupling one amino acid. By employing tentacle polymers with 8 μm diameter, the required time for a coupling cycle can again be reduced by one-tenth (i.e., 1–2 min). For these short times, however, there are no commercial synthesizers available.

The graft copolymers have been widely used for peptide and nucleotide synthesis. We have synthesized many peptides on PEG–PS graft polymers using continuous flow peptide synthesis.[30] As an example, Scheme 4 shows the synthesis of

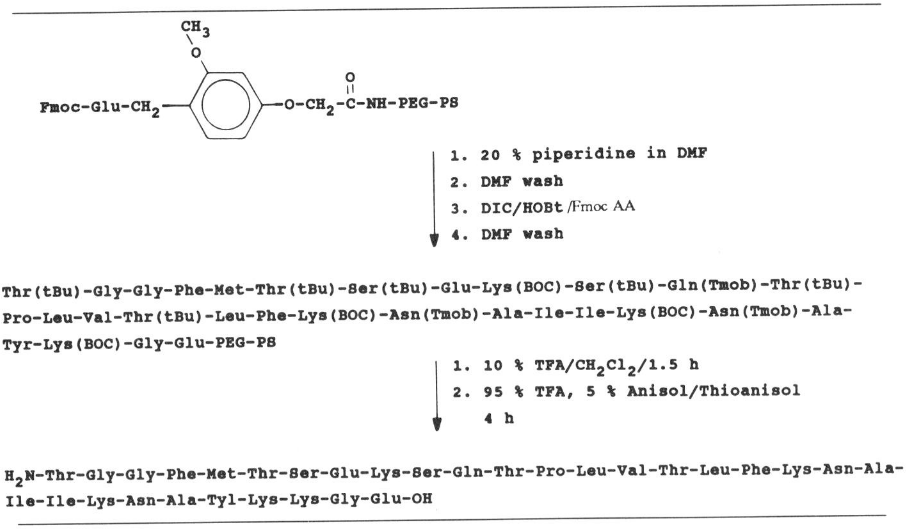

Scheme 4. Synthesis of human β-endorphin.

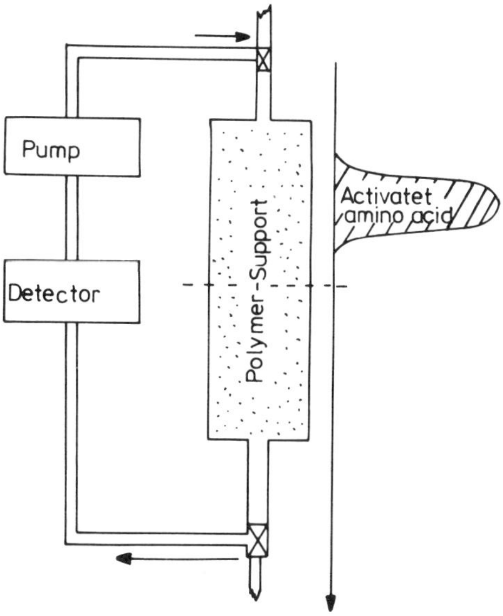

Figure 9. Moving band of the coupling components through the column during continuous flow synthesis.

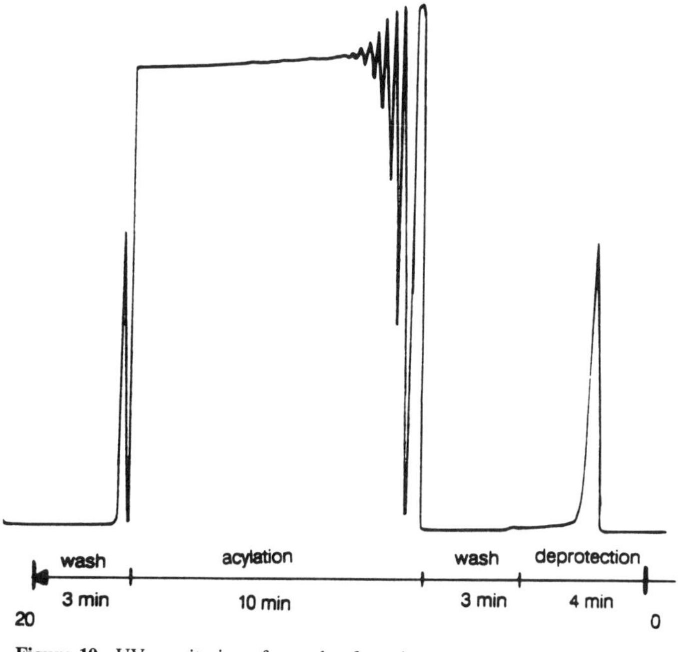

Figure 10. UV monitoring of a cycle of continuous flow peptide synthesis.

β-endorphine by using DIC/HOBt activation and single coupling for all amino acids. Typical acylation time was 8 min, total cycle time (washing, deprotecting, acylation, washing) was 20 min. By continuous UV monitoring the completeness and kinetics of coupling and deprotection can be followed, as shown in Figure 10. In the case of this synthesis, beads of 44–80 μm size have been used. Using monosized PEG–PS beads of 10 μm, a coupling cycle can be reduced to 1–2 min. Such fast coupling times will open the field to customary protein synthesis.

20.5. PEG–PS POLYMER PEPTIDES FOR *IN VIVO* AND *IN VITRO* FORMATION OF ANTIBODIES

The prediction of epitopes for the formation of antibodies and synthesis of these epitopes is nowadays an indispensable instrument in immunology. A multitude of overlapping peptides is required for a dependable prediction. Conjugates of peptides with carriers such as serum albumin or lipopeptides are currently employed for immunization.[32–34] It would be a tremendous methodological progress, if the peptides did not need to be cleaved from the polymeric support, and instead of a conjugate they could directly be applied for the *in vivo* and *in vitro* immunization and production of antibodies.[35–37] The tentacle polymers, due to the compatibility of the PEG chains with biological systems, are especially suitable in this respect.

This is demonstrated by the synthesis of an epitope region of the haemophilus influencae P6 protein, corresponding to the sequence 115–131, on TentaGel.[38] The synthesis of the sequence H_2N–Ala–Val–Leu–Gly–His–Asp–Glu–Ala–Ala–Tyr–Ser–Lys–Asn–Arg–Ala–Val–OH was carried out according to the Fmoc strategy with the following side-chain protecting groups: Arg(Mtr), Lys(Boc), His(Trt), Ser(tBut), Glu(OtBu), Asp(OtBu), and Asn(oNp). The side-chain protecting group of the polymer-bound peptide were removed under moderate acidic conditions to deliver the TentaGel-bound peptide. For comparison the PEG–peptide was cleaved from the PS backbone under strongly acidic conditions, generating the soluble peptide–PEG epitope conjugate.

Figure 11 shows the ion-spray mass spectrum of this PEG–peptide. It is interesting to note that with a single mass spectrum the exact molecular mass distribution of the peptide–polymer can be determined. This could otherwise only indirectly be achieved by means of exclusion chromatography and standard substances. The difference of the individual peaks in the mass spectrum of the threefold protonated polymer–peptide $(M + 3H)^+$ with average molecular mass 4850 corresponds exactly to one-third of the molecular mass of the CH_2—CH_2—O-monomeric unit of poly(ethylene glycol). For immunization, the soluble PEG–peptide as well as the insoluble immobilized PS–PEG–peptide were applied to mice, and after 4 weeks serum samples were taken for determination of antibodies, using the ELISA test. The titers of the ELISA in Figure 12 show that the polymer–peptides form antibodies as nicely as the serum albumin conjugate BSA–(PG 115-131).[38] Most

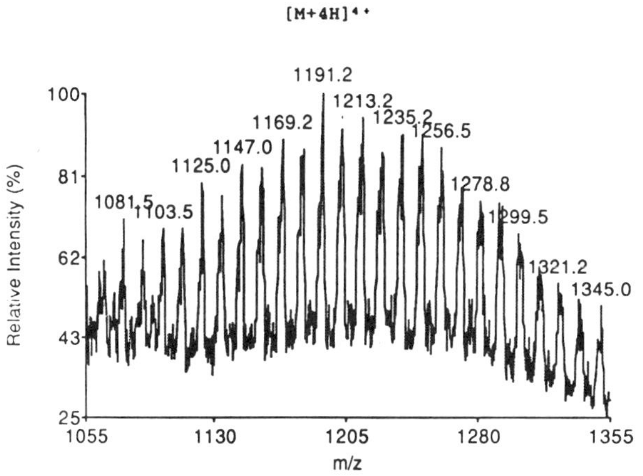

Figure 11. Ion spray mass spectrum of the $(M + 4H)^+$ peak of the PEG-bound peptide A–V–L–G–H–D–E–A–A–T–S–K–N–R–R–A–V–PEG_{3000}. Average molecular mass 4850.9; experimental: 4848.8.

important is the result that the insoluble immobilized PS–PEG–peptide shows no significant difference from the soluble cleaved PEG–peptide.

20.6. PEG–PS GRAFT COPOLYMERS AS STATIONARY PHASES FOR HPLC OF BIOMOLECULES

Pressure stability, small and equal bead size, and equal swelling ratios in different solvents render the PEG–PS polymers especially suitable as stationary phases for high performance liquid chromatography. A great advantage is the

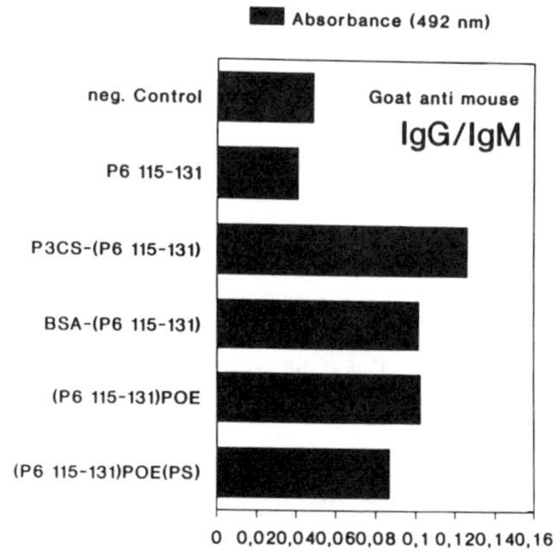

Figure 12. ELISA titers of serum after immunization with different conjugates of the sequence 115–131 of haemophilus influenzae P6 protein.

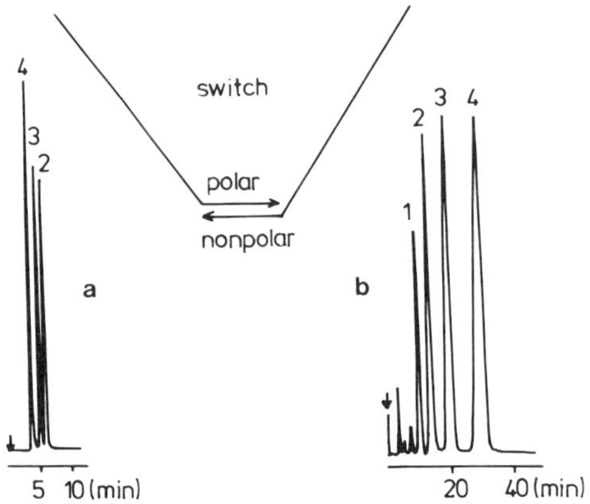

Figure 13. HPLC separation of benzene (1), toluene (2), ethylbenzene (3), propylbenzene (4), and n-butylbenzene on tentacle polymer PS–PEG: (a) normal-phase mode, (b) reversed-phase mode.

chemical stability in alkaline as well as strongly acidic (0.1 n H_2SO_4) solvents because these polymers have only C—C and C—O—C bonds. Up to 10% cross-linked monosized PS–PEG beads have been tested as HPLC column packings. Separation efficiencies of up to 36,000 plates per m column length can be generated.

An intriguing behavior of these packing materials is their "chameleon-like" behavior. Depending upon the choice of solvents, they can be used in the normal-phase or the reversed-phase (RP) mode of separation.[39]

This is attributed to the different conformations of the poly(ethylene glycol) in different solvents, shown in Figure 3. In apolar solvents like heptane, the PEG chains are covering the PS matrix as a dense layer. Since solvents cannot penetrate this polar layer, interaction of analytes follows polarity as is the case in normal-phase HPLC. In polar solvents the PEG chains are loosely packed. Analytes can penetrate the PEG layer to the PS matrix, and hydrophobic interaction is dominant as is the case in reversed-phase HPLC.

Figure 13a shows a chromatogram of benzene derivatives in normal-phase mode with heptane as mobile phase, and Figure 13b the reversed-phase mode separation with acetonitrile/water (35/65) as eluent. Also, with strongly polar analytes like amines, organic acids, peptides, and proteins, very narrow peaks without tailing can be obtained. Supports like unmodified silica gel normally cannot be used in normal-phase separations for polar analytes because of serious tailing. Even isomeric or closely related polypeptides and proteins can be separated.

20.7. PROTEIN IMMOBILIZATION FOR CATALYSIS AND AFFINITY COLUMNS

Besides the stepwise synthesis of peptides and small proteins directly on the resin, synthetic or natural peptides and proteins can be covalently attached to PEG–

PS support. The immobilization of enzymes stabilizes them and the enzyme is much more easy to handle.[40–42] Proteins can also be immobilized for affinity supports. The tertiary structure of the immobilized protein and therefore the catalytic activity is strongly influenced by matrix properties such as flexibility of the surface and hydrophobicity/hydrophilicity. A very strong and inflexible binding of the protein to a hard resin or inorganic surface damages the tertiary structure of the protein and also introduces strong hydrophobic interactions between matrix and protein. A great advantage of the tentacle polymer PEG–PS is its hydrophilicity and the flexibility of the PEG spacer.

The reactive sites on PEG–PS for attachment of the proteins are located at the end of the flexible PEG spacers and therefore the immobilized protein is allowed to move with the polymer chains. Because of the flexibility of the PEG spacers the tertiary structure of the protein is not influenced, and conformational changes during enzyme activity can occur that might be impossible for the attachment to hard and inflexible supports (Figure 14). It has been shown by CD measurements that peptide conformations are not changed if bound to PEG.[4,9–11]

The advantage of the hydrophobic and hydrophilic properties of PEG–PS is shown during functionalization and activation of the resin and the following immobilization reaction. In general the introduction of reactive sites into the matrix is performed in organic solvents while protein immobilization takes place in aqueous systems. Table 5 summarizes the functionalized supports used for protein immobilization.[44,45] After functionalization in organic solvents the activated resin is transferred directly into the aqueous system via a solvent gradient. The enzymes and proteins shown in Table 5 were immobilized in buffered aqueous solutions at pH 7.5–9.[44,45]

The amount of immobilized protein is independent of the activation method used. All protein loadings are in the range of 150–250 nmol protein per g resin. Low capacity indicates that the proteins are attached on the outer surface of the PEG–PS beads, because the porosity of the beads used was too small. Proteins and molecules larger than 20,000 dalton molecular mass cannot penetrate these gelatineous supports. However, by the attachment onto the surface there is no restriction of mass transport.

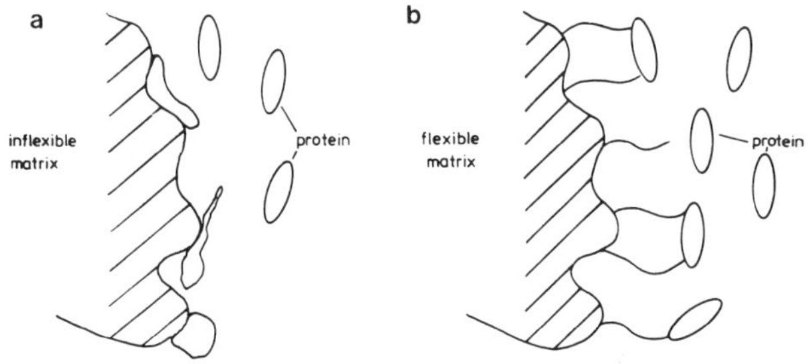

Figure 14. Schematic drawing of protein immobilization on (a) hard and (b) flexible surfaces.

Table 5. Functionalized PEG–PS Copolymers for Protein Immobilization

Starting resin	Activation	Functionalized resin	Immobilized protein
PEG–PS–NH_2	Thiophosgene	PSPEG–N=C=S	Trypsine
PEG–PS–OH	Bromoacetic acid	PSPEG–O–C(=O)–CH_2Br	Trypsine
PEG–PS–COOH	*N*-Hydroxysuccinimide	PSPEG–COO– 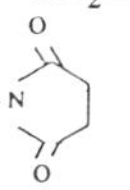	Chymotrypsine, albumine, acylase
PEG–PS–COOH	DIC/HOBt	PSPEG–COO– 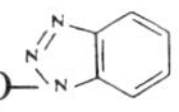	Histone
PEG–PS–NH_2	BOC–Hydrazido-succinyl crosslinker	PSPEG–CON_3	Trypsine, chymotrypsine, albumine, acylase

Via immobilization of Histon H1 onto the PEG–PS beads antiH1 Histon IgM antibodies could be received by affinity chromatography.[43] Using macroporous matrices with shorter length of the PEG spacers, the capacity of the resin increases by a factor of 10–30.[46]

20.8. CONCLUSION

The PEG–PS copolymers described are a new class of polymers with interesting properties. The immobilized PEG chains are relatively mobile and show a quasi-homogeneous behavior. Reactions of the terminal OH group show kinetics similar to those of nonimmobilized PEG in solution. Therefore, immobilized PEG can be used in many cases where normally soluble PEG is used and functionalized. Examples described in this chapter are peptide synthesis, antibody formation, and catalysis. The easy separation of immobilized PEG by simple filtration from educts and products is a great advantage.

Relatively equal swelling in different solvents and pressure stability of PEG–PS beads allows continuous column operation. There are many other conceivable fields in which these PEG polymers can be utilized.

REFERENCES

1. E. Bayer, *Nachr. Chem. Tech. 20*, 495 (1972).
2. M. Mutter, H. Hagenmaier, and E. Bayer, *Angew. Chem. 83*, 883 (1971); *Angew. Chem., Int. Ed. Engl. 10*, 811 (1971).
3. E. Bayer and M. Mutter, *Nature (London) 237*, 512 (1971).
4. M. Mutter and E. Bayer, in: *The Peptides* (E. Gross and J. Meienhofer, eds.), Vol. 2, p. 285, Academic Press, New York (1970).
5. M. Mutter, R. Uhmann, and E. Bayer, *Justus Liebigs Ann. Chem.*, 901 (1975).

6. E. Bayer, M. Mutter, R. Uhmann, J. Polster, and H. Mauser, *J. Am. Chem. Soc. 96*, 7333 (1974).
7. M. Mutter and E. Bayer, *Angew. Chem. 86*, 101 (1974); *Angew. Chem., Int. Ed. Engl. 13*, 149 (1974).
8. E. Bayer and M. Mutter, *Chem. Ber. 107*, 1344 (1974).
9. M. Mutter, *Macromolecules 10*, 1413 (1977).
10. M. Mutter, F. Maser, K.-H. Altmann, C. Toniolo, and G. M. Bonora, *Biopolymers 24*, 1057 (1985).
11. M. Mutter, H. Mutter, and E. Bayer, in: *Peptides: Chemistry, Structure and Biology* (M. Goodman and J. Meienhofer, eds.), p. 403, Wiley, New York (1977).
12. E. Bayer, M. Mutter, and G. Holzer, in: *Peptides: Chemistry, Structure and Biology* (R. Walter and J. Meienhofer, eds.), p. 426, Ann Arbor Sci. Publ., Ann Arbor (1975).
13. R. B. Merrifield, *J. Am. Chem. Soc. 85*, 2149 (1963).
14. A. Warshavsky, R. Kaliv, A. Doshe, H. Berkovitz, and A. Patchornik, *J. Am. Chem. Soc. 101*, 4249 (1979).
15. E. Tsuchida, H. Nishide, N. Shimidazu, A. Yamada, and M. Keneko, *Makromol. Chem., Rapid Commun. 2*, 621 (1981).
16. E. Bayer, B. Hemmasi, K. Albert, W. Rapp, and M. Dengler, in *Peptides: Structure and Function* (V. J. Hruby and D. H. Rich, eds.), p. 87, Pierce Chem. Comp. (1983).
17. E. Bayer and W. Rapp, German Patent DOS 3714258 (1988).
18. E. Bayer, M. Dengler, and B. Hemmasi, *Int. J. Pept. Protein Res. 25*, 178 (1985).
19. E. Bayer and W. Rapp, in *Chemistry of Peptides and Proteins* (W. Voelter, E. Bayer, Y. A. Ovchinnikov, and V. I. Ivanov, eds.), p. 3, Walter de Gruyter, Berlin (1986).
20. E. Bayer, H. Hellstern, and H. Eckstein, *Z. Naturforsch. 42c*, 455 (1987).
21. W. Rapp, L. Zhang, R. Häbich, and E. Bayer, in *Peptides 1988* (G. Jung and E. Bayer, eds.), p. 199, Walter de Gruyter, Berlin (1989).
22. E. Bayer, K. Albert, H. Willisch, W. Rapp, and B. Hemmasi, *Macromolecules 23*, 1937 (1990).
23. Tentacle polymers are supplied as TentaGel^{O} by Rapp Polymere, Tübingen.
24. J. Ugelstadt and P. C. Mork, *Adv. Colloid Interface Sci. 13*, 101 (1980).
25. F. K. Hansen and J. Ugelstadt, *J. Polym. Sci. 16*, 1953 (1978); *17*, 3033 (1979).
26. E. Bayer and W. Rapp, unpublished.
27. J. Rudinger and P. Buetzer, in *Peptides 1974* (J. Wolman, ed.), p. 211, Wiley, New York (1975).
28. T. Hori, W. Rapp, and E. Bayer, Proceedings of the 30th Symposium on Chemistry of Dying, Osaka, Japan, p. 42 (July 1988).
29. T. Hori, W. Rapp, and E. Bayer, unpublished results.
30. E. Bayer, *Angew. Chem. 103*, 117 (1991); *Angew. Chem., Int. Ed. Engl. 30*, 113 (1991).
31. L. Zhang, W. Rapp, and E. Bayer, in *Peptides 1990* (E. Giralt, ed.), p. 196, Epson, Amsterdam.
32. H. Zeppezauer, W. Rapp, and E. Bayer, in preparation.
33. E. Pfaff, H.-J.Thiel, H.-O. Böhm, and J. Leban, *Arzneim.-Forsch. Drug Res. 37*, 82 (1987).
34. H. Küpper, W. Keller, G. Kurz, S. Forss, H. Schaller, R. Franze, K. Strohmaier, O. Marquardt, V. G. Zuslavsky, and P. H. Hofschneider, *Nature (London) 289*, 555 (1989).
35. S. Modrow and H. Wolff, *J. Immunol. Methods 118*, 1 (1989).
36. J. T. Sparrow, D. A. Sparrow, Z. Li Xin, M. Kovar, W. Li, and R. B. Arlinghaus, in *Peptides: Chemistry, Structure and Biology* (J. E. Rivier and G. R. Marshall, eds.), p. 714, Escon, Leiden (1990).
37. M. Flegel, D. Pichova, P. Minarik, and R. C. Sheppard, in *Peptides 1990* (E. Giralt, ed.), p. 837.
38. W. Rapp, E. Bayer, L. Zhang, A. Beck-Sickinger, K. Deres, K.-H. Wiesmüller, and G. Jung, in preparation.
39. E. Bayer and W. Rapp, presented at the 14th International Symposium on Column Liquid Chromatography, 20–25th May, Boston (1990).
40. J. Chibata, in *Immobilized Enzymes*, pp. 9–142, Wiley, New York (1978).
41. G. Manecke, E. Ehrenthal, and J. Schlünsen in *Characterization of Immobilized Biocatalysts* (K. Buchholz, ed.), p. 49, Schön u. Wetzel, Frankfurt (1979).
42. W. H. Scouten (ed.), *Solid Phase Biochemistry*, Wiley, New York (1983).
43. M. Zeppezauer, unpublished results.
44. W. Rapp, Ph.D. thesis, University of Tübingen, Germany (1985).
45. H. Schlecker, Ph.D. thesis, University of Tübingen, Germany (1989).
46. E. Bayer, W. Rapp, and H. Fritz, unpublished results.

21

Use of Functionalized Poly(Ethylene Glycol)s for Modification of Polypeptides

SAMUEL ZALIPSKY and CHYI LEE

21.1. INTRODUCTION

The unique properties of poly(ethylene glycol) (or PEG) and its general compatibility with polypeptide materials facilitated development of a variety of different applications of this polymer.[1–11] A marked proportion of these applications involve the use of covalently linked polypeptide–PEG adducts (reviewed elsewhere[5–11]). For example, a number of PEG-enzymes were shown to be useful as catalysts, soluble and active in organic solvents.[7] Due to the affinity to the upper phase of PEG/Dextran and PEG/salt two-phase systems, PEG-modified proteins were proven useful both as diagnostic tools[8] and in preparative separations of biological cells.[9]

Unquestionably, the properties of polymer–polypeptide conjugates *in vivo* were a significant reason for the substantial amount of research reported in the area during the last two decades. It has been repeatedly demonstrated that covalent attachment of multiple strands of PEG to proteins produces conjugates with dramatically reduced immunogenicity and antigenicity.[10] Such preparations also show great resistance to proteolytic digestion, and remain present in the bloodstream a considerably longer time than the parent polypeptides. These beneficial properties of modified polypeptides allowed development of a number of PEG-modified therapeutic proteins.[10] Slight to moderate modification of potent allergenic proteins with PEG can often be

SAMUEL ZALIPSKY and CHYI LEE • Enzon, Inc., South Plainfield, New Jersey 07080. *Present address for S. Z.*: Department of Chemistry, Rutgers–The State University of New Jersey, Piscataway, New Jersey 08855.

Poly(Ethylene Glycol) Chemistry: Biotechnical and Biomedical Applications, edited by J. Milton Harris. Plenum Press, New York, 1992.

sufficient to convert them into tolerance inducers and/or substantially reduce their allergenicity.[11]

Although, in some instances, hydroxyl end groups of the polymer can be used directly for covalent attachment of molecules of biological relevance,[5,12–14] in most cases suitable functionalization of the polymer prior to the conjugation is essential.[12]

Here we will review the methods for PEG functionalization and its covalent attachment to polypeptides. Attention will be given to comparison of different methods for preparation of PEG conjugates, properties of the linkages between the polymer and peptide components and their influence on biological/enzymatic activities of the conjugates. This review will not deal with formation of PEG–peptide conjugates that are built by stepwise addition of amino acid residues.[5] However, some of the methods for interlinking peptides and PEG, though originally devised for chemical and/or enzymatic protocols for peptide synthesis, are of general applicability and will be included in our discussion.

First, we will briefly go over the relevant properties of PEG itself, so that the rationale for its extensive use as a carrier for biological molecules will become apparent.

21.2. PROPERTIES OF PEG

Poly(ethylene glycol) is a polyether–diol with the general structure HO—$(CH_2CH_2O)_n$—H. For the purpose of modification of peptides and proteins the useful molecular weight range is 2,000–20,000 daltons.

The polymer has a wide range of solubilities. It is soluble in most organic solvents as well as in aqueous solutions.[5] For example, it is soluble in both water and dichloromethane, to such an extent that very concentrated solutions ($> 50\%$) can be prepared. Most polypeptides, as a result of their conjugation with PEG, in addition to retaining and in some cases enhancing their water solubility, also acquire solubility in some organic solvents.[7]

It was shown that ethylene oxide-based oligomers ($MW < 400$) can be oxidized *in vivo* into toxic diacid and hydroxy acid metabolites through a process initiated by alcohol dehydrogenase[15]; however, toxicity of PEGs of molecular weights above 1000 daltons is very low.[16] Extensive toxicity studies on PEG-4000 showed that this polymer can be safely administered intravenously in 10% solution to rats, guinea pigs, rabbits, and monkeys at a dose level of 16 g per kg body weight.[17] This corresponded to at least 1000-fold the amount of the polymer present in PEG-purified factor-VIII doses normally administered in humans. It was also reported that when administered intravenously to humans, PEGs of molecular weight 1000 and 6000 are readily excreted mainly via the kidney.[18] Biological activities of PEG conjugates are typically dominated by the non-PEG part of a molecule.[19]

The polymer by itself, even with a molecular weight as high as 5.9×10^6, is a very poor immunogen. PEG-recognizing antibodies (specific to 6–7 oxyethylene units) could be generated in rabbits by administration of PEG-modified allergenic proteins together with Freund's complete adjuvant.[20] In the absence of Freund's complete adjuvant, PEG-modified proteins do not induce formation of immunoglobulins against the polymer, and also reduce the response against antigenic determinants of a protein molecule.[10,21]

The polyether backbone of PEG is not degradable by mammalian enzymes. However, several types of bacteria can readily degrade the polymer up to Mn = 20,000. The topic of polyether biodegradation was recently reviewed by Kawai.[22]

The presence of PEG in protein solutions, even at high concentrations, in free or in conjugated form does not have any adverse effect on protein molecules.[2] The observation that a single point attachment of PEG to a peptide does not restrict the access of enzymes to a modified peptide residue was made by a number of investigators,[23–25] and was used for preparation of enzymatically altered polymer–peptide adducts. Generally, the conformation of peptides or proteins does not change as a result of covalent conjugation with PEG.[6,26]

It was recognized in one of the first papers dealing with PEG–proteins[27] that the key property of PEG as a protein modifier is its ability to bind molecules of water. It was also suggested[1] that the ability of the polymer to influence a structure of several molecular layers of water might be one of the causes of its appreciable exclusion effect, which might explain such phenomena as: (i) the ability of PEG to act as a protein precipitating agent,[2] (ii) repulsion of proteins from PEG-modified surfaces,[4] and (iii) reduced immunogenicity and antigenicity of PEG–protein conjugates.[10,21,27]

21.3. METHODS FOR COVALENT ATTACHMENT OF PEG TO POLYPEPTIDES

21.3.1. PEG Derivatives Reactive toward Amino Groups

Covalent attachment of PEG derivatives in the vast majority of cases has been achieved utilizing amino groups of polypeptide molecules as sites of modification. The first step in this process is substitution of the hydroxyl end-groups of the polymer by electrophile-containing functional groups. This process is often referred to as "activation." Derivatives of monofunctional polymers, capped on one end by methyl ethers (mPEG), were usually the reagents of choice for protein modifications.[10–12] Such modifications are expected to be free from crosslinking and result in the attachment of multiple strands of the polymer to the globular polypeptide core. A number of popular ways to activate the polymer are shown in Scheme 1.

21.3.1.1. Most Commonly Used PEG-Based Reagents

In the original work of Davis and co-workers[27] trichloro-*s*-triazine (cyanuric chloride) was reacted with the primary alcohol groups on PEG so that only one of the chlorides is displaced from the triazine ring (**1**) and the remaining chlorides can be used for subsequent reaction with the amino groups of a protein. This approach was adopted by other investigators[8,9,28–30] and is still one of the most popular. The synthesis of **1** was recently optimized to assure the reproducible and complete

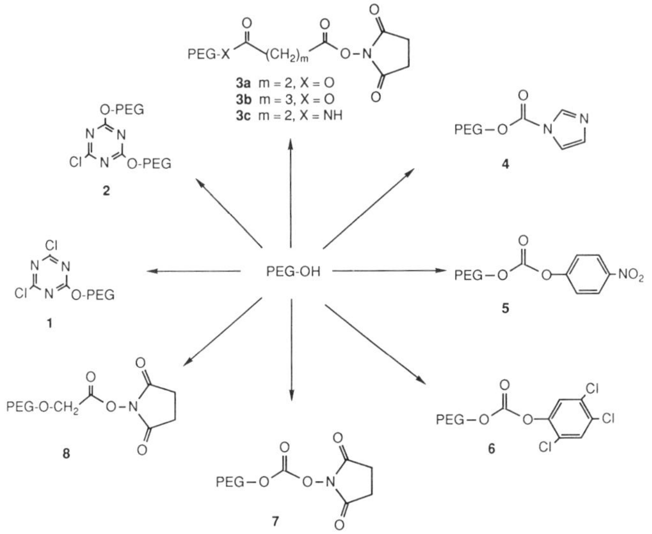

Scheme 1. Commonly used methods for activation of polyethylene glycol. For the purpose of this and the following schemes, the abbreviation PEG refers to the structures: RO—$(CH_2CH_2O)_n$—CH_2CH_2— or —$(CH_2CH_2O)_n$—CH_2CH_2—, where R is a simple alkyl residue, usually CH_3.

conversion of mPEG–OH to **1** as was proven by titration of chloride, ^{13}C-NMR, GPC, and elemental analysis.[31] Although this approach provides a very effective one-step activation of the polymer, it suffers from various disadvantages such as toxicity of cyanuric chloride derivatives and excessive reactivity of **1** toward nucleophilic functional groups other than amines (e.g., cysteinyl and tyrosyl residues). In fact, cyanuric halides are considered to be the least suitable reagents for selective protein modification.[32] Consequently, modification of some proteins with **1** is often accompanied by a substantial loss of biological activity.[29,33–35]

A variation of the cyanuric chloride method in which two of the three chlorides originally present on the trichloro-*s*-triazine molecule are displaced by mPEG chains (**2**), and the remaining chloride used for the reaction with amino groups of a protein, was developed by Inada and co-workers.[7] Unfortunately, the communications describing the synthesis of **2** failed to provide proof of the structure of this activated polymer.[36–38] A recent report from the laboratory of Sehon[39] convincingly demonstrated lack of reactivity of the third Cl group on the triazine ring toward ethanolamine and ovalbumin, which places serious doubt as to whether 2,4-bis-(mPEG-*O*-)6-chloro- *s*-triazine (**2**) is capable of reacting with proteins. In light of this conflicting evidence, the issue of reactivity of **2** clearly needs to be clarified. It is interesting to mention that reactivity toward sulfhydryl groups reported for **1** by Wieder *et al.*[33] was also found for reagent **2**.[40]

The shortcomings of cyanuric chloride activation were overcome by using PEG-succinimidyl succinate (SS–PEG, **3a**). This form of activated PEG was first used for crosslinking of proteins[41] and synthesis of peptides substituted by PEG chains at their terminals[42] but was later adapted for preparation of mPEG–protein conjugates.[43] The reagent (**3a**) is usually prepared by succinylation of the terminal hydroxyl groups of PEG[14] followed by dicyclohexylcarbodiimide-mediated condensation with *N*-hydroxysuccinimide.[42,43] It reacts with proteins within a short period of time under mild conditions (30 min, pH < 7.8, 25 °C), producing extensively modified proteins with well-preserved biological activities.[26,43–45] The ester linkage between the polymer and the succinic ester residue has limited stability in aqueous media,[11,25,46] which causes slow hydrolytic cleavage of PEG chains from SS–PEG-modified proteins under physiological conditions. Substitution of the succinate residue in **3a** by glutarate (**3b**)[47] produced an activated form of the polymer very similar to SS–PEG, but with slightly improved resistance to hydrolysis of the ester linkage. Replacement of the aliphatic ester in **3a** by an amide bond (**3c**) improves the stability of the reagent and its conjugates even further.[46,48] However, preparation of **3c**, starting from hydroxyl terminated mPEG, involves 4 to 5 synthetic steps.[48]

Formation of urethane (carbamate) linkages between the amino groups of a protein and PEG overcomes the problem of hydrolytic release of the polymer chains. In fact, it was demonstrated on radioactively labeled PEG derivatives that urethane linkages are completely stable under a variety of physiological conditions.[49] The attachment of mPEG chains to polypeptides through carbamate linkages was first accomplished using imidazolyl formate derivatives of the polymer (**4**).[50] Several versions of a one-step synthesis of **4** using carbonyldiimidazole were reported.[35,51] Complete transformation of the hydroxyl groups of PEG was confirmed by both elemental analysis and NMR data. The polymer activated in this manner has a rather mild reactivity, and therefore long reaction times were required for protein modification (48–72 h, pH 8.5). However, good preservation of activity in these modified enzymes was usually observed.[35,50,52,53]

The products of protein modification using phenylcarbonates of PEG (**5** and **6**) also have PEG chains grafted onto the polypeptide backbone through urethane linkages.[54] Single-step activation protocols with commercially available chloroformates of 4-nitrophenol and 2,4,5-trichlorophenol were used for the preparation of **5** and **6**, respectively. These activated PEGs seem to react faster than **4**. However, both 4-nitrophenol and 2,4,5-trichlorophenol produced during the PEG- attachment process are toxic and hydrophobic molecules with affinities toward proteins.

Succinimidyl carbonates of PEG (SC–PEG, **7**), developed in our laboratory, constitute a further improvement in urethane-forming PEG reagents.[55,56] SC–PEG and its bifunctional analogs (BSC–PEG) of different molecular weights were prepared in a one-pot procedure as shown by Scheme 2. Polymeric chloroformate, generated *in situ* by treatment of PEG with phosgene, was reacted with *N*-hydroxysuccinimide (HOSu). Purified preparations of the SC-activated polymers were determined to contain the theoretical amounts of active carbonate groups. To estimate the reactivity of SC–PEG and to compare it to SS–PEG, measurements of hydrolysis and aminolysis rates of both activated polymers derived from mPEG-5000 were performed. The results of these experiments showed that SC–PEG is slightly less

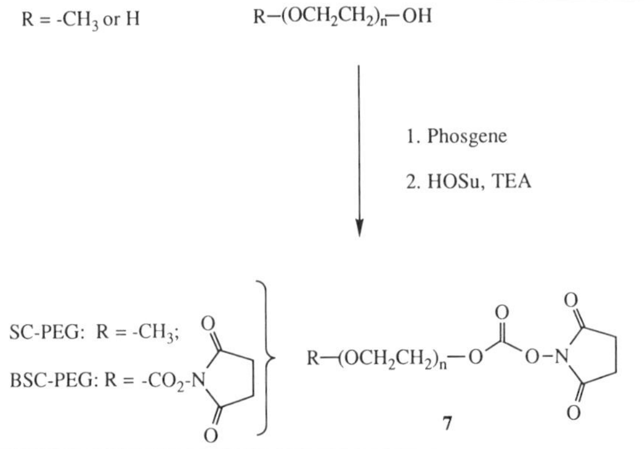

Scheme 2. Preparation of succinimidyl carbonate derivatives of PEG. *Abbreviations:* HOSu, *N*-hydroxysuccinimide; TEA, triethylamine.

reactive yet a more selective reagent. Protein modification reactions with **7** can be performed within short periods of time ($\approx$ 30 min) over a broad pH range, with the highest reactivity at pH $\approx$ 9.3. The HOSu released during polypeptide modification is a nontoxic material that is often used in bioconjugate and peptide chemistries as a leaving group residue. Our experience was that, unlike 4-nitrophenol and 2,4,5-trichlorophenol mentioned above, HOSu does not show affinity toward proteins and can be readily removed from the reaction solution by either dialysis or diafiltration. Under appropriate conditions BSC–PEG can be used as a homobifunctional cross-linker of proteins. It is also useful as a reactive macromonomer in polycondensation with diamines.[57]

The activated polymer **8** was prepared by carbodiimide-mediated esterification of carboxymethylated PEG with HOSu.[48,58,59] Carboxymethylation of PEG is one of the most straightforward ways to introduce carboxylic acid end groups onto the polymer. This is best accomplished by nucleophilic displacement of the bromide in ethyl bromoacetate with a PEG alkoxide, followed by saponification of the ester.[23,58–60] An alternative way to essentially the same derivative is by oxidation of the terminal hydroxyls of PEG,[48,61] but this process is often accompanied by polyether backbone degradation. A number of enzymes were modified using **8** with very good preservation of specific activity.[48,62,63] It was also used for modification of human hemoglobin.[46,59] The reactivity of **8** was reported to be one order of magnitude higher than that of **3**, as was estimated by hydrolysis ($t_{1/2} \approx 1$ min at pH 7.5, 27 °C) and aminolysis rates of the activated mPEGs.[64] The increased reactivity was explained by the presence of electron-withdrawing oxygen in one carbon-proximity to the carbonyl of the active ester in **8**. This result, combined with our own data (on SS– and SC–PEG)[55] and other literature sources,[54] allows us to estimate the order of reactivity of active acyl-bearing PEGs (assuming equality of all other variables, such as the molecular weight of the reagents) shown on Scheme 1:

$$\mathbf{8} >> \mathbf{3a} \approx \mathbf{3c} > \mathbf{3b} \geqslant \mathbf{7} > \mathbf{5} > \mathbf{6} > \mathbf{4}$$

One has to bear in mind that the higher the reactivity of a reagent, the less likely it is to be selective, and consequently the higher is the probability of side reactions.

21.3.1.2. Less Commonly Used PEG Reagents for Modification of Amino Groups

In some instances another activated form of the polymer, PEG-carboxymethyl azide (**9**), was generated *in situ* from PEG–carboxymethyl hydrazide and then reacted with proteins according to the reaction shown in Scheme 3.[21,41,65] Normally, the slow reacting acyl azide functionality gets an additional boost of reactivity due to the stronger electrophilicity of the carboxymethyl residue.

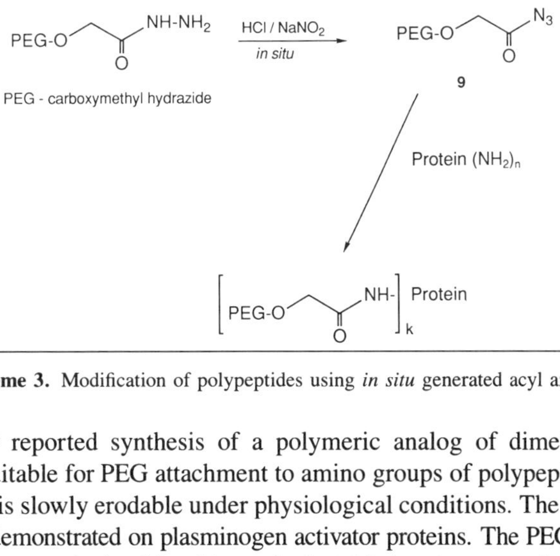

Scheme 3. Modification of polypeptides using *in situ* generated acyl azide.

Garman[66] reported synthesis of a polymeric analog of dimethylmaleic anhydride (**10**) suitable for PEG attachment to amino groups of polypeptides through a linkage which is slowly erodable under physiological conditions. The validity of this approach was demonstrated on plasminogen activator proteins. The PEG–plasminogen activator conjugates obtained by this method yielded active, completely deacylated proteins after 44 h incubation at pH 7.4, 37 °C.

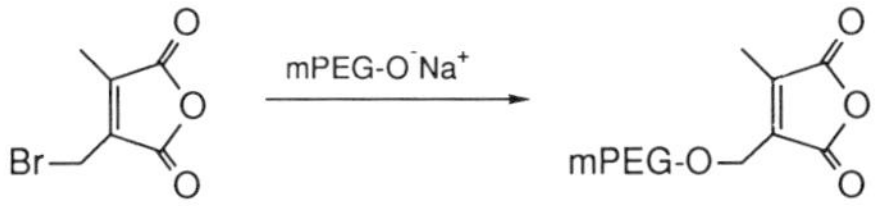

10

Polyethylene glycols can be quantitatively converted into dithiocarbonate (xanthate) derivatives,[67,68] **11**, which were shown to be useful for grafting PEG chains onto proteins via thionourethane linkages (Scheme 4). The composition of PEGoxythiocarbonylated proteins could be conveniently determined spectrophotometrically from the increase in absorption at 242 nm as compared with that of native protein. The conjugates showed good chemical stability in a variety of aqueous buffers. Because of the mild reactivity of **11**, the modification reactions with bovine serum albumin and

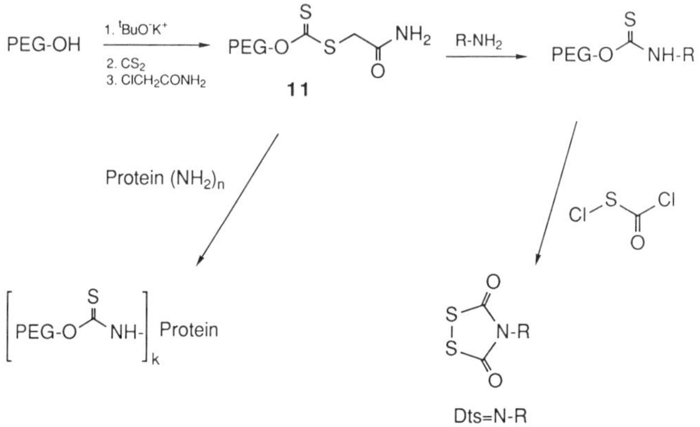

Scheme 4. Preparation and use of PEGoxy-*S*-carboxamidomethyldithiocarbonate. R represents a peptide or an amino acid residue.

ragweed antigen E were carried out at pH range 9–10 for ≈ 20 h. (All the common free amino acids reacted quite readily with **11** within 30 min.) The xanthate **11** derived from bifunctional PEG-2000 proved useful as a polymeric reagent for the introduction of dithiasuccinoyl (Dts) protecting group into amino acids and dipeptides following the sequence of reactions shown on Scheme 4.[68]

Esters of organic sulfonic acids and PEG, *p*-toluenesulfonates (tosylates) in particular, have been useful as starting materials for the preparation of a variety of functionalized polymers.[68,69] Tresylates (2,2,2-trifluoroethanesulfonates) of PEG (**12**, PEG–$OSO_2CH_2CF_3$) were shown to be sufficiently reactive toward amino groups (≈ 100-fold more reactive than tosylates) to be considered useful as protein modifying reagents.[70] Modification of bovine serum albumin (BSA) with mPEG-tresylate in 16-fold excess to the amino groups, performed at pH 7.5, was complete in approximately 1 h and resulted in attachment of mPEG to 18 amino groups per protein molecule.[71] The comparison of these results to our own data on BSA modifications[55,56] using **3** or **7** leads us to believe that PEG–tresylate is the less reactive reagent. As a result of protein modification with PEG– tresylate, the polymer chains become grafted onto the polypeptide through very stable secondary amine bonds. There are two important consequences to this: (i) the total charge of the protein does not change in the process of modification; (ii) the modified proteins could be conveniently characterized by quantitation of lysine by amino acid analysis. The chemical composition of such conjugates can also be determined by the fluorescamine assay, which specifically measures primary amines and is not interfered with by the presence of secondary amino groups.[72] The assay based on the use of trinitrobenzene sulfonate (TNBS), which was extensively used for determination of the extent of modification in PEG–proteins,[73] theoretically should not be suitable in the case of PEG–tresylate-modified proteins. TNBS reacts with both primary and secondary amines as well as with other nucleophiles.[32] In light of this, it is surprising that the extent of modification of alkaline phosphatase modified with **12** was determined by TNBS assay.[35]

Although amino acid analysis has been used routinely for characterization of PEG–peptides,[5] it has only recently been recognized as a powerful analytical tool for characterization of PEG–proteins.[28,74,75] A single amino acid analysis run of a protein conjugate can provide several valuable pieces of information: (i) protein concentration, (ii) detection of possible side reactions that may occur during protein modification, and (iii) determination of the conjugate composition, provided that the linker is designed for this purpose. For example, the polymeric active ester **13**

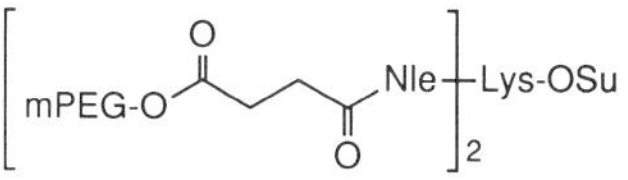

13

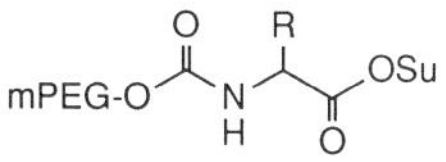

14 R - side-chain residue of an amino acid

composed of two mPEG-5000 chains and containing reference amino acid norleucine (Nle) was specifically designed for convenient determination of the amount of PEG in its conjugates by amino acid analysis.[74] The amount of Nle determined in hydrolysates of proteins modified with **13** provides an accurate measure of the extent of modification. A similar rationale was used with regard to activated PEGs of general structure **14**.[75] Synthesis of PEGoxycarbonyl–amino acids, needed for preparation of **14**, can be readily accomplished by using active carbonates **5–7**[75] or by reacting isocyanato derivatives of amino acids with hydroxyls of PEG–OH.[68,76]

Imidoesters of mPEG (**15**) useful for protein modification were recently described in the patent literature.[77] These activated polymers were prepared by acid-catalyzed methanolysis of mPEG–cyanolkylethers and reacted with amino groups of proteins at pH 7–9 (Scheme 5). Making use of the fact that imidyl linkages survive the conditions of acidic protein hydrolysis, the extent of mPEG-amidination was determined by diminished lysine content in the conjugates. The advantage of this approach to protein modification is that the amidinated proteins possess the same net charge as the native ones.

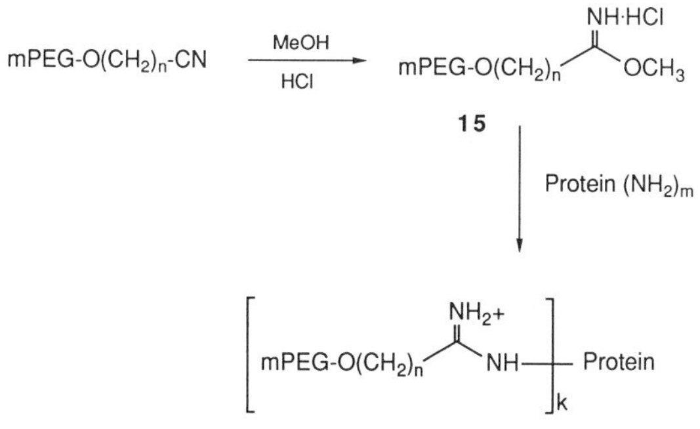

Scheme 5. Preparation and use of mPEG–imidoesters.

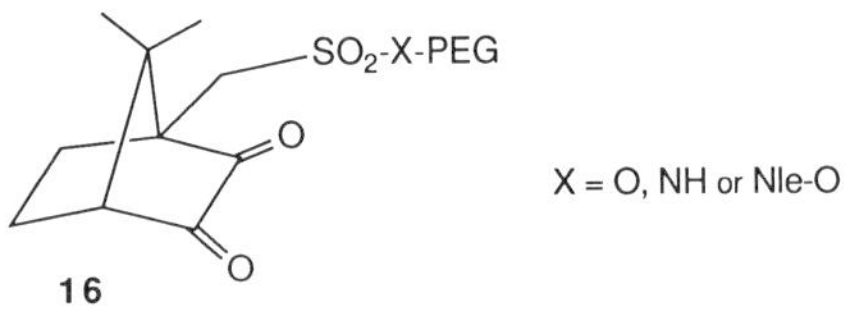

16

21.3.2. PEG Reagents Reactive toward Arginine Residues

Polymeric derivatives **16** for selective attachment of PEG to arginine residues were introduced to serve as carriers during semisynthesis of peptides in aqueous solutions.[78] The reagents were readily obtainable by reaction of camphorquinone-10-sulfonylchloride with PEG–OH, PEG–NH_2, or preferably with the norleucine ester of the polymer. The modified peptides were formed in borate buffer at pH 9.0, at 37 °C. They were stable to a variety of acidic and nucleophilic reagents, including hydroxylamine, the recommended agent for cleavage of cyclohexadione–arginine adducts. Active peptides could be released from the polymeric carrier by *o*-phenylenediamine. When norleucine was incorporated as a spacer between the camphorquinone moiety and the polymer, the extent of arginine modification was conveniently determined by amino acid analysis of hydrolysates of the PEG–peptides.

A variety of proteins were modified with a mPEG-5000 analog of phenylglyoxal (**17**),[79] a well-known reagent for modification of guanidino groups. The synthesis of **17** was performed according to the reactions depicted on Scheme 6. Readily obtainable mPEG–tosylate[68,69] was subjected to nucleophilic displacement by 4-hydroxyacetophenone, which was oxidized by selenium dioxide to yield mPEG–phenylglyoxal. Attachment of **17** to proteins through arginyl residues proceeded at pH range 5.5–9.3, at room temperature, and was measured by the decrease of arginine content in hydrolysates of modified polypeptides. The conjugates of **17** were claimed to possess very good stability and biological activity.

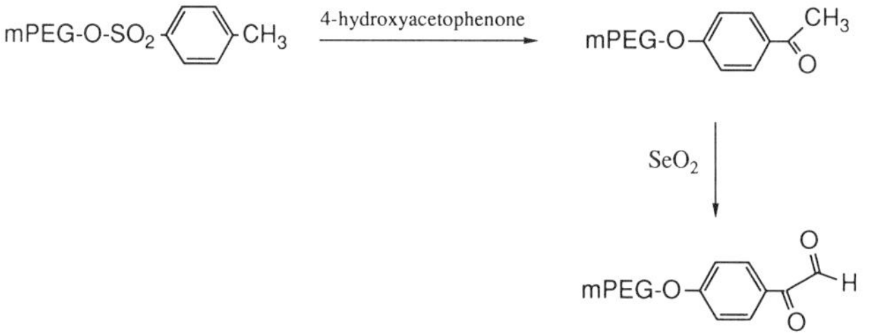

17

Scheme 6. Preparation of mPEG–phenylglyoxal.

21.3.3. PEG Reagents for Modification of Cysteine Residues

Maleimide-containing reagents are effective modifiers of free cysteinyl residues of polypeptides.[32] Preparation of PEG derivatives (**18**) bearing maleimido functional groups was reported using readily accessible[14,58] PEG–NH_2 and active esters of 6-maleimidohexanoic[80] and 3-maleimidopropionic[81] acids as starting materials (Scheme 7). Direct conversion of amino–PEG into a maleimido–PEG using maleic anhydride was also described.[21] A mutant protein of interleukin-2, in which cysteine at position −3 replaced the naturally occurring glycosylated threonine residue, was coupled with **18** (n = 5) to produce a well defined, fully bioactive PEG–polypeptide.[80] The reagent **18** (n = 2) was used for selective entrapment of free thiol-containing peptides on the polymer, in order to simplify purification of unsymmetrical cystine-peptides from a reaction mixture.[81]

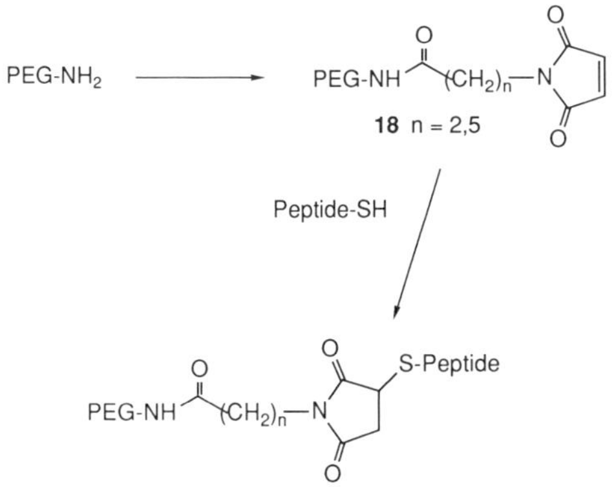

Scheme 7. Preparation and use of PEG–maleimides.

Glass and co-workers reported preparation of 4-phenoxy-3,5-dinitrobenzoylPEG (**19**).[24] This PEG derivative reacted rapidly and selectively at neutral pH with

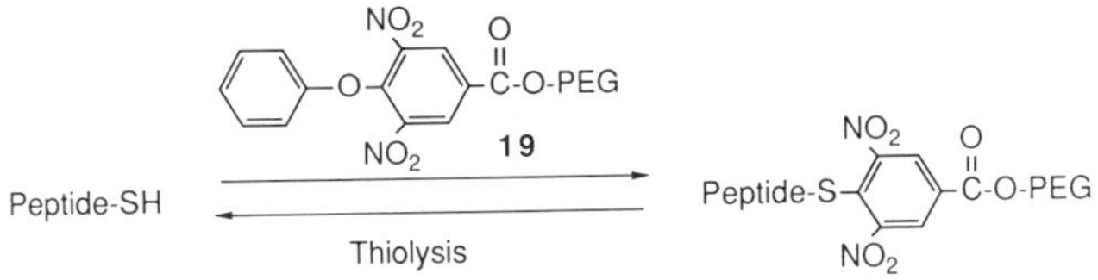

sulfhydryl groups of peptides to yield polymer–peptide adducts in which the components were linked by a thiol-sensitive dinitrophenylene linker. Thus it was possible to attach a peptide to the polymer and, after performing some chemical and/or enzymatic alterations on the conjugate, release the derivatized peptide from the PEG carrier.

21.3.4. Coupling of PEG Derivatives to Carboxylic Groups of Polypeptides

Amino–PEG was reacted with 1-ethyl-3-(3-dimethylaminopropyl)carbodiimide-activated carboxyl groups of trypsin and other proteins.[21] The selectivity of such reaction is rather poor, due to the fact that amino–PEG has similar reactivity to the lysiyl residues of proteins. In another variation of this reaction *p*-aminobenzylether of PEG was coupled to carboxyl groups of D-glucose 6-phosphate dehydrogenase by treatment with 1-ethyl-3-(3-dimethylaminopropyl)carbodiimide.[82] The crude conjugate retained 22% of the original enzymatic activity. The aromatic amino groups on the polymer have a lower pK_a than the amino groups of lysine residues and therefore would be expected to exhibit higher reactivity toward activated carboxyls under the acidic conditions of such reactions (pH 4.8–6.0). In both types of modification involving aliphatic and aromatic amino–PEGs, the composition of the conjugates was not determined.

In a recent report[83] bilirubin oxidase was modified by a two-step protocol. First, 1,4-diaminobutane was coupled to the carboxyl groups using water-soluble carbodiimide. Then, both the newly introduced amino groups and the original amines of the protein were modified with reagent **2**. The activity of PEG–bilirubin oxidase was approximately 30% that of the native enzyme. Unfortunately, the composition of the conjugate was not reported. When using carbodiimide-mediated coupling of an appropriate nucleophile to a protein, one has to be aware of possible side reactions, such as modification of tyrosyl and cysteinyl residues and the formation of *N*-acylurea derivatives.[32]

21.3.5. Coupling of PEG to Oligosaccharide Residue of Glycoproteins

We found only one reported case of this type of protein modification. In this example,[84] the specific reactivity of a carbohydrate residue of horseradish peroxidase was used for anchoring PEG chains. Mild oxidation of the carbohydrate portion of the enzyme molecule with sodium periodate resulted in the formation of six aldehyde groups. Bifunctional amino–PEG of molecular weight 20,000 was reacted with the oxidized peroxidase to form Schiff base links which were reduced by sodium borohydride. The conjugate produced retained 91% of the original specific activity and contained on average three PEG chains per glycoprotein molecule. In addition to water, it was soluble and active in tetrachloromethane, toluene, chloroform, and dimethylformamide.

21.4. RELATIONSHIP BETWEEN COUPLING CHEMISTRY AND BIOLOGICAL ACTIVITY OF PEG–POLYPEPTIDE CONJUGATES

Only a few attempts have been made to address the issue of the interrelationship between the chemistry of conjugation or activation of PEG and the biological activity of a particular PEG–polypeptide conjugate.[35,45,53,67,85] Therefore, in order to overview this topic we had to compare data published by a number of different research groups. We compiled the characteristics of selected PEG–protein conjugates in Table 1, showing the different functionalized polymers that were used to modify each protein. While the data in Table 1 are presented as reported in the primary sources, the methods used by the various authors to determine the chemical composition of the modified proteins were often different. For example, several variations of TNBS assay, often used to determine the extent of modification of proteins, had been employed by the researchers. While most authors used the procedure of Habeeb,[73] which measures the total number of amino groups on a protein, in some cases[50,52,54,62,87] other versions of the TNBS assay, yielding a measure of the readily accessible amino groups only, were employed. The reader also has to be aware that in many cases conditions for modification reactions and design of biological/enzymatic assays also differed from one laboratory to another. Despite the above-mentioned drawbacks, several conclusions can be drawn from the data summarized in Table 1, which also illustrates some of the interesting properties and applications of PEG–polypeptide conjugates.

Proteins modified with cyanuric chloride activated PEGs (**1** and **2**) almost always possessed lower enzymatic activity than the same enzymes modified using alternative chemistries. This is most likely due to excessive reactivity and thus lack of selectivity of these reagents, which results in modification of nucleophilic groups other than amines. The pattern of asparaginase inactivation as a result of exposure to **1** or **2** provides the clearest illustration of this phenomenon and is consistent with the known reactivity of cyanuryl halides toward tyrosyl residues.[32] Inactivation of asparaginase by tyrosyl-modifying reagents is well documented.[102]

King *et al.* reported that PEG conjugates of Antigen E obtained via use of **1** had approximately one order of magnitude lower antigenic activity compared to those conjugates derived from **11**,[28,67] even though about the same numbers of amino groups were modified with both reagents. One might speculate that attachment of mPEG chains to nucleophilic residues other than lysyls, in the case of reagent **1**, could be the reason for the substantial difference in the antigenic activities. The lower antigenicity could actually be advantageous for potential therapeutic use in allergic patients, since greater amounts of the modified allergen may be used safely.

Yoshinaga and Harris[35] examined the activity of PEG–alkaline phosphatase conjugates obtained by four different coupling methods. The best preservation of activity was observed when SS–PEG (**3a**) was used as a modifying reagent, and only **1** caused substantial loss of enzymatic activity. The same research group examined alkaline phosphatase activity of conjugates obtained using cyanuric chloride-activated mPEGs and a number of its bifunctional analogs.[86] Interestingly, in contrast to the pattern observed with cyanuric chloride-activated mPEGs, protein modifications with the bifunctional PEGs resulted in better preservation of enzymatic activity which was independent of the extent of the modification or the molecular weight of the polymeric reagent used. It is pertinent to note that alkaline phosphatase in its active form is known to be a dimeric enzyme,[103] present in equilibrium with the only slightly active monomeric form. Therefore, the reported improved preservation of enzymatic activity could be attributed to partial fixation of the active dimeric form of the enzyme by the crosslinking of two monomers of alkaline phosphatase by the activated

Table 1. Chemical and Biological Characteristics of Selected PEG–Protein Conjugates

Protein	Active PEG (MW)[a]	No. of PEGs per protein (% modified amino groups)	% Native enzyme activity (substrate)	Excerpts/applications	Ref.
Adenosine deaminase (ADA)	**1**	9(40)	28	No reaction between PEG-ADA and the antibodies raised against native ADA. PEG–ADA may be suitable for treating human ADA deficiency because of long circulating life and the lack of detectable antibody formation in mice.	34
	3a	—	—	PEG–ADA was used successfully for the treatment of two children suffering from adenosine deaminase deficiency. Neither toxic effects nor hypersensitivity reactions were observed.	44
	4	19(85[g])	76	Long circulating half-life and significant retention of enzymatic activity in mice. However, no reduction of the immunogenecity was reported.	52
	7	14(65[e])	51	Used as a model for evaluation of the activated PEG.	56
Alkaline phosphatase	**1**	19(88)	33	The reagent **1** caused significant loss of enzymatic activity. Best enzymatic activities were obtained in conjugates derived from **3a**.	35
		14(62)	67		
	3a	17(79)	93		
		13(61)	98		
	12	17(77)	86		
		16(73)	82		
	4	17(78)	77		
		12(56)	91		
	1	5(23)–18(82)	66–44	Modification with higher molecular weight mPEG gave more deactivation than did modification with the lower molecular weight mPEG. Modification with bifunctional PEG gave highly active protein conjugates and there was little dependence on molecular weight or degree of modification.	86
	1(m1900)	1(5)–19(86)	97–55		
	1(4000)	9(41)–17(77)	72		
	1(8000)	9(40)–16(72)	70		
	1(20000)	12(54)–18.5(84)	80		

Antigen E	**1**	8[f](44)	1.0[h]	PEG–antigen E with reduced allergenic activities, yet retaining the immunogenic properties of antigen E. Substantially lower antigenicity was observed for conjugates obtained with reagent **1**.	28
	1(m2000)	7[f](39)	2.4[h]		67
	11	5.6[i](31)	25[h]		
		7.8[i](43)	17[h]		
	11(m2000)	8.0[i](44)	15[h]		
Arginase	**6**	55(60)[g]	90	The PEG–enzyme had an increased structural stability, a decreased digestion by proteolytic enzymes, and an expanded clearance time in rats. It is potentially useful for the therapy of arginine-dependent tumors or of familial hyperargininemia.	87
	1	49(53)	65	Antisera from mice injected with native arginase reacted against arginase but not against PEG–arginase. In addition, antisera from animals injected with PEG–arginase reacted with neither native arginase nor PEG–arginase.	88
	7	56(61)	71	Used as a model for evaluation of the activated PEG.	56
Asparaginase (*E. coli*)	**1**	73(79)	7(Asn) 15(ANA)	The modification of asparaginase with PEG(m5000) showed the reduction of the antigenicity and had a resistivity against trypsin. Interestingly, the asparaginase modified with PEG of 750 and 1900 daltons did not show a substantial change of the immunogenic properties.	29
	1(m1900)	70(76)	14(Asn) 25(ANA)		
	1(m750)	77(84)	12(Asn) 20(ANA)		
	2	52(57)	8(Asn)	The PEG–asparaginase conjugate has the same enzymatic properties (K_m value and optimal pH) as the native enzyme. The half-lives of the modified and native enzymes in mice were 56 and 2.9 h, respectively.	89
	2	30–48 (33)–(52)	30–11 (Asn)	PEG–asparaginase conjugate retained 11% of the enzymatic activity but showed no binding activity toward anti-asparaginase serum.	36
	3a	59(64)	52(Asn)	PEG–asparaginase conjugates were active, stable, without immunosuppressive effect, and with extended half-lives in mice.	43
	7	62(68[e])	54(Asn)	Used as a model for evaluation of the activated PEG.	56
	8[c](4500)	—	8(Asn)	The modified asparaginase contained magnetite, which was used for facile removal of PEG–asparaginase from aqueous solution.	90

(continued)

Table 1. (*Continued*)

Protein	Active PEG (MW)[a]	No. of PEGs per protein (% modified amino groups)	% Native enzyme activity (substrate)	Excerpts/applications	Ref.
Chymotrypsin	**1**	16(95)	5(ATEE) 0(GFNA)	PEG–chymotrypsin retained catalytic activity and had increased solubility in organic solvents. It is useful for improving the coupling yield of peptide fragments and avoiding the risk of racemization.	85
		13(75)	25(ATEE) 10(GFNA)		
	5	9(55)	35(ATEE) 25(GFNA)		
	8	13(75)	50(ATEE) 30(GFNA)		
	5(m2000)	8.5–17 (50–100[g])	75–69 (BTNA)	The PEG–chymotrypsin was found to be soluble in benzene and DMF and catalyzed transesterification in cyclohexane. The enzymatic activity in organic solvents was decreasing with increasing the extent of modification and ranged 34–171% of the native activity.	91
	6(m2000)	2–7 (11–40[g])	77–64 (BTNA)		
	2	12.5(83)	57(ATEE)	The yields of ester and amide formations both were 90% in 1,1,1-trichloroethane and the reaction rate was linearly enhanced with increasing amount of the PEG–chymotrypsin conjugate.	92 38
	1	8(45)	60(BTNA)	Substrate specificity of PEG–chymotrypsin in organic solvents was altered since arginine and lysine esters were found to be as effective as substrates as derivatives of aromatic amino acids.	30
	7	9(54[e])	131(BTEE) 151(BTNA)	Used as a model for evaluation of the activated PEG.	56
		14(82[e])	122(BTEE) 161(BTNA)		
Gulonolactone oxidase	**3a**	18(47[e])	74	The modified enzyme was found to be more stable at 37 °C than native enzyme. The PEG–enzymes retained immunogenicity and reacted with preformed antibodies. The circulating half-life of the modified enzyme was not extended.	45
	1	15(38[e])	67		

Lipase (*Pseudomonas fragi*)	**2**	(49)	43[b]	The PEG–lipase catalyzed ester synthesis in organic solvents. The highest activity was observed in 1,1,1-trichloroethane.	93
	8[c](4500)	—	59–15[b]	The modified enzyme contained magnetite, which was used for convenient removal of PEG–lipase from reaction mixtures. The preparation was used for ester synthesis.	90
(*Pseudomonas fluorescens*)	**2**	(55)	80[b]	PEG–lipase catalyzed ester synthesis, transesterification and aminolysis reactions in organic solvents.	94
	2	(52)	67[d]	PEG–lipase was used in organic solvents with indoxyl acetate as a substrate; it was possible to determine the Michaelis-Menten constants for water.	95
	2	(60)	70[b]	Two types of lipase were modified with two different reagents (**2** & **8**). In contrast to the case of *Pseudomonas fluorescens*, the enzyme from *Candida cylindracea* when modified with **8** catalyzed ester synthesis form short-chain alcohols and α- or β-substituted carboxylic acids in benzene.	96
(*Candida cylindracea*)	**8**(m4500)	(47)	56[b]		
	8(m4500)	(95)	68[b]	The PEG–lipase catalyzed ester exchange reaction between dipalmitoyl phosphatidylcholine and eicosapentaeuoic acid.	63
Superoxide dismutase (SOD) (bovine erythrocyte)	**4**	18 or 19 (90 or 95)	>95	SOD coupled to PEG increased its plasma half-life from 3.5 minutes to 9 or more hours depending on the PEG derivative studied.	50
	8	3(15)–18(90)	90–72	SOD–PEG conjugate showed longer half-life in rats than native SOD.	48
	3c	3(13)–18(90)	90–70		
	1	12(60)–14(70)	100	The PEG-modified enzyme increased cellular enzyme activities and provided prolonged protection from partially reduced oxygen species.	97
	3a	—	100		
	1	19(95)	51	No evidence of an immune response to repetitive injections of PEG–SOD was observed.	98
	6	10(50[g])	80	The PEG-phenylcarbonate derivatives were stable in neutral aqueous solution and were reactive enough to modify proteins extensively in reasonable time periods.	54
	5	—	—		
	8	16(82[g])	75	No structural modification occurred at the metal active site region and, in fact, the metal binding was higher in PEG–SOD than native SOD. The biological life of PEG–SOD decreased in the order i.v.> i.p.> i.m.> s.c.	62
	3a	11(57)	47	PEG-modified SOD resulted in high heterogeneity and substantial changes in isoelectric point and hydrophobicity.	99

(continued)

Table 1. (Continued)

Protein	Active PEG (MW)[a]	No. of PEGs per protein (% modified amino groups)	% Native enzyme activity (substrate)	Excerpts/applications	Ref.
Superoxide dismutase (SOD) (*cont.*)					
(Human)	**17**	2.1 or 2.5		Arginine residues of SOD were selectively modified.	79
(Serratia)	**2**	5(24)	52	The PEG (m5000) modified SOD showed enhanced anti-inflammatory activities and radioprotective effects in mice.	100
	2(m1900)	10(48)	41		
	2(m750)	10(48)	41		
	2(m350)	10(48)	39		
Tissue plasminogen activator	**4**	10(44)	80	The reaction of rt-PA with activated-PEG **4** was much slower than the reaction with activated-PEG **3a**. The conjugate of PEG-rt-PA has a potential to be used as thrombolytic agent in human.	53
		13(60)	30		
		16.5(75)	20		
	3a	12(55)	36		
		14.5(66)	14		
		22(100)	0		

	10	13(59)	50–70	Reversible conjugation of t-PA has been achieved with the PEG-contianing maleic anhydride reagent.	66
	17	8(36) or 9(41)		Arginine residues of t-PA were selectively modified.	79
Trypsin	**1**	4(24) 9(59)	95(BAEE) 150(BAEE)	Proteolytic activity of the conjugate was markedly reduced. PEG–trypsin conjugate (59% modified) dissolved soft blood clots at one-fourth the rate of trypsin.	101
	7	7(46[e])	95(BAEE) 224(ZAPA)	Used as a model for evaluation of the activated PEG.	56
		12(78[e])	92(BAEE) 326(ZAPA)		
	8	12.5(83)	110(BAEE)	Used as a model for evaluation of the activated PEG.	48

Abbreviations: ANA, Aspartic acid β-*p*-nitroanilide; Asn, Asparagine; ATEE, Acetyl tyrosine ethyl ester; BTEE, N^{α}-benzoyl-L-tyrosine ethyl ester; BTNA, N^{α}-benzoyl-L-tyrosine-*p*-nitroanilide; GFNA, Glutaryl phenyl alanine *p*-nitroanilide; ZAPA, N^{α}-benzyloxycarbonyl-L-arginine-*p*-nitroanilide; rt-PA, Recombinant tissure plasminogen activator; i.v., intravenous; i.p., intraperitoneal; i.m., intramuscular; s.c., subcutaneous.

[a]Molecular weight given only for PEG derivatives different from mPEG-5000. Letter m appearing in parentheses prior to the number indicates mPEG derivative.

[b]Hydrolysis of olive oil in emulsified aqueous system.

[c]Bifunctional PEG-4500 was activated in presence of magnetite.

[d]Indoxyl acetate hydrolysis in emulsified aqueous solution.

[e]Fluorescamine assay (Ref. 72).

[f]Amino acid analysis.

[g]TNBS assay: version measuring only the readily accessible amino groups.

[h]% Antigenic activity of antigen E.

[i]The number of lysine residues modified were determined by a spectrophotometrical method and amino acid analysis.

bifunctional PEGs. Thus, the degree of this intermonomeric crosslinking, and not the number of modified amino groups, could have been the dominant factor determining enzymatic activity of a given preparation.

Superoxide dismutase (SOD) has been modified with PEG in a number of laboratories. With one exception,[99] the attachment of PEG through amide (reagents **3**, **8**) or urethane (reagents **4**, **6**) linkages caused only minimal inactivation of the enzyme. In the work of McGoff *et al.*,[99] which is the exception, the modified and the native enzymes were from two different sources. Consequently, the comparison of the two can hardly be valid. Interestingly, while Pyatak *et al.*[98] reported a 50% loss of SOD activity after modification with **1**, Beckman *et al.*[97] observed complete preservation of enzymatic activity of PEG–SOD derived from the same reagent. Unfortunately, no activity was reported for SOD derivatives obtained by attachment of **17** to arginyl residues of the protein.[79]

Preparation of functionally active, yet extensively modified, PEG conjugates derived from proteins having large-size substrates proved more difficult than with enzymes acting on low molecular weight substrates. For example, several PEG-tissue plasminogen activator (tPA) conjugates were prepared using succinimidyl succinate (**3a**) and imidazolyl formate (**4**) mPEGs as modifying reagents.[53] Preference was given to mPEG–imidazolyl formate-derived conjugates, due to their somewhat higher fibrinolytic activities. Regardless of the activated PEG employed, the activity of the conjugates decreased with increased extent of modification. Similarly to the case of tPA, proteolytic activity of PEG-modified trypsin was also drastically reduced[101] in contrast to the well preserved and in some cases even enhanced activity towards low molecular weight substrates (Table 1). Other proteolytic enzymes have shown similar behavior. Using far-ultraviolet circular dichroism and intrinsic protein fluorescence, Pasta *et al.*[26] showed that the serine protease, subtilisin, modified with **3a** maintains its native secondary structure and thus the integrity of its catalytic site. It is generally believed that steric hindrance is responsible for the diminished proteolytic activity of PEG-modified proteases. We believe that the well-documented ability of PEG to exclude proteins from its environment[1–4] is also partially responsible for this phenomenon.

From the examples shown in Table 1 the choice of the "best" performing activated PEG is not obvious. Overall, the acylating reagents (**3**–**8**) performed comparably well. In some cases the ease of preparation and shelf-life of reagents are very important considerations.[55] Based on these two criteria the urethane-forming functionalized PEGs (**4**–**7**) are clearly superior.

21.5. FUTURE PERSPECTIVES

Activity in the area of PEG-modified polypeptides has increased over the last two decades, as evidenced by growing numbers of research groups that have joined this field, as well as by the total number of relevant publications and patents. We expect this trend to continue during the nineties. It is clear that recombinant proteins, which have become more available and in many cases have potential for therapeutic use, can benefit from the increased stability, resistance to proteolysis, and extended plasma lifetime that conjugation with PEG is almost certain to provide. More sophisticated PEG-based reagents, that modify selective sites or residues of polypeptide molecules, will certainly emerge, accompanied by the development of the new analytical methods for the characterization of PEG–polypeptide conjugates. Superior quality of commercially available PEGs and their functionalized derivatives are already being developed by a number of companies. For example, the undesirable presence of bifunctional PEG contaminants in mPEG preparations[104] will have to be dramatically reduced to minimize the possibilities for crosslinking and heterogeneity of PEG–protein preparations. Some recently developed methods for selective introduction of one reactive functional group per polymer chain[60] as well as for synthesis of heterobifunctional PEG derivatives using readily available PEG–diols as starting materials[76] might minimize such complications and facilitate a more controlled and rationale design of PEG–polypeptide conjugates and their applications.[105]

REFERENCES

1. I. N. Topchieva, *Usp. Khim. 49*, 494; *Russ. Chem. Rev. 49*, 260 (1980).
2. K. C. Ingham, *Methods Enzymol. 104*, 351 (1984).
3. H. Walter and G. Johansson, *Anal. Biochem. 155*, 215 (1986).
4. E. Merrill and E. W. Salzman, *Am. Soc. Artif. Internal Organs J. 6*, 60 (1983).
5. M. Mutter and E. Bayer, in: *The Peptides* (E. Gross and J. Meienhofer, eds.), Vol. 2, p. 285, Academic Press, New York (1979).
6. V. N. R. Pillai and M. Mutter, *Acc. Chem. Res. 14*, 122 (1981).
7. Y. Inada, K. Takahashi, T. Yoshimoto, A. Ajima, A. Matsushima, and Y. Saito, *Trends in Biotechnol. 4*, 190 (1986).
8. B. Mattiason, *Methods Enzymol. 92*, 498 (1983).
9. D. E. Brooks, K. A. Sharp, and S. J. Stocks, *Makromol. Chem., Macromol. Symp. 17*, 387 (1988).
10. A. Abuchowski and F. F. Davis, in: *Enzymes as Drugs* (J. Holsenberg and J. Roberts, eds.), p. 367, Wiley, New York (1981).
11. S. Dreborg and E. B. Åkerblom, *Crit. Rev. Therap. Drug Carrier Syst. 6*, 315 (1990).
12. J. M. Harris, *J. Macromol. Sci., Rev. Macromol. Chem. Phys. C25*, 325 (1985).
13. S. Zalipsky, C. Gilon, and A. Zilkha, *J. Macromol. Sci. Chem. A21*, 839 (1984).
14. S. Zalipsky, C. Gilon, and A. Zilkha, *Eur. Polym. J. 19*, 1177 (1983).
15. D. A. Herold, K. Keil, and D. E. Bruns, *Biochem. Pharmacol. 38*, 73 (1989).
16. H. F. Smyth, Jr., C. P. Carpenter, and C. S. Weil, *J. Am. Pharm. Assoc. 39*, 349 (1950).
17. A. J. Johnson, M. H. Karpatkin, and J. Newman, *Br. J. Hematol. 21*, 21 (1971).
18. C. B. Shaffer and F. H. Critchfield, *J. Am. Pharm. Assoc. 36*, 152 (1947).
19. E. C. Dittmann, *Naunyn-Schmeideberg's Arch. Pharmacol. 276*, 199 (1973).
20. A. W. Richter and E. Åkerblom, *Int. Arch. Allergy Appl. Immunol. 70*, 124 (1983).
21. F. F. Davis, T. Van Es, and N. C. Palczuk, U.S. Patent 4,179,337 (1979).
22. F. Kawai, *CRC Crit. Rev. Biotechnol. 6*, 273 (1987).
23. G. P. Royer and G. M. Ananthramaiah, *J. Am. Chem. Soc. 101*, 3394 (1979).
24. J. D. Glass, L. Silver, J. Sondheimer, C. S. Pande, and J. Coderre, *Biopolymers 18*, 383 (1979).
25. K. Ulbrich, J. Strohalm, and J. Kopecek, *Makromol. Chem. 187*, 1131 (1986).
26. P. Pasta, S. Riva, and G. Carrea, *FEBS Lett. 236*, 329 (1988).
27. A. Abuchowski, T. Van Es, N. C. Palczuk, and F. F. Davis, *J. Biol. Chem. 252*, 3578 (1977).

28. T. P. King, L. Kochoumian, and L. M. Lichtenstein, *Arch. Biochem. Biophys. 178*, 442 (1977); T. P. King, L. Kochoumian, and N. Chiorazzi, *J. Exp. Med. 149*, 424 (1979).
29. Y. Ashihara, T. Kono, S. Yamazaki, and Y. Inada, *Biochem. Biophys. Res. Commun. 93*, 385 (1978).
30. H. F. Gaertner and A. J. Puigserver, *Proteins 3*, 130 (1988).
31. S. G. Shafer and J. M. Harris, *J. Polym. Sci., Polym. Chem. Ed. 24*, 375 (1986).
32. M. Z. Atassi, in: *Immunochemistry of Proteins* (M. Z. Atassi, ed.), Vol. 1, p. 1, Plenum Press, New York (1977).
33. K. J. Wieder, N. C. Palczuk, T. Van Es, and F. F. Davis, *J. Biol. Chem. 254*, 12579 (1979).
34. S. Davis, A. Abuchowski, Y. K. Park, and F. F. Davis, *Clin. Exp. Immunol. 46*, 649 (1981).
35. K. Yoshinaga and J. M. Harris, *J. Bioact. Compatible Polym. 4*, 17 (1989).
36. A. Matsushima, H. Nishimura, Y. Ashihara, Y. Yokota, and Y. Inada, *Chem. Lett.*, 773 (1980).
37. H. Nishimura, K. Takahashi, K. Sakurai, K. Fujinuma, Y. Imamura, M. Ooba, and Y. Inada, *Life Sci. 33*, 1467 (1983).
38. K. Takahashi, A. Ajima, T. Yoshimoto, M. Okada, A. Matsushima, Y. Tamaura, and Y. Inada, *J. Org. Chem. 50*, 3414 (1985).
39. C. C. Jackson, J. L. Charlton, K. Kuzminski, G. M. Lang, and A. H. Sehon, *Anal. Biochem. 165*, 114 (1987).
40. A. Yoshimoto, S. G. Chao, Y. Saito, I. Imamura, H. Wada, and Y. Inada, *Enzyme 36*, 261 (1986).
41. M. Rubinstein, U.S. Patent 4,101,380 (1978).
42. M. Joppich and P. L. Luisi, *Makromol. Chem. 180*, 1381 (1979).
43. A. Abuchowski, G. Kazo, C. R. Verhoest, T. Van Es, D. Kafkewitz, M. L. Nucci, A. T. Viau, and F. F. Davis, *Cancer Biochem. Biophys. 7*, 175 (1984).
44. M. S. Hershfield, R. H. Buckley, M. L. Greenberg, A. L. Melton, R. Schiff, C. Hatem, J. Kirtzberg, M. L. Markert, R. H. Kobayashi, A. L. Kobayashi, and A. Abuchowski, *N. Engl. J. Med. 316*, 589 (1987).
45. K. B. Hadley and P. H. Sato, *Enzyme 42*, 225 (1989).
46. K. Iwasaki, Y. Iwashita, and T. Okami, U.S. Patent 4,670,417 (1987).
47. V. N. Katre, M. J. Knauf, and W. J. Laird, *Proc. Natl. Acad. Sci. U.S.A. 84*, 1487 (1987).
48. E. Boccu, R. Largajolli, and F. M. Veronese, *Z. Naturforsch. 38c*, 94 (1983).
49. D. Larwood and F. Szoka, *J. Labelled Compd. Radiopharm. 21*, 603 (1984).
50. C. O. Beauchamp, S. L. Gonias, D. P. Menapace, and S. V. Pizzo, *Anal. Biochem. 131*, 25 (1983).
51. L. Tondelli, M. Laus, A. S. Angeloni, and P. Ferruti, *J. Controlled Release 1*, 251 (1985).
52. C. Beauchamp, P. E. Daddona, and D. P. Menapace, in: *Adv. Exp. Med. Biol.* (C.H.M.M. DeDruyn, H. A. Simmons, and M. Muller, eds.), **165A**, 47 (1984).
53. H. Berger and S. V. Pizzo, *Blood 71*, 1641 (1988).
54. F. M. Veronese, R. Largajolli, E. Boccu, C. A. Benassi, and O. Schiavon, *Appl. Biochem. Biotechnol. 11*, 141 (1985).
55. S. Zalipsky, Patent pending; S. Zalipsky, R. Seltzer, and S. Meron-Rudolph, *Biotechnol. Appl. Biochem. 15*, 100 (1992).
56. S. Zalipsky, R. Seltzer, and K. Nho, *Polym. Prepr., Am. Chem. Soc., Div. Polym. Chem. 31(2)*, 173 (1990); in: *Polymeric Drugs and Drug Delivery Systems* (R. L. Dunn and R. M. Ottenbrite, eds.), p. 91, ACS Symposium Series 469, Washington, DC (1991).
57. A. Nathan, S. Zalipsky, and J. Kohn, *Polym. Prepr., Am. Chem. Soc., Div. Polym. Chem. 31(2)*, 213 (1990).
58. A. Buckmann, M. Morr, and G. Johansson, *Makromol. Chem. 182*, 1379 (1981).
59. M. Leonard, J. Neel, and E. Dellacherie, *Tetrahedron 40*, 1581 (1984).
60. S. Zalipsky and G. Barany, *J. Bioact. Compat. Polym. 5*, 227 (1990).
61. G. Johansson, *J. Chromatogr. 368*, 309 (1986).
62. F. M. Veronese, P. Caliceti, A. Pastorino, O. Schiavon, L. Sartore, L. Banci, and L. M. Scolaro, *J. Controlled Release 10*, 145 (1989).
63. T. Yoshimoto, M. Nakata, S. Yamaguchi, T. Funada, Y. Saito, and Y. Inada, *Biotechnol. Lett. 8*, 771 (1986).

64. D. E. Nitecki, L. Aldwin, and M. Moreland in: *Peptide Chemistry 1987* (T. Shiba and S. Sakakibara, eds.), p. 243, Protein Res. Foundation, Osaka (1988).
65. K. Shimizu, T. Nakahara, and T. Kinoshita, U.S. Patent 4,495,285 (1985); K. Shimizu, T. Nakahara, T. Kinoshita, J. Takatsuka, and M. Igarashi, U.S. Patent 4,640,835 (1987).
66. A. J. Garman and S. B. Kalindjian, *FEBS Lett. 223*, 361 (1987); A. J. Garman, U.S. Patent 4,935,465 (1990).
67. T. P. King and C. Weiner, *Int. J. Peptide Protein Res. 16*, 147 (1980).
68. S. Zalipsky, F. Albericio, U. Slomczynska, and G. Barany, *Int. J. Peptide Protein Res. 30*, 740 (1987).
69. J. M. Harris, E. C. Struck, M. G. Case, M. S. Paley, M. Yalpani, J. M. Van Alstine, and D. E. Brooks, *J. Polym. Sci., Polym. Chem. Ed. 22*, 341 (1984).
70. K. Nilsson and K. Mosbach, *Methods Enzymol. 104*, 56 (1984).
71. C. Delgado, J. N. Patel, G. E. Francis, and D. Fisher, *Biotechnol. Appl. Biochem. 12*, 119 (1990).
72. P. Bohlen, S. Stein, W. Dairman, and S. Udenfriend, *Arch. Biochem. Biophys. 155*, 213 (1973); S. J. Stocks, A. J. M. Jones, C. W. Ramey, and D. E. Brooks, *Anal. Biochem. 154*, 232 (1986).
73. A. S. F. A. Habeeb, *Anal. Biochem. 14*, 328 (1966).
74. N. Yamasaki, A. Matsuo, and H. Isobe, *Agric. Biol. Chem. 52*, 2125 (1988).
75. L. Sartore, P. Caliceti, O. Schiavon, P. Ferruti, E. Ranucci, and F. M. Veronese, *Proc. Int. Symp. Control. Rel. Bioact. Mater. 17*, 208 (1990).
76. S. Zalipsky and G. Barany, *Polym. Prepr., Am. Chem. Soc., Div. Polym. Chem. 27(1)*, 1 (1986).
77. H. Ueno and M. Fujino, Eur. Patent Appl. 0 236 987 (1987).
78. C. S. Pande, M. Pelzig, and J. D. Glass, *Proc. Natl. Acad. Sci. U.S.A. 77*, 895 (1980); J. D. Glass, R. Miller, and G. Wesolowski, in: *Peptides Structure and Function* (D. Rich, ed.), p. 209, Pierce Chemical Co., Rockford, IL (1981); J. D. Glass, M. Pelzig, and C. S. Pande, in: *Peptides 1978* (I. Z. Siemion and G. Kupryszewski, eds.), p. 235, Wroclav University Press, Poland (1979).
79. A. Sano, H. Maeda, Y. Kai, and K. Ono, Eur. Patent Appl. 0 340 741 (1989).
80. R. J. Goodson and N. V. Katre, *Bio/Technology 8*, 343 (1990).
81. S. Romani, W. Gohring, L. Moroder, and E. Wunsch, in: *Chemistry of Peptides and Proteins* (W. Voelter, E. Bayer, Y. A. Ovchinnikov, and E. Wunsch, eds.), Vol. 2, p. 29, Walter de Gruyter, Berlin (1984).
82. A. Pollak and G. M. Whitesides, *J. Am. Chem. Soc. 98*, 289 (1976).
83. M. Kimura, Y. Matsumura, Y. Miyauchi, and H. Maeda, *Proc. Soc. Exp. Biol. Med. 188*, 364 (1988).
84. M. Urrutigoity and J. Souppe, *Biocatalysis 2*, 145 (1989).
85. M.-T. Babonneau, R. Jacquier, R. Lazaro, and P. Viallefont, *Tetrahedron Lett. 30*, 2787 (1989).
86. K. Yoshinaga, S. G. Shafer, and J. M. Harris, *J. Bioact. Compatible Polym. 2*, 49 (1987).
87. C. Visco, C. A. Benassi, F. M. Veronese, and P. A. Maglioli, *Il Farmaco 42*, 549 (1987).
88. K. V. Savoca, A. Abuchowski, T. Van Es, F. F. Davis, and N. C. Palczuk, *Biochem. Biophys. Acta 578*, 47 (1979).
89. Y. Kamisaki, H. Wada, T. Yaguara, A. Matsushima, and Y. Inada, *J. Pharmacol. Exp. Ther. 216*, 410 (1981).
90. T. Yoshimoto, T. Mihama, K. Takahashi, Y. Saito, Y. Tamaura, and Y. Inada, *Biochem. Biophys. Res. Commun. 145*, 908 (1987).
91. C. Pina, D. Clark, and H. Blanch, *Biotechnol. Tech. 3*, 333 (1989).
92. A. Matsushima, M. Okada, and Y. Inada, *FEBS Lett. 178*, 275 (1984).
93. T. Nishio, K. Takahashi, T. Tsuzuki, T. Yoshimoto, Y. Kodera, A. Matsushima, Y. Saito, and Y. Inada, *J. Biotechnol. 8*, 39 (1988).
94. Y. Inada, H. Nishimura, K. Takahashi, T. Yoshimoto, A. R. Saha, and Y. Inada, *Biochem. Biophys. Res. Commun. 122*, 845 (1984).
95. A. Matsushima, M. Okada, K. Takahashi, T. Yoshimoto, and Y. Inada, *Biochem. Int. 11*, 551 (1985).
96. Y. Kodera, K. Takahashi, H. Nishimura, A. Matsushima, Y. Saito, and Y. Inada, *Biotechnol. Lett. 8*, 881 (1986).

97. J. S. Beckman, R. L. Minor, C. W. White, J. E. Repine, G. M. Rosen, and B. A. Freeman, *J. Biol. Chem. 263*, 6884 (1988).
98. P. S. Pyatak, A. Abuchowski, and F. F. Davis, *Res. Commun. Chem. Pathol. Pharmacol. 29*, 113 (1980).
99. P. McGoff, A. C. Baziotis, and R. Maskiewicz, *Chem. Pharm. Bull. 36*, 3079 (1988).
100. K. Miyata, Y. Nakagawa, M. Nakamura, T. Ito, K. Sugo, T. Fujita, and K. Tomoda, *Agric. Biol. Chem. 52*, 1575 (1988).
101. A. Abuchowski and F. F. Davis, *Biochem. Biophys. Acta 578*, 41 (1979).
102. U. Bagert and K.-H. Rohm, *Biochem. Biophys. Acta 999*, 36 (1989).
103. H. Neumann and A. Lustig, *Eur. J. Biochem. 109*, 475 (1980).
104. M. Leonard and E. Dellacherie, *Makromol. Chem. 189*, 1809 (1988).
105. S. Zalipsky, F. Albericio, and G. Barany, in: *Peptides: Structure and Function* (V. J. Hruby and K. H. Kopple, eds.), p. 257, Pierce Chemical Co., Rockford, IL (1985).

22

Synthesis of New Poly(Ethylene Glycol) Derivatives

J. MILTON HARRIS, M. R. SEDAGHAT-HERATI, P. J. SATHER, DONALD E. BROOKS, and T. M. FYLES

22.1. INTRODUCTION

The chapters of this book describe the synthesis and use of a variety of active PEG derivatives designed to couple PEG to other materials. Despite the availability of these derivatives, there remains a need for new derivatives with presently unavailable properties and work continues in this area. Desirable properties include selectivity, stability, and ease of preparation. For example, it would be desirable to have derivatives that react with nucleophilic groups on proteins, but which do not react with water. Derivatives of a wide range of reactivities are always in demand. Similarly, there would be advantages to having derivatives that react with groups other than the commonly used amino groups. And, of course, the need for derivatives that can be prepared cheaply and easily in large quantity is critical for commercialization of the many biomedical and biotechnical applications of PEG chemistry.

Synthesis of new PEG derivatives is an on-going effort in our laboratories,[1–5] and here we describe three recent studies. The first applies solid-phase synthesis to prepare monodisperse PEGs, the second involves a new synthesis of PEG thiols and shows how this derivative can be used, and the third describes lab-scale synthesis of

J. MILTON HARRIS • Department of Chemistry, University of Alabama in Huntsville, Huntsville, Alabama 35899. M. R. SEDAGHAT-HERATI • Department of Chemistry, Southwest Missouri State University, Springfield, Missouri 65804-0089. P. J. SATHER and DONALD E. BROOKS • Departments of Pathology and Chemistry, University of British Columbia, Vancouver, British Columbia V6T 2B5, Canada. T. M. FYLES • Department of Chemistry, University of Victoria, Victoria, British Columbia V8W 2A2, Canada.

Poly(Ethylene Glycol) Chemistry: Biotechnical and Biomedical Applications, edited by J. Milton Harris. Plenum Press, New York, 1992.

heterofunctional PEGs by ethoxylation of appropriate anionic initiators at atmospheric pressure. In the present chapter we provide only qualitative description of the synthetic routes used, in part because of space limitations, but primarily because the goal of this work is to illustrate a range of new PEG chemistries and potential biotechnical applications.

22.2. SOLID POLYMER SUPPORTED SYNTHESIS OF MONODISPERSE ETHYLENE GLYCOL OLIGOMERS

It is frequently of interest to have PEG oligomers of various molecular weights to allow systematic variation of the separation between groups attached to either end of the chain. Oligoethylene glycols up to $n = 7$ (where n is the degree of polymerization) are available commercially, although the price of hepataethylene glycol is about 1000 times that of tetraethylene glycol. The problem of producing long-chain oligoethylene glycols can be solved by solution chemistry and separation from mixtures. However, separation of individual oligomers from average molecular weight mixtures is very difficult, and solution-phase synthesis has proven, in our hands, to produce low yields of products that are difficult to purify.[6] To avoid these problems we have developed a polymer-supported synthesis of long-chain ethylene glycol oligomers.

The solid-phase synthesis proceeds by attaching an ethylene glycol oligomer to polystyrene (PS) beads, extending the chain by coupling another oligomer, and removing the oligomer from the support. For example, octaethylene glycol is synthesized via the following five steps: (1) attachment of monoacetyl–tetraethylene glycol to PS having active trityl chloride groups; (2) removal of the acetyl protecting group; (3) activation of the exposed hydroxyl group by conversion to the mesylate; (4) reaction with tetraethylene glycol; and (5) removal of the octaethylene glycol by reaction with hydrochloric acid in dioxane. Higher oligomers, such as dodecaethylene glycol, can be prepared by repeating steps 3 and 4 before removal of the oligomer from the solid support. A general reaction scheme is given in Figure 1. A brief description of the procedure follows.

22.2.1. Coupling of Protected Oligomer to Activated Support

The solid, supporting polymer used in this synthesis was PS, crosslinked with 1% divinyl benzene, with trityl chloride active sites. Crosslinked PS was obtained from Polysciences and was treated with butyl lithium followed by reaction with benzophenone to produce trityl alcohol groups. These sites were then chlorinated by reaction with acetyl chloride. Chloride titration showed the polymer to have 0.55 mmol of chloride per gram of polymer.

The first step in oligoethylene glycol synthesis is to attach the first tetraethylene glycol to the polymer (Figure 1). Because tetraethylene glycol itself has two hydroxyl groups, it is possible for the glycol to react with two different trityl chloride groups on

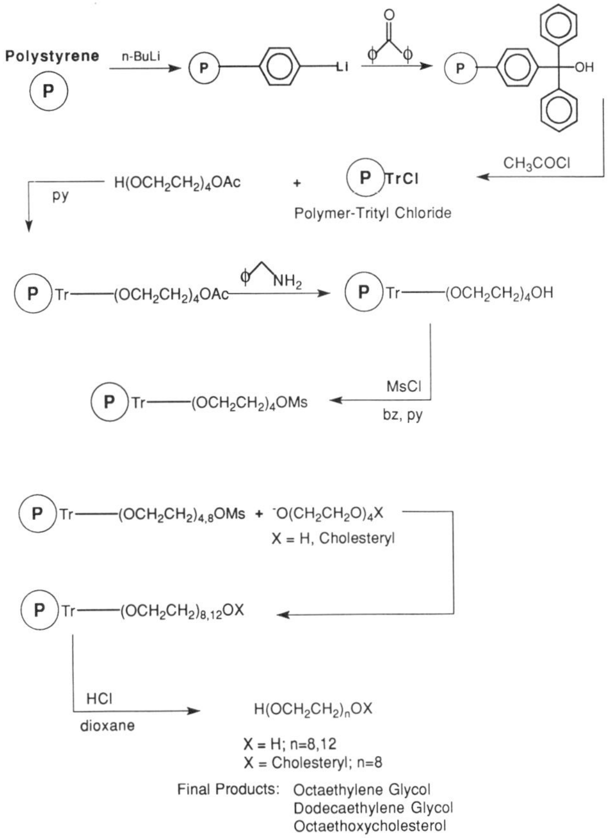

Figure 1. Polymer-supported synthesis.

the polymer, thus lowering overall yield. To avoid this problem, we prepared a monoprotected derivative.

The choice of the proper protecting group proved difficult. Before describing this search, we note that the hydroxyl group of the monoprotected oligomers reacted readily with the polymer trityl chloride groups in pyridine suspension. Acid cleavage after coupling showed that 0.5 mmol of the tetraethylene glycol monoacetate could be recovered for each gram of PS possessing 0.55 mmol of active chloride.

The key to the choice of hydroxyl protecting group turned out to be removal of the group. First we tried the commonly used *t*-butyl dimethyl silyl group. Monoprotected tetraethylene glycol was synthesized by reaction of the glycol with *t*-butyl dimethyl silyl chloride, and the resulting monoprotected derivative was isolated by distillation. Removal of the silyl group from the soluble derivative was straightforward under conditions generally used (25 °C, 0.5 M tetrabutyl ammonium fluoride in THF for 2 hours).[7] Unfortunately, these conditions were not effective for the polymer-supported oligomer. For example, reaction at 25 °C for 48 hours followed by reaction at 65 °C for 24 hours removed only 40% of the silyl groups (as shown by elemental analysis). Use of DMF solvent and the more reactive trimethylsilyl group proved similarly ineffective.

In the course of monitoring removal of the silyl group we found it effective to convert exposed hydroxyl groups to acetate by reaction with acetic anhydride in pyridine. The oligomer could then be removed from the polymer support by mild acid treatment (the trityl ether linkage is much more reactive with acid than the usual PEG-type ether linkage) and the number of acetate groups measured by proton NMR. This measurement provided a simpler, although indirect, means of following removal of silyl groups than did elemental analysis. Also, the presence of the acetate could be followed without removing the oligomer from the PS by preparing a KBr pellet of the sample and measuring the IR peak at 1730 cm^{-1} (silyl groups do not have a readily visible IR absorption). Since the acetate group is not removed by the acid conditions used to remove the oligomer from the PS support, we decided to examine use of acetate as the protecting group.

Monoacetyl tetraethylene glycol was made by reaction of the glycol with acetic anhydride in pyridine. The product was purified by distillation and analyzed by IR, NMR, mass spectrometry, and elemental analysis. These studies indicate that the distilled product contained 10% of diacetyl tetraethylene glycol as impurity. The presence of this impurity is not a problem here, as it does not react with the active polymer and is washed away after the coupling step.

As with the silyl derivative, removal of the acetate protecting group from the polymer-bound compound proved more difficult than expected from solution studies. After examining increasingly harsh reaction conditions, we determined that reaction for 65 hours at 160 °C with neat benzyl amine gave complete removal of the acetate group. Despite the harshness of these conditions, the PS support and the oligomer–PS linkage were not affected.

22.2.2. Chain Extension and Product Isolation

The next steps after removal of the protecting group are activation of the exposed hydroxyl group on the bound oligomer and coupling with another tetraethylene glycol. As indicated in Figure 1, activation was done by reaction with methanesulfonyl chloride to produce the methanesulfonate ester (mesylate). This reaction was monitored by IR analysis of the sulfur–oxygen stretch at 1350 cm^{-1}.

The anion of tetraethylene glycol is formed in solution by reaction with NaH in DMF and coupled to the active, bound oligomer. Mesylation of the hydroxyl terminus and extension of the oligomer chain could be repeated to provide higher oligomers. To demonstrate this possibility, we have prepared dodecaethylene glycol ($n = 12$) in this manner.

Cleavage of the oligomer from the chain was accomplished by reaction with hydrochloric acid in dry dioxane (0.5 M). Water slowed this reaction appreciably. The cleaved oligomers were isolated by filtration and solvent evaporation.

The isolated oligomers were purified by gel permeation chromatography in water on Fractogel 40S. This technique also proved useful analytically and showed that both the octaethylene glycol and the dodecaethylene glycol products were free of glycol impurities; baseline separation between oligomers four units apart could be achieved with this method. Overall yield was approximately 50%. Proton NMR and IR were used to confirm the structures. Molecular weights were confirmed with DCI mass spectroscopy.

22.2.3. Conclusions

The use of polymer-supported synthesis in the production of oligoethylene glycols offers several advantages over solution-phase synthesis. Work-up in polymer-supported synthesis involves simple filtration. Solvents, starting materials, and impurities are washed away. This is especially advantageous when removing excess tetraethylene glycol, since distillation of this compound tends to produce appreciable amounts of breakdown products. Difficult preparative separations are also avoided. And once the method is developed it can potentially be applied to other syntheses. One must take care, however, in that reaction rates on these polymer surfaces are often greatly reduced relative to those in solution.

22.3. SYNTHESIS AND APPLICATION OF PEG THIOL

PEG aldehydes are of interest because they are stable in water and remain active toward coupling with amines.[3] The propionaldehyde derivative, **1**, is especially of interest because it is more reactive than the benzaldehyde derivative and, unlike the acetaldehyde analog, is not susceptible to decomposition by aldol condensation. This compound was synthesized according to Eq. (1) (note that both difunctional PEG and monofunctional monomethyl ether, MPEG, have been used).[3]

$$HS(CH_2)_3S^-Na^+ + ClCH_2CH_2CH(OEt)_2 \rightarrow HS(CH_2)_3SCH_2CH_2CH(OEt)_2$$

$$(1) \qquad \xrightarrow[\text{3. Aq. acid}]{\text{1. NaOMe} \atop \text{2. MPEG–OTs}} \text{MPEG—S—}(CH_2)_3\text{—S—}CH_2CH_2\text{—CHO}$$

1

This route was devised primarily because a more straightforward route utilizing PEG alkoxide, PEG–O$^-$, and the chloroacetal gave extensive elimination in addition to the desired product (Eq. 2). One would expect that the softer thiolate anion PEG–S$^-$ would give less elimination than the harder alkoxide, and indeed this was found to be the case, with no elimination product detected from reaction 1 or reaction 4. The route of Eq. (1) was found to be more technically straightforward than that of Eq. (4), and thus this first route was used to prepare quantities of **1**.

$$(2)\ \text{MPEG–O}^{-1}Na^+ + ClCH_2CH_2CH(OEt)_2 \xrightarrow{\text{2. H}^+} \mathbf{1} + CH_2{=}CH\text{—}CH(OEt)_2$$

$$(3) \qquad HS(CH_2)_3S^-Na^+ + \text{MPEG–OTs} \rightarrow \text{MPEG–S—}(CH_2)_3\text{—SH}$$

2

(4) $$\mathbf{2} + \mathrm{ClCH_2CH_2CH(OEt)_2} \xrightarrow{\text{1. n-BuLi/2. H}^+} \mathbf{1}$$

It is evident from this work that PEG thiol, PEG–SH, would be a useful derivative. The thiol group is more nucleophilic than the hydroxyl group and more likely to give substitution rather than elimination reactions. Also, if one had a heterofunctional PEG with a hydroxyl group at one terminus and a thiol group at the other, then it would be feasible to conduct different reactions on the two terminae (next section). The synthesis via Eq. (3) with 1,3-propanedithiol proved unsatisfactory, primarily because of the stench of the dithiol and difficulty in removing this compound from the final product. Also, there is some concern that incorporation of the dithiol moiety into the PEG may subtly change the properties of the PEG by making the molecule terminus less hydrophilic. Thus we have examined other routes to preparation of PEG–SH.[8]

22.3.1. Synthesis of Thiol

We have utilized the following three routes to synthesize PEG thiol:

(5) $$\mathrm{MPEG{-}OTs} + \mathrm{NaSH} \xrightarrow{\text{water}} \mathrm{PEG{-}SH}$$

(6) $$\mathrm{MPEG{-}OTs} + \mathrm{H_2N{-}CS{-}NH_2} \xrightarrow{\text{water}} \mathrm{MPEG{-}\overset{+}{S}{=}C(NH_2)_2} \xrightarrow{\text{OH}^-} \mathrm{MPEG{-}SH}$$

(7) $$\mathrm{MPEG{-}OTs} + \mathrm{Na_2CS_3} \xrightarrow{\text{water}} \mathrm{MPEG{-}SCS_2^-Na^+} \xrightarrow{\text{H}^+} \mathrm{MPEG{-}SH}$$

Reaction 5 was attempted first. It is well known that thiols can be prepared from alkyl halides by reaction with hydrosulfide ion, HS^-, in alcoholic solvents.[9] We performed the reaction with MPEG–OTs in water as reaction solvent (where both reagents are soluble). The NMR spectrum of the resulting product in d_6-DMSO showed no trace of the tosylate group, indicative of completion of the reaction, and no PEG–OH peak at 4.57 ppm, indicating no tosylate hydrolysis.[5] However, there was also no —SH peak in the spectrum. Furthermore, GPC analysis of the product revealed the presence of only one peak with MW almost twice that of the starting MPEG. This result could either be due to formation of the corresponding sulfide from displacement of MPEG–SH on MPEG–OTs (i.e., MPEG–S–MPEG) or to oxidation of MPEG–SH to the disulfide (i.e., MPEG–S—S–MPEG).

The possibility of sulfide formation was excluded by reaction of MPEG–OTs with thiourea[9] and sodium trithiocarbonate[10] (Eqs. 6 and 7). GPC analysis of the products from these reactions showed two peaks, one with the same MW as the starting MPEG and the other corresponding to double this MW (the dimer peak for several batches varied from 25 to near 100%). Since these reactions would not be expected to give sulfide product, it became clear that the dimer was disulfide formed by oxidation of the desired thiol (although air had been excluded from the reaction mixture; measures to remove oxygen from solvents and reactants were not taken). To further address the existence of the disulfide, the product was reduced with lithium aluminum hydride in THF. Analysis of the reduction product by GPC and NMR showed that dimer disappeared and was replaced by MPEG–SH. H-NMR (d_6-DMSO in ppm): 2.26, t, —SH; 2.60, t, —CH_2S—; 3.21, s, CH_3O—; 3.51, s, backbone.

Reduction of disulfide with lithium aluminum hydride can be used to produce pure thiol, but a simpler method is available. Thiol oxidation is known to be catalyzed by trace amounts of transition metals such as iron. These metals can be rendered inert by complexation with the pentasodium salt of diethylenetriaminepentaacetic acid. Thus we have found that inclusion of 0.2% of this salt in reactions 5–7 gives clean production of the desired thiol, uncontaminated with disulfide.

22.3.2. Application of Thiol in Aldehyde Synthesis

Synthesis of MPEG–S—CH_2CH_2CHO from reaction of the thiol with 3-chloropropionaldehyde diethyl acetal via the route of Eq. (2) proved straightforward; for MPEG 2000, no alkene was formed and an isolated aldehyde yield of approximately 90% was achieved. This result illustrates one of the prime advantages of PEG thiolate as a nucleophile, which is that it is softer than PEG alkoxide and thus less likely to give elimination product. The thiolate is also a more powerful nucleophile and thus milder reaction conditions are required.

22.4. SYNTHESIS OF HETEROFUNCTIONAL PEG

There is a need in biotechnology for heterofunctional PEGs (i.e., X–PEG–Y) that provide coupling to a single terminus by having one active and one inert terminus, or that can be used to couple one substance to one PEG terminus and another substance to a second PEG terminus. The readily available monomethyl ethers (M–PEGs) are widely used for the first application, although the presence of contaminant diol PEG is frequently a problem in this case. The many available alkylated PEG surfactants have one terminus blocked with an inactive group and could potentially also be used for coupling at a single terminus. The second application, to have different reactive groups at the two terminae, is more chemically challenging. We know of only two examples from the literature of such compounds. In the first, polymerizaton of ethylene oxide onto an initiator containing a protected amino group was used to produce monoamino–PEG (i.e., NH_2–PEG–OH).[11,12] In the second example, PEG derivatives were synthesized that contain an ω-amino acid as one functional group and an amine or azide as the other or that contain hydroxyl and carboxyl terminal groups.[13–15] This synthesis starts with preformed PEG, and the key step is separation of the desired product from contaminants.

Synthesis by ethoxylation has the advantage that heterofunctional derivatives are produced directly, thus avoiding the requirement for highly effective separations of unmodified and modified polymers involved in beginning with preformed PEG.

$$(8)\quad A{:}^- + nCH_2OCH_2 \rightarrow A(CH_2CH_2O)_{n-1}CH_2CH_2\text{—}O^- \overset{H^+}{\rightarrow} \text{A–PEG–OH}$$

22.4.1. Synthesis of BzO–PEG–OH and NH_2–PEG–OH

The goal of the present work is to synthesize reactive, heterofunctional PEGs by rapid, anion-catalyzed ethoxylation at atmospheric pressure in glassware readily available in a typical organic chemistry laboratory. A disadvantage of this approach is that ethylene oxide is a toxic, potentially explosive gas, but we have encountered no difficulties by working with small quantities, at atmospheric pressure, in solution and in a well-ventilated hood.

The excellent solvation ability of DMSO and the high reactivity of anions dissolved in this solvent has promoted its widespread use for base-catalyzed reactions, and several authors have investigated epoxide polymerization in this medium using potassium *t*-butoxide as the initiator.[16–19] Our approach is to prepare a solution of dimsyl ion in DMSO (Eq. 9), then to add alcohols to produce the alkoxide anion, which acts as the anionic initiator (Eq. 10), where EO = ethylene oxide.

(9) $$K + CH_3\text{—}SO\text{—}CH_3 \rightarrow CH_3\text{—}SO\text{—}CH_2{}^-K^+ \text{ (dimsyl ion)}$$

(10) $$CH_3\text{—}SO\text{—}CH_2{}^-K^+ + R\text{—}OH \rightarrow R\text{—}O^-K^+ \xrightarrow[\text{2. aq. acid}]{\text{1. EO}} R\text{—}O\text{–}PEG\text{–}PH$$

Dimsyl ion is readily prepared[20] and its concentration can be measured by titration against standard hydrochloric acid. A nitrogen atmosphere and dry DMSO (3A molecular sieve) are used to avoid introduction of water and formation of OH^-, which would lead ultimately to production of contaminating HO–PEG–OH. In a typical run the required amount of ethylene oxide is bled from a cylinder into a round-bottomed flask, dried with calcium hydride, and then distilled into the reaction flask containing DMSO and the initiating alkoxide. Cooling was required to keep the reaction flask at approximately 25 °C. Product was obtained by addition of water to terminate polymerization, followed by removal of solvent under vacuum, solution in methylene chloride, and precipitation by addition to cold ethyl ether.

A variety of initiating alcohols work well for this reaction, producing polymers of MWs in the desired range of 2000–3000 g mol^{-1} in 3–5 hours. In contrast, polymerization in other solvents such as THF/benzene[12] or acetonitrile was much slower, requiring days for production of the desired MWs. Two alcohols of pertinence to the present consideration are benzyl alcohol and the Schiff base of 2-(2-aminoethoxy)ethanol and 4-methyl-2-pentanone, **3**.

$$(CH_3)_2CHCH_2C(CH_3)\text{=}N\text{—}CH_2CH_2\text{—}O\text{-}CH_2CH_2\text{—}OH$$

3

Initiation of polymerization with benzyl alcohol produces BzO–PEG–OH, polymer **4**. This polymer is of interest because the benzyl group is a hydroxyl protecting group, which can be readily removed by hydrogenation. Similarly, initiation with alcohol **3** produces the ethoxylated Schiff base **6** which, upon addition of water, hydrolyzes to give NH_2–PEG–OH.[21]

$C_6H_5—CH_2—O—PEG\text{-}OH$ (**4**) $NH_2–PEG–OH$ (**5**) $(CH_3)_2CHCH_2C(CH_3)=N—CH_2CH_2—O–PEG–O^-K^+$ (**6**)

22.4.2. Potential Applications of Heterofunctional PEGs

Heterofunctional polymers **4** and **5** can be used in a variety of ways. Schemes 1 and 2 illustrate some possible reactions of polymer **4**. In Scheme 1, polymer **4** is first converted to the electrophilic tresylate (other electrophilic leaving groups such as carbonyl imidazole or succinimide could also be used) and then reacted with an aminated surface. Aminated surfaces of interest include, for example, those prepared by silanation of glass, modification of an active polystyrene, or radiofrequency glow discharge treatment of an organic polymer surface. Once polymer **4** is attached to the surface, the benzyl group can be removed by treatment with hydrogen, and the exposed hydroxyl group can be converted to another tresylate. This tresylate can then be substituted by a variety of interesting amines such as antibodies, enzymes, peptides, heparin, etc.

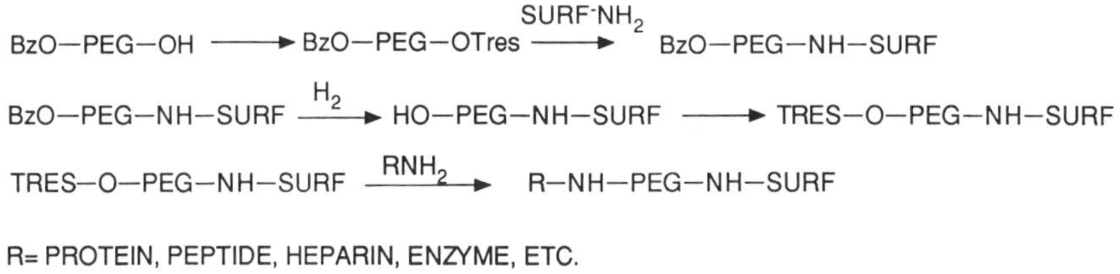

Scheme 1

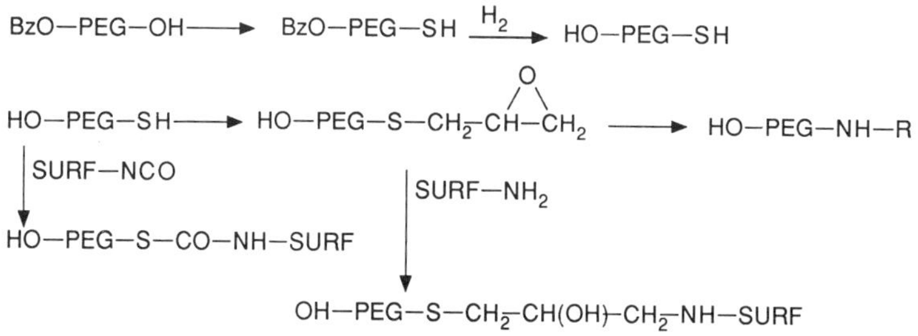

Scheme 2

The advantage of this route for reaction with surfaces, as compared to reaction with active, difunctional PEG (e.g., ditresylate), is that loop formation is avoided.

An alternative route to modified surfaces is shown in Scheme 2. A key difference in this approach is that the benzyl protecting group of polymer **4** is removed before attachment to the surface. First, polymer **4** is converted to the thiol, using methods of the preceding section of this chapter, and deprotected to give the monothiol HO–PEG–SH. This monothiol has a variety of potential applications deriving from the great difference in nucleophilicity between the thiol and hydroxyl groups. The monothiol can be reacted directly with an electrophilically activated surface (such as the isocyanate shown), and the surface-attached PEG (HO–PEG–Surf) used as in the preceding paragraph. Alternatively, the monothiol can be converted to the epoxide and reacted with an aminated surface. Similarly it could be reacted directly with a soluble amine, R—NH_2, to bind PEG by a single terminus.

This last reaction in which PEG is bound by a single terminus is a common goal when the amine is a protein or peptide and an absence of crosslinking is desired (this goal is also the driving force behind the search for low diol MPEG). Probably the ideal route to a noncrosslinked PEG–protein is provided by polymer **5**, NH_2–PEG–OH, in Scheme 3. Removal of small amounts of diol PEG from polymer **4** (or from MPEG) is difficult and the best route is to minimize its formation (it cannot be totally avoided) during polymerization by careful drying of reactants and solvents. However, the small amount of diol PEG formed during synthesis of NH_2–PEG–OH is readily removed by ion-exchange chromatography. This purified polymer **5** can then be coupled with protein, as shown in Scheme 3, to give a product with no crosslinking.

Polymer **5** can also be readily attached to electrophilically activated surfaces with little crosslinking or loop formation because of the great difference in nucleophilicity between amino and hydroxyl groups (Scheme 3).

NH_2–PEG–OH + PRO–COOH ⟶ HO–PEG–NH–CO–PRO

SURF-NCO ⟶ HO–PEG–NH–CO–NH–SURF

Scheme 3

22.5. CONCLUSIONS

The preceding work has shown that much can still be achieved in the relatively mature field of PEG derivative synthesis by applying both conventional and nonconventional synthetic routes to produce novel PEG derivatives. The authors are confident that application of the power of modern polymer chemistry to the field of PEG derivatives will be of value in moving the concepts presented in the present volume into the marketplace.

ACKNOWLEDGMENT

The authors gratefully acknowledge financial support from the National Institutes of Health and from the National Aeronautics and Space Administration.

REFERENCES

1. J. M. Harris and M. Yalpani, in: *Partitioning in Aqueous Two Phase Systems* (H. Walter, D. E. Brooks, and D. Fisher, eds.), Chapter 12, Academic Press, New York (1985).
2. J. M. Harris, K. Yoshinaga, M. S. Paley, and M. R. Herati, in: *Separations Using Aqueous Phase Systems: Applications in Cell Biology and Biotechnology* (D. Fisher and I. A. Sutherland, eds.), Chap. 5.3, pp. 201–210, Plenum Press, London (1989).

3. J. M. Harris, J. M. Dust, R. A. McGill, P. A. Harris, M. J. Edgell, M. R. Sedaghat-Herati, L. J. Karr, and D. L. Donnelly, in: *Water Soluble Polymers*, (S. W. Shalaby, C. L. McCormick, and G. B. Butler, eds.), pp. 418–429, ACS Symp. Ser. 467 (1991).
4. J. M. Dust and J. M. Harris, *J. Polym. Sci., Polym. Chem. Ed. 28*, 1875 (1990).
5. J. M. Dust, Z.-H. Fang, and J. M. Harris, *Macromolecules 23*, 3742 (1990).
6. Y. Nakutsuji, N. Kameda, and M. Okahara, *Synthesis*, 280 (1987).
7. K. K. Olgivie and D. J. Iwacha, *Tetrahedron Lett.*, 317 (1973).
8. J. M. Harris and R. S. Herati, *Polym. Prepr. 32*(1), 154 (1991).
9. L. Field, *Synthesis*, 101 (1972).
10. D. J. Martin and C. C. Greco, *J. Org. Chem. 33*, 1275 (1968).
11. G. P. Speranza, US Patent 3,231,619 (1966).
12. M. Sepulchre, G. Paulus, and R. Jerome, *Makromol. Chem. 184*, 1829 (1983).
13. S. Zalipsky and G. Barany, *J. Bioact. Compatible Polym. 5*, 227 (1990).
14. S. Zalipsky and G. Barany, *Poly. Preprints 27*(1), 1 (1986).
15. I. N. Topchieva, A. I. Kuzaev and V. P. Zubov, *Eur. Polym. J. 24*, 899 (1988).
16. C. C. Price and D. D. Carmelite, *J. Am. Chem. Soc. 88*, 4039 (1966).
17. C. C. Price and R. J. Spector, *J. Am. Chem. Soc. 88*, 4171 (1966).
18. C. E. Bawn and L. A. McFarlane, *Polymer 8*, 484 (1967).
19. K. Bridger and B. Vincent, *Polymer J. 16*, 1017 (1980).
20. M. R. Sedaghat-Herati, S. P. McManus, and J. M. Harris, *J. Org. Chem. 53*, 2539–2543 (1988).
21. Y-H. Huang, Z.-M. Li, and H. Morawetz, *J. Polym. Sci., Polym. Chem. Ed., 23*, 795 (1985).

Index